SURFACE ACTIVE AGENTS
Their Chemistry and Technology

SURFACE ACTIVE AGENTS

THEIR CHEMISTRY AND TECHNOLOGY

By ANTHONY M. SCHWARTZ, HARRIS RESEARCH LABORATORIES, WASHINGTON, D. C.

and JAMES W. PERRY, MASSACHUSETTS INSTITUTE OF TECHNOLOGY, CAMBRIDGE, MASS.

1 9 4 9

INTERSCIENCE PUBLISHERS, INC., NEW YORK
INTERSCIENCE PUBLISHERS LTD., LONDON

INTERSCIENCE PUBLISHERS, INC.
250 Fifth Avenue, New York 1, N. Y.

For Great Britain and Northern Ireland:
INTERSCIENCE PUBLISHERS, LTD.
2a Southampton Row, London W. C. 1

PRINTED IN THE UNITED STATES OF AMERICA

PREFACE

This book is an attempt to summarize the achievements of the last three decades in developing a host of new surface active agents. The different types of products are reviewed with particular attention to their laboratory synthesis, commercial production, and characteristic properties. Special attention has been directed to the numerous practical applications of surface active agents and to their utilitarian effects—foaming, wetting, detergency, emulsification, spreading, etc. In order to provide background for a better understanding of the diverse practical applications and related effects, the fundamentals of the chemistry and physics of surface phenomena have been summarized separately.

It was our aim to present a well-integrated picture of the present state of development of surface active agents. It is hoped that this book may prove interesting and helpful to practicing chemists in general, to advanced students in chemistry and chemical technology, and particularly to specialists in the production, investigation, and application of surface active agents.

Because the mass of factual data pertinent to our subject is too large to be encompassed within a single volume, numerous references have been included both to the original literature and to reviews of various special aspects of the broad field.

We are indebted to Mrs. Dolli Schoenberg and Mrs. Jean Benedict for bibliographic assistance, to Miss Alice Perry for typing most of the manuscript, and to the publisher's staff for editorial aid.

September, 1948

ANTHONY M. SCHWARTZ
Washington, D. C.
JAMES W. PERRY
Cambridge, Mass.

CONTENTS

PART II

THE PHYSICAL CHEMISTRY OF SURFACE ACTIVE AGENTS
IN THEORY AND PRACTICE

PART III

PRACTICAL APPLICATIONS OF SURFACE ACTIVE AGENTS

INTRODUCTION

CHAPTER 1

General Considerations

I. Phase Interfaces

From the chemists' point of view a surface or interface is the boundary between two phases. In any heterogeneous system the boundaries are of fundamental importance to the behavior of the system as a whole. They are particularly important in technology. For example, the phenomenon of adhesion is an effect of the surface or interface between two solid phases, or in some instances between a liquid and a solid phase. Similarly emulsification, wetting, spreading, foaming, detergent effects, adsorption effects, and various combinations of these are conditioned by what goes on in the interfaces between two phases. Simple geometric considerations show that a single surface exists only between two phases. Three phases or more can have only a line in common, but not a surface. This fact simplifies the theoretical treatment of even the most complex systems, and also makes the concept of a surface active agent much easier to grasp.

There are five types of phase interface possible: solid–gas, solid–liquid, solid–solid, liquid–gas, and liquid–liquid. The solid–gas and solid–solid interfaces will not concern us greatly in a study of surface active agents. The other three types, all of which involve at least one liquid phase, are susceptible to the influence of surface active agents and they will be the ones in which we are primarily interested.

There are three fundamental facts about phase interfaces which must be borne in mind in any discussion of them. First, the transition from one phase to the other is a sharp one. Even though there is a tremendous kinetic travel of molecules across the boundary at ordinary temperatures, the boundary itself is statistically not more than one or two molecules thick. This picture of the interface is presented very well by Adam.[1]

The second fundamental fact about any interface between two phases is the existence of a definite quantity of free energy associated with every unit of interfacial area. Interfacial energy expressed in ergs per square centimeter is mathematically and dimensionally equivalent to an inter-

[1] Adam, *The Physics and Chemistry of Surfaces*. Oxford Univ. Press, London, 1941.

3

facial tension expressed in dynes per linear centimeter. As a consequence the terms *interfacial energy* and *interfacial tension* are sometimes used interchangeably. In speaking of the liquid–gas interface, it is general practice to use the term *surface tension*. When so used, this term refers, strictly speaking, to the surface of a liquid in equilibrium with its own vapor rather than with a gas phase containing air or other foreign gas. Surface tension arises from the geometrical unbalance of the force fields acting on the molecules in the interface, and will be discussed further in Chapter 12. As might be expected, a low value of interfacial energy characterizes phase interfaces which are relatively easy to form and phase pairs which are relatively less easy to separate from each other. In everyday experience low interfacial tension is associated with high adhesion between the phases. The converse is also true. Thus the interfacial tension between water and clean glass is low, whereas between mercury and clean glass it is high. Mercury is easily separated from a glass plate by merely allowing it to roll off. Water must be evaporated off or mechanically removed by wiping. Similarly the interfacial tension between water and air is high, whereas between dilute soap solution and air it is relatively low. It is easy therefore to create an extensive soap solution – air interface in the form of a foam. This cannot be done with pure water.

The third characteristic of phase interfaces is the general existence of an electrical potential difference across them, *i.e.*, one phase is charged with respect to the other. This is not always the case as two-phase systems are known where no appreciable potential difference exists across the interface. The drop in electrical potential across a phase interface is, however, an extremely important effect in many instances. It is most noticeable in suspensions and emulsions, where one of the phases is very finely subdivided and consequently a very large interfacial area exists. Here, if the particles of the suspended phase are charged, cataphoretic effects can be easily produced. That is, if two electrodes connected to a source of e.m.f. are dipped into the liquid system, the suspended particles will migrate to one of the electrodes. This effect has been extensively studied and used in the laboratory, particularly in the investigation of biological systems. The anodic deposition of rubber is an example of an important technical process utilizing cataphoresis.

The interfacial energy and interfacial charge characteristics of a system are governed by the usual thermodynamic variables, temperature and pressure, and primarily by the chemical nature of the components present in each phase. Charge characteristics may often be modified profoundly by the addition of simple soluble ionic components to the system. Salts of polyvalent metals, for example, are often effective in modifying the

interfacial charges in emulsion systems. Interfacial energies are similarly affected by changes in composition. In liquid–gas systems the nature of the gas phase usually has relatively little effect on the surface energy. This is to be expected since the molecules in a gas are relatively far apart, and their total attractive effect on molecules in the interface is much less than the effect of the tightly packed molecules in the liquid phase. The interfacial tension, however, varies markedly with composition of the liquid phases. In systems consisting of two condensed phases, *i.e.*, liquid–liquid or liquid–solid systems, changes in the composition of either phase will generally raise or lower the interfacial energy by a measurable degree. The addition of any solute will generally alter the surface energy of the liquid phase in which it is dissolved. Thus the surface energy of water, *i.e.*, interfacial energy against air, under standard conditions is 72.7 erg/cm.^{2}. That of a 1% NaOH solution is 73, and of a 10% NaOH solution 77.5. The surface tension of aqueous NaOH solutions as a function of composition is shown in Figure 1. It is seen that relatively high concentrations of NaOH are necessary to change the surface tension significantly.[2] The majority of solutes show about the same order of activity in altering the surface energy of their solution systems. The alterations may be positive or negative, *i.e.*, surface energy may be raised or lowered by increasing amounts of the solute. It should be noted at this point that changes in the surface energy of a solution at the solution–air interface are usually paralleled by changes in its interfacial energy against other liquids or solids. Thus the changes in surface tension of a liquid upon addition of a solute are a good index of the alterations to be expected in its general interfacial behavior.

II. Surface Activity and Surface Active Agents

Certain solutes, even when present in very low concentration, have the startling property of altering the surface energy of their solvents to an extreme degree. The effect is invariably a lowering rather than an increase of the surface energy. Solutes having such properties have come to be known as the *surface active agents*, and their unusual effect is known as *surface activity*. Possibly the best-known and oldest of the surface active agents is soap. The concentration – surface tension curve for the aqueous solutions of a typical soap is shown in Figure 2.[3] Surface activity is best known and has been most extensively studied in aqueous systems. Thus

[2] *International Critical Tables*. Vol. IV, McGraw-Hill, New York, p. 465.
[3] Johlin, *J. Phys. Chem.* **29**, 1129–39 (1925).

most of the known surface active agents are water-soluble substances. Surface activity however can also be exhibited in non-aqueous solutions. Oleic acid, for example, is distinctly surface active when dissolved in hydrocarbon oils. A wide variety of other oil-soluble, water-insoluble substances are known, and used technically for their surface active effect in oils. Many extreme-pressure lubricants, drawing lubricants, and cutting oil additives fall into this class.

A large number of surface active agents are characterized by a molecular structure which is essentially linear, *i.e.*, considerably longer than it is wide. One end of the linear structure is composed of radicals which are

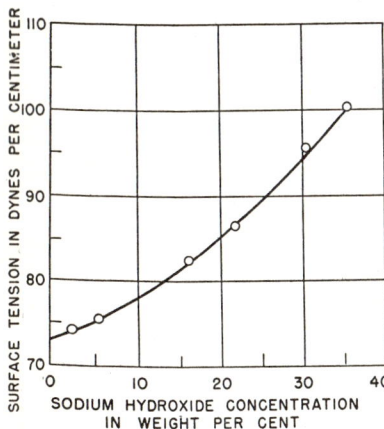

Fig. 1. Surface tension of concentrated aqueous sodium hydroxide solutions (measurements made at 18°C.).

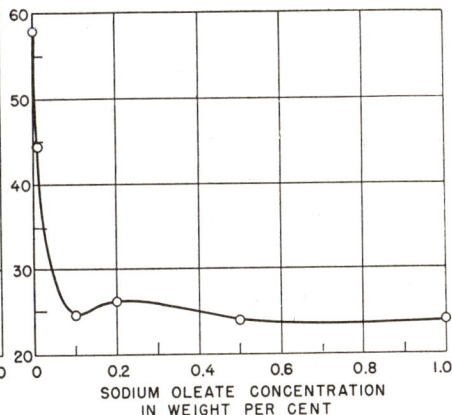

Fig. 2. Surface tension of dilute sodium oleate solutions (measurements made by capillary rise method at 25°C., 90 seconds after formation of fresh surface of solution).

compatible with the solvent system, and the opposite end of incompatible radicals. Usually one end comprises a hydrocarbon radical of hydrophobic nature characterized by weak residual valence forces, whereas the other end is of a hydrophilic nature with strong residual or secondary valence forces. It is these compounds which will form the main subject matter of this book.

The above description of the term "surface active agent" is somewhat arbitrary and is certainly narrower than the definition used in current technical terminology. There are many substances outside the scope of this definition which promote or effect lubrication, wetting, detergency, foaming, emulsification, and other effects associated with the term "surface activity." Some of these materials act definitely in the form of a third phase in the true macroscopic sense. Thus, finely divided clay or

bentonite is an excellent emulsifying agent under certain circumstances. As another example we have the water repellents used on textiles. These substances are used to deposit a true film of hydrophobic material constituting a separate phase on the fiber surface. They affect the fiber surface in the same sense that a coat of paint affects a wall surface. But in the colloid realm the distinction between a single-phase solution system and a multiphase dispersion system is not at all clean cut. If the hydrophobic material formed a monomolecular layer on the fiber surface, or even a layer a few molecules thick, it would fall well within the scope of our definitions.

Many of the soluble macromolecular substances have pronounced activity in emulsification, detergency, and other surface phenomena. As examples we have the starches, vegetable gums, polyvinyl esters and alcohols, cellulose derivatives, and proteins. These substances do not in general have the unbalanced hydrophilic structure, nor do they have the pronounced effect on surface tension characteristic of the soap-like surface active agents. Consequently they can logically be considered in a different class. Their mode of behavior in surface phenomena will be discussed in the theoretical part of this book for purposes of completeness.

We have included solubility as a necessary characteristic of a surface active agent. There is, however, a large class of insoluble materials which are capable of spreading on the surface of water (or other appropriate liquid) to form monomolecular films. These films do not constitute a third phase in any macroscopic sense. Their presence is detected largely by measurements of the energy and electrical characteristics of the treated surface. In this sense they are certainly surface active, at least in the special circumstance where they are in the form of a monomolecular layer on the base liquid surface. As we shall see, the true soluble surface active agents are active by virtue of a very similar type of surface film which forms on their solutions. This film is formed by a migration and orientation of molecules to the surface from within the bulk of the solution.

There are several large classes of compounds which are true surface active agents, but which are not used industrially for their effect in lowering surface tension. The well-known permanent water repellents are an excellent example. Zelan (du Pont), one of the most widely used products of this class, has the structure:

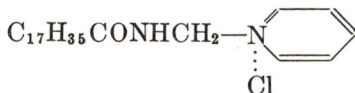

$$C_{17}H_{35}CONHCH_2-N \overset{Cl}{\underset{}{\bigcirc}}$$

It is a typical surface active compound of the cation active class, soluble in water, and forming foamy solutions of low surface tension. Its technical value, however, has nothing to do with these characteristics. Zelan is applied to fabrics and then chemically altered by heating to produce a thin, highly adherent layer of a water-repellent product on the fiber surface. The fact that the parent substance is surface active is immaterial in this application. The layer of hydrophobic material deposited on the fibers is directly observable under the microscope (with suitable techniques) and, being in no sense monomolecular, is therefore not within our limited definition.

In many technical operations such as the manufacture of glue, certain dyestuffs, etc., the presence of foam is troublesome. It has been found that certain substances such as spreading oils will prevent the foaming when added in small amounts. From a practical point of view, they are certainly surface active since they change a foaming liquid into a non-foaming one. Actually they may operate by several different physicochemical mechanisms, some anti-foaming agents being surface active and others being merely mechanical coverings.

The foregoing discussion illustrates the difficulty of defining surface active agents precisely, or in a manner satisfactory to all persons who will deal with them. The term surface active agent is used rather loosely to designate any substance whose presence in small amounts markedly alters the surface behavior of a given system. In a more restricted sense surface active agents are *soluble substances* whose presence makes the surface properties of a solution markedly different from those of the pure solvent. Almost all solutes are surface active to a certain extent, and therefore the distinction between a surface active and a non-surface active solute is quantitative rather than qualitative in nature. The borderline is not clean cut, nor is it of particular importance. The term is essentially descriptive rather than categorical, and the surface activity of any particular substance is best defined by the measurements of its various surface properties. Before proceeding to discuss either the theory or the applications of surface active agents, it is advisable to consider the various molecular structures characterized by surface activity, *i.e.*, the organic chemistry of surface active agents.

Soap is still the most widely used of all surface active agents. Within the past two decades, .however, the newer synthetic products have received an increasingly wider recognition. The chemistry and applications of soap have been the subject of many excellent texts, papers, and monographs over a long period of years, and will not be discussed in detail. The outstanding disadvantage of soap is its instability to hard or acidic water. Most of the early synthetic surface active agents were developed specifically to overcome this defect of soap, particularly as applied to the textile industry. The earliest surface active agents to be developed for this purpose were probably the sulfonated oils, used as dyeing and wetting assistants. The first real impetus was given to the modern development of surface active agents, however, by World War I. The critical soap shortage in Germany at that time led to the development of the lower alkylated naphthalene sulfonates for use as soap substitutes. Although they were poor general-purpose detergents they did have the one great advantage of being completely resistant to hard water and even to acidulated solutions. This virtue was immediately recognized, and stimulated the search for other products which would have increased detergent powers while retaining the hard water stability. During the next twenty years a tremendous number of new detergents were made and patented. At the same time the development of surface active agents for non-aqueous systems was beginning to take place. The whole development pattern of the surface active agents was similar to that of the synthetic plastics and polymeric materials which occurred during the same years. Current developments are aimed largely at producing specific agents to meet very rigid specifications for specialized applications. They are also aimed at producing the general-purpose agents on a larger scale and at lower cost. It is estimated that the total production of synthetic detergents in the United States for 1945 was about 92 million pounds. It has been estimated that facilities installed and projected will have brought this figure to about 300 million pounds by the end of 1947. The present annual production of soap (all types) is about 3 billion pounds.

III. Characteristic Features of Surface Active Agents

It has already been noted that most surface active agents are characterized structurally by an elongated portion of low residual affinity, and one end of high residual affinity. Chemically, there are two great classes, the ionic or ionogenic, and the non-ionic. The non-ionic class (a class of growing importance) has non-ionizable high-affinity end groups, usually containing a number of oxygen, nitrogen, or sulfur atoms in non-ionizing configurations. This class is of primary importance in non-aqueous systems but a number of non-ionic surface active agents have been widely and successfully used in aqueous systems, both as emulsifiers and as straight detergents.

The ionogenic class of agents has two main divisions. If the elongated, low-affinity portion of the molecule is included in the anion in aqueous solution, the substance is called anion active or simply *anionic*. Sodium stearate is a typical anionic surface active agent since it ionizes in solution to form Na^+ and the long-chain stearate anion, $C_{17}H_{35}COO^-$, which we may consider responsible for the surface activity. The cation active or *cationic* surface active agents form a cation containing the elongated low-affinity portion of the molecule. Cetylpyridinium chloride is an example of this class. In aqueous solution it ionizes according to the equation:

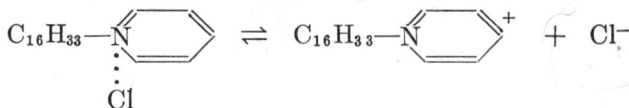

A third class of lesser importance exists wherein the molecule as a whole forms a zwitterion. Cetylaminoacetic acid, for example, comes to the following equilibrium in aqueous solution:

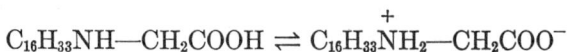

$$C_{16}H_{33}NH—CH_2COOH \rightleftharpoons C_{16}H_{33}\overset{+}{N}H_2—CH_2COO^-$$

We shall term this the ampholytic class of surface active agents. It has not been widely recognized as a separate class although numerous individual substances which fall into it have been described. This is probably because few of them have as yet attained widespread use. They are extremely interesting from the theoretical standpoint since the charge on the molecule as a whole varies with the pH of the medium, as in the case of any other ampholytic substance.

The anionic and cationic classes can be subdivided according to the nature of the ionogenic group. In the anionic class the most important ionogenic groups are the carboxy (—COOH), sulfonic acid (—SO$_3$H), and sulfuric ester (—OSO$_3$H). There are numerous others of lesser importance which will be considered later. The free bond indicated in the various group formulas is attached directly or indirectly to the elongated low-affinity portion of the molecule. In the cationic class the most prevalent ionogenic groups are the primary, secondary, and tertiary amino groups and the quaternary ammonium groups. Phosphonium and sulfonium groups are sometimes encountered but they are of far less importance.

The two essential groupings of a surface active molecule noted above are often linked directly together, as for example in the sulfated fatty alcohols or the alkyl pyridinium halides. In many cases, however, there is present an intervening group whose main purpose is simply to connect the hydrophobic and hydrophilic groups. Igepon T, for example, has the structure:

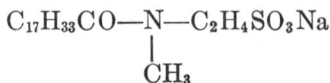

$$C_{17}H_{33}CO—N—C_2H_4SO_3Na$$
$$\overset{|}{C}H_3$$

The oleyl group is the essential hydrophobic portion and —SO₃Na the hydrophilic. They are connected by the acylmethyltaurine structure, a convenient and economical structure for the organic chemist to use. Many of the patented novel features in the field of surface active agents have been concerned with these connecting structures or links.

We have elected to classify the surface active agents according to the nature of the solubilizing group. They could in many cases be classified according to the connecting links, and many patents have been granted (in those nations where the patent law permits it) which cover both anionic and cationic detergents characterized by a definite type of linkage.

Classification on the basis of the nature of the hydrophobic group is also possible. The number of sources of hydrophobic groups, however, is limited, and therefore many substances of greatly differing properties would be grouped together under this system. For instance, if we classed together all the surface active agents having a cetyl radical as the hydrophobic group we would have such obviously disparate compounds as sodium palmitate, cetyl sulfate, cetylpyridinium chloride and cetyl ether of nonaethylene glycol all in one grouping. It is important to bear in mind, however, that a single convenient hydrophobic group can often be used to synthesize a wide range of surface active agents. This fact is of considerable economic importance. The discovery of new inexpensive hydrophobic groupings which can be utilized for making surface active agents invariably marks the start of a period of great advancement in the technology of these products. The really widespread interest in synthetic surface active agents began when it was found possible to use the cheap kerosene fractions of petroleum as hydrophobic groups.

The surface active agents are often loosely classified according to the purposes for which they are used. Thus the various products might be listed as wetting agents, detergents, emulsifying agents, hydrotropic substances, dyeing assistants, lime soap dispersers, etc. It is obvious that a single substance might fulfill several of these different functions quite efficiently, and this is in fact normally the case. Accordingly the use classification is of limited value.

IV. Nomenclature of Surface Active Agents

In the majority of technical surface active agents the long-chain low-affinity portion of the molecule is a mixture of homologous radicals rather than a clearly defined individual. For example, the highly important "sulfated lauryl alcohol" is the sodium salt of the half sulfuric ester of a mixture of fatty alcohols derived from coconut oil. These alcohols range from C_8 to C_{18} and are especially rich in the C_{12} and C_{14} members of the series   Another important type of surface active agent is made by chlorinating a paraffinic petroleum fraction in the kerosene range, condensing it with an aromatic hydrocarbon and sulfonating the resultant complex mixture of alkylated aromatic hydrocarbons. For the sake of convenience and brevity a variety of trivial names and trade names have become attached to these mixed radicals and to a number of the various detergent types. These will be indicated in the text as they appear and will be used thereafter. In general we shall refer to the elongated low-affinity portion of the molecule as the *hydrophobic* portion, and designate it by the generic symbol R. The portion of high residual affinity will be referred to variously as *hydrophilic, solubilizing,* or *polar.*

The newer synthetic surface active agents are most often referred to, for brevity's sake, as detergents and/or wetting agents. The generality implied in these terms and the implied distinction between them is apt to be misleading. Both detergent and wetting effects are often highly specific. They depend not only on the surface active agent which is used but also on the nature of the material to be cleaned or wet. Nevertheless the terms "detergent" and "wetting agent," are convenient, and much less cumbersome than "surface active agent." The reader will encounter these terms in their less rigorous sense in some places throughout the book where there is no danger of ambiguity.

The development of the synthetic surface active agents as products has taken place almost entirely in industrial laboratories. This fact has resulted in an inevitable arbitrariness with respect to their names. The same product (with possible variations in purity and concentration) may be known under a score of different names. Furthermore, products which have no chemical relationship to each other sometimes have very similar names. Smaller manufacturers will sometimes introduce a new composition under the same name borne by a discontinued product, a situation which has tended to correct itself as the industry has expanded. It is possible that in the future authoritative lists of surface active agents, giving their compositions and main properties, will appear. These would serve the same purpose that Rowe's *Colour Index* and Schulz's *Farbstofftabellen* serve in the dyestuff industry. From time to time limited lists of this nature are published.[4] One of the most useful of these is published by the American Association of Textile Chemists and Colorists in their year book.[5] In lists of this type the general structure is often indicated, but

[4] Van Antwerpen, *Ind. Eng. Chem.* **33**, 16–22 (1941); **35**, 126–130 (1943). McCutcheon, *Chem. Inds.* **61**, 811 (1947).

[5] *Year Book, American Association of Textile Colorists and Chemists.* Howe Publishing Co., New York.

only seldom is the complete structure, the percentage concentration, or the nature of the diluents and additives given.   In view of the frequency with which the commercial names of surface active agents, as well as their compositions, are being changed it is usually advisable for the user to obtain the fullest possible information concerning his product directly from the manufacturer.

V. Classification System for Surface Active Agents

Surface active agents may be classified on the basis of the uses to which they are put, on the basis of physical properties (water-soluble or solvent-soluble, etc.), or on the basis of chemical structure.   None of these systems is entirely satisfactory.   Certainly in discussing the chemical constitution of surface active agents a system based on differences in structure is desirable.   We have chosen to arrange the surface active agents primarily according to the nature of the solubilizing or hydrophilic group and secondarily according to the way in which the hydrophilic groups and hydrophobic groups are joined, i.e., directly or indirectly, and if indirectly according to the nature of the linkage.   The tertiary basis for classification is the nature of the hydrophobic group.   A separate class has been established to include a variety of structures used primarily in non-aqueous systems.   This classification is admittedly not logically connected with the rest of the system, and it includes a number of compounds which are already classified in other sections of the system.   This has been done in an effort to point out the newer fields to which the principles of surface activity have been applied rather than in any attempt at completeness.   The technology and theory of surface activity in non-aqueous systems has developed largely in the lubricants industry and the paint, varnish, and printing ink industries.   Technologists in these fields are concerned with what happens at the interface between oils and solids.   The solids may be metal bearings or finely divided pigment systems.   In either event it is known that relatively small amounts of certain solutes present in the oils may have a profound effect on phenomena at the interface.   No attempt has been made in this book to survey these broad fields completely.   Such a survey would require a book in itself.   Only a few of the highlights have been pointed out, and only those compounds mentioned which are reasonably closely related to the surface active agents used in aqueous systems.

Another separate class has been made to include the water-insoluble products which are used primarily as emulsifying agents.   This has been done because a certain amount of confusion exists with regard to emulsifying agents.   The novice in this specialized field is often surprised to find such dissimilar products as cholesterol, dodecylbenzene sulfonate, stearyl alcohol, and soap all grouped together as emulsifying agents. This class also includes many types of chemical structures and contains many compounds which are already separately classified according to structure. Again it is by no means complete, but includes only those substances which are reasonably closely related to the main group of water-soluble agents which form the subject of the book. Thus such well-known W/O (water-in-oil) emulsifying agents as ethylcellulose and the oil-modified alkyd resins are not included.

The inclusion of an ampholytic group is open to some question. It is apparent that many of the patentees who have developed ampholytic surface active agents did not have their ampholytic character specifically in mind. Nor do they refer to this ampholytic character as conferring any special properties to the product. Nevertheless they do represent a distinct chemical group, and in certain instances the fact that they exhibit the properties of both their anionic and cationic groups has been definitely pointed out. It is possible that future work in the field may bring out certain effects and properties of the ampholytes as surface active agents, which will distinguish them from the other classes much more clearly than is now the case. There are few if any commercially important products in the ampholytic class at present. Perhaps that is the reason they have hardly received enough attention to have been recognized as a separate class.

There are several important surface active agents which contain two different hydrophilic groups. The sulfated fatty acids, for example, contain a carboxyl group and a sulfuric ester group on the same molecule. The half amides and half esters of sulfosuccinic acid contain a carboxyl group and an alkane sulfonic acid group. No separate class has been made for this type of compound. Products falling in this category have been classified according to the most important (functionally) of the solubilizing groups present. The sulfated fatty acids resemble their allied sulfuric esters much more closely than they resemble the carboxylic soaps. Accordingly they are classed with the sulfuric esters. For the same reason the half amides and esters of sulfosuccinic acid are classed with the alkane sulfo acids.

The main classes of surface active agents which we shall consider are: (I) anionic, (II) cationic, (III) non-ionic, (IV) ampholytic, (V) water-insoluble emulsifying agents, and (VI) substances that are surface active in non-aqueous systems.

Under each of these main headings, subclasses can be established logically, using structure as the basis of classification as shown below. Accordingly, we have prepared a rather idealized listing of subheadings which we have not followed in full detail in our discussion of various surface active agents because of considerations of convenience.   In some instances, the economic and processing relationships are such that we have felt compelled to group together certain products which, from a strictly logical standpoint, might be classified separately.   Another important consideration leading to deviation from the idealized scheme presented below is the fact that some of the subclasses listed are virtually empty.   We believe, however, that the classification scheme as presented will serve to give the reader a bird's-eye view of the field even though it would have been awkward to follow the scheme rigidly in developing our discussion.

I. Anionic

A. Carboxylic Acids

1. Carboxyl joined directly to the hydrophobic group (subclassification on basis of the hydrophobic group, *e.g.*, fatty acid soaps, rosin soaps, etc.).
2. Carboxyl joined through an intermediate linkage.
   a. Amide group as intermediate link.
   b. Ester group as intermediate link.
   c. Sulfonamide group as intermediate link.
   d. Miscellaneous intermediate links, ether, —SO₂—, —S—, etc.

B. Sulfuric Esters (Sulfates)

1. Sulfate joined directly to hydrophobic group.
   a. Hydrophobic group contains no other polar structures (sulfated alcohol and sulfated olefin type).
   b. Sulfuric esters with hydrophobic groups containing other polar structures (sulfated oil type).
2. Sulfate group joined through intermediate linkage.
   a. Ester linkage (Arctic Syntex M. type).
   b. Amide linkage (Xynomine type).
   c. Ether linkage (Triton 770 type).
   d. Miscellaneous linkages (*e.g.*, oxyalkylimidazole sulfates).

C. Alkane Sulfonic Acids

1. Sulfonic group directly linked.
   a. Hydrophobic group bears other polar substitutents ("highly sulfated oil" type).   Chloro, hydroxy, acetoxy, and olefin sulfonic acids (Nytron type).
   b. Unsubstituted alkane sulfonic acids (MP 189 type; also cetane sulfo acid type).
   c. Miscellaneous sulfonic acids of uncertain structure, *e.g.*, oxidation products of sulfurized olefins, sulfonated rosin, etc.
2. Sulfonic groups joined through intermediate linkage.
   a. Ester linkage
      1. RCOO—X—SO₃H (Igepon AP type).
      2. ROOC—X—SO₃H (Aerosol and sulfoacetate type).
   b. Amide linkage
      1. RCONH—X—SO₃H (Igepon T type).
      2. RNHOC—X—SO₃H (sulfosuccinamide type).

I. Anionic (*continued*)

 c. Ether linkage (Triton 720 type).
 d. Miscellaneous linkages and two or more linkages.

D. Alkyl Aromatic Sulfonic Acids

 1. Hydrophobic group joined directly to sulfonated aromatic nucleus (subclasses on basis of nature of hydrophobic group. Alkyl phenols, terpene, and rosin-aromatic condensates, alkyl aromatic ketones, etc.)
 2. Hydrophobic group joined to sulfonated aromatic nucleus through an intermediate linkage.
  a. Ester linkage (sulfophthalates, sulfobenzoates).
  b. Amide and imide linkages.
   (1) R—CONH—ArSO$_3$H type.
   (2) Sulfobenzamide type.
  c. Ether linkage (alkyl phenyl ether type).
  d. Heterocyclic linkage (Ultravon type, etc.)
  e. Miscellaneous and two or more links.

E. Miscellaneous Anionic Hydrophilic Groups

 1. Phosphates and phosphonic acids.
 2. Persulfates, thiosulfates, etc.
 3. Sulfonamides.
 4. Sulfamic acids, etc.

II. Cationic

A. Amine Salts (Primary, Secondary, and Tertiary Amines)

 1. Amino group joined directly to hydrophobic group.
  a. Aliphatic and aromatic amino groups.
  b. Amino group is part of a heterocycle (Alkaterge type).
 2. Amino group joined through an intermediate link.
  a. Ester link.
  b. Amide link.
  c. Ether link.
  d. Miscellaneous links.

B. Quaternary Ammonium Compounds

 1. Nitrogen joined directly to hydrophilic group.
 2. Nitrogen joined through an intermediate link.
  a. Ester link.
  b. Amide link.
  c. Ether link.
  d. Miscellaneous links.

C. Other Nitrogenous Bases

 1. Non-quaternary bases (classified as guanidine, thiuronium salts, etc.).
 2. Quaternary bases.

D. Non-nitrogenous Bases

 1. Phosphonium compounds.
 2. Sulfonium compounds, etc.

III. Non-ionic

A. *Ether Linkage to Solubilizing Groups*
B. *Ester Linkage*
C. *Amide Linkage*
D. *Miscellaneous Linkages*
E. *Multiple Linkages*

IV. Ampholytic

A. *Amino and Carboxy*

   1. Non-quaternary.
   2. Quaternary.

B. *Amino and Sulfuric Ester*

   1. Non-quaternary.
   2. Quaternary.

C. *Amino and Alkane Sulfonic Acid*
D. *Amino and Aromatic Sulfonic Acid*
E. *Miscellaneous Combinations of Basic and Acidic Groups*

V. Water-Insoluble Emulsifying Agents

A. *Ionic Hydrophilic Group*
B. *Non-ionic Hydrophilic Group*

VI. Non-aqueous Systems

A. *Acids*   C. *Esters*   E. *Halogen Compounds*
B. *Salts*   D. *Hydroxy Compounds*   F. *Miscellaneous and Mixed Groups*

Classification of Hydrophobic Groups

The above classification scheme is in general accord with the trends implicit in American patents and publications. Many German patentees, however, will disclose a new method of obtaining a certain hydrophobic residue, and in a single basic patent describe its combination with a wide variety of solubilizing groups. A typical illustration of this is the use of the diisobutylphenyl group. Diisobutene, an inexpensive olefinic hydrocarbon, condenses very readily with phenol to give a *p-tert*-octylphenol:

$$C_8H_{17}-\!\!\langle\ \rangle\!\!-OH$$

By synthetic methods which will be more or less "obvious to one skilled in the art," as patent terminology has it, the following products may be made (R stands for the tertiary octyl phenyl radical, C_8H_{17}—C_6H_4—; the class symbol refers to the system given above):

$$R—O—CH_2COOH \dots \dots \text{I-}A\text{-2-}d$$

$$R—O—CH_2CONHCH_2COOH \dots \dots I A\text{-2-}a \text{ and } d$$

$$R—O—CH_2CONHC_2H_4SO_3H \dots \dots \text{I-}C\text{-2-}b \text{ and } c$$

$$R—O—C_2H_4SO_3H \dots \dots \text{I-}C\text{-2-}c$$

$$R—O—(C_2H_4O)_nH \dots \dots \text{III-}A$$

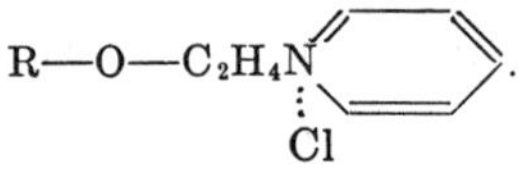

$$R—O—C_2H_4N\langle\ \rangle \dots \dots \text{II-}B\text{-2-}c$$
$$\overset{\vdots}{Cl}$$

$$R—O—C_2H_4NHC_2H_4SO_3H \dots \dots \text{IV-}C$$

These are only a few of the structures that can be built around the di-isobutylphenyl radical as a hydrophobic group. Not all of them are mentioned in any single patent, but in this and similar cases as many as three or four different classes are mentioned in one patent. Some of the older patent claims are sufficiently broad to cover even more classes.

In view of these considerations it is important to examine some of the more usual types of hydrophobic groups and the ways in which they are obtained.

(1) The straight alkyl chains of eight to eighteen or more carbon atoms are derived from the natural fatty acids. These may be used as such or may be converted to acid chlorides for use in acylation reactions. The acids or their esters may also be reduced to the corresponding fatty alcohols or through the nitriles to the fatty amines. These in turn can be used as intermediates in a variety of reactions to introduce the long alkyl chain. The natural fatty acids are possibly the most important raw material for hydrophobic groups. A wider variety of detergents has been derived from them, directly or indirectly, than from any other single source.

(2) The lower alkyl groups of three to eight carbon atoms are frequently attached to aromatic nuclei such as benzene or naphthalene, the combination forming a useful type of hydrophobic group. The sources of lower alkyl groups are usually the alcohols. The lower alkyl groups are sometimes used by themselves as hydrophobic groups, as in the case of the dialkyl sulfosuccinates:

$$ROOC—CH_2—\underset{\underset{SO_3H}{|}}{CH}—COOR$$

In this important series of wetting agents the R groups used commercially vary from C_4 to C_8.

The octyl group, C_8H_{17}—, is widely encountered in surface active agents, both by itself and condensed with aromatic nuclei. The three major sources of octyl groups are (1) capryl alcohol (octanol-2) obtained by the pyrolysis of ricinoleates; (2) 2-ethylhexanol, a pure synthetic product built up from acetaldehyde; (3) diisobutene, which is made by the dimerization of butene or by treating isobutanol or tertiary butanol with strong sulfuric acid.

(3) Propene, isobutene, and some of the isomers of pentene and hexene can be readily polymerized to a low degree yielding branched-chain mono-olefins of eight to twenty or more carbon atoms. These olefins have been widely used as hydrophobic group sources. The di-, tri- and tetraiso-butenes are best known in this country. One of the widely used detergents in Germany during the recent war was an alkyl benzene sulfonate in which the alkyl group was a tetramer of propene, i.e., a twelve-carbon olefin. This was condensed with benzene and the product subsequently sulfonated.

(4) Petroleum hydrocarbons in the range of eight to twenty or more carbon atoms are a very important raw material for hydrophobic groups. The kerosene, light oil, and paraffin wax fractions are commonly used and may be treated in a variety of ways: (a) They may be chlorinated and subsequently condensed with aromatic nuclei or dehydrohalogenated to give olefins. (b) They may be converted directly to sulfonyl chlorides by reaction with sulfur dioxide and chlorine (the Reed reaction). (c) They may be oxidized to carboxylic acids, which in turn can be used in the same ways as the natural fatty acids.

(5) Certain types of petroleum yield naphthenic acids in the normal course of refining. These are surface active in themselves and have also been used as a source of hydrophobic groups in derived surface active agents. In the refining of petroleum oils by washing with sulfuric acid the surface active petroleum sulfonates are obtained.

(6) The higher alcohols and hydrocarbons which are obtained by the reaction of carbon monoxide or carbon dioxide with hydrogen in the Fischer-Tropsch synthesis, or as by-products in the methanol synthesis have been widely used (particularly in Germany) as sources of hydrophobic groups.

(7) The rosin acids are most often used as such for surface active agents, but they have also been used to make up the hydrophobic groups of derived detergents and wetting agents. For this purpose they can be used in the same manner as the fatty acids.

(8) The terpenes and terpene alcohols have also been utilized, largely as alkylating agents for aromatic nuclei. The alkyl aromatic compounds thus formed are in some instances excellent hydrophobic groups.

The above are among the more important sources of hydrophobic groups although several others have been described.

The methods of linking together the hydrophobic and hydrophilic groups are numerous and some are startling in their ingenuity. The most commonly used structures are as follows:

(1) Direct linkage, as in the n-alkyl sulfuric esters, alkane sulfonic acids, and alkyl benzensulfonates.

(2) Linkage through an ester group. The hydrophobic group bears either a carboxyl or an alcoholic hydroxyl and the solubilizing group is attached to an intermediate structure which also bears a hydroxyl or a carboxyl. The latter groups are esterified with their counterpart on the hydrophobic chain.

(3) Linkage through an acid amide group. This is analogous to the ester linkage.

(4) Linkage through an ether group. This is quite common in the nonionic series; less common among the anionics and cationic series.

(5) Linkage through two or more intermediate groups; the ether-ester and ether-amide combinations are encountered fairly frequently.

The above types of linkage account for a very high percentage of all commercial surface active agents.

VI. Patent Situation

The source literature on the chemical structure and preparation of surface active agents consists almost entirely of patents. The majority of those patents which refer to products of well-established structure and most of the important patents relating to commercially used products have been included in the references. These cover only a fraction of the total number of patents issued in the field. There are numerous patents in which the products are of uncertain structure, and are defined in terms of the process by which they are made. Some of these cover important products. The majority of them, however, are of little practical interest. There is also a vast number of patents covering mixtures or compositions comprising various surface active agents and adapted to special purposes. Relatively few of these are of general importance. Still other groups of patents cover minor details in the manufacturing processes for previously disclosed products. Some refer to methods of obtaining specific physical forms such as beads, flakes, gels, clear solutions, etc.

There is also, unfortunately, a considerable number of patents in which it is doubtful if the disclosed structures represent the actual product of the described processes. Finally there is the large group of patents (mostly foreign) which cover various applications and uses of specific surface active agents. Some of these are important in introducing the whole range of surface active agents to new industrial fields, and they will be referred to in the last part of this book (Chapters 17–28).

The writers have made a serious effort to include most of the patents which have contributed significantly to the art. No attempt at exhaustive coverage has been made, however, and many patents and references which might be of great historical or legal interest have doubtless been omitted.

PART I

PROCESSES FOR SYNTHESIZING AND MANUFACTURING SURFACE ACTIVE AGENTS

1. Anionic Surface Active Agents

CHAPTER 2

Carboxy Acids

I. Carboxy Groups Linked Directly to the Hydrophobic Group

(a) *Natural Fatty Acids*

The oldest, best-known, and by far the most important surface active agents are the soaps. Since soaps are, strictly speaking, salts of naturally occurring fatty acids, it is outside the scope of this book to discuss their ramified technology and chemistry in any detail. The subject of soaps would of itself require a separate volume, and, in fact, several excellent specialized treatises are available. A large portion of our theoretical knowledge of surface activity is based on experimental work with soaps. Since this will be considered at length in subsequent chapters, and since soaps are the commonest surface active agents, a very brief survey of their chemistry seems amply justified.

Various types of soap-forming fatty acids occur in nature as glycerol esters (the fats and oils) and as esters of the higher aliphatic alcohols (the waxes). The fatty acids present in these natural products include straight-chain and branched-chain structures, both saturated and unsaturated, and also certain hydroxy substituted acids such as the ricinoleic acid of castor oil and certain acids present in wool fat.

The number of carbon atoms in a fatty acid is of decisive importance in determining whether or not its salts exhibit soap-like properties. Thus on passing from the lower to the higher members of the homologous series of straight-chain, saturated fatty acids, soap-like properties begin to appear as low in the series as eight carbon atoms. Above about 20 to 22 carbon atoms, the soaps become too insoluble to be generally useful. The most important soap-forming fatty acids are, from a practical point of view, the C_{12} to C_{18} straight-chain, saturated acids and oleic acid, the C_{18} straight-chain, singly unsaturated acid.

Although soaps of the alkali metals are usually made by direct saponification of the fats and oils, an increasing amount of soap is being made by neutralization of the free fatty acids which are prepared from fats and oils in separate manufacturing processes. The fatty acids are often distilled or otherwise purified to remove unsaponifiable matter and other undesirable impurities present in the oil. Some manufacturers offer the individual members of the series in a relatively high state of purity, obtained by fractional distillation. Most often, however, the mixture which is obtained by simple hydrolysis (supplemented at most by a simple and incomplete separation) is marketed. Thus the mixed fatty acids of tallow, coconut oil, palm oil, etc. are common raw stock for the soap maker. Tallow acids are usually separated by filtration or pressing into "red oil," which is largely oleic acid, and "stearic acid." This "stearic acid" of commerce is sold as single-, double-, or triple-pressed, depending on the extent to which the oleic acid has been removed. It is largely a mixture of stearic and palmitic acids rather than straight stearic acid.

The suitability of an oil for soap making depends largely on the nature of the fatty acid mixture which it contains. A high content of diene or polyene acids is undesirable in soaps for most purposes. The reader is referred to Hilditch's excellent treatise for a compilation of the fatty acid composition of various fats and oils.

Besides the alkali metal bases a wide variety of organic bases and ammonia itself are used to make soaps for special purposes. These are all made by neutralization of the fatty acid with the base. Mono-, di-, and triethanolamine, the isopropanolamines and morpholine are widely used to make solvent-soluble soaps and emulsifying agents. The alkanolamines made from the nitroparaffins are similarly used.[1] These products are seldom encountered as such. They are often made in situ, in admixture with solvents or other compositions of which they are a component.

Soaps made by the saponification of waxes are likewise seldom isolated. Such isolation would be difficult, since it would involve a separation of the soap from the water-insoluble, high-boiling fatty alcohols, hydrocarbons, sterols, etc. normally present in the wax. Large amounts of such waxes as candellila, beeswax, lanolin, and spermaceti are saponified with alkali in the manufacture of emulsifying agents and cosmetic creams. The resultant product is an intimate mixture of a water-soluble surface active agent (the soap) and a water-insoluble surface active agent (the fatty alcohol, which is surface active in many non-aqueous systems). This mixture is a better emulsifying agent in many cases than pure soap alone. Some of these waxes, lanolin for example, contain branched-chain acids whose properties are appreciably different from those of the commoner soap formers.

The salient disadvantage of the soaps is their instability toward heavy-metal ions, particularly the calcium and magnesium found in hard water, and toward acids. The calcium and magnesium salts of the fatty acids as

[1] U. S. Pat. 2,247,106, Vanderbilt to *Purdue Research Foundation*.

well as the free acids themselves are quite insoluble in water. This short-coming was possibly the most important single factor in spurring the development of the newer synthetic surface active agents. To offset this disadvantage soap has at least two major points of superiority, low cost and high detergent powers in most of the cleaning operations encountered practically.

(b) Other Natural Carboxy Acids

Another important group of carboxylic surface active agents is that of the rosin acids. The most important acidic ingredient of rosin is abietic acid.

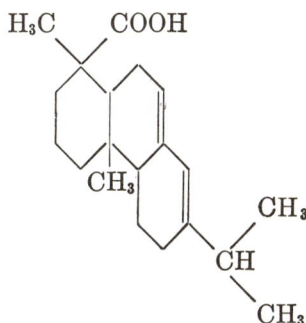

Other acids are also present, most of which appear to be isomers of abietic acid in which the double bonds are located in different positions. With metallic ions rosin forms soaps which resemble the fatty acid soaps. They are used largely in admixture with the fatty soaps to impart increased solubility and foaming power. They are especially useful in compositions for washing in sea water.[2]

Rosin soaps, however, have certain disadvantages. Like the fatty acid soaps, they are not resistant to lime or acid. Rosin soaps also have a marked tendency toward discoloration and instability on aging. In order to counteract this instability various modifying processes have been developed. On a commercial scale, rosin is modified in three ways: by hydrogenation, dehydrogenation, or polymerization. Hydrogenation elim-inates the double bonds and destroys colored or color-forming impurities. The resulting di- and tetrahydroabietic acids have good germicidal proper-ties. Dehydrogenation aromatizes the compounds and gives a product of increased resistance to oxidative deterioration, suitable as a raw material for making surface active agents. Rosin is readily polymerized, and the

[2] Van Zile and Borglin, *Oil and Soap* **21**, 164 (1944).

soaps of this polymerized material are more stable than raw rosin soaps. They are also stated to have better emulsifying properties.

Studies of the behavior of soap mixtures with the above types of modified rosins have been made by Van Zile and Borglin.[3] The rosins increase the solubility of the soaps, and, therefore, considerably increase the speed of lathering. They also increase the wetting power and do not seriously affect the detergency in concentrations of less than about 50%. Mixtures of hydrogenated rosin soaps with fatty acid soaps are suitable for laundry and toilet purposes,[4] and were widely used during World War II.

Soaps from partially or completely hydrogenated rosin may be further modified, e.g., by heating.[5]

More complex modifications of the rosin carboxylic acids have been effected by hydrogenating condensates of rosin, formaldehyde, and acetic anhydride. The product is an acetylated carbinol of hydrogenated rosin and forms soaps of high wetting power.[6]

Tall oil, a by-product of the sulfate process for making wood pulp consists largely of a mixture of fatty acids and resin type acids. The resin acids present in tall oil appear to be different from abietic acid in many of their properties but to have the same partly saturated phenanthrene carbon skeleton. Raw tall oil is very dark in color and has an extremely disagreeable odor. The resin acids tend to crystallize out but the separation thus effected is incomplete. Considerable work has been done on the refining of tall oil and particularly on the separation of the resin and fatty components. Numerous patents have been granted on various methods of refining, improving color and odor, and separating the acids. Soaps made from tall oil or the various refined tall oils are among the lowest-cost surface active agents. They have become more popular as their color and odor have been improved, and are used as detergents and emulsifying agents to replace or supplement the more expensive soaps.

The *naphthenic acids*, present in certain types of petroleum from which they can be extracted by alkaline treatment, form soaps of high surface activity. Crude naphthenic acids are usually quite dark in color and contain a large percentage of unsaponifiable matter. Refined grades, usually prepared by vacuum distillation, are much lighter in color and may contain as little as 5–10% unsaponifiable (hydrocarbon oil). Even the refined acids have a characteristic unpleasant odor.

Structurally the naphthenic acids are saturated aliphatic monocarboxylic

---

[3] Van Zile and Borglin, *Oil and Soap* **22**, 331–4 (1945).
[4] U. S. Pat. 2,285,333, Humphrey to *Hercules Powder Co.*
[5] U. S. Pat. 2,327,132, Schantz to *Hercules Powder Co.*
[6] U. S. Pat. 2,383,289, Bried to *Hercules Powder Co.*

acids containing alkyl-substituted cyclopentane or cyclohexane rings. The carboxy group may be attached to a ring, but is usually in the terminal position on one of the side chains, which range in size from six to about thirty carbon atoms. The larger naphthenic acid molecules contain two, three, or four rings and commercial mixtures usually range in average molecular weight from 200–300. Naphthenic acids are mainly used in the form of their heavy-metal salts. The copper salt is widely used for mildewproofing fabrics, since it is an excellent fungicide, easily applied, and quite stable to leaching. The lead salt has been used in extreme-pressure lubricants. The lead, cobalt, manganese, and zinc salts are widely used as driers in the paint and varnish industry.

The water-soluble soaps, however, have also been studied and proved to be excellent surface active agents. Levinson and Minich[7] examined the potassium, sodium, and ethanolamine soaps of a highly purified naphthenic acid of molecular weight 255. These are highly soluble and do not show the tendency to gel which is characteristic of fatty acid soaps, nor are they easily salted out. They are good foamers, detergents, and emulsifying agents. They also have marked hydrotropic properties and have a high solvent power for hydrocarbon oils. In contrast with the fatty acid soaps, they are potent germicides.[8] The water-soluble naphthenates do not appear to be used at present on any considerable commercial scale.

Among the most interesting naturally occurring surface active agents are the bile acids. These substances occur in bile to the extent of about 6–7%. In the alkaline environment of the intestine they exist as the sodium salts and their function is the emulsification of fats in the food and the facilitation of their digestion. Structurally they are closely related to cholesterol. Cholic acid from ox bile has the structure:

$$H_3C \qquad CH_2CH_2COOH$$

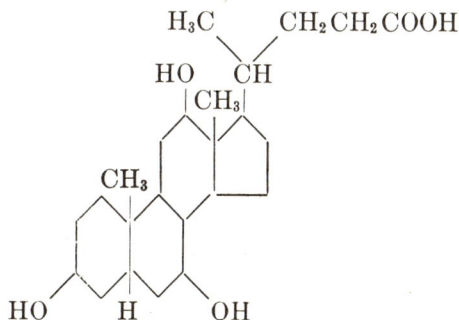

---

[7] Levinson and Minich, *Soap* **14**, No. 12, 24–6, 69–70 (1938); *Textile Colorist* **60**, 698–9, 705 (1938).

[8] Bachrach, *Soap* **10**, No. 11, 21–3 (1934).

It exists conjugated with glycine as glycocholic acid and with taurine as taurocholic acid. Other bile acids, present in smaller amounts, have similar carbon skeletons.

(c) Synthetic Carboxy Acids

Under normal economic conditions the natural fatty acids of the fats and oils are the most economical raw materials for ordinary soaps, and relatively little effort has been spent in developing other raw materials. In Germany, however, the national plans for economic self-sufficiency called for the development of detergents based on materials other than the natural fats. The production of synthetic aliphatic carboxy acids suitable for soap was accordingly developed to a high degree. The most important process involves the direct oxidation of aliphatic hydrocarbons (usually paraffin wax) or hydrocarbons from the Fischer-Tropsch synthesis to a mixture of oxygenated compounds including alcohols, ketones, and acids. The acids are then separated from the unsaponifiable material. The oxidation is usually carried out by blowing air through the liquid hydrocarbon at temperatures ranging from 100–180°C. in the presence of a catalyst. Potassium permanganate or other compounds of manganese are good catalysts for this reaction. Numerous others have been mentioned. The patent literature on both the oxidation and separation stages is fairly extensive.[9] With certain waxes good results are obtained[10] without any catalyst.

The processes used commercially in Germany during the war are described by Sheely[11] as follows. The preferred raw materials are: (1) scale

---

[9] Brit. Pat. 433,305, to *I. G. Farbenindustrie*; Ger. Pat. 627,808, Franzen to *I. G. Farbenindustrie*; Ger. Pat. 621,979, Dietrich v. Luther to *I. G. Farbenindustrie*; French Pat. 852,597, to *Vereinigte Ölfabriken Hubbe & Farenholtz*; Ger. Pat. 702,143, Keunecke and Becke to *I. G. Farbenindustrie*; Ger. Pats. 704,428 and 706,951, to *Markische Seifen-Industrie*; Brit. Pat. 494,853, to *I. G. Farbenindustrie*; Brit. Pat. 497,170, to *I. G. Farbenindustrie*; Ger. Pat. 684,968, Harder to *I. G. Farbenindustrie*; U. S. Pat. 1,943,427, Franzen, Beller, and Luther to *I. G. Farbenindustrie*; U. S. Pat. 2,193,321, Leithe to *I. G. Farbenindustrie*; U. S. Pat. 2,230,582, Beller to *Jasco, Inc.*; U. S. Pat. 2,274,632, Owen to *Standard Oil Development Co.*; Brit. Pat. 507,521, to *Hubbe u. Farenholtz*; Ger. Pat. 695,647, to *I. G. Farbenindustrie*; U. S. Pat. 1,999,184, Ellis to *Standard Oil Development Co.*; Bauschinger, *Fette u. Seifen* **45**, 629 (1938). U. S. Pat. 2,216,238, Harder to *Jasco, Inc.* Review articles include: Strauss, *Fettchem. Umschau* **41**, 45–51 (1934); Heublyum, *Seifensieder-Ztg.* **62**, 421 (1935); Imhausen, *Kolloid-Z.* **85**, 234–46 (1938); Neu, *Pharm. Zentralhalle* **84**, 239–44 (1943).

[10] See Berlin and Kuznetzova, *Groznenskiĭ Neftyanik* **5**, No. 1–2, 52–6 (1935); *C.A.* **30**, 3975; Dovankov, *Trans. VI Mendeleev Congr. Theor. Applied Chem.* **1932**, **2**, Pt. 1, 867–76 (1935); *Chem. Abstracts* **30**, 3975.

[11] Sheely, *PB 2422*, Office of Technical Services, Dept. of Commerce, Washington,

wax, (*2*) wax produced by the low-temperature hydrogenation of lignite, (*3*) wax produced by the Fischer-Tropsch process. If necessary the wax is pretreated to remove unsaturates, sulfur, and other objectionable impurities. This is done by heating and agitating with $AlCl_3$ at 100°C. for one to two hours, settling and separating the sludge, and finally filtering the molten wax through Tonsil clay. The purified wax is charged into oxidation towers constructed of aluminum. The catalyst, consisting of 0.1–0.15% $KMnO_4$ and 0.05–0.15% Na_2CO_3, is added in the form of a concentrated aqueous solution. The temperature is raised to 130–150°C. and air is blown in at atmospheric pressure or slightly above (1.2 atmospheres). As the exothermic reaction proceeds, it is necessary to cool the towers with external sprays and the temperature is lowered to 80–120°C. The time of heating and blowing at the higher temperature varies from fifteen minutes to five hours, at the lower temperature from ten to thirty hours. The exit gases from the reactors contain lower-molecular-weight acids, which are recovered by washing with water. The oxidized wax, containing about 35% fatty acids, is washed with water to remove catalyst salts and the remaining lower acids. It is then saponified with 25% aqueous NaOH at about 90°C. and the resulting mixture is allowed to separate into two layers. The upper layer consisting of unreacted wax and non-acidic oxidized material is separated for recycling. Since the lower layer still contains considerable unsaponifiable matter, a further separation is effected under pressure at 150–170°C. The soap layer is finally flash distilled to remove water and remaining volatile material. The anhydrous soap is then made up to a 10% aqueous solution and acidified to recover the fatty acids. The crude acids have an acid number of 230–245 and a saponification number of 250. They are subsequently fractionally distilled into cuts suitable for various purposes.

Vapor phase oxidation of paraffin hydrocarbons has also been described. The vapor mixture contains atmospheric oxygen in proportions lower than the explosive limit, and various catalysts are stated to be effective.[12]

An indirect method of obtaining carboxy acids from petroleum sources involves reacting olefins in the surface active range of molecular weight with carbon monoxide and hydrogen at elevated temperatures and in the presence of catalysts.[13] Saturated aldehydes are formed which can readily be oxidized to carboxylic acids.

D. C. This report also contains the translation of an excellent review paper with complete references on paraffin wax oxidation by Ludwig Mannes, *Die Chemie* **51**, No. 1/2, 16 (Jan. 8, 1944).

[12] U. S. Pats. 2,215,472–4, to King and Sheely.

[13] Ger. Pat. 734,219, Mannes and Köhler to *Henkel et Cie.*

α-Alkoxy fatty acids with alkoxy groups of five to ten carbon atoms are said to be obtained as by-products in paraffin oxidation.[14]   They are highly surface active.

The electrolytic oxidation of paraffin wax to fatty acids has been described by Nabuco de Araujo.[15]   The anolyte is an emulsion of paraffin in 20% H_2SO_4 together with a salt of cerium, chromium, vanadium, or titanium as a catalyst.   The same investigator[16] reports that simple boiling of an aqueous solution of ceric sulfate with paraffin gives a good yield of fatty acids.

A variety of other synthetic carboxy acids in which the carboxyl group is attached directly to the hydrophobic group have been described, and their alkali metal salts suggested as surface active agents.   Few if any of these appear to have attained commercial significance.

Branched-chain aliphatic acids of six to ten carbon atoms, such as 1,3-dimethylvaleric acid have been disclosed[17] as wetting agents, particularly in mercerizing baths which contain 15–40% NaOH.

The normal soap-forming fatty acids which have been substituted on the α-carbon atom with a variety of hydrophilic groups such as:

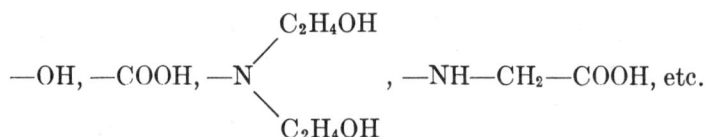

$$\text{—OH, —COOH, —N}\begin{matrix} \diagup\ C_2H_4OH \\ \diagdown\ C_2H_4OH \end{matrix}\text{, —NH—CH}_2\text{—COOH, etc.}$$

have been described and are stated to have better wetting properties than the parent compounds.[18]

Polyhalogenated fatty acids have been sulfurized and the resulting thiol compounds oxidized to disulfides to produce detergents and emulsifying agents.[19]

Emulsifying agents have been produced by heating polyhalogenated fatty acids with alkalis whereby a mixture of hydroxylated and/or unsaturated acids are formed.[20]

Soap-forming carboxylic acids have been made from twelve- to eighteen-carbon-atom monoolefins (such as the olefinic products of the hydrogena-

[14] Ger. Pat. 738,445, Kirstahler to *Deutsche Hydrierwerke A.-G.*

[15] Nabuco de Araujo, *Chimica (Brazil)* **1**, 281–2 (1933). *Chem. Abstracts* **28**, 3309.

[16] Nabuco de Araujo, *Chimica e Industria (Brazil)* **2**, 4–6 (1934). *Chem. Abstracts* **28**, 3310.

[17] French Pat. 768,355 to *E. I. du Pont de Nemours & Co.*, U. S. Pat. 2,033,819, Downing and Clarkson to *du Pont de Nemours & Co.*

[18] French Pat. 789,004, to *I. G. Farbenindustrie.*

[19] Ger. Pats. 539,449; 545,693; Keller to *I. G. Farbenindustrie.*

[20] U. S. Pat. 1,959,478, Keller to *General Aniline Corp.*

tion of carbon monoxide) by first oxidizing these to glycols, using H_2O_2, per salts, permanganate, or other suitable oxidizing agent. The glycols are then fused with caustic alkali at 250–280°C. forming[21] the carboxylic acid salt. A similar[22] process starts with unsaturated fatty alcohols and fuses them with caustic at 200–280°C. under a pressure of 5 atmospheres.

An interesting series of carboxylic detergents has been made by taking advantage of the ability of maleic anhydride to undergo addition and condensation reactions on its double bond. Maleic acid or anhydride reacts with olefins such as di- or triisobutylene or with alkyl halides of five to sixteen carbon atoms to form an alkylene succinic acid:

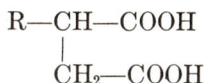

$$R—CH—COOH$$
$$|$$
$$CH_2—COOH$$

where R contains a double bond. The alkali metal salts of this acid, or of its monoesters or monoamides, are typical surface active agents.[23]

A hydrophobic chain can also be introduced into maleic acid by using diolefins (produced by the action of NaOH on chlorinated paraffin) in a Diels-Alder type addition reaction.[24]

Surface active aralkylcarboxy acids have been produced by condensing a long-chain olefin or alkyl halide with phenylacetic acid or its analogs, a typical example being the condensation[25] of a monochlorinated C_{14} synthetic hydrocarbon (from Fischer-Tropsch synthesis) with phenylacetic acid in the presence of a zinc catalyst. The alkylation takes place on the benzene nucleus and the carboxy group is unaffected.

Surface active aromatic carboxy acids have been prepared by starting with a long-chain alkyl benzene (such as dodecylbenzene or dibutylbenzene) and condensing with acetyl chloride in a Friedel-Crafts reaction to give the long-chain acetophenone. The methyl ketone group is then oxidized with hypochlorite to give the corresponding carboxylic acid.[26]

Alkylated phenols with side chains in the hydrophobic range have been converted to the corresponding salicylic acids by reaction with carbon dioxide according to the classical Kolbe method. Thus, p-isooctylphenol yields p-isooctylsalicylic acid, a surface active carboxy acid.[27]

[21] Brit. Pat. 492,595, to *Deutsche Hydrierwerke A.-G.*

[22] U. S. Pat. 2,159,700, Hennig to *Deutsche Hydrierwerke A.-G.*

[23] U. S. Pats. 2,283,214; 2,360,426; 2,380,699; Kyrides to *Monsanto Chemical Co.* U. S. Pat. 2,182,178, Pinkernelle to *I. G. Farbenindustrie.*

[24] Brit. Pat. 524,521, to Crawford, George, and *Imperial Chemical Industries Ltd.*

[25] Brit. Pat. 493,109, to *I. G.Farbenindustrie.*

[26] U. S. Pat. 2,195,198, Balle, Wagner, and Nold to *I. G. Farbenindustrie.* Brit. Pat. 497,353, to *I. G. Farbenindustrie.*

[27] French Pat. 847,716, to *I. G. Farbenindustrie.*

II. Carboxy Acids with Intermediate Linkage between the Carboxy and the Hydrophobic Groups

It has long been recognized that the lime resistance (and consequent hard-water serviceability) of a carboxylic detergent can be increased by increasing the total number of hydrophilic residues in the molecule. One of the most practical and effective ways of increasing the hydrophilic character of the surface active molecule is linkage of the hydrophobic chain to the carboxy-bearing radical through an intermediate linkage of a polar or hydrophilic nature. There has been considerable investigation along these lines and at least three product types of this general nature have attained commercial significance in Germany, one of which (Lamepon) has also been successfully produced and marketed in the United States. One very fruitful type of synthesis involves the condensation of a higher fatty acid chloride with a lower primary or secondary amino acid to form a detergent having an acyl amido group interposed between the hydrophobic groups and the solubilizing carboxy groups.

The carboxylic detergents, whether lime resistant or not, have certain valuable properties which are not shared by the commoner sulfonate or sulfate detergents. Among these are the high detergent efficiency on cotton goods, a mildness on the human skin which makes them desirable for personal use, and their own characteristic adsorption and emulsification behavior.

(a) Intermediate Acid Amide Linkages

One of the early patents[28] in this field describes the condensation of a higher fatty acid chloride with aspartic acid to produce a surface active agent suitable for wetting out in mercerizing baths. The condensation is carried out in dilute aqueous alkaline solution and proceeds according to the equation:

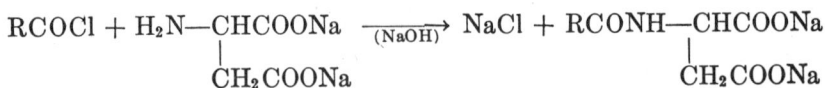

$$\text{RCOCl} + \text{H}_2\text{N—CHCOONa} \xrightarrow{\text{(NaOH)}} \text{NaCl} + \text{RCONH—CHCOONa}$$
$$\qquad\qquad\qquad | \qquad\qquad\qquad\qquad\qquad\qquad\qquad\qquad | $$
$$\qquad\qquad\quad \text{CH}_2\text{COONa} \qquad\qquad\qquad\qquad\qquad\quad \text{CH}_2\text{COONa}$$

R stands for a higher aliphatic radical.

Another early patent[29] describes a detergent and wetting agent of high

---

[28] Ger. Pat. 546,942, Hentrich, Keppler, and Hintzmann to *I. G. Farbenindustrie*.
[29] U. S. Pat. 1,973,860, Ulrich and Saurwein to *I. G. Farbenindustrie*.

lime resistance and acid resistance made by condensing a higher fatty acid chloride with N-β-hydroxyethylglycine. The product has the formula:

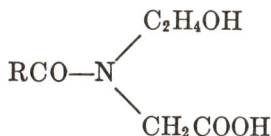

$$RCO-N\begin{cases} C_2H_4OH \\ CH_2COOH \end{cases}$$

A series of products which is reported to have been extensively used in Germany during the war is made by condensing a fatty acid chloride with sarcosine (N-methylglycine). The products[30] have the generic formula:

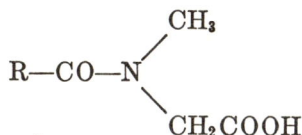

$$R-CO-N\begin{cases} CH_3 \\ CH_2COOH \end{cases}$$

They are marketed under the name Medialan. The fatty acid radical may be derived from coconut oil or oleic acid (the two most popular natural acids for this and many other synthetic detergents) or from the synthetic fatty acids produced by hydrocarbon wax oxidation. Sarcosine, once classed as a rare chemical, has become quite inexpensive in the German chemical economy. It is made from methylamine, formaldehyde, and hydrogen cyanide, all of which are inexpensive tonnage chemicals. Sarcosine is used extensively in certain other phases of German chemical industry, notably as an intermediate in the Rapidogen series of fast cotton dyestuffs. In choosing the fatty acyl sarcosides for large scale development it is possible that the low cost was as important a factor as the efficacy of the final detergent. The Medialans have good solubility and detergent power. They are sufficiently lime resistant to be fully effective in all but the hardest waters. They have been particularly favored for personal use because of their mild feel and good foaming properties.

More complex products closely related to the Medialans have been made by condensing surface active isocyclic carboxy acids, in the form of their acid chlorides, with lower aminocarboxy acids. A representative product[31] of this type is p-$tert$-octylphenoxy acetylglycine:

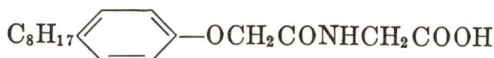

$$C_8H_{17}\langle\underline{\quad}\rangle-OCH_2CONHCH_2COOH$$

[30] Ger. Pat. 635,522, Hentrich, Keppler, and Hintzmann to *I. G. Farbenindustrie.* Brit. Pats. 459,039; 461,328; 456,142, to *I. G. Farbenindustrie.*
[31] U. S. Pat. 2,215,367, Balle and Schild to *I. G. Farbenindustrie.*

Similar products[32] have been made by etherifying fatty acyl ethanol-amides with chloroacetic acid:

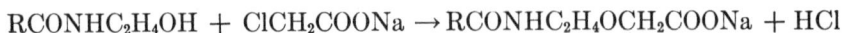

$$RCONHC_2H_4OH + ClCH_2COONa \rightarrow RCONHC_2H_4OCH_2COONa + HCl$$

Beside the individual amino acids mentioned above, mixtures of amino acids and polypeptides made by hydrolyzing proteins have been used to condense with fatty acid chlorides.

Lamepon is the best-known surface active agent of this class. It is made by hydrolyzing waste protein (usually scrap leather from the chrome tanning process, low-grade glue, or other easily hydrolyzable collagen) with alkali, thereby obtaining a hydrolyzate which contains amino acids and lower peptides. This mixture is then condensed with the desired fatty acyl halide. Although the dominating[33] patents in this group mention and stress the use of the lower polypeptides, it appears that in making commercial Lamepon the protein is largely hydrolyzed down to the individual amino acids. As with most products of this class the condensation is carried out by slowly feeding the acid chloride into a solution of the amino acids, maintaining efficient stirring and adding caustic soda to keep the mixture continuously alkaline. The temperature is kept at about 25–30°C. and may be raised to 60–70° to finish the reaction. The product may be isolated by acidification, since the free acids are only slightly soluble in water. Both the oleic and the coconut fatty derivatives are marketed in the form of viscous brownish liquids. These are aqueous solutions containing about 35% active ingredient and very small amounts of inorganic salt. The Lamepons are unusually resistant to lime and are good dispersing agents for lime soaps. They have good protective-colloid and emulsifying properties as well as a softening effect on fabrics. They have been recommended for use in cosmetic preparations, particularly shampoos, as well as in the textile industry.

Modifications of the Lamepon type have been disclosed[34] in which the amino acids or peptides are monobenzylated or otherwise aralkylated before condensing with the fatty acid chloride.

[32] French Pat. 794,012, to *I. G. Farbenindustrie*.

[33] U. S. Pat. 2,015,912, Sommer to *Chemische Fabrik Grünau, Landshoff u. Meyer A.-G.*; 2,041,265, Orthner and Meyer to *I. G. Farbenindustrie*. For compositions and applications, see: Ger. Pat. 692,070, to Grünau A.-G.; Ger. Pat. 697,324, Orthner and Meyer to *I. G. Farbenindustrie*; Brit. Pat. 413,016, to *Chemische Fabrik Grünau, Landshoff u. Meyer A.-G.*; Brit. Pats. 435,481; 450,467; and French Pats. 772,585; 770,636, to *I. G. Farbenindustrie*.

[34] U. S. Pat. 2,119,872, Wiegand to *Chemische Fabrik Grünau, Landshoff u. Meyer A.-G.* Ger. Pats. 670,096–7, to *Chemische Fabrik Grünau A.-G.*

Another modification of this type makes use of alkylated arylcarboxy acid chlorides of generic formula, R—Ar—R′—COCl, where R is a medium-sized hydrophobic group such as isooctyl (from diisobutene), or is an aromatic nucleus (benzene or naphthalene) and R′ is a connecting link, —CH$_2$—, —O—CH$_2$, —S—CH$_2$—, etc. These are condensed[35] with amino acids such as sarcosine or protein hydrolyzates. A typical product is:

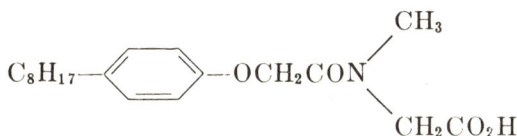

$$C_8H_{17}\text{—}\left\langle\bigcirc\right\rangle\text{—OCH}_2\text{CON}\begin{smallmatrix}\nearrow CH_3\\ \searrow CH_2CO_2H\end{smallmatrix}$$

They are claimed to have high detergent power.

Aromatic amino acids have been used instead of the aliphatics already mentioned to condense with fatty acid chlorides. The oleyl and stearyl derivatives of 3-aminobenzoic acid are stated[36] to be good dyeing assistants, softeners, and emulsifying agents.

Surface active agents containing an intermediate amide linkage have also been made by condensing a long-chain (or otherwise hydrophobic) amine with one carboxy group of a dicarboxylic acid.[37] Examples of this type are:

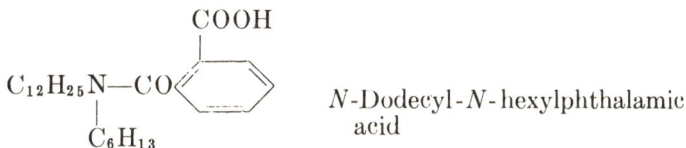

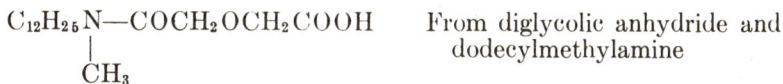

$C_{18}H_{35}NH\text{—COCOOH}$ Oleyloxamic acid

$$C_{12}H_{25}N\text{—CO}\left\langle\bigcirc\right\rangle\overset{\text{COOH}}{\underset{\displaystyle C_6H_{13}}{|}}$$

N-Dodecyl-N-hexylphthalamic acid

$$C_{12}H_{25}N\text{—COCH}_2\text{OCH}_2\text{COOH}\quad\underset{\displaystyle CH_3}{|}$$

From diglycolic anhydride and dodecylmethylamine

These products are good detergents and wetting agents and have remarkably high resistance to lime and acid. They do not appear to have attained any commercial significance, probably because of their inherent high cost.

[35] U. S. Pat. 2,215,367, Balle *et al.* to *I. G. Farbenindustrie.* Brit. Pat. 498,136, to *I. G. Farbenindustrie* French Pat. 839,600, to *I. G. Farbenindustrie.* For preparation of the R—Ar—R′—COOH, see Brit. Pat. 490,416, to *I. G. Farbenindustrie.*

[36] Ger. Pat. 548,816, Hentrich, Keppler, and Hintzmann to *I. G. Farbenindustrie.*

[37] U. S. Pat. 2,191,738, Balle to *I. G. Farbenindustrie.* French Pat. 795,662, to *I. G. Farbenindustrie.* Brit. Pat. 510,308, to *I. G. Farbenindustrie.*

Intermediate urea[38] linkages and urethan[39] linkages have also been described.　To produce the former type a hydrophobic primary or secondary amine is converted to the carbamyl chloride by reacting, for example, with phosgene.

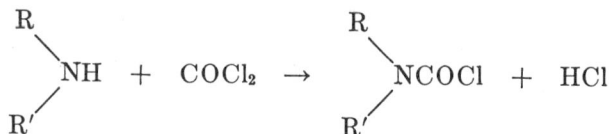

$$\begin{array}{c} R \\ \diagdown \\ NH \\ \diagup \\ R' \end{array} + COCl_2 \rightarrow \begin{array}{c} R \\ \diagdown \\ NCOCl \\ \diagup \\ R' \end{array} + HCl$$

This is then condensed with an amino acid or mixture thereof such as glycine, sarcosine, or partly or totally hydrolyzed protein.　A typical product would be:

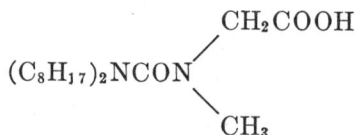

$$(C_8H_{17})_2NCON \begin{array}{c} \diagup CH_2COOH \\ \\ \diagdown CH_3 \end{array}$$

where C_8H_{17} represents the α-ethylhexyl radical.　The urethan type is made by condensing a hydrophobic alcohol such as lauryl alcohol with phosgene to form the chloroformate:

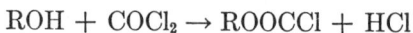

$$ROH + COCl_2 \rightarrow ROOCCl + HCl$$

This is then condensed with the amino acid, producing the surface active agent of generic formula:

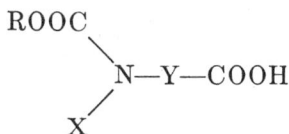

$$\begin{array}{c} ROOC \\ \diagdown \\ N-Y-COOH \\ \diagup \\ X \end{array}$$

where X represents hydrogen or R' and Y represents a hydrocarbon linking unit such as $-CH_2-$.　These products are obviously capable of wide variation in properties depending on the specific hydrophobic groups used.　They have been suggested for wetting, washing, and emulsification and

[38] Brit. Pat. 510,310, to *I. G. Farbenindustrie.*　U. S. Pat. 2,251,892, Orthner *et al.* to *I. G. Farbenindustrie.*　U. S. Pat. 2,143,490, Meyer *et al.* to *General Aniline Corp.*
[39] Ger. Pat. 585,161, to *I. G. Farbenindustrie.*　U. S. Pat. 2,157,362, Ulrich and Koerding to *I. G. Farbenindustrie.*

also as textile softeners. A similar product has been made[40] by condensing a primary or secondary fatty amine:

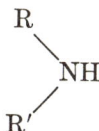

$$\begin{array}{c} R \\ \diagdown \\ \qquad NH \\ \diagup \\ R' \end{array}$$

with a cyclic carbamic-carboxylic anhydride of formula:

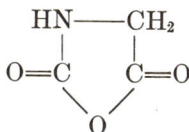

$$\begin{array}{c} HN\text{——}CH_2 \\ | \qquad\quad | \\ O\!=\!C \qquad C\!=\!O \\ \diagdown \quad \diagup \\ O \end{array}$$

or a homologue in which one of the H atoms may be replaced by methyl.

(b) Ester Intermediate Linkages

Carboxylic surface active agents containing an intermediate ester linkage have been made in a number of ways. None of them appear to have had any outstanding commercial success, nor to be possessed of any specific advantages over the more common and less expensive types.

The higher fatty alcohols such as lauryl alcohol have been reacted with dicarboxylic anhydrides such as phthalic, succinic, and maleic anhydrides to produce monoesters of the type ROOC—X—COOH which are stated[41] to be useful as cleansing and softening agents. Wetting agents and dyeing assistants have been made[42] by esterifying, for example, the dodecyl ether of diethylene glycol with maleic acid to give:

$$R\text{—}OC_2H_4OC_2H_4OOCCH\!=\!CHCOOH.$$

Fatty acid esters of hydroxycarboxy acids, such as the stearic acid ester of tartaric acid have been made by reacting the fatty acid chloride in dry pyridine with the hydroxycarboxy acid.[43] They are said to be useful in a wide variety of applications requiring a surface active agent. A more complicated type of structure has been made[44] by starting with a fatty acid amide of an amino alcohol, e.g., lauric monoethanolamide, $RCONHC_2$-

[40] Brit. Pat. 555,129, to *Imperial Chemical Industries Ltd.*
[41] French Pat. 772,538, to *Ciba Co.*
[42] U. S. Pat. 2,183,858, Haussmann, Scheuffer, and Kaupp to *I. G. Farbenindustrie.*
[43] U. S. Pat. 2,025,984, to Benjamin R. Harris.
[44] U. S. Pat. 2,371,097, Cahn and Harris to *The Emulsol Corp.*

H_4OH, and esterifying it with one carboxy group of an acylated hydroxy-polycarboxy acid, *e.g.*, diacetyltartaric acid, the product having the formula:

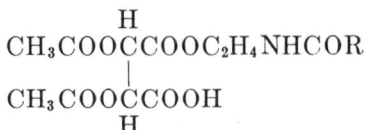

$$CH_3COOCCOOC_2H_4NHCOR$$
$$\overset{H}{|}$$
$$CH_3COOCCOOH$$
$$\underset{H}{}$$

These products are claimed to be quite resistant to hard water. This would be expected from the multiplicity of hydrophilic groups concentrated near the polar end of the molecule. Similar products may be made[45] in which the position of the monoethanolamine residue is inverted. The product of formula R—$COOC_2H_4NH$—CO—CH=CH—$COOH$ from the lauric ester of monoethanolamine, and maleic anhydride is an example.

(c) Sulfonamide Intermediate Linkages

One of the earlier and broader patents on the Lamepon-type detergent also discloses[46] the condensation of hydrophobic sulfonyl chlorides with amino acids (completely or partially hydrolyzed proteins) to yield a carboxylic detergent with a sulfonamide linkage interposed between the hydrophobic and hydrophilic groups. The general reaction follows the equation:

$$RSO_2Cl + H_2N—X—COOH \rightarrow RSO_2—NH—X—COOH$$

The long-chain sulfonyl chlorides used in this type of reaction are most readily derived by the Reed reaction[47] of sulfur dioxide and chlorine on an aliphatic hydrocarbon. Alkyl aromatic sulfonyl chlorides have also been used. These may be made by sulfonating[48] long-chain alkyl aromatic hydrocarbons with chlorosulfonic acid, or indirectly from the corresponding sodium sulfonates. Sulfonyl chlorides for this purpose have also been described which are sulfamic derivatives of the structure R—NH—SO_2Cl or RCO—NH—SO_2Cl. These are made from the amine or acid amide by reacting with sulfuryl chloride. They may be condensed with the simple primary or secondary amino acids or the lower polypeptides to give lime-resistant surface active agents.

Another method of making this type of product is the reaction of a long-

---

[45] U. S. Pat. 2,322,783, Katzman and Epstein to *The Emulsol Corp.*
[46] U. S. Pat. 2,041,265, Orthner and Meyer to *I. G. Farbenindustrie.*
[47] U. S. Re. 20,968 to Reed.
[48] U. S. Pats. 2,373,602–3, Rust and Spialter to *Montclair Research Corp.*

chain sulfonamide, made from the sulfonyl chlorides and ammonia or a primary amine, with a reactive halogenated acid such as chloroacetic acid.[49] A product of this nature was made on a large scale in Germany during the recent war under the name Emulphor STH.[50] It is made by reacting a C_{14}-C_{16} aliphatic hydrocarbon mixture with sulfur dioxide and chlorine to give the sulfonyl chloride. This is then converted to the amide with ammonia and the amide is reacted with chloroacetic acid to give the final product, RSO_2NH—CH_2COOH. Emulphor STH is reported to be an excellent emulsifying agent for oils and waxes, giving emulsions which are stable to hard water. It also has a specific high affinity for metal surfaces, and exhibits rust-preventing properties when applied on steel. It was found to be particularly useful, therefore, in preparing the lubricating emulsions used in drawing both steel and non-ferrous metals.

(d) Ether, Sulfide, and Sulfone Intermediate Linkages

A variety of surface active agents of the general formula R—X—R'—COOH have been described where R stands for a hydrophobic radical, X represents —O—, —S— or —S— and R' is a short hydrocarbon link
$$\underset{O_2}{—S—}$$
usually —CH_2—. These products are stated to have the advantages over soap of improved wetting, solubility, and emulsifying power and lime resistance. Typical examples are: (1) lauryloxyacetic acid[51] (R—O—CH_2COOH), which may be made from the sodium derivative of lauryl alcohol and chloroacetic acid, or from lauryl chloromethyl ether through the nitrile synthesis, (2) laurylmercaptoacetic acid, (R—S—CH_2COOH), lauryl bromide and thioglycollic acid or lauryl mercaptan and chloroacetic acid,[52] and (3) laurylsulfoneacetic acid, (R—SO_2—CH_2COOH) made by the oxidation of laurylmercaptoacetic acid.[53] The sodium salt of dodecylmercaptoacetic acid is said[54] to be an excellent foamer and wetting agent. Higher alkoxyacetic acids made from the higher alcohols obtained in the

[49] U. S. Pat. 2,225,960, Orthner et al. to General Aniline Corp.

[50] L. F. Hoyt, Report on Emulsifying Agents, PB 3867, Office of Technical Services, Dept. of Commerce, Washington, D. C.

[51] We shall use the term "lauryl" to indicate the mixed aliphatic radicals (mainly $C_{12} + C_{14}$) derived from coconut oil. The term dodecyl shall be reserved to indicate the true C_{12} radical. The two terms are often used ambiguously in the patent technical literature.

[52] Urquhart and Conner, J. Am. Chem. Soc. 63, 1483 (1941).

[53] Ger. Pat. 706,122, Elbel and Kirstahler to Deutsche Hydrierwerke A.-G. U. S. Pat. 1,987,526, to Henkel et Cie. French Pat. 763,206, to Henkel et Cie. G.m.b.H.

[54] U. S. Pat. 2,050,169, Elbel and Muller to Henkel et Cie.

hydrogenation of carbon monoxide have been described[55] as surface active agents. Perhydrogenated alkylphenoxyacetic acids have also been claimed as textile scouring agents and dyeing assistants. An example is the perhydrogenated p-benzylphenoxyacetic acid:[56]

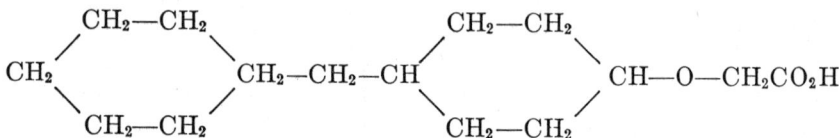

$$\begin{array}{c}
CH_2{-}CH_2 \qquad\qquad CH_2{-}CH_2 \\
CH_2 \Big\backslash \qquad\qquad \Big/ \qquad\qquad \Big/ \\
\quad CH_2{-}CH_2{-}CH \qquad\qquad CH{-}O{-}CH_2CO_2H \\
CH_2{-}CH_2 \qquad\qquad CH_2{-}CH_2
\end{array}$$

The alkylated phenoxyacetic acids have been prepared by alkylating phenoxyacetic acid itself. The diisobutyl derivative, for example, has been made[57] by reacting diisobutene and phenoxyacetic acid in carbon tetrachloride at 60°C. with a boron trifluoride catalyst. They have also been prepared[58] by reacting such compounds as p-$tert$-octylphenol with chloroacetic acid.

Two or more ether groups in the molecule increase the solubility greatly. Thus $R{-}(OC_2H_4)_n{-}OCH_2COOH$, where $n = 1$ or more, is claimed[59] as a wetting agent for mercerizing lyes. The high caustic soda concentration of these baths will salt out the usual wetting agents.

Another interesting type of ether-linked detergent has been made by adding lauryl alcohol across the double bond of diethyl maleate and subsequently saponifying:[60]

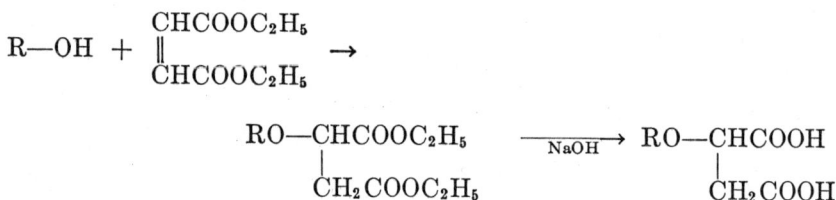

$$R{-}OH + \begin{array}{l} CHCOOC_2H_5 \\ \parallel \\ CHCOOC_2H_5 \end{array} \rightarrow$$

$$\begin{array}{l} RO{-}CHCOOC_2H_5 \\ \quad\mid \\ \quad CH_2COOC_2H_5 \end{array} \xrightarrow{\text{NaOH}} \begin{array}{l} RO{-}CHCOOH \\ \quad\mid \\ \quad CH_2COOH \end{array}$$

An unusual sulfone-substituted product has been made by reacting p-toluene sodium sulfinate with an alpha-bromo fatty acid:[61]

[55] U. S. Pat. 2,333,726, to Alien Property Custodian.
[56] Brit. Pat. 446,498, to *Henkel et Cie*.
[57] Brit. Pat. 497,487, to *I. G. Farbenindustrie*.
[58] French Pat. 808,852, to *Henkel et Cie*.
[59] French Pat. 848,529, to *Sandoz*.
[60] U. S. Pat. 2,377,246, Kyrides to *Monsanto Chemical Co*.
[61] U. S. Pat. 2,010,754, Felix and Albrecht to *Ciba Co*.

$$CH_3(CH_2)_9\underset{\overset{|}{COOH}}{CHBr} + NaO_2S\langle\rangle CH_3 \rightarrow$$

$$CH_3(CH_2)_9\underset{\overset{|}{COOH}}{CH}-O_2S\langle\rangle CH_3 + NaBr$$

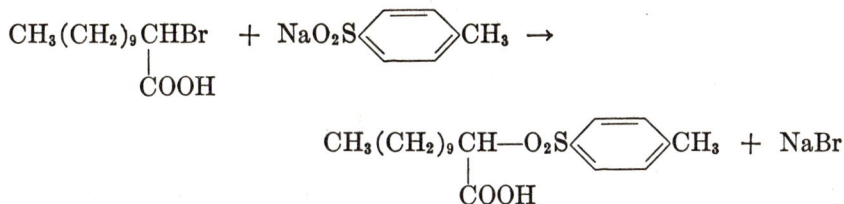

Compounds containing both an amide and a thioether link have also been described.[62]   As an example N-(hydroxymethyl)-lauramide is heated with thioglycollic acid to give the product of formula:

$$C_{11}H_{23}CONH-CH_2-S-CH_2CO_2H$$

[62] Swiss Pats. 208,530; 210,959; to *Ciba Co.*

CHAPTER 3

Sulfuric Esters

I. General Considerations

The surface active agents which are half esters of sulfuric acid include a very wide variety of commercially important products. On the basis of structure they can be divided into two large classes: class 1—products containing the sulfate group attached directly to the hydrophobic group; class 2—products containing the sulfate group attached to the hydrophobic group through an intermediate linkage. Class 2 has been subclassified in accordance with the nature of the connecting link, as was done in the case of the carboxylic detergents. Class 1 can, however, also be subdivided into two major sections. The first section includes those compounds in which the sulfate group is located near the center of the hydrophobic group and the hydrophobic group itself bears another polar group which only contributes in a minor measure to the surface active properties of the molecule. We shall refer to these products as the "sulfonated oil type." Most of them are in fact derived directly or indirectly from the natural oils, particularly the oleic and ricinoleic derivatives. Most of the so-called sulfonated oils are really sulfuric esters and should more strictly be called sulfated oils. The term "sulfonation" is often used in this field to indicate a reaction with sulfuric acid, whether the end product is a true sulfonic acid or a sulfuric ester. The reader will thus sometimes encounter in technical literature the terms "sulfonated" and "sulfonation" where, strictly speaking, "sulfated" and "sulfation" should be used. The reverse of this terminology, *i.e.*, calling a true sulfonic acid a "sulfate," or true sulfonation "sulfation," is almost never encountered. The second subdivision of class 1 includes those compounds in which the sulfate group is located near one end of the chain or, if near the middle, is the only polar-type group in the structure. We shall call this the "sulfated alcohol" or "sulfated olefin" type.

There are two important general reactions by which the sulfuric esters may be formed.[1] The first is the reaction of an alcoholic hydroxyl with sulfuric acid or its equivalent:

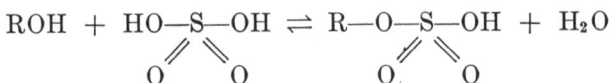

$$\text{ROH} + \text{HO—S—OH} \rightleftharpoons \text{R—O—S—OH} + \text{H}_2\text{O}$$

[1] An excellent detailed discussion of the formation of sulfates from alcohols and olefins is given in Suter, *The Organic Chemistry of Sulfur*, Wiley, New York (Chapman & Hall, London), 1944.

This is a reversible reaction, like most other esterifications, and the esterification is favored by removal of water from the reaction mixture. In place of sulfuric acid a variety of sulfating compounds or mixtures are often used, examples being oleum, chlorosulfonic acid, sulfur trioxide complexes with pyridine or ethers, acetylsulfuric acid, etc. Most of these are used to overcome the reaction-reversing effect of water. The stoichiometric reactions with sulfur trioxide and with chlorosulfonic acid, for example, are:

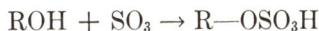

$$ROH + SO_3 \rightarrow R\!-\!OSO_3H$$

and

$$ROH + ClSO_3H \rightarrow R\!-\!OSO_3H + HCl$$

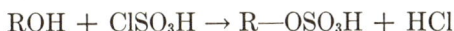

In the first case there is direct addition to form the final product. In the second case gaseous hydrogen chloride is formed and escapes from the reaction mixture. When sulfuric acid itself is used as the sulfating agent, it is usually necessary to use a large excess over the stoichiometric quantity. This serves to take up the water of reaction. The "spent acid" in this type of reaction, *i.e.*, sulfuric acid left over when the product is ready to be isolated, will often run 85% or higher in H_2SO_4.

The second general reaction for forming sulfuric esters is the direct addition of sulfuric acid to an olefinic linkage:

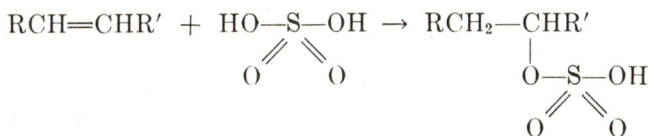

$$RCH\!=\!CHR' + HO\!-\!\underset{\underset{O}{\diagup \diagdown}O}{S}\!-\!OH \rightarrow RCH_2\!-\!\underset{\underset{\underset{\underset{O}{\diagup \diagdown}O}{S}}{|}}{CHR'}\!-\!OH$$

The product is the sulfuric ester of a secondary or tertiary alcohol. The anhydrides and chlorides of sulfuric acid sometimes react to form true sulfonic acids rather than sulfates. These reactions and their products will be considered in the chapter on sulfonic acids.

II. Sulfuric Ester Group Linked Directly to the Hydrophobic Group

(a) Sulfonated-Oil Type[2]

The natural fats and oils are triacid esters of glycerol of the general formula:

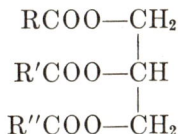

$$RCOO\!-\!CH_2$$
$$R'COO\!-\!CH$$
$$R''COO\!-\!CH_2$$

[2] For a complete discussion of sulfonated oils the reader is referred to Burton and Robertshaw, *The Sulfated Oils and Allied Products*, Harvey, London, 1939 (Chemical Pub. Co., New York, 1940).

The fatty acid components of these glycerides may be saturated, mono-olefinic, or polyolefinic. One important acid, ricinoleic acid, which is the main constituent of castor oil, contains one hydroxyl group and one double bond. It is 12-hydroxy-9,10-octadecenoic acid. Most natural oils on saponification will yield appreciable amounts of several different acids. It is now recognized that these exist in the unsaponified oil in the form of mixed glycerides as well as symmetrical glycerides. For example, consider a hypothetical oil which on saponification yields equimolar quantities of oleic, stearic, and palmitic acids. The original oil in this case would contain not only triolein, tristearin, and tripalmitin but also the esters:

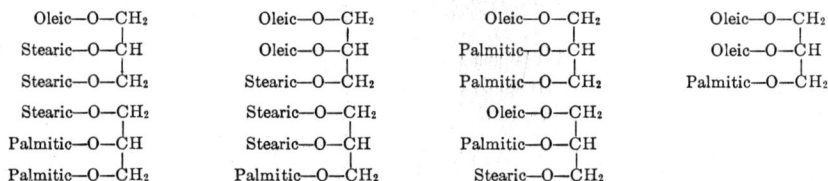

Oleic—O—CH₂	Oleic—O—CH₂	Oleic—O—CH₂	Oleic—O—CH₂
Stearic—O—CH	Oleic—O—CH	Palmitic—O—CH	Oleic—O—CH
Stearic—O—CH₂	Stearic—O—CH₂	Palmitic—O—CH₂	Palmitic—O—CH₂

Stearic—O—CH₂	Stearic—O—CH₂	Oleic—O—CH₂
Palmitic—O—CH	Stearic—O—CH	Palmitic—O—CH
Palmitic—O—CH₂	Palmitic—O—CH₂	Stearic—O—CH₂

There are geometric isomers present also due to the difference between the α- and β-hydroxyl groups of glycerol. Of the ten structural isomers possible, six contain at least one oleic acid residue, and are therefore capable of being sulfated. Assuming a random distribution of fatty acids among the available hydroxyl groups this means that 60% of an oil is sulfatable when only 33% of the fatty acid content is sulfatable. On this basis a large number of natural oils are capable of being sulfonated to technically useful products. A high content of dienoic or polyenoic acids is undesirable in oils to be sulfonated. These compounds tend to polymerize and oxidize, forming colored and gummy products. The desired oils for sulfonation are therefore sufficiently rich in oleic (or ricinoleic acid or other monoenoid acid) as to contain relatively few completely saturated glyceride molecules. Preferably they should also have a relatively low content of highly unsaturated acids.

The first sulfonated oils were produced about 1850 and were made from inedible grades of olive oil. They were used as emulsifying and wetting agents and dyeing assistants in textile processing, a use which is still very important today. The less expensive castor oil was introduced shortly thereafter. Since then a great many different oils have been sulfonated and it is largely a matter of economics which ones enjoy commercial favor. Castor and olive oils are widely used. Sulfonated tallow is favored as a textile finish. Teaseed oil has been used as a substitute for olive oil. Sulfonated rice oil is stated[3] to have the great advantage of high resistance

[3] U. S. Pat. 2,180,256, Printon to *National Oil Products Co.*

to oxidation. Neat's foot, cottonseed, rapeseed, and corn oils are also often used as raw materials.

The sulfonation of most oils gives products which are liquid at room temperature. Certain glycerides, however, notably mixed glycerides with a relatively low content of unsaturated acids (such as lard, oleostearin, etc.) give solid products.[4] The use of sulfated hydrogenated castor oil (glyceride of 12-hydroxystearic acid) in preparing ointment bases has been described by Fiero and Loomis.[5]

Few of the true sulfated glycerides are trade-named or protected by patent monopolies at the present time. They do not require elaborate equipment or a highly skilled operating force for their manufacture. Hence, they are made by a large number of companies, including many of very modest size, and are usually sold on specification.[6] The sulfonated oils today are widely used as emulsifying agents. They are also used to a certain extent as detergents, particularly in shampoos and hand cleansers. One of their main uses is in the finishing of textile materials, where they act as a surface lubricant for the fibers.

The sulfation of the double bonds and/or hydroxyl groups of oils is usually accompanied by side reactions such as hydrolysis of the glyceride ester linkages. Furthermore, the reaction of sulfation as carried out commercially is usually incomplete. A high degree of sulfation is often desirable, particularly where a completely water-soluble product of high wetting, foaming, and detergent power is desired. The highly sulfated oils are also more stable toward lime and acid. In some other applications, lubricants for example, high sulfonation is not required and may even be undesirable. A usual method of sulfonating oils is to treat the oil with ordinary 66° Bé. H_2SO_4 at temperatures not substantially above room temperature. After standing for a suitable reaction time the mass is washed one or more times with a strong aqueous solution of salt or sodium sulfate, which washes out the spent sulfuric acid and leaves the sulfonated oil as a supernatant layer. This acidic oil is then neutralized with alkali. The final product is a yellow to dark-brown viscous liquid. Depending on the type of oil used and the method of sulfation, it may dissolve in water to give clear foaming solutions or it may form cloudy oily dispersions. Beside the true sulfated glyceride it may contain varying amounts of water, salt, fatty acid, sulfated

[4] U. S. Pat. 2,029,168, to Benjamin R. Harris.

[5] Fiero and Loomis, *J. Am. Pharm. Assoc.* **34**, 218 (1945).

[6] For reviews of preparation, properties, and uses of sulfonated oils, see: A. E. Sunderland, *Soap* **11**, No. 10, 61–4, 71; No. 11, 61–4; No. 12, 67–9 (1935); A. Hecking, *Fette u. Seifen*, **45**, 628–9 (1938).

fatty acid, and unsulfated glyceride.   The analysis of sulfonated oils has accordingly received considerable attention.[7]

The journal and patent literature on the sulfation of oils is voluminous.[8] The patent literature on the preparation of sulfated oils, acids and esters is very extensive.   The sulfation of the free unsaturated or hydroxylated fatty acids is often considered at the same time since the products are used for the same purposes.   A majority of the reported studies are concerned with increasing the degree of sulfation,[9] which increases the solubility and general surface activity.   The sulfation of neat's foot oil,[10] olein,[11] peanut oil,[12] and castor oil[13] have been given particular attention.   The chemical constituents of typical sulfated oils have ably been discussed by Koppenhoefer,[14] and a number of Japanese investigators[15] have also made exhaustive studies of this problem.

Tall oil esters have also been sulfated[16] and stable products are obtained if strong sulfating conditions are used.   In this case the rosin acids present in the tall oil are probably sulfated as well as the fatty acids.

High degrees of sulfation have been achieved by taking an oil sulfated by usual methods and extracting the unsulfated material from it with appro-

---

[7] For discussions and methods of analysis, see: Ralph Hart, *Am. Dyestuff Reptr.* **20**, 565–6 (1931); **21**, 1–2, 33 (1932); **22**, 695–704 (1933); **23**, 290–9 (1934); *Ind. Eng. Chem. Anal. Ed.* **7**, 137–40 (1935); **9**, 177–80 (1937); *Am. Dyestuff Reptr.* **25**, 122–5 (1936).   Albert H. Grimshaw, *Am. Dyestuff Reptr.* **21**, 87–9 (1932).   H. Jahn, *Seifensieder-Ztg.* **59**, 45–7 (1932).   C. Riess, *Fettchem. Umschau* **41**, 199–200 (1934).   H. Finken and H. Holters, *Fette u. Seifen* **46**, 70 (1939).

[8] The patent literature on the preparation of sulfated oils, acids, and esters is very extensive.   For typical examples see: U. S. Pats. 1,796,801; 1,822,977–9; 1,836,487; Münz to *General Aniline Works.*   U. S. Pats. 1,918,372; 1,967,655; 2,032,313–4; Bertsch to *H. Th. Böhme A-G.*   U. S. Pat. 2,352,698; J. T. Eaton and P. W. Volk to *E. F. Houghton Co.*   U. S. Pats. 2,344,154, Lighthipe to *National Oil Products Co.*   Ger. Pats. 591,196; 670,962; to *I. G. Farbenindustrie.*   Ger. Pats. 643,052; 640,997; to *Böhme Fettchemie G.m.b.H.*   Ger. Pat. 645,608, to *Ciba Co.*   Brit. Pats. 272,967; 296,999; 303,917; 306,052; 298,599; 344,828; to *I. G. Farbenindustrie.*   Brit. Pats. 313,160; 315,832; to *H. Th. Böhme A.-G.*

[9] C. Steiner, *Fettchem. Umschau* **42**, 201–5 (1935).

[10] Welwart, *Seifensieder-Ztg.* **63**, 717–8 (1936).

[11] Welwart, *Seifensieder-Ztg.* **63**, 549 (1936).

[12] Gallent, *Am. Dyestuff Reptr.* **33**, 148 (1944).   See also Glicher, *Petroleum (London)* **8**, 130, 232 (1945).

[13] Schulze, *Textil-Betrieb*, **18**, No. 12, 48 (1938).

[14] Koppenhoefer. *J. Am. Leather Chemists' Assoc.* **34**, 622–39 (1939).

[15] Winokuti, Igarasi, and Yagi, *J. Soc. Chem. Ind. Japan*, **35**, Suppl. binding, 159 (1932).   Winokuti and Yagi, *ibid.* 161.   Winokuti, *ibid.* 163.   Nishizawa, *ibid.* **39**, Suppl. binding, 234 (1936).

[16] Brit. Pat. 340,272, to *I. G. Farbenindustrie.*

priate solvents such as the lower alcohols or halogenated hydrocarbons.[17] Separation of unsulfated matter by partial neutralization has also been described.[18] Sulfation at low temperature, 0°C. or below, with excess sulfuric acid, oleum, or chlorosulfonic acid is another method of obtaining highly sulfated products.[19] Solid carbon dioxide[20] or solid fats may be added to aid in maintaining low temperature.[21] The sulfation of oils may be carried out in a series of stages[22] using sulfonating agents of increasing energy in each stage to achieve a high degree of sulfation. The use of dehydrating solvents such as acetic anhydride, acetyl chloride, acetylsulfuric acid, particularly in conjunction with oleum or chlorosulfonic acid has also been widely recommended.[23] Liquid sulfur dioxide has also been proposed[24] as a solvent, using oleum or sulfur trioxide as the sulfating agent.

Highly purified sulfated fatty oils may be obtained[25] by extracting a highly sulfated oil with chlorinated solvents such as trichloroethylene. Sulfonated oils free of inorganic salts have been prepared by precipitating out the latter with a solvent such as ethyl alcohol or isopropanol.[26] They may be bleached by the ordinary bleaching agents such as hydrogen peroxide, hypochlorite, or sulfites, or with ozonized air.[27] Sulfonated oils of low carboxy content have been produced[28] by re-esterifying the free carboxyl groups with glycerol or glycol after sulfation.

Sulfonated free fatty acids as well as the oils are also quite widely pre-

[17] U. S. Pats. 2,280,118, Dombrow to *National Oil Products*; 2,203,524, Dombrow and Beach to *National Oil Products*.

[18] U. S. Pat. 2,186,308, Charles L. Schuttig to *Arkansas Co.*

[19] French Pat. 690,022, to *Soc. Anon. pour l'Ind. Chim. à Saint Denis.* Ger. Pats. 608,692, to *Rudolf and Co.*; 623,632, to *Fettsaure u. Glyzerin Fabrik G.m.b.H.*

[20] Ger. Pat. 622,728, to *Hellmuth Jahn.*

[21] Can. Pat. 354,961, Kaplan to *Richards Chemical Works Ltd.*

[22] French Pat. 796,463, to *Hansawerke Lürman, Schütte and Co.*

[23] U. S. Pat. 1,980,342, Kern to *Chemische Fabrik R. Baumheier.* French Pat. 688,637, to *N. V. Chemische Fabriek Servo and Meindert D. Rozenbroek.* Ger. Pats. 564,759 and 614,347, to *Farb- und Gerbstoffwerke Carl Flesch Jr.*; 666,828, to *Oranienburger Chemische Fabrik A.-G.*

[24] Brit. Pats. 346,945, to *I. G. Farbenindustrie;* 404,364; Greenhalgh to *Imperial Chemical Industries Ltd.*

[25] U. S. Pats. 1,906,924, H. Bertsch to *Böhme Fettchemie G.m.b.H.*, 2,328,931, Steik to *National Oil Products Co.*

[26] U. S. Pats. 2,285,337, Kapp, Mosch, and Wo ds to *National Oil Products Co.*; 2,328,931, Steik to *National Oil Products Co.* Fr. Pat. 852,077, to *National Oil Products Co.*

[27] Ger. Pat. 541,090, to *H. T. Böhme, A.-G.*

[28] U. S. Pat. 1,955,766, to Edward Pohl.

pared and used. They are characterized by greater solubility than the glycerides, and have apparently found considerable favor in Germany, particularly as lime soap dispersing agents in textile processing. The highly sulfated oils and fatty acids are best known there under the trade names Prestabitol and Intrasol.

Ricinoleic acid on heating forms inner esters between the 12-hydroxy group and the carboxy group. These products are called estolides. The sulfated estolides constitute a large portion of the so-called Monopole soap.[29]

Beside the natural glycerides some of the natural waxes are commercially sulfated. The two which have been most widely used are sperm oil and spermaceti. Sperm oil consists largely of the esters of higher fatty acids with higher fatty alcohols, and a considerable proportion of both the acids and alcohols are unsaturated. The sulfation takes place on these double bonds. The product is used largely as a finishing oil for textiles and as a constituent of lubricants.

Spermaceti consists largely of cetyl palmitate, a saturated compound. In the sulfation of this wax the ester is hydrolyzed and the cetyl alcohol is sulfated. The neutralized product is largely a mixture of sodium palmitate and sodium cetyl sulfate, and accordingly is not structurally related to the sulfated oil type of surface active agent but rather to the sulfated alcohol type (see Section II, d, p. 53).

The hydroxyl group of ricinoleic acid (or its glyceride) may be acylated and the product is still capable of being sulfated on the double bond. The acylating agent is usually acetic anhydride. The acylation[30] may be carried out before sulfation, or simultaneously by using a mixture of acetic anhydride, glacial acetic acid, or acetyl chloride together with an appropriate sulfating agent.

Phosphatides have been sulfated to produce water-soluble surface active compositions.[31] Oleyl alcohol is usually sulfated on the hydroxyl group to give a commercially important member of the sulfated alcohol series (see Section II, d). The hydroxyl group may be blocked by acylation or etheri-

[29] Sulfated oleic acid and its various applications are described in Brit. Pat. 400,587, to *Fettsaure u. Glyzerin Fabrik G.m.b.H.* A review and tabulation of commercial German sulfonated oil preparation is given in *Seifensieder-Ztg.* **63**, 242–3 (1936).

[30] French Pat. 719,901, to *Imperial Chemical Industries Ltd.* Ger. Pats. 591,196; 629,182; to *I. G. Farbenindustrie.* U. S. Pat. 1,986,808, Richard Greenhalgh to *Imperial Chemical Industries Ltd.* Brit. Pat. 357,670, to *Imperial Chemical Industries Ltd.*

[31] Ger. Pat. 691,486, to *Chemische Fabrik Stockhausen and Co.*

fication and the products[32] sulfated on the double bond to give surface active agents similar to the sulfated oil type. Oleyl alcohol, castor oil and ricinoleic acid have been esterified on the hydroxyl groups with naphthenic acid, and the products sulfated.[33]

(b) Sulfated Esters and Acids

When ricinoleic or oleic acid is esterified with a low-molecular-weight alcohol and the product is then sulfated a very useful type of surface active agent is produced. These products differ from the sulfated glycerides in having much greater foaming and wetting power. They are an important class of wetting agents and have found great favor in the textile industry. They are particularly useful because of their high re-wetting power combined with their lubricant effect on textile fibers. This makes them well adapted for finishing cotton piece goods which are to be preshrunk by the Sanforizing process. They also differ from many other wetting agents in maintaining their wetting power in hot solutions of dilute caustic soda such as are used in kier boiling and in some of the continuous bleaching processes.

The propyl, butyl, and amyl esters of both ricinoleic and oleic acids are all used as raw materials for the sulfation. Probably the most widely used of these is butyl oleate. The oleates are prepared by the simple esterification of red oil or the elaine oils (technical oleic acids) with the appropriate alcohol. The ricinoleates are often prepared by a transesterification reaction. Thus castor oil is reacted with butyl alcohol, for example, in the presence of a mineral acid catalyst to form glycerol and butyl ricinoleate.

The esters are sulfated by the same procedures used to produce the highly sulfated oils. The products usually contain somewhat less than the theoretical sulfur trioxide content for complete sulfation, but this is true of a great number of commercial surface active agents. For most practical purposes they can be regarded as completely sulfated. They are marketed as clear yellow-to-brown viscous solutions containing up to 50% or more water and often up to 10% or more of pine oil or terpineol. They form clear stable easily foaming solutions in water. This type of product was introduced in Germany as Avirol AHX. It is currently produced in this country by several manufacturers and sold under their trade names as

[32] U. S. Pat. 2,163,133, Walther Schrauth to *Unichem*, describes sulfated oleyl acetate. U. S. Pat. 2,275,413, Heinrich Bertsch to *American Hyalsol Corp.*, describes sulfated oleyl butyl ether, etc.

[33] U. S. Pats. 2,203,641-2, Kapp to *National Oil Products Co.*

Surfax W. O. (*Houghton*), Phi-Sol (*Onyx Oil and Chemical Co.*), Parapon (*Arkansas Co.*), etc. Wide variations in manufacturing procedure and raw materials are described[34] in the patent literature.

An interesting process[35] for making sulfated oleic esters consists in sulfating oleic acid and then esterifying the carboxy group with an alkyl sulfate such as diethyl sulfate. In the exactly opposite procedure,[36] where the desired product is a sulfated oleic acid, fatty oils have been first sulfated and the glyceride ester groups subsequently hydrolyzed.

α-Hydroxy saturated fatty acids such as α-hydroxystearic acid have been sulfated[37] to produce surface active agents of the type:

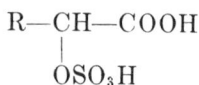

$$R—CH—COOH$$
$$\underset{OSO_3H}{|}$$

Similarly the fatty acids of lanolin may be oxidized (presumably to effect hydroxylation) and subsequently sulfated to yield surface active sulfates[38] containing hydroxyl groups. An interesting variation[39] of this type uses an ether alcohol of the Cellosolve type (monoethyl ether of ethylene glycol) in place of the usual butyl or amyl alcohol. In another variation[40] a glycol such as ethylene or diethylene glycol is esterified with two molecules of oleic acid and the product is sulfated.

The transesterification of sperm oil with methanol or ethanol to produce a sulfatable mixture of fatty esters and higher fatty alcohols has also been described.[41] Products of this general class have also been described[42] in which ricinoleic acid is converted to a hydroxydicarboxy acid, and subsequently esterified and sulfated.

(c) Sulfated Amides

Instead of being esterified the free carboxy group of oleic or ricinoleic acids may also be converted to an amide by using ammonia, or primary or

[34] French Pat. 721,070, to *Farb-u. Gerbstoff-Werke Carl Flesch, Jr.* U. S. Pats. 1,823,- 815; 1,974,007; 2,032,313–4, Bertsch to *H. Th. Böhme A.-G.* Ger. Pats. 625,637; 633,082; 655,942; 659,528; 676,343; 681,441, Bertsch to *Böhme.*

[35] Brit. Pat. 343,989, to *I. G. Farbenindustrie.*

[36] U. S. Pat. 2,344,154, Lighthipe to *National Oil Products Co.*

[37] U. S. Pat. 2,199,398, to *E. I. du Pont de Nemours & Co.* See also Ger. Pat. 546,142, to *H. Th. Böhme A.-G.*

[38] Ger. Pat. 539,625, to *I. G. Farbenindustrie.*

[39] Brit. Pat. 351,456, to *H. Th. Böhme A.-G.* U. S. Pat. 2,371,284, Cook *et al.* to *Arkansas Co.*

[40] U. S. Pat. 2,136,379; Can. Pat. 376,873; Flett to *National Aniline and Chemical Co.* U. S. Pat. 2,192,721, Toone to *National Aniline and Chemical Co.*

[41] Japanese Pat. 111,447; *Chem. Abstracts* **30**, 2289.

[42] U. S. Pats. 2,007,492; 1,947,673; Bertsch to *H. Th. Böhme A.-G.*

secondary aliphatic, aromatic, or heterocyclic amines. These substances on sulfation yield products[43] similar in their general properties to the sulfated esters. They are considered to be even more stable to high temperatures, excess acid or alkali, and hard water. Products of this type have been marketed in Germany under the trade name Humectol. They are also made in this country although they do not appear to have won the widespread acceptance accorded to the sulfated fatty esters.

Humectol CX, the sulfated diisobutyl amide of oleic acid,[44] achieved widespread popularity in Germany during the war. It is stated to be an excellent wetting agent and levelling agent for direct colors. It is also used as a wetting agent in bleaching processes and for boiling off yarns.

The analogous sulfated oleic monoethylanilide was produced under the name Dismulgan IV, and was used primarily as an emulsifying agent.

Analogous compounds have also been made[45] in which the amino group is attached to the hydrophobic chain. A typical example is made by sulfating benzoyloleylamide:

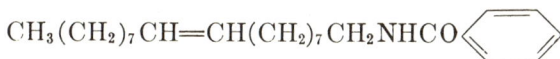

$$CH_3(CH_2)_7 CH{=}CH(CH_2)_7 CH_2 NHCO\left\langle\!\!\!\bigcirc\!\!\!\right\rangle$$

(d) Sulfated Alcohols

The half sulfuric esters of the normal primary aliphatic alcohols have been known since the early days of organic chemistry. Sodium cetyl sulfate, $C_{16}H_{33}OSO_3Na$, was first prepared by Dumas[46] in 1836.

The alkyl sulfates of eight to eighteen or more carbon atoms are typical surface active compounds. The lower ones show their maximum activity at room temperatures. With increasing molecular weight, higher temperatures are needed to attain maximum detergent and wetting effects. These compounds are among the most important commercial surface active agents. Their introduction, about the year 1930, signalled the start of intense interest and activity in the whole field of synthetic detergents. Their properties have probably been more thoroughly studied from a scientific point of view than those of any other group of synthetic detergents. They are still among the most widely used in spite of the advent of the less expensive, large-tonnage alkyl aromatic sulfo acids and alkane sulfo acids.

[43] Ger. Pats. 595,173; 628,828; U. S. Pat. 1,918,363; French Pat. 718,393, Bertsch to *H. Th. Böhme A.-G.* Ger. Pats. 671,085; 678,731, to *I. G. Farbenindustrie.* U. S. Pat. 2,372,786, Kelley and Abramowitz to *National Oil Products Co.*

[44] Brit. Pats. 343,899; 341,053, to *I. G. Farbenindustrie.*

[45] French Pat. 787,505, to *I. G. Farbenindustrie.* U. S. Pat. 2,086,690, Zerweck and Gofferje to *General Aniline Works.*

[46] Dumas and Peligot, *Ann. Pharmazie* **19**, 293 (1836).

The hydrophobic groups of the normal primary aliphatic alcohol sulfates are identical with those of the fatty soaps. The surface active properties of these two classes are therefore similar in many respects, and the users of soaps have readily accepted the sulfates as improved soap substitutes. The advantages over soap are apparent and yet the products are quite obviously "soapy" in nature. Beside the normal primary alcohols, however, almost any olefin or alcohol of suitable configuration and molecular weight can be sulfated to give a surface active product. The olefins and secondary or tertiary alcohols give secondary or tertiary sulfates. As pointed out in the discussion of sulfated oils, the more energetic sulfating agents also give a certain proportion of true alkane sulfonic acid (carbon joined directly to sulfur) during the sulfation reaction. Since sulfation processes, even the more complicated and ingenious ones, are well known, easy to work out, and usually inexpensive, the primary concern of the investigators in these fields has been with cheap sources of suitable alcohols and olefins. It was not until inexpensive reduction processes had been developed for the production of fatty alcohols that their sulfates could be successfully introduced into industry. We shall therefore consider first the various alcohols and olefins which have been used as raw materials for sulfation, and their sulfation products. Special methods of sulfation and purification, many of which are applicable to a variety of alcohol or olefin types, will be discussed separately.

The two most important fatty alcohol sulfates are made from oleyl alcohol (9,10-octadecenol-1) and from the mixture of fatty alcohols made by reducing the mixed fatty acids of coconut oil. This mixture consists of about 15% mixed C_8 and C_{10} (octyl and decyl) alcohols, 40% C_{12} (lauryl or dodecyl) alcohol, 30% C_{14} (myristyl or tetradecyl) alcohol, and 15% mixed C_{16} and C_{18} (cetyl, stearyl, and oleyl) alcohols. In trade circles the mixture itself is called "coconut fatty alcohol" or simply "lauryl alcohol." It should be noted that the term, "lauryl alcohol," is also widely used by organic chemists as a trivial name for the chemical individual dodecanol-1.

In the group are also included some of the alcohol fractions obtained from coconut fatty alcohol by fractional distillation. Thus coconut fatty alcohol is sometimes distilled into two or three fractions, cuts being made between C_{10} and C_{12} and C_{14} and C_{16} to give three fractions, or simply between C_{14} and C_{16} to give a lower and a higher fraction. Relatively pure cetyl and stearyl alcohols are often separated and their sulfates prepared. These alcohols are more often prepared, however, from fats which are richer in the C_{16} and C_{18} acids.

This group of products[47] is marketed in this country under the names Gardinol, Modinal, Duponol, Orvus (*du Pont Co.* and *Procter and Gamble Co.*), Aurinol, Maprofix (*Onyx Oil and Chemical Co.*), also Tergavon (*Ciba Co.*), Sandopan (*Sandoz*), Cyclopon, Cyclanon, (*General Aniline Corp.*), Dreft and Drene, (for consumer's use). The suffix letters after the trade names are important in indicating the particular alcohol mixture which is present. Thus Gardinol WA contains sulfated coconut fatty alcohol, Duponol D contains sulfated oleyl alcohol, etc. On the Continent the trade names Gardinol, Modinal, Cyclanon, Sapidan, Ocenol, and Sandopan are the best known for these products. In England they are produced by *Imperial Chemical Industries* as the Lissapol series. Usually included in this group are the sulfates[48] of the more highly unsaturated fatty alcohols derived from fish oil acids or the acids of the more unsaturated vegetable oils. For brevity's sake, and to distinguish this group from other alkyl sulfates, we shall refer to the above products as *Gardinol type* detergents.

The alcohols from which Gardinol type detergents are made are obtained by two general methods. The first method consists of saponification of certain natural waxes, notably spermaceti and sperm oil, which consist mainly of esters of higher fatty alcohols with fatty acids.[49] The saponification is usually carried out with caustic alkali or lime forming a mixture of fatty alcohols and fatty acid soaps. The alcohols are then separated by distillation with superheated steam or vacuum, or by extraction processes. Sometimes the waxes themselves are sulfonated without previous saponification. This has been done[50] with sperm oil, spermaceti, wool fat (lanolin), and beeswax in particular. During the sulfation the ester linkage is at least partially split and the product is a mixture of the sulfated alcohol and fatty acid. This may be neutralized and used without separation or the fatty acid portion may be separated by solvent extraction.[51]

The second and most important way of obtaining fatty alcohols is by the reduction of fatty acids, either directly or indirectly. The fatty acids

[47] U. S. Pats. 1,968,793, 4, 5, 7; Can. Pat. 356, 113; U. S. Pat. 2,114,042; Bertsch to *American Hyalsol Corp.* Brit. Pats. 350,432; 350,080; 351,452, to *H. Th. Böhme A.-G.* Ger. Pat. 542,048, Schrauth to *Deutsche Hydrierwerke A.-G.*

[48] French Pat. 776,044, to *Deutsche Hydrierwerke A.-G.*

[49] U. S. Pat. ·1,962,941, Schrauth to *Unichem.* Ger. Pat. 616,765, Schrauth to *Deutsche Hydrierwerke A.-G.* U. S. Pats. 2,021,926, Sexton and Ward to *Imperial Chemical Industries Ltd.*; 2,245,538, Thurman to *Refining, Inc.* See also, U. S. Pat. 2,114,043, Bertsch to *American Hyalsol Corp.*

[50] Swiss Pats. 174,511-4; 171,359; to *Chemische Fabrik vorm. Sandoz.* U. S. Pat. 1,980,414, K. Lindner to *Oranienburger Chemische Fabrik, A.-G.*

[51] Brit. Pat. 354,217, to *I. G. Farbenindustrie.*

used may be purified individual acids, but more often they are mixtures obtained by saponification of natural fats followed by a partial separation of less desirable components. Thus commercial stearic acid is reduced to give a mixture of cetyl and stearyl alcohols. Sometimes the mixture of acids occurring in a single fat is reduced without separation. This is done when the mixed alcohols give a sulfate of particularly desirable properties as in the case of coconut oil and palm oil. The sulfated mixed alcohols from certain other natural fats have been specifically patented.[52] Catalytic hydrogenation of the lower alkyl esters of the fatty acids is a well-known method[53] of obtaining the fatty alcohols. The higher esters (waxes) can be reduced in a similar manner. The naturally occurring glyceride esters can also be reduced satisfactorily by this method. Nickel or copper chromite catalysts are particularly effective.[54] The free carboxylic acids themselves can be reduced to give the fatty alcohols in good yields. Copper chromite, copper on kieselguhr, zinc–copper, and copper chromite – iron oxide appear to be the preferred catalysts.[55] The mixed anhydrides of fatty acids and weak oxygenated inorganic acids, such as boric-stearic anhydride have also been catalytically hydrogenated[56] to fatty alcohols. Under usual operating conditions the processes of catalytic hydrogenation reduce olefinic double bonds as well as carboxy or ester groups. Oleic acid, for example, is reduced to stearyl alcohol rather than to oleyl alcohol. By selecting the proper catalyst and conditions, however, the ester or acid group can be reduced without affecting the double bond. A zinc–vanadium or cadmium–vanadium catalyst at 250–350°C. and 50 atmospheres pressure has been used[57] for this purpose.

Reduction of fatty acid esters by sodium and a lower alcohol, the Beauvalt-Blanc method, is a very important method of producing fatty alcohols. In some plants it is economically more feasible than catalytic hydrogenation. It is particularly useful for producing oleyl alcohol since it does not

[52] U. S. Pat. 2,195,345, to Ozren and Georg Stefanovic. Brit. Pat. 531,194, Greenhalgh and McKay to *Imperial Chemical Industries Ltd.*

[53] U. S. Pat. 2,091,800, Adkins, Folkers, and Connor to *Rohm and Haas Co.*

[54] Brit. Pat. 433,549, to *Imperial Chemical Industries Ltd.* and Green. Swiss Pat. 178,814, to *Chemische Fabrik vorm. Sandoz.* U. S. Pat. 2,023,383, Schrauth and Böttler to *Unichem.*

[55] U. S. Pat. 1,839,974, Wilbur Lazier to *E. I. du Pont de Nemours & Co.* Ger. Pat. 670,832, Rittmeister to *Deutsche Hydrierwerke A.-G.* Brit. Pat. 440,934, to *Sandoz.* Ger. Pats. 594,481; 617,542; French Pat. 718,394, to *H. Th. Böhme A.-G.* Also U. S. Pats. 2,322,095–99, Schmidt to *General Aniline and Film Corp.*

[56] U. S. Pat. 1,987,558, Hintermaier to *Henkel et Cie G.m.b.H.*

[57] U. S. Pat. 2,374,379, Rittmeister to *American Hyalsol Corp.*

reduce the double bond.   It was for a long time, and possibly still is, the most important method[58] of producing this alcohol.

An isomer of oleyl alcohol has been made by dehydrating 1,12-octadecandiol to the olefinic primary alcohol.   1,12-Octadecandiol is obtained[59] by hydrogenating ricinoleic acid or its esters.

Among the more unusual methods which have been described for producing the fatty alcohols is a reduction[60] of the corresponding ester with activated aluminum and absolute alcohol.   Methanol at 200–500°C. in the presence of a barium – chromium oxide catalyst has been used[61] to reduce fatty esters to alcohols.   Fatty alcohols have also been made by treating the corresponding fatty primary amines with nitrous acid.   The amines are made[62] by reduction of the corresponding fatty nitriles.

The saturated fatty alcohols are normally sulfated with concentrated sulfuric acid, an excess of sulfuric acid being used.   The product is neutralized and may then be separated from the excess sodium sulfate.   Often the sodium sulfate is left in and the whole mixture dried.   Chlorosulfonic acid is often used as a sulfating agent.   In this case only one mole is needed per mole of alcohol, the reaction taking place according to the equation:

$$ROH + ClSO_3H \rightarrow ROSO_3H + HCl$$

The HCl which is formed escapes as a gas.   The resultant sulfuric ester may be neutralized directly giving a product of very low inorganic salt content.

When oleyl alcohol is sulfated with sulfuric acid both the double bond and the terminal hydroxyl group react.   The reaction rates are of about the same order over a fairly wide temperature range, and consequently a mixture of sulfated products is obtained.   The terminal sulfate with unaffected double bond is a much more effective and prized surface active agent than its isomer.   In order to obtain the desired isomer as the sole product, or at least as the predominant component, certain special sulfation methods are used:

[58] U. S. Pat. 2,019,022, Scott and Hansley to *E. I. du Pont de Nemours & Co.*

[59] U. S. Pat. 2,086,713, Grün to *H. Th. Böhme A.-G.*

[60] Russian Pat. 31,431; *Chem. Abstracts* **28**, 3079.

[61] U. S. Pat. 2,156,217, Andrews and Fenske to *Rohm and Haas Co.*

[62] Brit. Pat. 494,666, to *Armour and Co.*

[63] U. S. Pat. 2,098,114, Suter to *Procter and Gamble Co.*   Also U. S. Pat. 2,099,214. McAllister to *Procter and Gamble Co.*

[64] Ger. Pat. 648,448, Kern to *Chemische Fabrik R. Baumheier A.-G.*   U. S. Pat. 2,079,347, Hailwood to *Imperial Chemical Industries Ltd.*

(*1*) Sulfating[63] with the SO_3-dioxane complex.

(*2*) Sulfating[64] with the SO_3-pyridine complex which may be made from pyridine and chlorosulfonic acid in an inert solvent.

(*3*) Sulfating[65] with sodium or potassium chlorosulfonate.

(*4*) Sulfating[66] with H_2SO_4 and urea, or other lower acid amide, or[67] with the addition complex of $ClSO_3H$ and urea or other lower acid amide.

(*5*) Sulfating[68] with the addition complex of SO_3 and $NaNO_2$.

These methods can, of course, all be applied to the saturated alcohols as well as the unsaturated. Usually sulfation with sulfuric acid is incomplete and accompanied by a certain amount of dehydration and olefin formation.

A number of methods for obtaining a more satisfactory sulfation reaction have been described. Thus, sulfation in the presence of acetic acid or acetyl chloride has been used.[69] Ethers such as dimethyl ether have also been used[70] as solvents for the sulfation reaction. Low-temperature sulfation with sulfuric acid has been suggested,[71] particularly for the low-melting unsaturated alcohols. Heating fatty alcohols with NH_4HSO_4 to 100–150°C. effects sulfation.[72] Another high-temperature method involves heating the fatty alcohol with the salt of a lower alkyl sulfate such as potassium ethyl sulfate.[73] Sulfation in a mixture of liquid sulfur dioxide and petroleum ether has also been described.[74] Sulfation of the boric acid esters of the fatty alcohols has been disclosed. This probably results in the formation of the sulfuric esters mixed with boric acid.[75]

An interesting continuous sulfation process[76] involves the flowing together of streams of fatty alcohol and sulfating agent, and arresting the reaction by neutralization or dilution.

Purification. Unless the fatty alcohols used for sulfation are free of un-

---

[65] U. S. Pats. 2,049,670; 2,075,914–5; Snoddy and Martin to *Procter and Gamble Co.* 2,214,254, Mills and Wood to *Procter and Gamble Co.*

[66] U. S. Pat. 2,147,785, Cupery and Shipp to *E. I. du Pont de Nemours & Co.*

[67] Brit. Pat. 506,049, to *I. G. Farbenindustrie.*

[68] U. S. Pat. 2,231,979, Wolter to *Procter and Gamble Co.*

[69] Ger. Pat. 669,955, Bertsch to *H. Th. Böhme A.-G.*

[70] Ger. Pat. 701,403, Petz and Wappes to *I. G. Farbenindustrie.*

[71] U. S. Pat. 2,044,919, Schrauth and Hueter to *Unichem.* See also *Matumoto Prefectural Inst. Avancement Ind. Tokyo* 3, 28 (1940); *Chem. Abstracts* 35, 7938.

[72] U. S. Pat. 2,150,557, MacMullen to *Rohm and Haas Co.*

[73] Ger. Pats. 606,083; 557,428; Brodersen and Quaedvlieg to *I. G. Fardenindustrie.*

[74] U. S. Pat. 2,235,098, Brandt and Ross to *Colgate-Palmolive-Peet Co.*

[75] U. S. Pat. 2,042,952, Mauersberger to *Richards Chemical Works.* See also U. S. Pat. 2,195,418, Mauersberger to *American Hyalsol Corp.*

[76] U. S. Pat. 2,187,244; Brit. Pat. 499,373; Mills to *Procter and Gamble Co.*

sulfatable impurities these impurities will remain unchanged throughout the reaction and will be present in the final neutralized product. Insoluble organic matter may also be present in the form of unsulfated fatty alcohol or olefin. Small amounts of these impurities (up to about 5% of the total active sulfate present) generally have no detrimental effect and may even be beneficial in some cases. In many instances, however, it is desirable to remove them, particularly when they are present in large amounts. Inorganic salts are also normally present in the finished product, due to neutralization of excess sulfating agent. For some special purposes (such as cosmetic detergents, shampoos, oil-soluble emulsifiers, etc.) it is often desirable to remove these also.

Removal of unsulfated matter (which term includes unsulfatable matter) is usually accomplished by extracting an aqueous solution of the neutralized product with a lower chlorinated solvent such as carbon tetrachloride or a lower aliphatic hydrocarbon solvent such as petroleum ether or gasoline. These substances are solvents for the impurities and non-solvents for the fatty alkyl sulfates. The extraction process is often complicated by the formation of emulsions of the extracting solvent in the aqueous detergent. To help break these emulsions a water-miscible oxygenated solvent such as the lower alcohols, acetone, or dioxane may be added to the system. The emulsions may also be broken by centrifuging. Water-immiscible oxygenated solvents such as diethyl ether have also been used to extract unsulfated matter from detergent solutions.[77] In many instances the unsulfated matter is volatile enough so that a satisfactory amount can be removed by spray drying the product.[78] The spray-dried form of the Gardinol type detergents is very popular among both industrial and household consumers and was described early in the course of the development.[79]

Fatty alkyl sulfates substantially free of inorganic salts may be obtained in several different ways. When sulfuric acid is used as the sulfating reagent, the reaction mixture may be separated into two layers by the addition of a solvent such as butyl alcohol or carbon tetrachloride. The layer of spent acid is withdrawn, leaving the alkyl sulfuric ester dissolved in the solvent. After neutralization and evaporation of the solvent the product contains only minor amounts of inorganic salt.[80]

The separation of inorganic salts is usually effected by the use of one or

[77] U. S. Pat. 2,149,265, to Beller and Owen. Brit. Pat. 448,350, to Ehrhart Franz. Ger. Pat. 682,195, Luther to *I. G. Farbenindustrie*. Brit. Pat. 407,990, to *I. G. Farbenindustrie*.

[78] U. S. Pat. 2,047,612, Bertsch to *American Hyalsol Corp.*

[79] Ger. Pat. 546,807; French Pat. 718,395, to *H. Th. Böhme A.-G.*

[80] U. S. Pat. 2,081,865, Elbel to *Henkel et Cie.* Ger. Pat. 640,681, Wenzel to *Rudolf and Co.*

more of the low-molecular-weight oxygenated solvents in which the detergents are soluble and the inorganic salts insoluble. In analyzing the products for salt content, for example, the dried product is extracted with alcohol, leaving the salt as an insoluble residue. Separation of salt on a large scale, however, is usually effected by taking a concentrated aqueous solution of the neutralized product and adding a solvent to it. Appropriate solvents include most of those which are normally water miscible but can be salted out and some which are only slightly soluble in water but have good solvent power for the detergent. Isopropanol, n-propanol, the butanols, amyl, hexyl, and heptyl alcohols, terpineol, pine oil, acetone, and methyl ethyl ketone, the Cellosolve and Carbitol solvents (monoalkyl ethers of ethylene and diethylene glycol), butyl acetate, glycols and glycerol have all been used for this purpose. Of these isopropanol, pine oil, and the Cellosolves are probably the most widely used. The effect of the added solvent is to form a two-phase liquid system in which the lower layer consists mainly of inorganic salt and water. The upper layer contains most of the detergent and the added solvent. This procedure, which is often called "splitting out," is widely used[81] in the manufacture of many sulfated and sulfonated surface active agents besides the Gardinol type. One process of removing inorganic salt involves heating the product, with high-boiling solvents if necessary, to form an anhydrous melt from which the salts can be filtered.[82] The addition of glycerol, glycol, or their sulfuric esters after or during sulfation is said to facilitate the separation of excess sulfating agent from the mixture before neutralization.[83]

Forms and Salts. The Gardinol type surface active agents are usually marketed as the sodium salt in dry form or as an aqueous paste. The dry products are furnished in grades containing from 25 to 90% active detergent, the remainder being inorganic salt. The paste products normally contain about 25% active detergent and less than 5% salt, the remainder being water. Large amounts of the triethanolamine salt are made, mostly for use in shampoos. Water-soluble salts with a number of organic bases have been described.[84] The calcium salts of all the Gardinol type fatty sulfates except cetyl and stearyl are sufficiently soluble so that they are

[81] U. S. Pat. 2,152,162, Tulleners to *Shell Development Co.* Brit. Pat. 538,375, Miles *et al.* to *Colgate-Palmolive-Peet Co.* Ger. Pat. 670,297, Reibnitz to *I. G. Farbenindustrie.*

[82] Ger. Pat. 683,206, to *E. I. du Pont de Nemours & Co.*

[83] French Pat. 791,690; Swiss Pat. 177,938, to *Chemische Fabrik vorm. Sandoz.*

[84] Ger. Pats. 622,640; 627,055; U. S. Pat. 2,256,877, Bertsch to *H. Th. Böhme A.-G.* and *American Hyalsol Corp.* U. S. Pats. 2,139,277, Lehner and Arnold to *E. I. du Pont de Nemours & Co.*; 2,286,364, Jayne and Day to *American Cyanamid Co.*

effective in the hardest waters. The cetyl and stearyl calcium sulfates are only sparingly soluble but they do not have the curd-forming, greasy properties of the calcium fatty acid soaps and are accordingly much less objectionable. The magnesium salts of the fatty alkyl sulfates are sufficiently surface active and soluble to be claimed as detergent components.[85]

Gardinol type detergents are generally quite resistant to hydrolysis in alkaline solution, but hydrolyze rapidly in hot strong acid. Desseigne has recently reported on their stability under varying conditions of acidity.[86]

Miscellaneous Alcohols. Beside the higher primary normal alcohols there is a large number of other alcohols whose sulfuric esters are surface active. Most of those which have been described have not attained great commercial significance but some are produced and used on a large scale.

Cholesterol and other sterols give surface active sulfates.[87] By the catalytic reduction of lanolin there is obtained a mixture of sterols and fatty alcohols which can be sulfated to give a detergent.[88] These products appear to have little commercial significance. Sulfated mixtures of cholesterol and oleyl alcohol have also been described.[89]

Rosin acid esters have been reduced to abietyl and hydroabietyl alcohols whose sulfates[90] are highly surface active. Products of this nature have been offered in the market in this country but do not appear to have achieved any great volume up to the present time.

The naphthenyl alcohols, prepared by reducing naphthenic acids or their esters, also give sulfates[91] of high surface activity. Elaidyl alcohol, the *trans* isomer of oleyl alcohol, is stated to give a sulfate[92] of strong detergent properties. Sulfated acyloins, such as those derived[93] from capric or caprylic acids have been claimed as surface active agents.

[85] Ger. Pat. 690,628, to *Böhme-Fettchemie G.m.b.H.*

[86] Desseigne, *Industries Corps Gras* 1, 136 (1945). *Chem. Abstracts* 40, 4901.

[87] Sobel and Spoerri, *J. Am. Chem. Soc.* 63, 1259 (1941).

[88] Japanese Pat. 132,708, to K. Mugisima; *Chem. Abstracts* 35, 3366.

[89] French Pat. 797,679, to E. A. Mauersberger.

[90] U. S. Pat. 2,021,100; Brit. Pat. 430,578, to *E. I. du Pont de Nemours & Co.* See also Can. Pat. 377,543, Zimmerman to *Can. Aniline and Extract Co.*; U. S. Pat. 2,203,339, Parmelee to *E. I. du Pont de Nemours & Co.*

[91] U. S. Pat. 2,000,994, Schrauth to *Unichem.* See also Turkiewicz, *Kolloid-Z.* 92, 208 (1940).

[92] U. S. Pat. 1,981,901, Bunbury and Baldwin to *Imperial Chemical Industries Ltd.*

[93] U. S. Pat. 2,174,127, Henke and Richmond to *E. I. du Pont de Nemours & Co.*

A number of higher alcohols which give surface active sulfates[94] have been produced by hydrogenating substituted phenols. The resulting products are substituted cyclohexanols. The various alkylated or arylated phenols used as starting materials are relatively inexpensive to produce and it would seem that these products could be made to sell at a competitive price. They have not, however, appeared in the market so far as the writers are aware. Representative compounds of this nature include 4-octyl-2-methylcyclohexanol, perhydrogenated benzylcresols, perhydrogenated amylnaphthols, etc. Phenol-aldehyde condensates of the "pre-Bakelite" type have also been perhydrogenated to give sulfatable alcohols. The well-known intermediate from acetone and phenol is an example:

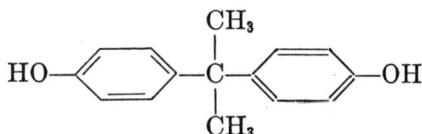

Alkylated cyclohexanols suitable for sulfation[95] to wetting agents have also been produced by reduction of isophorones. The latter compounds are made by the trimolecular condensation of methyl ketones. Other alcohols which contain aromatic rings, but have the hydroxyl group attached to a non-cyclic carbon atom, have also been sulfated. Examples of these are the alkylated cyclohexylethanols:

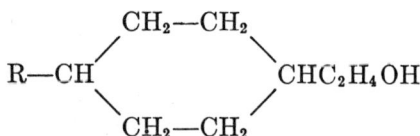

and perhydrodiphenylcarbinol.[96]

Sulfated alcohols containing halogen have been mentioned by a number of investigators. The halogen may be introduced before or after sulfation. Cetyl alcohol, for example, has been chlorinated and subsequently sulfated.[97] Sperm oil can be treated with hypochlorous acid to form a chlorohydrin, and the resulting product sulfated.[98] Hydrogen chloride has been

---

[94] U. S. Pats. 2,283,437–8, Hentrich *et al.* to *Procter and Gamble Co.* Ger. Pats. 681,782; 683,316, to *Henkel et Cie.* French Pats. 798,263, to *Henkel et Cie*; 794,422, to *Henkel et Cie.* Brit. Pat. 461,957, to *Henkel et Cie.* French Pat. 757,725, to *Ciba Co.* See also Ger. Pat. 696,904, to *Böhme Fettchemie A.-G.*

[95] U. S. Pat. 2,148,103, Bruson to *Rohm and Haas Co.*

[96] French Pats. 772,302; 796,980, to *Henkel et Cie.*

[97] Brit. Pat. 418,139, to *Chemische Fabrik Stockhausen.*

[98] Japanese Pat. 133,774, T. Maruyama *et al.*; *Chem. Abstracts* **35,** 4127.

added to oleyl sulfate to give a monochlorooctadecyl sulfate.[99] Olefins in the surface active range have been converted to chlorine-substituted alcohols by adding hypochlorous acid, and these have been sulfated.[100] Sulfated fatty alcohols containing further inorganic ester substituents on the alkyl chain have been made, for example, by sulfating dichlorooleyl alcohol. Oleyl alcohol which has been sulfated or phosphated at the double bond may also be sulfated at the hydroxyl group to produce[101] surface active agents of this class.

Synthetic higher alcohols which are probably mixtures of isomers have been made by condensing[102] formaldehyde with higher olefins, by condensing[103] a higher alcohol with cyclohexanone and metallic sodium, and by reducing[104] the condensation product of an alkyl phenol with a halogenated lower ketone.

Higher tertiary alcohols suitable for sulfation[105] to surface active agents have been made by treating a fatty acid ester such as ethyl oleate or a glyceride such as palm oil with a Grignard reagent such as CH_3MgBr or $C_6H_5CH_2MgCl$.

Higher normal secondary alcohols and their sulfates have been disclosed in several patents.[106] They are usually made by reducing the corresponding ketones either catalytically or with sodium and an alcohol. The ketones can be obtained by heating fatty acids or mixtures of fatty acids to high temperatures with lime or thoria.

Methylheptadecenylcarbinol has been sulfated to form surface active agents.[107] Like oleyl alcohol, this product can be sulfated either at the double bond or at the hydroxyl group, or in both positions.

The sulfation of the normal secondary alcohols with sulfuric acid evidently results in some isomerization. Baumgarten[108] has studied this phenomenon and reports that the hydroxyl group can shift progressively down the chain through the formation of intermediate olefin linkages.

Higher alcohols which can be sulfated to give wetting agents have also

[99] Brit. Pat. 400,986, to *Deutsche Hydrierwerke A.-G.*

[100] Ger. Pat. 687,462, to *Böhme Fettchemie G.m.b.H.* See also Brit. Pat. 418,139, to *Chemische Fabrik Stockhausen.*

[101] U. S. Pat. 2,264,737, Bertsch to *American Hyalsol Corp.*

[102] Ger. Pat 672,370, Haussman and Dimroth to *I. G. Farbenindustrie.*

[103] Ger. Pat. 683,569, Machon to *Böhme Fettchemie A.-G.*

[104] Brit Pat. 516,879, to *Deutsche Hydrierwerke A.-G.*

[105] Brit. Pat. 422,804; 424,891; to *Henkel et Cie.*

[106] U. S. Pat. 2,163,651; 2,326,270; to *E. I. du Pont de Nemours & Co.* U. S. Pat. 2,321,020, Dreger and Ross to *Colgate-Palmolive-Peet Co.* Ger. Pat. 589,946, Lommel and Schröter to *I. G. Farbenindustrie.*

[107] U. S. Pat. 2,229,649, Guenther *et al.* to *General Aniline and Film Corp .*

[108] Baumgarten, *Ber,* **76B,** 213 (1943).

been made by condensing triply unsaturated acyclic terpenes such as allooci-
mene with crotonaldehyde and hydrogenating the condensate. Products
such as trimethylbutylhexahydrobenzyl alcohol are obtained.[109]

Several mono- and disulfated higher glycols have been described as surface active agents. Thus, 7,18-stearic glycol results[110] from the hydrogenation of ricinoleic esters. Glycols suitable for sulfation to surface
active agents have been made[111] by treating higher olefins with H_2O_2 and
have also been obtained[112] from sperm oil. α,ω-Glycols whose sulfates[113]
are surface active have been made by reducing the polymerized esters of
polyene fatty acids. These acids have been developed and studied in
connection with the development of the rubbery polyester Norepol.

Just as the natural fatty acids can be reduced to alcohols, so the acids
made by oxidizing paraffin wax or Fischer-Tropsch hydrocarbons can be
reduced to alcohols suitable for sulfation. These products[114] appear to
have attained commercial importance in Germany but are of little, if any,
importance in the United States.

The reduction of carbon monoxide or carbon dioxide with hydrogen is
carried out under varied conditions to produce substances ranging from
methanol to higher wax-like hydrocarbons. Under most circumstances
there may be recovered, from the mixed reaction products, alcohols and
olefins in the surface-active range. A number of patents describe[115] the
sulfates of these products. The extent to which they are used is difficult to
estimate. There are no trade-named detergents of which the writers are
aware which are composed wholly of these alkyl sulfates. It is probable,
however, that some of them, particularly the sulfated by-product alcohols
of the methanol synthesis, are mixed with other surface active agents.

Higher primary alcohols have been produced on a large scale by the
action of carbon monoxide and hydrogen on higher olefins. The usual
Fischer-Tropsch cobalt–thoria catalyst is used. The carbon monoxide
evidently reacts with the olefin to form an aldehyde, which is then reduced

[109] U. S. Pat. 2,339,818, Rummelsburg to *Hercules Powder Co.*

[110] U. S. Pat. Re. 19194, Guenther and Saftien to *I. G. Farbenindustrie*. See also
U. S. Pat. 2,007,492, Bertsch to *H. Th. Böhme A.-G.*

[111] U. S. Pat. 2,138,917, Grun to *American Hyalsol Corp.* See also Ger. Pat. 670,-
556, Schrauth and Hueter to *Deutsche Hydrierwerke A.-G.*

[112] U. S. Pat. 2,014,782, Schrauth and Hueter to *Unichem.*

[113] U. S. Pat. 2,347,562, Johnston to *American Cyanamid Co.*

[114] U. S. Pats. 2,341,218, to J. H. James; 2,151,106, Hentrich *et al.* to *Henkel et Cie.*
Ger. Pat. 684,927, to *I. G. Farbenindustrie*. Brit. Pat. 489,863, to *Henkel et Cie.*
See also Ger. Pat. 682,195, to *I. G. Farbenindustrie.*

[115] U. S. Pat. 2,042,747, Ulrich *et al.* to *I. G. Farbenindustrie*. Ger. Pat. 674,751,
to *Deutsche Gold u. Silber-Scheideanstalt*. U. S. Pat. 1,979,303, Woodhouse to *E. I.
du Pont de Nemours & Co.* See also Brit. Pat. 499,542, Snoddy to *Procter and Gamble
Co.*

to the primary alcohol. This process was operated in Germany during the war and was called the Oxo process.[116] The higher alcohols produced were used for sulfation to produce detergents.

In the oxidation of paraffin wax and Fischer-Tropsch hydrocarbons (for the primary purpose of producing carboxylic acids) a certain proportion of by-product olefins and alcohols as well as ketones are obtained. After separation, these alcohols can be sulfated[117] to surface active agents.

An important method of synthesizing alcohols in the surface active range is by means of the aldol or ketol condensation reaction, followed by dehydration and reduction. For example two molecules of butyraldehyde can be condensed in the presence of a mild alkaline catalyst to form an aldehyde alcohol:

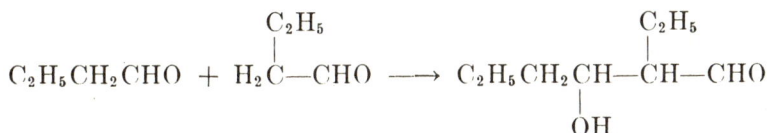

$$\underset{}{C_2H_5CH_2CHO} \;+\; \underset{\underset{}{\overset{|}{H_2C}}}{\overset{C_2H_5}{}}\!-CHO \;\longrightarrow\; C_2H_5CH_2CH-\underset{\underset{OH}{|}}{\overset{\overset{C_2H_5}{|}}{CH}}-CHO$$

By dehydration this product yields:

$$C_2H_5CH_2CH=\underset{}{\overset{\overset{C_2H_5}{|}}{C}}-CHO$$

which in turn is readily reduced to 2-ethylhexanol, a C_8 alcohol. The sulfate of 2-ethylhexanol is an excellent wetting agent. It is best known under the trade name Tergitol 08 (*Carbide and Carbon Chemicals Corp.*), which is an aqueous solution of the sodium salt containing about 25% active ingredient and a small amount of non-aqueous solvent.[118]

By using various aldehydes and ketones in the aldol condensation a wide range of higher alcohols can be synthesized.[119] The only sulfated products of this nature which have achieved prominence in this country are the Tergitols of *Carbide and Carbon Chemicals Corp.* Besides Tergitol 08

[116] See editorial report *Chem. Inds.* **60**, 232 (1947). U. S. Pat. 2,327,066, Roelen vested in Alien Property Custodian. *U. S. Bur. Mines, Information Circ.* No. 7370, p. 87.

[117] U. S. Pat. 2,020,453, Beller *et al.* to *I. G. Farbenindustrie.* Brit. Pat. 494,616 to *I. G. Farbenindustrie* and *Standard Oil Development Co.* U. S. Pat. 2,085,501, to J. H. James. French Pat. 806,990, to *I. G. Farbenindustrie.* Ger. Pats. 577,428; 589,511; 626,521; to *I. G. Farbenindustrie.*

[118] U. S. Pat. 2,052,027, to Benjamin R. Harris. U. S. Pat. 2,161,857, Davidson to *Carbide and Carbon Chemicals Co.*

[119] Brit. Pat. 446,026, Wickert and Freure to *Carbide and Carbon Chemicals Corp.* French Pats. 782,835; 786,734; 789,406; to *Carbide and Carbon Chemicals Corp.* U. S. Pat. 2,088,019, Wickert to *Union Carbide and Carbon Corp.* Brit. Pats. 407,187; 437,869; French Pat. 832,072; to *I. G. Farbenindustrie.*

mentioned above, similar products include Tergitol 4, the sulfate of the C_{14} alcohol, 7-ethyl-2-methylundecanol-4, $C_4H_9CH(C_2H_5)C_2H_4CH(OH)-CH_2CH(CH_3)_2$, and Tergitol 7 made from the C_{17} alcohol, 3,9-diethyl-tridecanol-6, $C_4H_9CH(C_2H_5)C_2H_4CH(OH)C_2H_4CH(C_2H_5)_2$.

Higher alcohols can also be synthesized by the Guerbet reaction. This long-known but infrequently used reaction involves heating a lower alcohol with caustic soda to a high temperature. Two molecules of the alcohol condense with each other eliminating water and forming an alcohol with double the number of carbon atoms. Capryl alcohol has been converted to dicapryl alcohol by this method. Sulfated dicapryl alcohol has been produced recently on a moderate commercial scale by a number of manufacturers in this country. It resembles cetyl and myristyl sulfates in its properties.

(e) Sulfated Olefins

The olefins corresponding to the higher aliphatic alcohols yield sulfation products having surface active properties substantially identical to those of the corresponding sulfated alcohols. This fact is of considerable practical importance as in many instances it is easier to obtain an olefinic raw material than a hydroxylated one. A wide variety of methods are used for obtaining olefins suitable for sulfation to surface active products.

Squalene, a highly unsaturated hydrocarbon occurring in shark liver oil and other marine oils, having the formula:

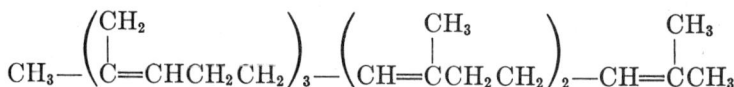

$$CH_3 - \left(\begin{array}{c} CH_2 \\ | \\ C \end{array} = CHCH_2CH_2 \right)_3 - \left(\begin{array}{c} CH_3 \\ | \\ CH \end{array} = CCH_2CH_2 \right)_2 - CH = \begin{array}{c} CH_3 \\ | \\ C \end{array} CH_3$$

has been sulfated[120] to produce an emulsifying agent.

Cetene-1 has been produced by the dry distillation of spermaceti. The cetyl palmitate which is the main ingredient of spermaceti pyrolyzes to cetene and palmitic acid. Cetene-1 on sulfation yields cetyl-2-sulfuric ester.[121] The 1-olefins can in general be made by dehydrating the corresponding fatty alcohols. The olefins yield[122] secondary sulfates in contrast to the primary sulfates obtained from the alcohols.

Olefins for sulfation have been produced by the decarboxylation of unsaturated fatty acids. Oleic acid, for example, can be decarboxylated by[123]

[120] Brit. Pat: 354,417; U. S. Pat. 1,961,683; Bunbury et al. to Imperial Chemical Industries Ltd.

[121] U. S. Pat. 2,027,896, Bertsch to H. Th. Böhme A.-G.

[122] French Pat. 716,178; Brit. Pat. 343,872; Ger. Pat. 705,179; to I. G. Farbenindustrie.

[123] Ger. Pat. 594,093, Platz to I. G. Farbenindustrie.

pyrolysis to a mixture of heptadecenes. This reaction is not smooth and high yields of clean products are evidently not obtained, although catalysts improve the process considerably.

A variety of natural terpenes and terpene alcohols have been sulfated to produce surface active agents.[124] Modified terpenes, such as their reduced condensation products with aldehydes, have also been used,[125] as well as dimerized terpenes.[126] Both abietyl alcohol and the olefin made by dehydrating hydroabietyl alcohol give surface active sulfates.[127]

Olefins of eight to twenty or more carbon atoms which give surface active sulfates have been made by polymerizing suitable lower olefins. Isobutylene dimers, trimers, and tetramers are the most thoroughly studied. Propene, isoamylene, and isohexene and butadiene have also been polymerized to a low degree for sulfation. The polymers are often obtained directly by the action of sulfuric acd or other acidic dehydrating agent on the corresponding lower alcohol. Mixed di-, tri-, and tetraisobutylenes are obtained by treating tertiary butyl alcohol with sulfuric acid. The sulfated products[128] are in general excellent wetting agents and detergents but the sulfation process is usually difficult to carry out. It is possibly for this reason that the products have not been produced in large tonnages.

Olefins and alcohols for sulfation have been prepared[129] from saturated petroleum hydrocarbons in the kerosene range by chlorinating them and then treating the chlorinated product with caustic alkali. An interesting study of this process, starting with a kerosene fraction boiling at 95–100°C. and carrying it through to the final sulfated wetting agent, has been made by Padgett and Degering.[130]

Chlorine-containing olefins, such as 1-choro-9,10-octadecene, obtained by the action of thionyl chloride on oleyl alcohol, have been sulfated. Chlorine may also be introduced into an alkyl sulfate after sulfation. Whether the halogen adds anything to the efficiency of the surface active agent is not clearly established.[131]

[124] See, for example, Ger. Pat. 699,655, to *Ciba Co.*

[125] U. S. Pat. 2,354,774, Rummelsberg to *Hercules Powder Co.*

[126] U. S. Pats. 2,344,833, Rummelsberg to *Hercules Powder Co.*

[127] U. S. Pat. 2,203,339, Parmalee to *E. I. du Pont de Nemours & Co.* Brit. Pat. 463,569, to *I. G. Farbenindustrie.*

[128] Ger. Pat., 550,242; Brit. Pat. 329,622; U. S. Pat. 1,950,287; to *Chem. Fabrik Pott and Co.* U. S. Pat. 2,160,343, Ross to *Colgate-Palmolive-Peet Co.*; 2,166,981, Rosen *et al.* to *Standard Oil Development Co.*; 2,139,394, van Peski to *Shell Development Co.*

[129] Ger. Pat. 622,296; Brit. Pat. 344,829; to *I. G. Farbenindustrie.*

[130] Padgett and Degering, *Ind. Eng. Chem.* **32**, 204 (1940).

[131] Brit. Pat. 498,008; French Pat. 838,527; to *I. G. Farbenindustrie.* U. S. Pat. 2,267,731, Guenther *et al.* to *General Aniline and Film Corp.*

The most interesting source of olefins from the economic point of view is petroleum. Many efforts have been made to prepare and sulfate olefins of the proper molecular weight, configuration, and purity from this source. One of the early attempts in this direction was the sulfation of the so-called Edeleanu extracts. In the Edeleanu process for refining oils the oil is extracted with liquid sulfur dioxide. This solvent dissolves out aromatics and unsaturates, which can be recovered by allowing the solvent to evaporate. This extract can be sulfated[132] (the aromatic portion is simultaneously sulfonated) to give surface active products. These products are said to resemble the "petroleum sulfonates" obtained in the refining of oils with sulfuric acid more than they do the completely sulfated water-soluble surface active agents.

The major line of development in this field has involved subjecting paraffin wax or petrolatum to a high-temperature cracking process and fractionally distilling the products to obtain suitable cuts of olefins. Products boiling in the range about 150–300°C. and having a high percentage of monoolefins give sulfation products of desirable properties. The practical difficulties appear to lie mainly in the sulfation process and particularly in the separation of unsulfatable and other insoluble material from the final product. The olefins as obtained by fractionating the cracking products contain a significant proportion of these undesirable constituents, and the sulfation reaction itself is incomplete, producing varying amounts of alcohols, polymers, and other water-insoluble substances. Most of the development work on this class of products appears to have been done mainly by two groups, one at the *Standard Oil Development Co.*,[133] and the other at *Shell Development Co.*[134] The commercial wetting agent Teepol, recently introduced by the *Shell Oil Co.*, is probably a product of this nature.

The general methods for sulfating olefins parallel those used for alcohols. Sulfuric acid of varying strengths, oleum, and chlorosulfonic acid have all been used. Solvents may be used[135] either to accelerate the reac-

---

[132] Ger. Pat. 545,968, Kirch to *Chemische Fabrik Pott and Co.*

[133] U. S. Pat. 2,139,669, Buc to *Standard Oil Development Co.* Can. Pat. 381,193, Fasce and Rolfs to *Standard Oil Development Co.* U. S. Pats. 2,157,320, Buc to *Standard Oil Development Co.* U. S. Pat. 2,153,286; Can. Pat. 430,059; Sweeney to *Standard Oil Development Co.* See also U. S. Pat. 1,999,128, MacLaren to *Standard Oil Co. of Indiana.*

[134] U. S. Pats. 2,078,516; 2,139,393; 2,152,292; 2,155,027; 2,339,038; Tulleners *et al.* to *Shell Development Co.* See also Brit. Pat. 504,977; Dutch Pat. 46,546; French Pat. 838,367; to *N. V. de Bataafsche Petroleum Maatschappij.*

[135] See for example: French Pat. 773,656, to *Henkel et Cie*; Ger. Pat. 677,463, Günther and Hetzer to *I. G. Farbenindustrie.*

tion or to separate excess acid. Special reagents such as the SO_3-dioxane complex have also been used[136] with great success in sulfating olefins. As a rule, the optimum conditions for obtaining complete sulfation and good separation from insoluble residues vary from one olefin to another, and must be worked out in each individual case.

The alkyl sulfuric esters are in general quite stable in alkaline solutions even at the boiling point and in the presence of free caustic soda. They are rather readily hydrolyzed, however, by hot mineral acids, forming sulfuric acid and the alcohol or its corresponding olefin.[137] This affords a convenient method of analysis, since the resulting alcohol or olefin is usually insoluble in the aqueous hydrolyzing mixture and can be separated by some physical process for weighing. The alkyl sulfates themselves are usually soluble in the lower alcohols such as ethanol or isopropanol and can be separated from inorganic salt by extraction with these solvents. They form insoluble salts with benzidine and can be estimated by precipitating with this reagent and subsequently titrating with sodium hydroxide.[138]

III. Sulfuric Esters with Intermediate Linkage between the Hydrophobic and the Sulfate Groups

(a) Ester Intermediate Linkage

It has been pointed out that one of the economic disadvantages of the normal primary alkyl sulfates is the relatively high cost of producing the fatty alcohols. Considerable effort has accordingly been spent in developing less expensive substances containing both a hydrophobic group and an aliphatic hydroxyl group capable of sulfation. One very effective way of doing this is the esterification of a fatty acid of eight or more carbon atoms with one hydroxyl group of a low-molecular-weight glycol leaving the other hydroxyl group free to be sulfated. Similarly polyhydric alcohols may be partially esterified with fatty acid to yield an intermediate product containing one or more free hydroxyl groups susceptible to sulfation.

The most important product of this class is the sulfated monoglyceride of coconut fatty acids, $RCOO—CH_2CHOHCH_2OSO_3Na$. It is produced on a very large scale in this country by *Colgate-Palmolive-Peet Co.* and sold as Arctic Syntex M, Arctic Syntex L, and for household use as Vel.

The mixed fatty acid – sulfuric acid esters of glycerol were studied as far back as 1909, but their practical value as surface active agents was not

---

[136] U. S. Pat. 2,135,358, Suter to *Procter and Gamble Co.*

[137] See Seck, *Fettchem. Umschau* **41**, 61 (1934).

[138] Kertess, *Textile Mfr.* **60**, 336 (1934). Snell, *Ind. Eng. Chem., Anal. Ed.* **7**, 234 (1935).

realized at that time.[139]   The modern development of this type of product in the United States is based on the disclosures in two broad patents, U. S. 2,023,387 and U. S. 2,023,388 to Benjamin R. Harris.   These have been reissued as Re. 20,636 and 21,322, respectively.   These patents describe a wide range of structures in which one or more fatty acid radicals are esterified with the corresponding number of hydroxyl groups on a central polyhydric alcohol.   At least one of the remaining hydroxyl groups is esterified with sulfuric acid.   There are two general procedures for making the Syntex type of product.   One is the preparation of the fatty monoglyceride first and then its sulfation.   The other is effecting a simultaneous esterification of glycerol with both the fatty acid and sulfuric acid.

Fatty monoglycerides may be prepared by the direct reaction of a fatty acid with an excess of glycerol under esterification conditions.   They may also be prepared by heating a fatty triglyceride with an excess of glycerin to a high temperature.   This latter procedure is normally used on a large scale in the preparation of oil-modified alkyd resins and is accordingly a well-known industrial-scale process.   The reactions in both cases produce mixtures of mono-, di-, and triglycerides, the relative yields of each depending on the reaction conditions, catalysts, etc.[140]   The fatty monoglycerides may actually be isolated[141] by distillation in a high vacuum.   The charging stock in such a process may be the crude monoglyceride mixture obtained by heating a fatty oil with glycerin in the presence of an alcoholysis catalyst.

The fatty monoglycerides may be sulfated by the usual sulfation procedures already discussed.   Products of low salt content may be made by using reagents such as sodium chlorosulfonate.[142]   They may also[143] be made by the usual procedure of extractng with alcohol, leaving the inorganic salts undissolved.

Beside the fatty monoglycerides a wide range of other partially esterified polyhydric alcohols have been sulfated to form surface active agents of this class.   Among these are the partial esters of ethylene[144] and propylene glycols and polyglycols, pentaerythritol, hexitols, and trimethylol propane.

The most interesting processes for preparing the Syntex type detergent

[139] Thieme, *J. prakt. Chem.* **85**, 284; *Koninkl. Akad. Wetenschap. Amsterdam* **10**, 855 (1909); *Chem. Abstracts* **4**, 754 (1910).

[140] See Brit. Pat. 440,888, Hilditch and Rigg to *Imperial Chemical Industries Ltd.* French Pat. 757,763, to *Procter and Gamble Co.* See also Kawai and Nobori, *J. Soc. Chem. Ind. Japan* **43**, Suppl. binding, 59 (1940); *Chem. Abstracts* **34**, 4598.

[141] U. S. Pat. 2,383,581, Arrowsmith and Ross to *Colgate-Palmolive-Peet Co.*

[142] Brit. Pat. 494,870, to *Procter and Gamble Co.* and A. O. Snoddy.

[143] U. S.Pat. 2,316,719, Russell to *Colgate-Palmolive-Peet Co.*

[144] Brit. Pat. 499,144, to *I. G. Farbenindustrie.*

are those in which fatty triglyceride, glycerol, and sulfuric acid or fatty acid, glycerol, and sulfuric acid are all reacted in one stage in the proper molecular proportions. The ideal reaction using fatty triglyceride as starting material is as follows:[145]

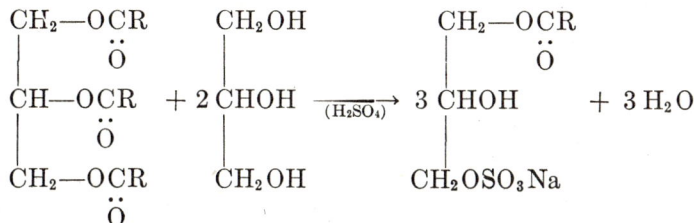

$$
\begin{array}{ccc}
\text{CH}_2\text{—OCR} & \text{CH}_2\text{OH} & \text{CH}_2\text{—OCR} \\
\quad\ \ \ddot{\text{O}} & | & \quad\ \ \ddot{\text{O}} \\
\text{CH—OCR} \ + \ 2 \text{ CHOH} \xrightarrow[(\text{H}_2\text{SO}_4)]{} 3 \text{ CHOH} \ + \ 3\,\text{H}_2\text{O} \\
\quad\ \ \ddot{\text{O}} & | & | \\
\text{CH}_2\text{—OCR} & \text{CH}_2\text{OH} & \text{CH}_2\text{OSO}_3\text{Na} \\
\quad \ddot{\text{O}} & &
\end{array}
$$

This process has been made continuous.[146] Because of the simplicity of processing and cheapness of the raw materials the Syntex type is potentially one of the least expensive synthetic surface active agents. The raw material costs are considerably less than for straight soap. As is the case with most of the synthetic detergents produced on a large scale a variety of different salts and applications have been patented.[147]

The Syntex type detergent, in common with other ester-linked sulfate detergents, is relatively unstable in hot acids or alkalis. The alkalis saponify the carboxylic ester linkage and the acids hydrolyze the sulfuric ester linkage. They are sufficiently stable for all ordinary purposes in solutions near the neutral point such as are used in household cleaning and washing and in most commercial processes involving surface active agents.

It is theoretically possible to sulfate the free hydroxyl group (or groups) of any product of the general formula:

$$\text{RCOO—X—(OH)}_n$$

and surface active products will result provided R is any typical hydrophobic radical. Products of the general formula:

$$\text{RCOO—X—CH=CH}_2$$

can, of course, also be sulfated. Among the examples of these types which have been described are:

(1) Sulfated pentaerythritol esters of hydrogenated rosin (dihydro- or perhydroabietic acid).[148] (2) Products obtained by reacting lauric acid

---

[145] See Ger. Pats. 702,598; 689,511; Brodersen and Quaedvlieg to *I. G. Farbenindustrie.*

[146] U. S. Pat. 2,242,979, Muncie to *Colgate-Palmolive-Peet Co.*

[147] See for example, U. S. Pat. 2,187,144, Bell *et al.* to *Colgate-Palmolive-Peet Co.* Canadian Pats. 426,100; 426,101; 426,103; to *Colgate-Palmolive-Peet Co.*

[148] U. S. Pat. 2,362,882, Carson to *Hercules Powder Co.*

with ethylene glycol, formaldehyde, and H_2SO_4.   Presumably an intermediate hydroxyether ester:

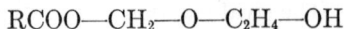

$$RCOO—CH_2—O—C_2H_4—OH$$

is formed which undergoes sulfation.[149]   (3) Sulfated fatty acid esters of methallyl alcohol:[150]

$$RCOO—CH_2—C(CH_3)=CH_2$$

(4) Sulfated monoglycerides and glycol esters[151] of naphthenic acids, such as sulfated ethylene glycol naphthenate.

A number of other ester-linked sulfates have been described, none of which has attained great industrial importance.

Phosphatides such as egg lecithin and soy bean lecithin have been sulfated to form water soluble surface active products[152] having good lime resistance.   Lecithin is the naturally occurring phosphoric ester of a bis fatty ester of glycerol.   Presumably sulfation takes place either by replacing the phosphate group, or more likely, by reaction with the double bonds of the fatty acid chain.

α,ω-Chlorohydrins such as 6-chloro-1-hexanol can be reacted with a fatty acid sodium salt to form, e.g., 6-fatty acylated 1,6-hexanediol, which can then be sulfated.[153]

Polyhydroxycarboxylic acids such as tartaric acid may be converted to products[154] in which one hydroxyl group is acylated with a fatty acid and another hydroxyl group is sulfated.   The resultant product contains both carboxyl group(s) and sulfuric ester group(s) as solubilizing radicals.

Another interesting product[155] of this type has been made by sulfating the tetrahydrofurfuryl ester of a fatty acid.   The sulfating agent splits the ether linkage in the ring:

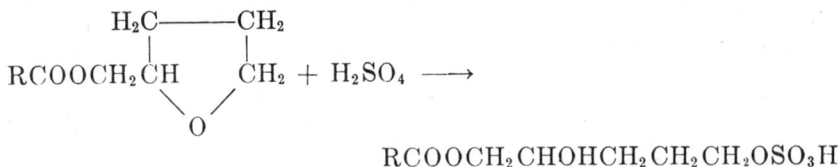

$$\begin{array}{c} H_2C\text{————}CH_2 \\ | \qquad\quad | \\ RCOOCH_2CH \qquad CH_2 + H_2SO_4 \longrightarrow \\ \diagdown \qquad \diagup \\ O \end{array}$$

$$RCOOCH_2CHOHCH_2CH_2CH_2OSO_3H$$

---

[149] U. S. Pat. 2,366,738, Loder and Gresham to *E. I. du Pont de Nemours & Co.*

[150] U. S. Pat. 2,341,060; Brit. Pat. 545,415-6; Price and Kapp to *National Oil Products Co.*

[151] U. S. Pat. 2,293,965, Mikeska to *Standard Oil Development Co.*   Ger. Pat. 713,-853, to *I. G. Farbenindustrie.*

[152] U. S. Pat. 2,079,973 Strauch to *Chemische Fabrik Stockhausen.*   Ger. Pats. 660,-736; 667,085; to *Chemische Fabrik Stockhausen.*

[153] Brit. Pat. 444,239 to *Deutsche Hydrierwerke A.-G.*

[154] U. S. Pat. 2,285,773, Harris to *Colgate-Palmolive-Peet Co.*

[155] U. S. Pat. 2,235,534, Russell and Bell to *Colgate-Palmolive-Peet Co.*

Instead of using a dihydroxy compound as the intermediate link, an hydroxycarboxylic acid may be used. In this case the carboxyl group is esterified with a long-chain fatty alcohol and the hydroxyl group is sulfated. The generic reaction is:

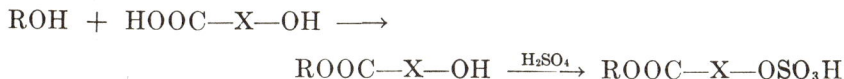

$$ROH + HOOC\text{—}X\text{—}OH \longrightarrow$$

$$ROOC\text{—}X\text{—}OH \xrightarrow{H_2SO_4} ROOC\text{—}X\text{—}OSO_3H$$

As an example, lactic acid may be esterified with stearyl alcohol and the resulting octadecyl lactate sulfated.[156] In a modification[157] of this general procedure hydroxybutyric or hydroxyvaleric lactones may be treated with a sulfating agent to give the corresponding sulfuric ester of the carboxylic acid. The carboxy group is then esterified with a higher fatty alcohol.

Ester-linked surface active sulfuric esters containing two hydrophobic groups have been made[158] by sulfating intermediates such as dioctyl maleate:

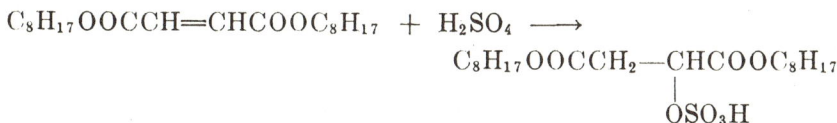

$$C_8H_{17}OOCCH\text{=}CHCOOC_8H_{17} + H_2SO_4 \longrightarrow$$

$$C_8H_{17}OOCCH_2\text{—}CHCOOC_8H_{17}$$
$$\underset{\displaystyle OSO_3H}{\mid}$$

Other long-chain esters of olefinic or hydroxylated di- or polycarboxylic acids such as didodecyl citrate, octadecyl malate, etc. may also be sulfated. An unusual product[159] of this type has been made by sulfating the esters of cyclohexenedicarboxy acid with alcohols of six or more carbon atoms.

(b) Amide Intermediate Linkage

Another widely used method of producing hydroxyl compounds suitable for sulfation to surface active agents involves the use of a low-molecular-weight aminohydroxy compound and acylation of the amino group with a higher fatty acid. The generic reaction may be represented as:

$$RCOOH + H_2N\text{—}X\text{—}OH \rightarrow RCONH\text{—}X\text{—}OH$$

The amino group in the linking compound may be primary or secondary. There may also be more than one hydroxyl group present.

By far the most widely used product of this class is made by sulfating the monoethanolamide of coconut fatty acids. It has the formula:

$$RCONHC_2H_4OSO_3Na$$

[156] Brit. Pat. 348,040, to *I. G. Farbenindustrie.*
[157] Ger. Pat. 623,948, Russe to *Oranienburger Chemische Fabrik A.-G.*
[158] U. S. Pats. 2,104,782; 2,121,617, Werntz to *E. I. du Pont de Nemours & Co.*
[159] Ger. Pat. 708,429, Hopff and Rapp to *I. G. Farbenindustrie.*

The intermediate amide is usually made by heating equimolar quantities of fatty acid and monoethanolamine to about 170–180°. Water is eliminated in the reaction and the acid number of the mixture gradually falls. When it has reached a value of about 5 or less, the reaction is considered substantially complete. The amide is then sulfated with H_2SO_4 at about 30°C. or less. It may also be sulfated with chlorosulfonic acid or oleum with or without the aid of solvents.[160] Detergents of this type have also been made[161] by condensing preformed sulfuric ester of ethanolamine in alkaline solution with a fatty acid chloride:

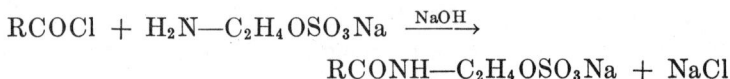

$$RCOCl + H_2N—C_2H_4OSO_3Na \xrightarrow{\text{NaOH}}$$
$$RCONH—C_2H_4OSO_3Na + NaCl$$

The reaction of fatty acid with monoethanolamine results in the formation of a small amount of the ester, $RCOOC_2H_4NH_2$, which is unsulfatable and remains in the product. It detracts considerably from the detergent, wetting, and foaming properties of the final sulfated material. An improved product[162] is obtained if the amide is made by reacting coconut fatty acid chloride with monoethanolamine. The ethanolamides may also be made[163] by reacting a fatty ester with monoethanolamine, and may be purified by washing.

Although coconut fatty acid is most widely used in preparing these substances, other acids such as palmitic, stearic, oleic, palm oil acids, etc. have been used. The product[164] made from a mixture of caprylic and capric acids is an excellent foaming agent.

The sulfated ethanolamide of coconut fatty acids is made by a number of firms in this country and sold under many trade names. Among these are Alframine, Xynomine, Alrosene, Hytergen, Sulframine, Miranol, Emcol, and others. Because of the simplicity of the operations involved and the lack of troublesome by-products, it is particularly favored by smaller-scale manufacturers. The properties of this product vary remarkably with the degree of purity. Pure preparations are excellent foamers and detergents,

[160] U. S. Pat. 1,981,792 to John W. Orelup. Note that this type of product is covered by the claims of U. S. Pat. 1,918,373, Bertsch to *H. Th. Böhme A.-G.*, but the disclosure does not mention the products specifically.

[161] U. S. Pat. 1,932,180, Guenther and Haussman to *I. G. Farbenindustrie*. See Examples 9 and 10. This is one of the most important patents in the whole field of surface active agents. It is generally considered the basic patent on detergents of the Igepon T type. (Compare Chapter 4, Section III, *b*.) Its voluminous disclosure and broad claims, however, cover a much wider field.

[162] U. S. Pat. 2,355,503; Can. Pat. 427,726; Bertsch to *The Hydronaphthene Corp.*

[163] Brit. Pat. 523,466, Greenhalgh to *Imperial Chemical Industries Ltd.* U. S. Pat. 2,167,931, McAllister to *Shell Development Co.*

[164] U. S. Pat. 2,353,081, Robinson and Davis to *National Oil Products Co.*

being equal in these respects to the Gardinol type detergents. Many of the commercial products, however, contain sufficient unsulfated material to detract remarkably from their surface active properties. This is particularly noticeable at lower temperatures. In high-temperature baths, such as are used in many textile scouring operations, the unsulfated matter has much less effect, and even the less pure products foam and wash very satisfactorily.

The sulfated fatty acyl ethanolamines are slowly hydrolyzed on storage when they are in aqueous paste form, so that commercial preparations of this nature should be used within a few weeks of shipment. After about three or four months of storage under ordinary conditions they will often be hydrolyzed to such an extent as to interfere seriously with their efficiency. The dried product, however, is apparently stable indefinitely. This product was used in Germany during the war under the name Waschmittel E.[165]

Many other sulfated hydroxy amides of fatty acids have been described.[166] Diacylated ethanolamines have been sulfated to give surface active agents, as have been the lower aldehyde condensates of the monoacyl ethanolamines.

Amines other than ethanolamine have been used as intermediate links. Among the most frequently mentioned[167] are 1-amino-2,3-dihydroxypropane (known also as glycerolamine or aminoglycerol), trimethylolaminomethane, H_2N—$C(CH_2OH)_3$, and diethanolamine.

The product made by sulfating the fatty acyl derivative of 4-aminobutanol-2:

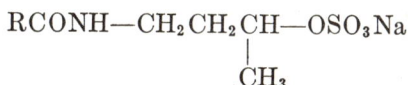

$$RCONH-CH_2CH_2CH-OSO_3Na$$
$$\vert$$
$$CH_3$$

was used in Germany under the name Igepon B.[168] Its stability characteristics are similar to those of the monoethanolamine analogue. m-Aminocyclohexanol has been acylated with higher fatty acids and sulfated to form surface active agents.[169] The sulfated ethanolamides of blown or bodied castor oil acids are described as good emulsifying agents.[170]

[165] W. Baird, *Textile Auxiliary Products Manufactured by I. G. Farbenindustrie at Ludwigshafen*, PB 28754, Office of Technical Services, Dept. of Commerce, Washington, D. C.

[166] Brit. Pat. 515,882; French Pat. 838,169; U. S. Pat. 2,185,817, Mauersberger to *Alframine Corp.*

[167] Brit. Pat. 414,403, to *Imperial Chemical Industries Ltd.*

[168] W. Baird, *loc. cit.*

[169] Swiss Pat. 187,421; 189,302, to *Ciba Co.*

[170] U. S. Pat. 2,340,112, Davis and Abramowitz to *National Oil Products Co.*

In place of a fatty acylated hydroxyamino compound, a fatty acylated olefinic amino compound may also be sulfated to give surface active agents. The allyl or methallyl amide of coconut fatty acids, for example, may be sulfated to give products[171] very similar to those obtained from the mono-ethanolamide.

Synthetic acids containing an aromatic nucleus have been used in producing the sulfated hydroxy amide type of detergent, an example being the amide[172] formed from isooctylphenylacetic acid and aminoethylsulfuric ester. The aromatic nucleus may also form part of the intermediate linking portion of the molecule as, for example, in the compound:[173]

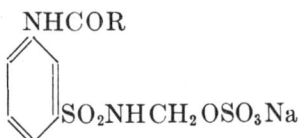

$$\underset{\text{SO}_2\text{NHCH}_2\text{OSO}_3\text{Na}}{\overset{\text{NHCOR}}{\bigcirc}}$$

Surface active agents have also been described in which the amide linkage is "reversed." For example, coconut fatty amine, RNH_2, can be reacted with the lactone of hydroxyethoxyacetic acid to form the hydroxy amide:

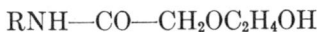

$$RNH—CO—CH_2OC_2H_4OH$$

This substance is then sulfated at the hydroxyl group to yield a product[174] containing both an amide and an ether linkage between the hydrophobic and hydrophilic groups.

The lactic acid derivatives of the higher amines have been sulfated to surface active agents. They may be made, as in a typical example,[175] by reacting butyl lactate with dodecylamine to give the sulfatable product:

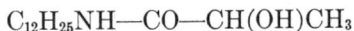

$$C_{12}H_{25}NH—CO—CH(OH)CH_3$$

Long-chain amines of the general type $R—CH(NH_2)—R'$, where R and R' together contain eleven or more carbon atoms, and the closely related secondary amines, $R—CH(NHR'')—R'$, have been acylated with sulfatable acids such as those which contain a hydroxyl group or a double bond, crotonic acid being frequently mentioned. The resulting acyl amides are

[171] Ger. Pats. 677,601; 640,581; to *I. G. Farbenindustrie.* U. S. Pat. 2,367,010, Davis *et al.* to *National Oil Products Co.* These products are also disclosed in U. S. Pat. 1,932,180, Guenther and Haussman to *I. G. Farbenindustrie* examples 12 and 22.

[172] U. S. Pat. 2,193,944, Steindorff *et al.* to *I. G. Farbenindustrie.*

[173] French Pat. 785,475, to *Ciba Co.*

[174] Ger. Pat. 675,723; U. S. Pat. 2,139,037; Rosenhauer to *Henkel et Cie.*

[175] Ger. Pat. 640,581, to *I. G. Farbenindustrie.*

then sulfated to form products[176] having the generic formula:

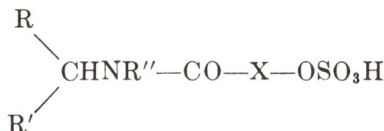

$$\begin{array}{c} R \\ \diagdown \\ CHNR''{-}CO{-}X{-}OSO_3H \\ \diagup \\ R' \end{array}$$

where R'' may be hydrogen or a hydrocarbon radical.

The substituted diamides of malic acid or maleic acid have also been sulfated to form surface active agents. A typical example of this type of product is the sulfate of bis(N-methyl-N-2-ethylhexyl)malamide:[177]

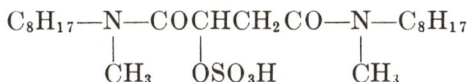

$$C_8H_{17}{-}\underset{\underset{CH_3}{|}}{N}{-}CO\underset{\underset{OSO_3H}{|}}{CH}CH_2CO{-}\underset{\underset{CH_3}{|}}{N}{-}C_8H_{17}$$

A similar type of product is the hydroxyethylurethan of a fatty alcohol:

$$ROOC{-}NH{-}C_2H_4OH$$

which may be made by reacting the chloroformic ester of the fatty alcohol with monoethanolamine. This substance can also be sulfated[178] at the hydroxyl group. More complex products of this general nature, containing multiple intermediate links, have also been made. An example[179] is the series:

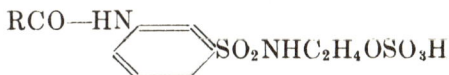

$$RCO{-}HN{-}\langle\!\!\langle\ \rangle\!\!\rangle{-}SO_2NHC_2H_4OSO_3H$$

(c) Ether Intermediate Linkage

Except in some special cases the use of an intermediate ether linkage for the purpose of forming sulfatable hydroxyl compounds offers little economic advantage. Many representatives of this type have, nevertheless, been described, and some have special advantages which offset their higher cost.

The simplest type of ether-linked sulfate with surface active properties is the sulfuric ester[180] of a higher alkyl ether of either ethylene glycol or polyethylene glycol:

$$R{-}OC_2H_4{-}OSO_3H \qquad \text{or} \qquad R{-}(OC_2H_4)_n{-}OSO_3H$$

[176] U. S. Pat. 2,225,082, Orthner et al. to General Aniline and Film Corp.
[177] U. S. Pat. 2,192,906, Hanford and Henke to E. I. du Pont de Nemours & Co.
[178] U. S. Pat. 1,964,654, Ulrich et al. to I. G. Farbenindustrie.
[179] U. S. Pat. 2,036,932, Felix et al. to Ciba Co.
[180] Brit. Pat. 463,624, to I. G. Farbenindustrie.

Furthermore, the α-fatty alcohol ethers of glycerol and other polyhydric alcohols have been sulfated to produce surface active agents.[181]

The glycerol and glycol ethers may be made by heating the fatty alkyl halide with glycerol or glycol in the presence of NaOH. The sulfuric esters of β-glyceryl fatty ethers have been specifically disclosed and claimed.[182]

Ether alcohols for surface active sulfates have also been prepared[183] by reduction of the fatty alkoxy acetic acids, which in turn are made from a fatty alcohol and chloroacetic acid.

A kerosene-derived ether alcohol has been described[184] which is made by treating chlorinated kerosene with a solution of metallic sodium in glycol. Glycerol ethers of alkylated phenols, such as p-tert-octylphenyl glycerol ether, have also been sulfated to produce surface active agents.[185]

One of the most practical methods of producing an ether alcohol is the reaction of a hydroxyl compound (either alcohol or phenol) with an alkylene oxide. The general reaction, illustrated with ethylene oxide, is:

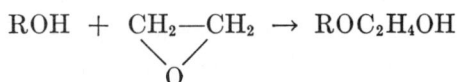

$$\text{ROH} + \underset{\diagdown\!\!\diagup}{\underset{O}{CH_2\!-\!CH_2}} \rightarrow ROC_2H_4OH$$

This can react further with ethylene oxide to form ethers of the polyethylene glycols. The number of molecules of ethylene oxide added can be controlled and products of the formula $R-(OC_2H_4)_n-OH$, where n is a known and controllable integer, which may vary from 1 to 20 or more, are formed. If R is a typical hydrophobic group, compounds of this type where n is above about 10 are water soluble and form an important class of non-ionic detergents. Where n is a lower number, they may be sulfated to give valuable anionic surface active agents.[186] Products of this nature have been developed and used mainly in Germany where they are sold as members of the Igepal, Alipal, and Leonil series. These trade names, however, include detergents of other structures.

The most widely used products[187] of the sulfated ether type are those made from ethylene or propylene oxide and an alkylated phenol or naph-

---

[181] Ger. Pat. 675,934; French Pats. 762,910; 768,554; to *Henkel et Cie.* U. S. Pat. 2,155,899, Harris to *Colgate-Palmolive-Peet Co.*

[182] Brit. Pats. 436,143; 436,209, to *Imperial Chemical Industries Ltd.*

[183] French Pat. 850,702, to *Böhme Fettchemie G. m. b. H.*

[184] Ger. Pat. 672,710, Keller *et al.* to *I. G. Farbenindustrie.*

[185] U. S. Pats. 2,167,325–6, Steindorff *et al.* to *I. G. Farbenindustrie.*

[186] Ger. Pat. 705,357; French Pat. 842,184; Brit. Pats. 443,559; 443,631–2; to *I. G. Farbenindustrie.*

[187] U. S. Pat. 2,203,883, Balle *et al.* to *I. G. Farbenindustrie.*

thol. The alkyl phenols and naphthols are relatively inexpensive and often more available than the fatty alcohols. Probably the only important members of this whole series manufactured and sold in the United States are the products made by first condensing di- or triisobutylene with phenol to give, respectively, p-octylphenol and p-dodecylphenol. This is then converted to the monoether of di- or triethylene glycol by the use of ethylene oxide or dichloroethyl ether and caustic soda. The phenolic ether-alcohol is then sulfated. A typical product is:

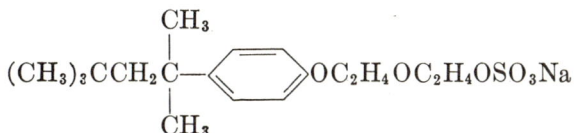

$$(CH_3)_3CCH_2 \overset{\overset{\displaystyle CH_3}{|}}{\underset{\underset{\displaystyle CH_3}{|}}{C}} -\left\langle \bigcirc \right\rangle OC_2H_4OC_2H_4OSO_3Na$$

which is a member of the Triton series of trade-named surface active agents (*Rohm and Haas Co.*).

The di- or triisobutyl phenols are very readily made[188] in high yields from the corresponding olefin and phenol in the presence of sulfuric acid. The number of $-OC_2H_4-$ groups between the aromatic nucleus and the sulfuric ester group can vary from one to ten or more.[189]

Ether-linked sulfates based on an alkyl phenol hydrophobic group have been made by etherifying the phenolic hydroxyl with a halogenated ketone and reducing the keto group to a sulfatable hydroxyl group. For example:[190]

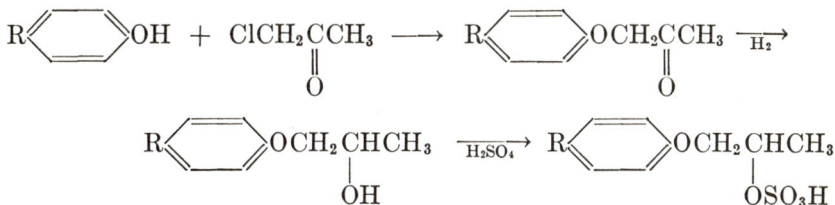

$$R\left\langle \bigcirc \right\rangle OH + ClCH_2\underset{\underset{\displaystyle O}{\|}}{C}CH_3 \longrightarrow R\left\langle \bigcirc \right\rangle OCH_2\underset{\underset{\displaystyle O}{\|}}{C}CH_3 \xrightarrow{H_2}$$

$$R\left\langle \bigcirc \right\rangle OCH_2\underset{\underset{\displaystyle OH}{|}}{C}HCH_3 \xrightarrow{H_2SO_4} R\left\langle \bigcirc \right\rangle OCH_2\underset{\underset{\displaystyle OSO_3H}{|}}{C}HCH_3$$

The sulfatable ether-linked group in this general type of product may also be an olefinic double bond. For example, the allyl ethers of higher alkyl phenols have been sulfated to yield surface active agents.[191] The typical product:

$$C_{12}H_{25}\left\langle \bigcirc \right\rangle OCH_2-CH=CH_2$$

[188] U. S. Pat. 1,987,228, Bruson to *Resinous Products and Chemical Co.* U. S. Pats. 2,008,017, Hester to *Rohm and Haas Co.*; 2,008,032, Niederl to *Rohm and Haas Co.*
[189] U. S. Pats. 2,075,018; 2,143,759; 2,178,829; Bruson *et al.* to *Rohm and Haas Co.* French Pats. 781,350; 835,406; to *I. G. Farbenindustrie.*
[190] U. S. Pat. 2,320,181, Hentrich *et al.* to *Unichem.*
[191] French Pat. 847,549, to *I. G. Farbenindustrie.*

may be made by reacting allyl chloride with dodecylphenol in alkaline media. It can be readily sulfated with sulfuric acid and acetic anhydride.

Alkylated phenol, alkylated benzene, or alkylated naphthalene hydrophobes have been chloromethylated and etherified with glycerol to give hydroxylated compounds suitable for sulfation to surface active agents.[192]

Using octylphenol, for example, the successive reactions are as follows:

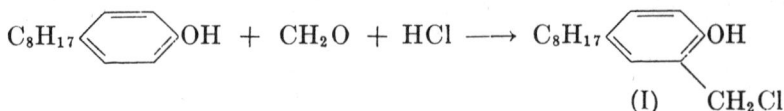

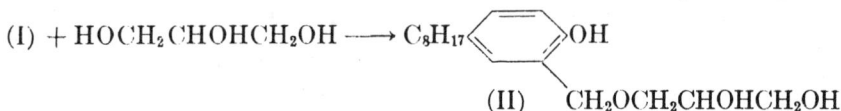

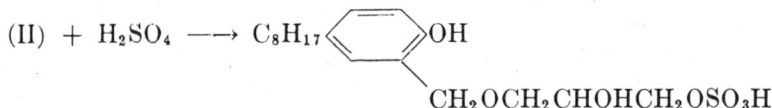

$$C_8H_{17} \langle \overline{} \rangle OH \;+\; CH_2O \;+\; HCl \;\longrightarrow\; C_8H_{17} \langle \overline{} \rangle OH$$

$$(I) \qquad CH_2Cl$$

$$(I) + HOCH_2CHOHCH_2OH \;\longrightarrow\; C_8H_{17} \langle \overline{} \rangle OH$$

$$(II) \qquad CH_2OCH_2CHOHCH_2OH$$

$$(II) + H_2SO_4 \;\longrightarrow\; C_8H_{17} \langle \overline{} \rangle OH$$

$$CH_2OCH_2CHOHCH_2OSO_3H$$

An interesting ether – alkyl sulfate has been described[193] in which a fatty alcohol is condensed with cyclohexene oxide to give the alkyloxy cyclohexanol which is then sulfated:

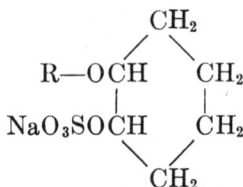

$$
\begin{array}{c}
\quad\quad CH_2 \\
\quad\;\diagup \quad \diagdown \\
R\!-\!OCH \quad\; CH_2 \\
\;\;\mid \qquad\quad \mid \\
NaO_3SOCH \quad CH_2 \\
\quad\; \diagdown \quad \diagup \\
\quad\quad CH_2
\end{array}
$$

Fatty acid ethanolamides have been condensed with ethylene oxide to give products of the type $RCONHC_2H_4$—$(OC_2H_4)_n$—OH, which are then sulfated.[194] The allyl and crotyl ethers of alkylated phenols have also been sulfated to produce detergents[195] similar to those made from the hydroxyalkyl ethers of alkylated phenols. Sulfation in this case take place at the olefinic double bond.

(d) Miscellaneous Linkages

Several other intermediate linkages have been used in synthesizing sulfatable hydroxyl compounds. Glucosides of the fatty alcohols have been

[192] U. S. Pat. 2,268,126, Orthner and Sönke to *I. G. Farbenindustrie.*

[193] U. S. Pats. 2,174,131, Lubs to *E. I. du Pont de Nemours & Co.*; 2,197,105, Holt to *E. I. du Pont de Nemours & Co.*

[194] U. S. Pat. 2,002,613, Orthner and Keppler to *General Aniline Works.*

[195] Ger. Pat. 707,023, Michel and Buschmann to *I. G. Farbenindustrie.*

sulfated to form surface active agents.[196] The fatty alkyl mercaptans, R—SH, have been converted to hydroxy thioethers by reacting with compounds such as ethylene oxide, glycerol chlorohydrin, or ethylene chlorohydrin, and subsequently sulfated.[197] The sulfones formed by oxidizing the above hydroxy thioethers have also been sulfated to give surface active compounds.[198] The sulfuric ester of dodecyl-β,γ-dihydroxypropyl sulfone is a typical example:

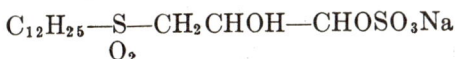

$$C_{12}H_{25}—\underset{O_2}{S}—CH_2CHOH—CHOSO_3Na$$

Alkane sulfonamides of the type RSO_2NH_2, which are easily prepared from products of the Reed reaction (see Chapter 4, Section I) have been reacted[199] with ethylene oxide to give sulfatable compounds of the type:

$$RSO_2NHC_2H_4O(C_2H_4O)_xH$$

None of these products characterized by miscellaneous intermediate linkages appear to have had any commercial use up to the present time.

[196] U. S. Pat. 1,951,784; Brit. Pat. 404,684; to *H. Th. Böhme A.-G.*

[197] Ger. Pat. 671,546, Elbel and Kirstahler to *Henkel et Cie.* Brit. Pat. 435,039, to Baldwin and Piggott. French Pat. 786,625, to *Imperial Chemical Industries Ltd.*

[198] Ger. Pat. 676,273, Kirstahler and Kaiser to *Henkel et Cie.* U. S. Pat. 2,017,004, Kirstahler and Kaiser to *Henkel et Cie.*

[199] Ger. Pat. 738,703, Orthner *et al.* to *I. G. Farbenindustrie.*

CHAPTER 4

Alkane Sulfonates

I. Sulfonate Group Linked Directly to the Hydrophobic Group

The normal primary alkane sulfonic acids R—SO_3H, where R varies from C_8 to C_{20} or higher, are typical anionic surface active agents.

The physical properties of the purified individual sulfonic acid salts from octane to octadecane have been studied by Reed and Tartar.[1] Cetanesulfonic acid and a large number of its salts have been studied[2] by Flaschenträger and Wannschaff. Although primary alkane sulfonic acids have received considerable attention in purely scientific studies they have had little if any industrial use. One company has recently offered these substances on a limited commercial scale primarily for research and exploratory uses. They are inherently expensive and apparently offer few, if any, important advantages over a number of less expensive products. The higher members of the series, moreover, are relatively insoluble, particularly in the cold, and their lime resistance is considerably poorer than that of the corresponding sulfuric esters.

The classical method[3] of preparing the alkane sulfonic acids is the reaction of the corresponding alkyl halides (usually the bromide is preferred) with Na_2SO_3:[4]

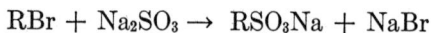

$$RBr + Na_2SO_3 \rightarrow RSO_3Na + NaBr$$

Ammonium sulfite may also be used. The reaction is usually carried out in aqueous alcohol and in a closed vessel. Unsaturated higher alkane sulfonic acids suitable as wetting agents and detergents have been made by reacting a higher allyl type halide, i.e., a halide with the halogen atom in the alpha position to a double bond, with sulfites. Allyl type halides undergo metathetical reactions more readily than normal saturated halides. They may be made by halogenating the appropriate olefin at high temperatures, as exemplified by the high-temperature chlorination of propene to give allyl chloride.[5]

---

[1] Reed and Tartar, *J. Am. Chem. Soc.* **57,** 570 (1935); **58,** 322 (1936).

[2] Flaschenträger and Wannschaff, *Ber.* **B67,** 1121 (1934).

[3] Heimilian, *Ann.* **168,** 145 (1873); Zuffanti, *J. Am. Chem. Soc.,* **62,** 1044 (1940).

[4] U. S. Pats. 2,053,424, Davidson to *Imperial Chemical Industries Ltd.*; 2,171, 117, Schirm and Schrauth to *Unichem.*

[5] U. S. Pats. 2,278,064, O'Conner and DeSimo to *Shell Development Co.*; 2,077,382,

The normal fatty alkane sulfonic acids may also be made[6] by reacting the corresponding sulfuric ester with a sulfite, using conditions similar to those used with the halides:

$$ROSO_3Na + Na_2SO_3 \rightarrow R—SO_3Na + Na_2SO_4$$

Beside the general reaction of an alkyl inorganic ester (halide or sulfate) with an inorganic sulfite, a variety of other syntheses have been used to produce alkane sulfonic acids in the surface active range. One general synthetic method is the use of various oxidizing agents, *e.g.*, bichromate, permanganate, peroxides, or halogens[7] to convert into sulfonic acids the corresponding mercaptan, RSH, or the di- or polysulfides:

$$R—S—S—R \text{ or } R—S—S—R$$
$$|$$
$$(\dot{S})_x$$

These sulfur derivatives may be obtained from the halides by treatment with alkali sulfides or polysulfides.[8] The mercaptans in the C_8–C_{18} range have been extensively used as modifiers in the production of GR—S synthetic rubber. They can be made by the addition of H_2S to olefins of the di- and triisobutene type, or by reaction of an alkyl halide with NaSH. A very recent and interesting process for preparing these products starts with an olefin such as pentadecene-7. This is heated with elemental sulfur producing a sulfurized product corresponding to the formula $C_{15}H_{30}S_3$. This sulfurized intermediate is then hydrogenated in the presence of one of the newly developed sulfur-resisting hydrogenation catalysts, such as cobalt sulfide, forming pentadecyl mercaptan, $C_{15}H_{31}SH$. The mercaptan is finally oxidized to the sulfonic acid[9] with aqueous nitric acid. Secondary mercaptans may be used,[10] in which case the product is a secondary sulfonic acid. Sulfonic acids of the type:

$$\begin{array}{c} R \\ \diagdown \\ \qquad CH—SO_3H \\ \diagup \\ R' \end{array}$$

Engs and Redmond to *Shell Development Co.*; 2,130,084, Groll, Hearne, Burgin, and LaFrance, to *Shell Development Co.*

[6] U. S. Pat. 2,170,380, Holsten to *I. G. Farbenindustrie*. French Pat. 716,705, to *I. G. Farbenindustrie*. Brit. Pat. 433,312, Davidson to *Imperial Chemical Industries Ltd.*

[7] Hilditch *et al.*, *J. Soc. Chem. Ind.* **52**, 272 (1933).

[8] U. S. Pat. 1,966,187, Schirm to *Unichem*. Flaschenträger and Wannschaff, *Ber.* **B67**, 1121 (1934).

[9] U. S. Pat. 2,402,587, Alvarado, Lazier, and Werntz to *E. I. du Pont de Nemours & Co.*

[10] U. S. Pats. 2,142,162; 2,187,338–9; Werntz to *E. I. du Pont de Nemours & Co.*

have been made[11] by oxidizing the disulfides:

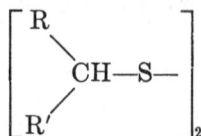

$$\left[\begin{array}{c} R \\ {\diagdown} \\ CH{-}S{-} \\ {\diagup} \\ R' \end{array} \right]_2$$

with nitric acid.

The complex sulfide type products formed by sulfurizing higher olefins have been oxidized[12] to form surface active alkane sulfonic acids. The olefins must contain at least one H atom on each ethylenic carbon atom. About 2.5–3.5 atoms of sulfur are used for each molecule of olefin, and nitric acid is used as the oxidizing agent.

Xanthates may be used as intermediates in place of sulfides or mercaptans. An alkyl halide can be reacted with potassium ethyl xanthate to form the alkyl ethyl dithiocarbonate RS—CS—OEt. This can then be oxidized[13] directly, for example, with hypohalites, to the alkane sulfonic acid, RSO_3H. It is probable that hydrolysis to the mercaptan occurs as an intermediate step in the reaction:

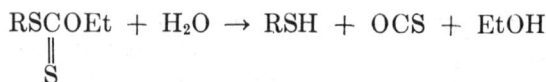

$$RSCOEt + H_2O \rightarrow RSH + OCS + EtOH$$
$$\overset{\|}{S}$$

The S-alkyl pseudothioureas, made by reacting an alkyl halide with thiourea, can be directly chlorinated with aqueous chlorine solution to give the alkane sulfonyl chloride. ·This can be hydrolyzed readily to the sulfonic acid. Chlorinated kerosene may be used[14] as the alkyl halide. The alkyl disulfides may be similarly reacted with aqueous chlorine to form alkane sulfonyl chlorides.[15]

Sulfonic acids containing a hydroxyl group have been made by adding hypochlorous or hypobromous acid across the double bond of an olefin in the surface active range and treating the resultant halohydrin with a sulfite.[16] Naphthenesulfonic acids, made from naphthenic acids through the corresponding alcohols and halides, as well as a number of close analogues of purely synthetic origin, have been studied and are said[17] to have remarkably high surface activity.

[11] U. S. Pat. 2,204,210, Farlow to *E. I. du Pont de Nemours & Co.*
[12] U. S. Pat. 2,338,829–30, Werntz to *E. I. du Pont de Nemours & Co.*
[13] Stone, *J. Am. Chem. Soc.* **62,** 571 (1940).
[14] U. S. Pat. 2,142,934, Bruson and Eastes to *Rohm and Haas Co.*
[15] U. S. Pat. 2,277,325, Hueter to *Unichem.*
[16] Ger. Pat. 687,462, to *Böhme Fettchemie G.m.b.H.*
[17] Pilat and Turkiewicz, *Ber.* **B72,** 1527 (1939).

Direct sulfonation of an aliphatic hydrocarbon with sulfuric acid is difficult to accomplish except in special positions, such as at a double bond or a tertiary carbon atom, although normal hydrocarbons of medium molecular weight have been sulfonated[18] using sulfur trioxide in the vapor phase. The saturated fatty acids and ketones, however, can be sulfonated much more readily. Sulfonation takes place at the alpha position. Energetic sulfonating agents, for example sulfur trioxide in liquid sulfur dioxide solvent, give the best results. The fatty acid esters or anhydrides or their sodium salts may also be directly sulfonated[19] to produce α-sulfonic acids. These products apparently have never been introduced on a commercial scale even though they would appear to be essentially inexpensive.

Saturated aliphatic ketones can be sulfonated directly by the use of energetic sulfonating agents such as oleum and acetic anhydride. Palmitone, stearone, lauryl ethyl ketone, and other similar ketones have been sulfonated[20] in this way to give surface active agents. The sulfonic acid group presumably enters alpha to the keto group. Ketones of this type may also be converted[21] to sulfonic acids by first halogenating and then replacing the halogen atom by reacting with sodium sulfite.

Fatty acids containing a sulfonic acid group near the middle of the chain have been made[22] by reacting sodium or ammonium sulfite with the adducts of hydrogen chloride, hypochlorous acid, or chlorine to oleic or ricinoleic acids.

The α-sulfo saturated fatty acids have been made[23] by treating an α-halogenated acid with sodium sulfite. Their derivatives in which the carboxy group is esterified with a lower alcohol are more resistant to lime and are stated[24] to be more effective surface active agents. A typical product would be the ethyl ester of α-sulfolauric acid:

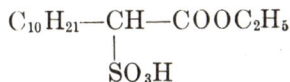

$$C_{10}H_{21}-CH-COOC_2H_5$$
$$|$$
$$SO_3H$$

[18] U. S. Pat. 2,383,752, Sveda to *E. I. du Pont de Nemours & Co.*

[19] U. S. Pats. 2,195,186-8, Moyer to *Solvay Process Co.*; 2,195,145, Crittenden to *Solvay Process Co.*; 2,195,088, Keppler and Schroeter to *I. G. Farbenindustrie.*

[20] U. S. Pats. 2,268,443, Crowder to *Solvay Process Co.*; 2,037,974, Guenther and Holsten to *I. G. Farbenindustrie.* French Pats. 850,753; 853,353, to I. G. Farbenindustrie.

[21] French Pat. 785,561 *Compagnie nationale de matières colorantes & manufactures de produits chimiques du nord réunies établissements Kuhlmann.*

[22] U. S. Pat. 1,949,837, Kalischer and Keller to *General Aniline Co.* See also 1,851,-102, Kalischer and Keller to *General Aniline Co.*

[23] Brit. Pat. 353,475, to *H. Th. Böhme, A.-G.*

[24] U. S. Pat. 2,043,476; French Pat. 721,794; Ger. Pat. 608,831; Guenther *et al.* to *I. G. Farbenindustrie.*

The addition of sodium bisulfite to a double bond to form a sodium sulfonate takes place quite readily when the double bond is in the alpha position to a carbonyl group. Accordingly long-chain α-alkene ketones can be converted to surface active sulfonates by the following reaction[25]:

$$\underset{\overset{\|}{O}}{R\text{—}CH\text{=}CHCR'} + NaHSO_3 \rightarrow \underset{\overset{|}{SO_3Na} \quad \overset{\|}{O}}{R\text{—}CH\text{—}CH_2CR'}$$

Addition of bisulfite proceeds with difficulty, however, in the case of unsubstituted olefins. This addition reaction can be made to give practical yields if a peroxide type catalyst is used. Thus, unsubstituted or remotely substituted olefins such as the polyisobutenes and the unsaturated fatty esters have been converted[26] to surface active agents by adding sodium bisulfite across the double bond in the presence of peroxides.

A method of producing surface active agents which is superficially similar but actually quite different in mechanism has been described[27] very recently. It consists in heating ethylene and sodium bisulfite in the presence of a promoter such as CCl_4 to about 115–125°C. under a pressure of 700–1000 atmospheres. The product is a mixture of formula $H(CH_2\text{—}CH_2)_nSO_3H$, where n is large enough so that the products are in the surface active range, i.e., about 4 to 12. The reaction is described as a "telomerization." This is the name applied to a general type of reaction wherein a polymerizable ethylenic compound, $\overset{\cdot}{—C}\text{=}\overset{\cdot}{C}—$, (called the taxogen) and a compound of type Y—Z (called the telogen) are reacted under polymerization conditions to form the product $Y\text{—}(\overset{\cdot}{—C}\text{—}\overset{\cdot}{C}\text{—})_n\text{—}Z$ (called a telomer). A limited number of molecular types can act as telogens.

One of the most important methods of making alkane sulfonic acids is by the action of energetic sulfonating agents such as sulfur trioxide or chlorosulfonic acid on olefins or alcohols. Under mild conditions the main products of reaction are the sulfuric esters, but it has been known for a long time that high yields of true sulfonic acids can be obtained under the proper conditions. The true sulfonic acids are stable in hot mineral acid solutions, whereas the sulfuric esters hydrolyze and are therefore less widely applicable. This simple test serves to differentiate between the two types. The so-called "highly sulfated" oils are made by treating oleic acid or ricinoleic acid or their derivatives with oleum or chlorosulfonic acid usually in

[25] U. S. Pat. 2,308,841, Werntz to *E. I. du Pont de Nemours & Co.*

[26] U. S. Pat. 2,318,036, Werntz to *E. I. du Pont de Nemours & Co.* Ger. Pats. 551,424; 672,491; to *I. G. Farbenindustrie.* Ger. Pat. 545,264, Ott and Mauthe to *I. G. Farbenindustrie.*

[27] U. S. Pat. 2,398,426, Hanford to *E. I. du Pont de Nemours & Co.*

the presence of a solvent. Chlorosulfonic acid adds to a double bond to form an α-chloro-β-sulfonic acid. This is hydrolyzed to the hydroxy-sulfonic acid in subsequent aqueous treatments:

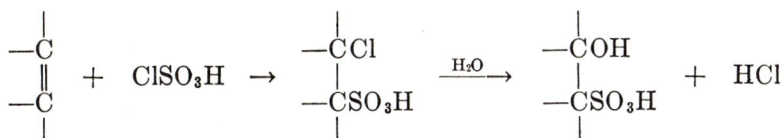

$$
\begin{array}{c}
-\overset{|}{\underset{\parallel}{C}} \\
-\overset{}{\underset{|}{C}}
\end{array}
+ \; ClSO_3H \; \rightarrow \;
\begin{array}{c}
-\overset{|}{\underset{}{C}}Cl \\
-\overset{}{\underset{|}{C}}SO_3H
\end{array}
\xrightarrow{\;H_2O\;}
\begin{array}{c}
-\overset{|}{\underset{}{C}}OH \\
-\overset{}{\underset{|}{C}}SO_3H
\end{array}
+ \; HCl
$$

Oleum or sulfur trioxide may act directly to sulfonate one of the olefinic carbon atoms. More probably this reaction takes place through the intermediate formation of a carbyl sulfate type of compound:

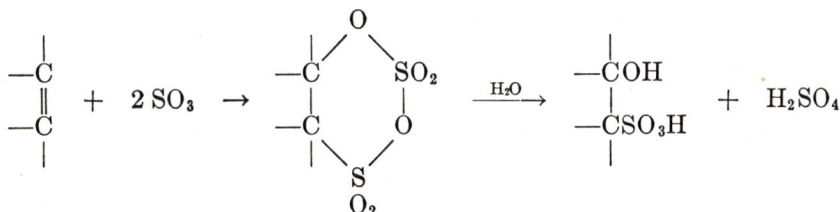

$$
\begin{array}{c}
-\overset{|}{\underset{\parallel}{C}} \\
-\overset{}{\underset{|}{C}}
\end{array}
+ \; 2\,SO_3 \; \rightarrow \;
\begin{array}{c}
-C \\
| \\
-C \\
|
\end{array}
\begin{array}{c}
O \\
\diagdown \diagup \\
SO_2 \\
| \\
O \\
\diagup \diagdown \\
S \\
O_2
\end{array}
\xrightarrow{\;H_2O\;}
\begin{array}{c}
-\overset{|}{\underset{}{C}}OH \\
-\overset{}{\underset{|}{C}}SO_3H
\end{array}
+ \; H_2SO_4
$$

Ethyl chlorosulfonate, $C_2H_5ClSO_3$, made from sulfuryl chloride and ethanol, has also been used[28] to sulfonate oleic acid.

Beside the oleic acid derivatives[29] several other olefins and alcohols have been converted to true sulfonic acids by energetic sulfonating agents. Among them are 1-hexadecene,[30] lauryl alcohol and homologous normal primary fatty alcohols,[31] the higher tertiary alcohols made by reacting fatty esters with Grignard reagents[32] and oleone, the C_{35} ketone made by the pyrolysis of oleic acid.[33] Olefins and alcohols from petroleum, which may be obtained either directly or by chlorination and subsequent treatment with alkalis have also been sulfonated to form true sulfonic acids.[34]

When olefins in the hydrophobic range having non-terminal double bonds

[28] U. S. Pat. 1,931,491, Haussman to *I. G. Farbenindustrie*.

[29] See: Pomeranz, *Seifensieder Ztg.* **59**, 3, 79 (1932); Bauer, *ibid.* **59, 34** (1932); Meyer, *Allgem. Oel- u. Fett-Ztg.* **37**, 165, 207 (1940).

[30] U. S. Pat. 2,061,617, Downing and Clarkson to *E..I. du Pont de Nemours & Co.*

[31] U. S. Pat. 1,968,796, Bertsch to *American Hyalsol Corp.* Note that this patent is one of the "Gardinol group" but its claims and disclosure relate to true sulfonic acids rather than sulfuric esters.

[32] French Pat. 778,373, to *Henkel et Cie.*

[33] Brit. Pat. 343,098, to *I. G. Farbenindustrie*.

[34] Brit. Pat. 344,829, to *I. G. Farbenindustrie*. Ger. Pat. 550,243, Haussmann to *I. G. Farbenindistrie*. Brit. Pat. 516,735, to *Colgate-Palmolive-Peet Co.* U. S. Pat. 2,192,713, Mottern to *Standard Oil Development Co.*

are sulfonated with acetic anhydride and oleum, or with ethyl ether and chlorosulfonic acid, true sulfonic acids rather than alkyl sulfates are formed.[35]

Tertiary olefins having free hydrogen on a carbon in the alpha position can be sulfonated[36] directly without affecting the double bond, using the SO_3-dioxane complex as the sulfonating agent:

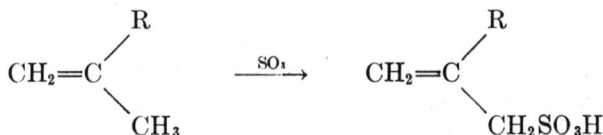

$$CH_2{=}C{<}^{R}_{CH_3} \xrightarrow{SO_3} CH_2{=}C{<}^{R}_{CH_2SO_3H}$$

The SO_3-thioxane complex may also be used to effect this sulfonation. If olefins such as di- or triisobutylene are used, surface active products are formed[37] directly. Surface active alkane sulfonic acids have been made[38] by the metathetical reaction between a higher alcohol and methanedisulfonic acid. One of the sulfonic groups reacts and the other remains in the final product.

By far the most important of the directly linked alkane sulfonate detergents are those made from aliphatic hydrocarbons (the kerosene and white oil fractions of paraffin base petroleums are usually used) by the Reed reaction and its various modifications. This reaction, named after its discoverer, Cortes F. Reed, takes place between an aliphatic hydrocarbon, sulfur dioxide, and chlorine to produce an alkane sulfonyl chloride, RSO_2Cl:

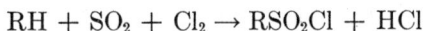

$$RH + SO_2 + Cl_2 \rightarrow RSO_2Cl + HCl$$

The reaction is activated by visible light of the shorter wave lengths such as is obtained from an ordinary tungsten filament bulb. The sulfonyl chlorides can be readily hydrolyzed to sulfonic acids and their salts. When the starting hydrocarbon is in the surface active range the products are typical surface active agents.[39] A wide variety of individual hydrocarbons, including members of the alicyclic and terpene series, have been converted to sulfonic acids by this reaction. The compounds of medium molecular weight in the range of eight to twelve carbon atoms are useful[40] as dispersing agents and for wetting out cotton in mercerizing baths. Superior

[35] U. S. Pat. 2,267,731, Guenther *et al.* to *General Aniline and Film Corp.*
[36] U. S. Pat. 2,365,783, Suter to *Procter and Gamble Co.*
[37] U. S. Pat. 2,335,193, Nawiasky and Sprenger to *General Aniline and Film Corp.*
[38] Brit. Pat. 406,889, to *Flesch-Werke A.-G.* U. S. Pat. 2,029,073, to Huttenlocher and Hess.
[39] U. S. Pats. Re 20,968; 2,174,110; 2,263,312; to Cortes F. Reed.
[40] U. S. Pats. 2,174,505–9, Fox *et al.* to *E. I. du Pont de Nemours & Co.*

general-purpose detergents are obtained from highly refined white oils boiling in the range 250–350°C. These are treated in the presence of actinic light with a mixture of sulfur dioxide and chlorine in a molar ratio which may vary from 1:1 to 4:1 at temperatures in the range 40–100°C. When the reaction is complete the resulting mixture is hydrolyzed with caustic soda and the unreacted hydrocarbon may be separated by adding water and/or by extraction with a suitable solvent.[41] Various process improvements, including a continuous-flow process, and separation of the sulfonyl chloride from unreacted hydrocarbon before hydrolyzing to the alkane sulfonate, have been described in a series of patents[42] which are continuing to be issued.

The mechanism of the Reed reaction has been studied[43] in Germany by Asinger and co-workers and several reviews on the subject have appeared[44] in that country.

Sulfuryl chloride, SO_2Cl_2, can be used to replace the SO_2-Cl_2 mixture in the Reed reaction. Actinic light catalyzes this reaction also. The addition of relatively small amounts of nitrogen bases such as pyridine improves the yield of sulfonyl chloride and shortens the time of reaction considerably.[45] Sulfuryl chloride will also react under similar conditions with fatty acids and other saturated carboxylic acids to form sulfonyl halides. The —SO_2Cl group apparently replaces one of the hydrogens on the carbon atom beta to the carboxy group.

In the reaction of sulfuryl chloride with aliphatic hydrocarbons a certain amount of alkyl chloride is formed along with the alkane sulfonyl chloride. Furthermore a single molecule of hydrocarbon may be substituted by both

[41] U. S. Pats. 2,197,800; 2,202,791, Henke et al. to E. I. du Pont de Nemours & Co.

[42] U. S. Pats. 2,193,824, Lockwood and Richmond to E. I. du Pont de Nemours & Co.; 2,212,786, McQueen to E. I. du Pont de Nemours & Co.; 2,333,568, Henke and Schofield to E. I. du Pont de Nemours & Co.; 2,333,788, Holbrook et al. to E. I. du Pont de Nemours & Co.; 2,392,841, Detrick and Hamilton to E. I. du Pont de Nemours & Co. Canadian Pat. 427,348, to E. I. du Pont de Nemours & Co. Ger. Pat. 742,741, to I. G. Farbenindustrie (uses FeCl₂ catalyst). U. S. Pat. 2,334,764, Henke and Lockwood to E. I. du Pont de Nemours & Co. Brit. Pats. 545,541, to E. I. du Pont de Nemours & Co.; 553,467, to Imperial Chemical Industries Ltd.; 549,512, Henke and Lockwood to E. I. du Pont de Nemours & Co.; 548,276, to Colgate-Palmolive-Peet Co. U. S. Pats. 2,337,552, Henke to E. I. du Pont de Nemours & Co.; 2,228,598, Fox et al. to E. I. du Pont de Nemours & Co.; 2,321,022, 2,346,568–9, 2,319,121, Fox to E. I. du Pont de Nemours & Co.

[43] Asinger, Ber. B77, 191 (1944).

[44] Schumacher and Stauff, Die Chemie 55, 341 (1942). Möllering, Chem. Ztg. 67, 224 (1943).

[45] U. S. Pats, 2,335,259, Calcott to E. I. du Pont de Nemours & Co.; 2,383,319–20, Kharasch to E. I. du Pont de Nemours & Co. Kharasch and Read, J. Am. Chem. Soc. 61, 308 (1939).

Cl and one or more —SO₂Cl groups. The alkyl chlorides are separated together with unreacted hydrocarbon during the purification process but the halogenated sulfonic acids and polysulfonic acids remain in the final product. The amounts of these by-products depend on reaction conditions. Increased yields of sulfonyl chlorides have been claimed when chlorine gas, sulfur dioxide, sulfur, carbon monoxide, or thionyl chloride is used[46] as an activating agent. These activating agents are particularly effective in the presence of strong actinic light such as is obtained from a mercury arc lamp.

In contrast to this use of strong illumination a process has been described in which sulfuryl chloride is reacted with refined paraffinic or alicyclic hydrocarbons in the dark. Organic peroxides such as benzoyl peroxide are used as catalysts.[47] In this process polycyclic aromatic hydrocarbons organic sulfides and mercaptans are strong anti-catalysts. Higher alkyl-atated monocyclic aromatic hydrocarbons, on the contrary, react in the same manner as the paraffins to form sulfonyl chlorides.

The alkane sulfonic acids have been produced on a pilot plant scale in Germany during the war directly from the paraffin hydrocarbons, sulfur dioxide, and oxygen. This eliminates the necessity for using chlorine to form the intermediate sulfonyl halide. In one process ozone and actinic light are used as activating agents. Another process uses organic per acids as activators. This development has been reported in Belgian and German patents[48] and has been confirmed by American technologists investigating German wartime industrial developments.

Alkane sulfonate detergents made by the Reed reaction are manufactured and sold in this country by the *du Pont Co.* under the trade name MP-189. They are extremely stable to hydrolysis in either acid or alkaline solutions and fully resistant to lime. Since they are essentially among the less expensive detergents to produce, and are highly efficient in a wide variety of applications, they are already a very important group and are rapidly gaining in popularity.

The industrial status of this type of detergent in Germany during the war has been described by Hoyt.[49] A fraction of saturated Fischer-Tropsch hydrocarbons averaging fifteen carbon atoms was used as the principal raw material. This was converted to the crude sulfonyl halide with sulfur dioxide and chlorine. Two types of crude sulfonyl chloride were

[46] Brit. Pat. 548,276, to *Colgate-Palmolive-Peet Co.*

[47] U. S. Pat. 2,374,193, Grubb and Tucker to *Standard Oil Co. of Indiana.*

[48] Belgian Pats. 443,658; 444,546; 445,312; 445,349; Ger. Pat. 735,096; to *I. G. Farben-industrie.*

[49] Hoyt, *PB 3868*, Office of Technical Services, Dept. of Commerce, Washington, D. C.

produced. Mersol D (which contained about 50% monosulfonyl chloride, 30% disulfonyl chloride, and 20% unreacted hydrocarbon) and Mersol H (which contained 50% monosulfonyl chloride and 50% unreacted hydrocarbon). These products were largely distributed to soap makers who hydrolyzed them with caustic soda and separated the unreacted hydrocarbon to whatever degree was most practical. A saponified product comparatively free of hydrocarbon material was also made by *I. G. Farbenindustrie* under the name Mersolat H. The Mersols, *i.e.*, crude sulfonyl chlorides, were also used as intermediates for condensing with sarcosine, taurine, etc. to make other types of surface active agents. The volume of these condensation products, however, was relatively small, compared with the amount of Mersol which was directly saponified. The annual production of Mersols reached the amazing total of 70,000 metric tons. This is conservatively equivalent to about 75,000,000 pounds of 100% active sodium alkane sulfonate. The German manufacturers have avowedly copied the art as revealed in American patents and have contributed very little in the way of fundamental improvements. Belgian patents 443,917 and 445,473 and French patent 853,686 (all assigned to *I. G. Farbenindustrie*) disclose some improvements in the process of saponifying the sulfonyl chloride and separating the residual hydrocarbon material.

Although aliphatic hydrocarbons are almost universally used as raw materials, the Reed reaction may also be carried[50] out with alkyl aromatic hydrocarbons. In the simplest case of toluene the product is benzylsulfonic acid, and in general an "exo-" sulfonyl chloride group is introduced. With starting materials such as the amylnaphthalenes the products are surface active.

An unusual and novel type of surface active agent, which may be classed with the alkane sulfonates but which is really a complex mixture, is made[51] by adding nitrosyl chloride to an olefin and treating the resultant addition product with sodium sulfite or bisulfite. The olefins used are derived from petroleum but the usual array of other olefin sources in the surface active range, such as cetene, oleic esters, Fischer-Tropsch olefins, etc., may also be used. The reaction of nitrosyl chloride with the olefin follows the equation:

$$-\overset{|}{\underset{}{C}}=\overset{|}{\underset{}{C}}- \ + \ NOCl \ \rightarrow \ -\overset{|}{\underset{\underset{Cl}{|}}{C}}-\overset{|}{\underset{\underset{NO}{|}}{C}}-$$

[50] U. S. Pats. 2,321,022; 2,351,674; Fox to *E.I. du Pont de Nemours & Co.*

[51] U.S. Pats. 2,265,993; 2,336,387; 2,343,362; 2,354,359; 2,371,418; 2,373,643; Beckham to *Solvay Process Co.* U. S. Pat. 2,370,518, Beekhuis to *Solvay Process Co.*

forming a saturated nitroso halide. This product is then treated with an aqueous alcoholic mixture of sodium sulfite and bisulfite, whereupon several different reactions may take place. The chlorine can be replaced by a —SO_3Na group. The nitroso group may be partly reduced and hydrolyzed to produce a keto group. It may be reduced completely to an amino group, or it may undergo a reduction and sulfitation to form a sulfamic acid group. Products which contain a double bond, a sulfonic acid group, and a sulfamic acid group may also be formed. These compounds are all present in varying amounts in the final product. After nitrosation and sulfitation the reaction products are separated from residual hydrocarbon material and may be further purified[52] by extraction or steam distillation processes. Products of this type are made by the *Solvay Process Co.* and have been recently introduced under the name Nytron.

Surface active agents of the alkane sulfonate series, containing no intermediate linkage but having an aromatic ring in the hydrophobe chain have been recently described by Suter.[53] An alkyl aromatic hydrocarbon in the surface active range, such as dodecylbenzene or nonylnaphthalene, is condensed with an alkene sulfonic acid. Boron trifluoride is the preferred catalyst. Isobutenesulfonic acid, which may be made by reacting methallyl chloride with sodium sulfite or by the direct sulfonation of isobutene with SO_3-dioxane, is used in a typical example:

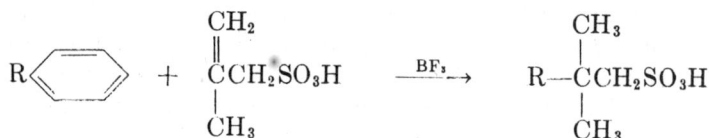

$$R\langle\rangle \ + \ \begin{matrix} CH_2 \\ \| \\ CCH_2SO_3H \\ | \\ CH_3 \end{matrix} \quad \xrightarrow{BF_3} \quad R{-}\begin{matrix} CH_3 \\ | \\ CCH_2SO_3H \\ | \\ CH_3 \end{matrix}$$

II. Petroleum Sulfonates

Another type of surface active agents which may be classed with the alkane sulfonates—but which chemically is a complex mixture—is represented by the products generally known industrially as "petroleum sulfonates." These were once regarded as worthless by-products of petroleum refining but they have in recent years assumed major importance in at least three fields: as emulsifying agents in cutting oils and textile spinning lubricants, and as sludge-dispersing agents in engine oils. The name "petroleum sulfonates" may be applied to any of the various chemical compounds containing sulfonic acid or sulfuric ester radicals which are obtained by the direct action of a strong sulfonating agent on a suitable petroleum stock. Sulfuric acid is widely used in the refining of many

[52] U. S. Pats. 2,381,658; 2,383,120; Fessler to *Solvay Process Co.*
[53] U. S. Pats. 2,366,133; 2,365,783; Suter to *Procter and Gamble Co.*

different petroleum products deriving from different crude oils. In most cases whatever sulfonates are formed are discarded in the sludge and removed from the raffinate by percolation through clays, washing, etc., without any attempt at recovery. The petroleum sulfonates which are recovered for use are obtained largely in the drastic refining of white oils, deodorized kerosenes, or lubricating stocks. In these processes large quantities of strong sulfuric acid or oleum are used. The petroleum sulfonates vary widely in chemical constitution and physical properties depending on the nature of the stock being refined. They may be classified roughly into two types, those which are water soluble (the so-called green acids) and those which are hydrocarbon soluble (the mahogany acids). The latter are the more important class although in some cases the two types are worked up together.

Various methods of separating petroleum sulfonates in relatively purified forms have been described. In refining operations such as the preparation of white oils, where it is desired to recover the sulfonates, the crude stock is often given a preliminary treatment with a small amount of sulfuric acid or with one of the refining solvents. This serves to remove asphaltic matter, easily polymerizable and oxidizable hydrocarbon constituents, and many of the sulfur and nitrogen compounds present. The oil is then treated with the main charge of acid, usually oleum, and allowed to settle. The lower layer or sludge is drawn off leaving an upper layer consisting of a solution of mahogany sulfonates in oil. The sulfonates may then be separated by washing with alkali or by extraction with solvents such as methyl or ethyl alcohol, as described in numerous patents.[54] Oleum can form true sulfonic acids by reaction with saturated branched-chain or

---

[54] Illustrative of the various refining procedures which are aimed primarily at obtaining petroleum sulfonates are: U. S. Pats. 1,958,630, Limburg to *Patent and Licensing Corp.;* 2,036,469, Field to *Standard Oil of California;* 2,266,084, Sachanen *et al.* to *Socony-Vacuum Oil Co.;* 2,252,957, Averill and Claytor to *Petrolite Corp.* The following patents illustrate various methods of separating petroleum sulfonates in relatively purified forms: Ger. Pats. 510,403; 595,604; to *Sudfeldt and Co.* U. S. Pat. 1,960,828, Reddish and Meyers to *Twitchell Process Co.* U. S. Pat. 1,963,257, Bird and Rosen to *Standard Oil Development Co.* U. S. Pat. 2,035,106, Vesterdal and Carlson to *Standard Oil Development Co.* U. S. Pats. 2,140,263; 2,166,117; 2,396,673; Blumer *et al.* to *L. Sonneborn Sons, Inc.* U. S. Pat. 2,158,680, Retailliau to *Shell Development Co.* U. S. Pat. 2,246,374, Lohman and Myers to *Emery Industries Inc.* U. S. Pat. 2,381,708, Amott to *Union Oil Co.* U.S. Pats. 2,307,743; 2,355,310; Liberthson to *L. Sonneborn Sons, Inc.* U. S. Pats. 2,358,773-4, Blumer to *L. Sonneborn Sons, Inc.* U. S. Pat. 1,903,466, Limburg to *Flintkote Corp.* U. S. Pat. 2,214,037, Archibald to *Standard Oil Development Co.* U. S. Pat. 2,403,185, Lemmon *et al.* to *Standard Oil Co. of Indiana.* U. S. Pat. 2,395,774, Amott and Grebe to *Union Oil Co. of Cailfornia.*

naphthenic hydrocarbons, and even with normal paraffins. In the latter
case, the paraffin is apparently first oxidized to an olefin which then forms[55]
a sulfatosulfonate.

Although petroleum sulfonates are produced in large tonnage, relatively
small amounts have come into the market as such until quite recently. The
major proportion is blended by the manufacturers with mineral or vegetable
oils, solvents, and other ingredients to form self-emulsifiable textile lubricants, cutting oils, and other finished products. Some producers sell the
petroleum sulfonates themselves for formulation, compounding, or direct
use. They are tan-to-dark-brown viscous liquids and usually contain a
considerable proportion of unsulfonated oil. *They are outstanding for
their powerful emulsifying and oil-solubilizing properties.

Water-soluble surface active agents which are typical foamers, wetting
agents, and detergents have been produced by sulfonating the hydrocarbon
mixtures obtained by solvent-extraction of petroleum fractions. The
sulfonated products so obtained are without doubt closely related to the
petroleum sulfonates, although the latter name is usually reserved for
products obtained by the sulfuric acid treatment of petroleum fractions
prepared without benefit of solvent extraction. When a petroleum oil is
extracted with liquid sulfur dioxide (or certain other solvents such as
phenol, dichloroethyl ether, nitrobenzene, or furfural) the highly olefinic
and aromatic constituents are removed and may be recovered by evaporat-
ing the solvent. These can then be sulfonated and if the original stock
was of suitable molecular weight the sulfonated products are surface active.
When liquid sulfur dioxide is used as the solvent the sulfonation may be
effected[56] advantageously before the solvent is removed. The extent to
which this type of material is used is uncertain.

III. Alkane Sulfonates with Intermediate Linkages between the Hydro-phobic and the Sulfonate Groups

(a) Ester Intermediate Linkage

There are two general types of ester-linked surface active agents in the
alkane sulfonate class. In the first type the hydrophobic radical is derived
from a carboxylic acid. This acid is esterified with the hydroxyl group
of a low-molecular-weight aliphatic hydroxy sulfonate. The type reaction

[55] Burkhardt, *J. Chem. Soc.* **1930**, 2387.
[56] U. S. Pat. 1,955,859, Osborn and Craig to *Standard Oil of California.* U. S. Pats.
2,149,661–2; 2,179,174; French Pat. 831,100; Brit. Pat. 506,337; Brandt to *Colgate-Palmolive-Peet Co.*

is:

$$RCOOH + HO—X—SO_3H \rightarrow RCOO—X—SO_3H + H_2O$$

In the second type the hydrophobic radical is derived from a long-chain alcohol which is esterified with the carboxy group of a low-molecular-weight aliphatic carboxy sulfonate. The type reaction is:

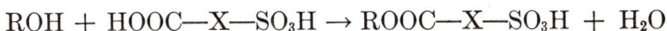

$$ROH + HOOC—X—SO_3H \rightarrow ROOC—X—SO_3H + H_2O$$

Both these types are represented by a number of commercially important products. The oldest, best-known, and most widely used products of the first type have the formula:

$$RCOOC_2H_4SO_3Na$$

where R is the radical of a natural fatty acid, most frequently oleic acid. These products are marketed in various physical forms and concentrations by the *I. G. Farbenindustrie* and *General Aniline and Film Corp.* as Igepon A, AP, and AP Extra. They are made[57] by the direct or indirect esterification of the fatty acid with isethionic acid, $HOC_2H_4SO_3H$. Processes which have been described include: (*a*) The reaction of a fatty acid chloride with the sodium salt of isethionic acid:

$$RCOCl + HOC_2H_4SO_3Na \rightarrow RCOOC_2H_4SO_3Na + HCl$$

(*b*) The reaction of fatty acid with carbyl sulfate. Carbyl sulfate is the anhydride of the sulfuric ester of isethionic acid:

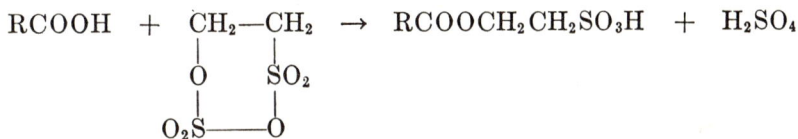

$$RCOOH + \begin{array}{c} CH_2—CH_2 \\ | \quad\quad | \\ O \quad\quad SO_2 \\ | \quad\quad | \\ O_2S———O \end{array} \rightarrow RCOOCH_2CH_2SO_3H + H_2SO_4$$

(*c*) The reaction of the sodium salt of a fatty acid with chloroethane sulfonic salts:

$$RCOONa + ClC_2H_4SO_3Na \rightarrow NaCl + RCOOC_2H_4SO_3Na$$

This reaction has been investigated independently, and is apparently best carried out[58] in the presence of mutual solvents of the lower acyl amine type at temperatures in the range of 150°C.

[57] U. S. Pats. 1,881,172; 1,916,776, Daimler and Platz to *I. G. Farbenindustrie.* Ger. Pats. 652,410; 655,999; 657,357; 657,404; 679,186; French Pats. 693,620; 720,590. Brit. Pats. 359,893; 366,916; 372,005; to *I. G. Farbenindustrie.*

[58] U. S. Pats. 2,289,391; 2,342,563; Tucker to *Procter and Gamble Co.*

(*d*) The reaction[59] of the chloroethyl ester of a fatty acid with sodium sulfite:

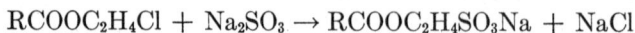

$$RCOOC_2H_4Cl + Na_2SO_3 \rightarrow RCOOC_2H_4SO_3Na + NaCl$$

(*e*) The transesterification reaction[60] utilizing a lower analogue of Igepon A such as acetylisethionic acid and a higher fatty acid.

Recent German practice, as reported by Hoyt,[61] utilizes the reaction of oleic acid chloride and sodium isethionate. Sodium isethionate is made by reacting ethylene oxide with sodium bisulfite in aqueous solution at 70–80°C., using sufficient pressure to retain the ethylene oxide. Oleic acid chloride is made in the usual manner from oleic acid and phosphorous trichloride. To make the detergent 1 M proportion of oleic acid chloride, 1.2 M proportions of dry sodium isethionate and about 0.1 M proportion of soda ash are heated together with good agitation to 115°C. Gaseous hydrogen chloride is evolved. As the reaction proceeds the doughy mass crumbles to granular flakes. It is neutralized and ground to form a finished product containing about 70% active material. Igepon AP has also been successfully made on a large scale by heating[61] sodium isethionate and oleic acid directly to 180–200°C., avoiding the necessity for using the acid chloride.

Igepon AP and products of analogous structure are relatively unstable in hot alkaline solutions since the carboxylic ester linkage is readily saponified. They are relatively inexpensive and efficient detergents, however, and quite satisfactory in neutral acid or mildly basic solutions.

Several analogous surface active agents have been described in which either the hydrophobic group or the sulfonate-bearing radical or both differ from those in Igepon AP. 1-Sulfo-2,3-propanediol, and the corresponding 2-methyl ether have been used as hydroxysulfonic acids. These are derived[62] from glycerol chlorohydrin by treating with sulfites. The compounds:

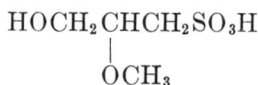

$$HOCH_2CHCH_2SO_3H$$
$$|$$
$$OCH_3$$

made from β-methylglycerolmonochlorohydrin and sodium sulfite, and:

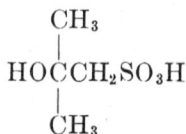

$$CH_3$$
$$|$$
$$HOCCH_2SO_3H$$
$$|$$
$$CH_3$$

[59] French Pat. 788,748, to *Henkel et Cie.*
[60] Ger. Pat. 731,393, to *I. G. Farbenindustrie.*
[61] Hoyt, *PB 3868*, Office of Technical Services, Dept. of Commerce, Washington.
[62] French Pat. 837,370, to *Colgate-Palmolive-Peet Co.*

made from isobutenechlorohydrin and sulfite have also been proposed[63] as practical intermediates for condensing with higher fatty acids. As another example of a variation in the intermediate linkage, 2,2'-dichlorodiethyl ether is reacted with one mole of sodium sulfite to form the product $ClC_2H_4OC_2H_4SO_3Na$. This is then converted to an Igepon AP type of detergent by heating[64] with a sodium soap. Compounds of similar type may also be made[65] by heating a fatty acid soap with a halogenated alkane sulfonic acid salt such as 2-chloroethanesulfonic acid sodium salt to about 140°C.

Beside oleic acid (or sometimes stearic, palmitic, coconut fatty or other mixed natural fatty acids), which furnishes the hydrophobic portion of Igepon AP, the usual array of hydrophobic fatty acyl nuclei has been described. The fatty acids from paraffin wax oxidation were widely used in Germany during the war. Abietic and hydroabietic acids have been used.[66] Alkoxyacetic acids such as lauryloxyacetic acid[67] alkylated phenoxyacetic acids,[68] and the acids of tall oil[69] have also been used as hydrophobe sources.

An entirely different method of obtaining closely related products of the ester-linked type has been described in which the alkyl ester of a fatty acid is sulfonated directly. Sulfonation only occurs if the alkyl group contains a tertiary carbon atom as is the case, for example, with isobutyl esters. Energetic sulfonating agents such as 100% sulfuric acid (generally called "monohydrate" since it is the monohydrate of SO_3), oleum, or chlorosulfonic acid are used.[70] The reaction follows the equation:

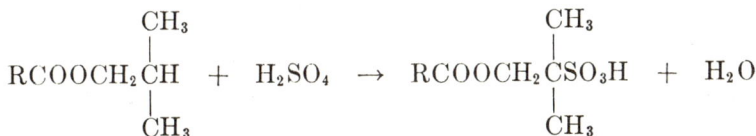

$$\underset{\overset{|}{CH_3}}{\overset{CH_3}{\overset{|}{RCOOCH_2CH}}} + H_2SO_4 \rightarrow \underset{\overset{|}{CH_3}}{\overset{CH_3}{\overset{|}{RCOOCH_2CSO_3H}}} + H_2O$$

It is doubtful, however, if this interesting process is used commercially or if it affords practical yields.

The second class of ester-linked alkane sulfonate surface active agents includes those products in which the hydrophobic portion is derived from an alcohol, and is esterified with the carboxyl group of an aliphatic carboxy-

[63] Brit. Pat. 514,053, to *Colgate-Palmolive-Peet Co.*

[64] U. S. Pat. 2,394,834, Young and Spitzmueller to *Industrial Patents Corp.*

[65] U. S. Pats. 2,342,562–3, Tucker to *Procter and Gamble Co.*

[66] U. S. Pats. 1,984,713–4; French Pat. 755,769, Weiland *et al.* to *E. I. du Pont de Nemours & Co.*

[67] French Pat. 763,743, to *Henkel et Cie.*

[68] U. S. Pat. 2,179,209, Daimler and Platz to *I. G. Farbenindustrie.*

[69] Ger. Pat. 596,510, to *H. Th. Böhme A.-G.*

[70] U. S. Pat. 2,093,576, Segessemann to *National Oil Products Co.*

sulfonic acid.   There are two general sub-groups in this class, each of which is of considerable commercial importance.   In the first sub-group the molecule has one hydrophobic group (and one carboxylic ester linkage) and one sulfonic acid group.   The long-chain alkyl sulfoacetates are representative of this type.   The second sub-group has only one sulfonic acid group but has two or more carboxylic ester groups linked to an equal number of hydrophobic residues.   The dialkyl sulfosuccinates are the best-known representatives of this second type.

A large number of surface active sulfoacetates have been described, having the general formula $ROOCCH_2SO_3H$.[71]   They are made by esterifying a hydroxyl compound bearing a hydrophobic group with chloroacetic acid or chloroacetyl chloride.   This chloro ester is then reacted with an alkali metal sulfite, replacing the chlorine by a sulfonate group:

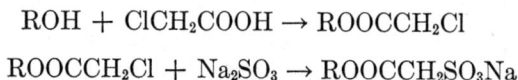

$$ROH + ClCH_2COOH \rightarrow ROOCCH_2Cl$$

$$ROOCCH_2Cl + Na_2SO_3 \rightarrow ROOCCH_2SO_3Na$$

The simplest type of ROH compounds which may be used is the long-chain fatty alcohol, such as coconut fatty, oleyl, cetyl, and stearyl alcohols. Coconut fatty sulfoacetate is sold under the trade name Nacconol LAL by National Aniline Division of *Allied Chemical and Dye Corp.*   It is an excellent washing and foaming agent, and it has an attractive colorless, clean appearance.   However, its relatively high cost restricts its use mainly to toothpastes, shampoo preparations, and cosmetics.

The same general type of reaction has been carried out[72] with many more complex types of ROH compounds.

Partially esterified polyhydric alcohols such as monostearin or monoolein have been used.   Partially esterified triethanolamine has also been used[73]:

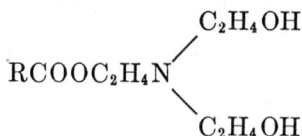

$$RCOOC_2H_4N \overset{\textstyle C_2H_4OH}{\underset{\textstyle C_2H_4OH}{\Big\langle}}$$

---

[71] U. S. Pats. 2,166,141–5; 2,145,443; 2,181,890; to Benj. R. Harris. See also U. S. Pats. 1,917,250; 1,917,255–6; to Harris. French Pat. 716,605; U. S. Pat. 2,169,998; Guenther and Saftien to *I. G. Farbenindustrie.*

[72] U. S. Pats. 2,166,141–5; 2,255,285; to Harris; and 2,221,377, Harris to *Emulsol Corp.*, have exceptionally complete disclosures of ether alcohols, ester alcohols, amide alcohols, etc., all of which bear hydrophobic groups and all of which can be converted to sulfoacetates through their chloroacetates.

[73] U. S. Pat. 2,178,139, Epstein and Katzman to *Emulsol Corp.*

Amides, such as monoethanolamide of lauric acid, $C_{11}H_{23}CONHC_2H_4OH$,[74] and ethers such as monolauryl glyceryl ether[75] have also been esterified with maleic anhydride or chloroacetic acid and then treated with sulfite to form sulfosuccinates and sulfoacetates, respectively. α-Hydroxy long-chain carboxylic acids such as α-hydroxylauric acid may also be used[76] as the ROH component in this reaction. Among other ROH components are the hydroxy esters of abietic and hydroabietic acids,[77] and chlorinated fatty alcohols.[78] It is doubtful if any of these more complex products, with the possible exception of the partially esterified glycerols and the fatty acyl monoethanol amides, have been converted to sulfoacetates on a commercial scale.

Sulfopropionates and sulfobutyrates of ROH type hydrophobes have been made as well as the sulfoacetates. They can be prepared through the corresponding halogenated acid in the same manner as the sulfoacetates. Another method of making these compounds, however, involves the preparation of an acrylic or crotonic ester, $ROOCCH{=}CH_2$ or $ROOCCH{=}CHCH_3$. These acrylates and crotonates having a double bond adjacent to the activating carboxylic ester group, are capable of adding $NaHSO_3$ across the double bond readily to form the saturated sulfonates:

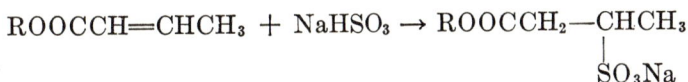

$$ROOCCH{=}CHCH_3 + NaHSO_3 \rightarrow ROOCCH_2{-}\underset{\underset{SO_3Na}{|}}{C}HCH_3$$

Among the esters of alkane dicarboxy sulfonates are included some of the best-known commercial wetting agents. Although the basic patents[79] covering this series disclose a large number of dicarboxy mono- and disulfonic acids which may be used as the hydrophilic portion of the molecule, sulfosuccinic acid is by far the most widely used. The sulfosuccinates are made indirectly by first esterifying maleic acid with the appropriate alcohol:

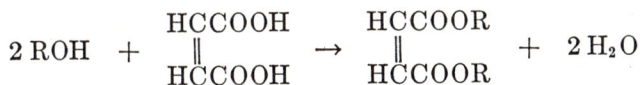

$$2\,ROH \ + \ \begin{matrix} HCCOOH \\ \| \\ HCCOOH \end{matrix} \ \rightarrow \ \begin{matrix} HCCOOR \\ \| \\ HCCOOR \end{matrix} \ + \ 2\,H_2O$$

The dialkyl maleate is then heated with a concentrated aqueous solution

---

[74] U. S. Pat. 2,236,529, Epstein and Katzman to *Emulsol Corp.*

[75] Swiss Pat. 162,732; Fr. Pat. 760,138; to *J. R. Geigy A.-G.*

[76] U. S. Pat. 2,185,541, Cahn to *Emulsol Corp.*

[77] French Pat. 841,487, to *Ciba Co.*

[78] U. S. Pat. 2,251,932, Harris and Cahn to *Emulsol Corp.*

[79] U. S. Pats. 2,028,091; 2,176,423; Brit. Pat. 446,568; Fr. Pat. 776,495; Jaeger to *American Cyanamid Co.*

of bisulfite, which adds to the double bond forming the sulfosuccinate:

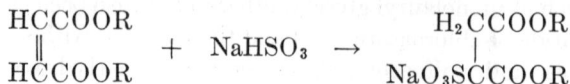

$$\begin{array}{c} \text{HCCOOR} \\ \parallel \\ \text{HCCOOR} \end{array} + \text{NaHSO}_3 \rightarrow \begin{array}{c} \text{H}_2\text{CCOOR} \\ \mid \\ \text{NaO}_3\text{SCCOOR} \end{array}$$

The monomaleates as well as the diamaleates may be prepared and sulfited. In this case the resulting product contains both a sulfo and a carboxy group. A typical example is monocetyl sulfosuccinate, made[80] from cetyl alcohol, maleic anhydride, and sodium sulfite.

The dialkyl sulfosuccinates are powerful wetting agents and are marketed by *American Cyanamid Co.* under the trade names Aerosol and Deceresol. Aerosol IB is the diisobutyl ester, AY is the diamyl ester, MA the dihexyl ester, and OT the dioctyl ester. In Aerosol OT 2-ethylhexanol is the octyl alcohol used. These products are sold in pure colorless form which makes them adaptable for many purposes (such as cosmetic preparations) where other less-attractive-appearing products would be undesirable. They are adequately stable under most conditions, but like most carboxylic esters they are hydrolyzed by hot alkaline solutions. A variety of salts, *e.g.*, guanidine and biguanide salts, and physical forms, mixtures, etc. of these products, have been disclosed[81] but the sodium salts are by far the most widely used. The general properties of these products have been discussed by Caryl and Ericks.[82]

The dialkyl sulfosuccinates become relatively insoluble and lose their wetting powers with alkyl groups larger than eight carbon atoms. With mixed alkyl esters using one lower group and one higher group a total of about sixteen or eighteen carbon atoms in both groups is the practical limit for maintaining surface active properties at ordinary temperatures. The same is true of the monoalkyl esters.

Beside the simple alcohols any number of more complex alcohols may be used to furnish the hydrophobic groups. Typical examples are the hydroxyethyl amides of fatty acids,[83] $RCONHC_2H_4OH$, and ricinoleic acid esters:[84]

$$CH_3(CH_2)_5CH(OH)CH_2CH{=}CH(CH_2)_7\text{---}COOCH_3$$

Esters of sulfotricarboxy aliphatic acids have recently been described[85]

[80] U. S. Pat. 2,316,234, Flett to *Allied Chemical and Dye Corp.*

[81] U. S. Pats. 2,181,087, Jaeger to *American Cyanamid & Chemical Corp.*; 2,265,944, Langhorst *et al.* to *American Cyanamid Co.*

[82] Caryl and Ericks, *Ind. Eng. Chem.* **31**, 44–7 (1939).

[83] U. S. Pat. 2,236,528, Epstein and Katzman to *Emulsol Corp.*

[84] U. S. Pat. 2,184,794, de Groote to *Petrolite Corp.*

[85] U. S. Pat. 2,315,375, Nawiasky and Sprenger, to *General Aniline and Film Co.* Also see U. S. Pat. 2,345,041, Ericks and Meincke to *American Cyanamid Co.* and Brit. Pat. 551,246, to *National Oil Products Co.*

as surface active agents. Nekal NS, a powerful wetting agent marketed by *General Dyestuffs Corp.*, is the trihexyl ester of sulfotricarballylic acid. It may be made from trihexyl citrate, first dehydrating to form trihexyl aconitate, then adding bisulfite across the double bond:

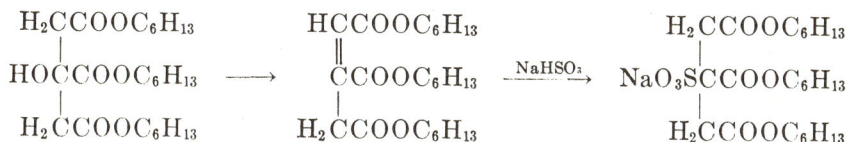

$$H_2CCOOC_6H_{13} \qquad HCCOOC_6H_{13} \qquad\qquad H_2CCOOC_6H_{13}$$
$$HOCCOOC_6H_{13} \longrightarrow CCOOC_6H_{13} \xrightarrow{NaHSO_3} NaO_3SCCOOC_6H_{13}$$
$$H_2CCOOC_6H_{13} \qquad H_2CCOOC_6H_{13} \qquad\qquad H_2CCOOC_6H_{13}$$

It should be noted that, although each hydrophilic group is so short (six carbon atoms) as to be ineffective by itself, the fact that there are three of them in the same spatial region of the molecule renders the compound strongly surface active.

An interesting series of surface active agents closely related to the sulfosuccinates has been made[86] by adding $NaHSO_3$ to the Diels-Alder adducts of maleic diesters with cyclopentadiene:

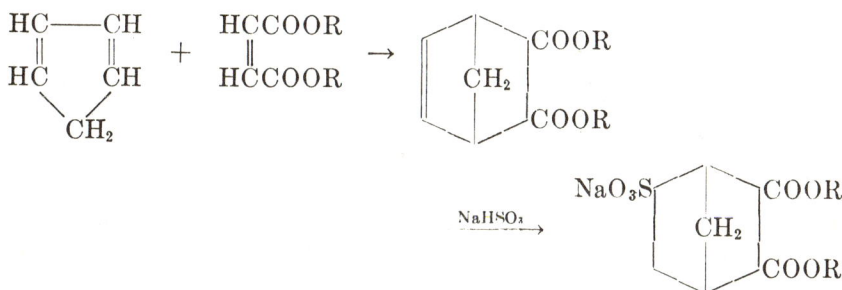

In a similar manner the Diels-Alder adduct of maleic anhydride and alloöcimine has been esterified and sulfited to give surface active agents.[87]

Surface active agents containing two ester links and based on a succinic acid structure have been recently described.[88] They may be made by sulfonating compounds such as lauryl allyl succinate with chlorosulfonic acid. On hydrolysis the product:

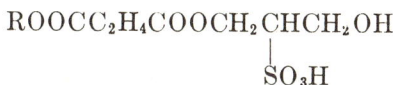

$$ROOCC_2H_4COOCH_2CHCH_2OH$$
$$SO_3H$$

is formed. The alkenyl succinic acids may be used in place of succinic acid itself. If milder sulfonating agents are used, alkyl sulfates rather than

[86] U. S. Pats. 2,314,846 and 2,345,539, McClellan and Bacon to *American Cyanamid Co.*

[87] U. S. Pat. 2,403,038, Aelony to *Monsanto Chemical Co.*

[88] U. S. Pat. 2,402,823, Kyrides to *Monsanto Chemical Co.*

alkane sulfonates are formed. A typical example from lauryl allyl tri-isobutenyl succinate would be:

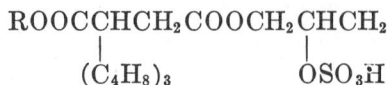

$$\underset{(C_4H_8)_3}{ROOCCHCH_2} \underset{OSO_3H}{COOCH_2CHCH_2}$$

(b) Amide Intermediate Linkage

There is a very close parallel between the various molecular types encountered among the amide-linked alkane sulfonate detergents and the types of ester-linked products considered above. The amide-linked products also fall into two general classes, those in which the hydrophobic group bears the carboxy group and those in which it bears the primary or secondary amino group when it enters into amide formation. Of those products in which a high-molecular-weight carboxy acid furnishes the hydrophobic group the best known is Igepon T[89] originated in Germany by *I. G. Farbenindustrie* about 1930. It is made by condensing oleic acid chloride with methyltaurine in aqueous alkaline solution:

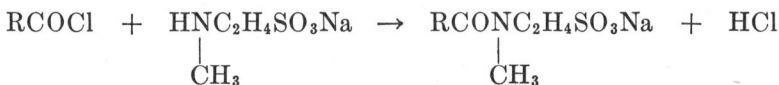

$$\underset{}{RCOCl} \ + \ \underset{CH_3}{HNC_2H_4SO_3Na} \ \rightarrow \ \underset{CH_3}{RCONC_2H_4SO_3Na} \ + \ HCl$$

Although this type of synthesis may be effected with either taurine, or methyltaurine, the latter is preferred, in spite of its higher price, since it yields a product having better detergent properties. This point will be discussed later, in the theoretical section of this book.

Igepon T is generally a more effective surface active agent than Igepon A, and is considerably more resistant to hydrolysis, particularly in alkaline solution. It is marketed in dry form, cut to standard strengths with sodium sulfate, and also in paste and gel form. This product has enjoyed very wide acceptance in the textile industry on account of its versatility and high efficiency. This has been particularly true in Germany.

The methyltaurine used in making Igepon T may be prepared in several ways. The reaction of chloroethane sodium sulfonate with methylamine has been used, as has the addition of methylamine to vinylsulfonic acid, and the reaction of methylamine with carbyl sulfate. Current large-scale practice as reported by Hoyt[90] involves making methyltaurine by reacting

---

[89] U. S. Pat. 1,932,180, Guenther, Münz, and Haussmann to *I. G. Farbenindustrie*. Ger. Pats. 584,703; 633,334; Brit. Pats. 341,053; 343,524; 360,982; 372,389; 389,543; to *I. G. Farbenindustrie*.

[90] Hoyt, *PB 3868*, Office of Technical Services, Dept. of Commerce, Washington, D. C.

sodium isethionate with methylamine at 270–290°C. under 200 atmospheres pressure. The yields run over 85% based on isethionic acid. The methyltaurine is recovered as a 25% aqueous solution which is used directly in the condensation step. Products of the Igepon T type may also be made[91] by reacting the metal derivative of a fatty amide with chloroethane sulfonates in inert media:

$$RCONHNa + ClC_2H_4SO_3Na \rightarrow RCONHC_2H_4SO_3Na$$

Beside oleic acid and methyltaurine a wide variety of higher carboxylic acids and aminoalkylsulfonic acids can be combined. Some of these have been produced commercially. Igepon KT is a high-foaming product made from coconut fatty acids and methyltaurine. Igepon 702 K is made from a mixture of myristic and palmitic acids (derived from palm oil) and methyltaurine. Alipon OT is made from oxidized paraffin wax acids and methyltaurine. These products have not been introduced in the United States, but were of considerable commercial importance in Germany.

Among numerous other hydrophobic carboxy acids which have been described for condensation with methyltaurine are the alkylated phenoxyacetic acids[92] and the acyloxy or alkoxy substituted natural fatty acids.[93] Compounds of the Igepon T type have been described[94] in which the hydrophobic carboxy acid is of the alkylated aryloxyalkanoic type. An example is the product obtained by condensing p-isoöctylphenoxyacetyl chloride with taurine:

$$C_8H_{17}—\langle\ \rangle—OCH_2CONHC_2H_4SO_3H$$

Fatty acid chlorides have been condensed[95] with amino methane sulfonates (made from ammonia or methylamine with formaldehyde and sodium bisulfite). These products are stated to be quite stable, which is surprising in view of the ease with which most analogous methylene derivatives hydrolyze.

In a similar disclosure[96] oleic amide is condensed with formaldehyde and methyltaurine, the reaction presumably being:

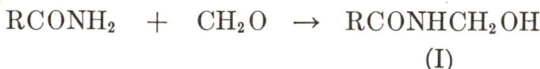

$$RCONH_2 + CH_2O \rightarrow RCONHCH_2OH$$
$$(I)$$

---

[91] Ger. Pat. 664,309, Hentrich et al. to I. G. Farbenindustrie.

[92] Brit. Pat. 501,004, to I. G. Farbenindustrie.

[93] Brit. Pat. 436,075, to N. V. Chemische Fabriek Servo.

[94] U. S. Pat. 2,213,984, Horst and Schild to General Aniline and Film Corp. See also French Pat. 841,681, to I. G. Farbenindustrie.

[95] French Pat. 721,988, to I. G. Farbenindustrie.

[96] Brit. Pat. 446,912, to I. G. Farbenindustrie.

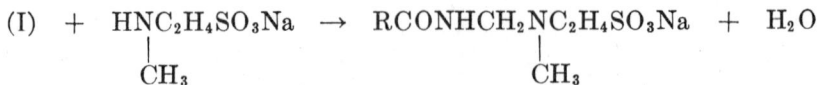

$$(I) \quad + \quad \underset{\underset{CH_3}{|}}{HNC_2H_4SO_3Na} \quad \rightarrow \quad \underset{\underset{CH_3}{|}}{RCONHCH_2NC_2H_4SO_3Na} \quad + \quad H_2O$$

β-Amino-β'-sulfodiethyl ether, $H_2NC_2H_4OC_2H_4SO_3Na$, and the corresponding β-methylamino derivative have been used[97] to condense with fatty acid chlorides. These products are made from dichlorodiethyl ether by successively replacing the chlorine atoms with amino (or methylamino) and sulfonate groups.

A closely related surface active agent has been described[98] in which hydroxymethyllauramide is reacted with 2-mercaptoethane sulfonate.

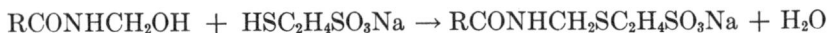

$$RCONHCH_2OH + HSC_2H_4SO_3Na \rightarrow RCONHCH_2SC_2H_4SO_3Na + H_2O$$

Surface active agents in which two or more amide linkages separate the hydrophobic and the hydrophilic groups are represented[99] by the example:

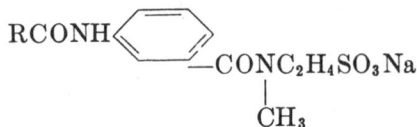

$$\underset{\underset{CH_3}{|}}{RCONH\langle\rangle\!\!-\!\!CONC_2H_4SO_3Na}$$

Urethan, carbamide, and thiocarbamide linkages have also been used to unite hydrophobic groups with aminoalkane sulfo groups. Lauryl chlorocarbonate, for example, may be condensed with methyltaurine to form the product:[100]

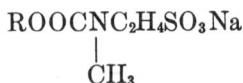

$$\underset{\underset{CH_3}{|}}{ROOCNC_2H_4SO_3Na}$$

Compounds of the type[101] $RNHCONHC_2H_4SO_3Na$ and[102] $RNHCSNHC_2$-H_4SO_3Na have also been described. The latter may be prepared by the addition reaction between a long-chain isothiocyanate R—N=C=S and the amino group of taurine.

Just as there is almost a complete correspondence in structure between the Igepon A and Igepon T classes, so there is a similar correspondence between the sulfoacetate class and the sulfoacetamide class. Long-chain

---

[97] Brit. Pat. 434,358; Ger. Pat. 677,600; to Waldmann and Chwala.

[98] Swiss Pat. 210,962, to *Ciba Co.*

[99] French Pat. 789,578, to *Henkel et Cie.* See also Brit. Pat. 442,135, to *Henkel et Cie.*

[100] U. S. Pat. 2,157,362, Ulrich and Koerding to *I. G. Farbenindustrie.*

[101] Ger. Pat. 671,352, to *I. G. Farbenindustrie.*

[102] U. S. Pat. 2,139,697, Salzberg to *E. I. du Pont de Nemours & Co.*

sulfoacetamides are made[103] by acylating a fatty amine with chloroacetyl chloride and subsequently sulfiting the product:

$$RNH_2 + ClCH_2COCl \rightarrow RNHCOCH_2Cl \xrightarrow{Na_2SO_3}$$
$$RNHCOCH_2SO_3Na + NaCl$$

Primary or secondary long-chain amines may be used, and chloro-propionyl chloride, etc., may replace the chloroacetyl chloride. They may also be prepared by reacting[104] the fatty amine with ethyl sulfoacetate sodium salt:

$$RNH_2 + C_2H_5OOC—CH_2SO_3Na \rightarrow RNHCOCH_2SO_3Na + C_2H_5OH$$

The products resemble the sulfoacetates in their general physical properties, but apparently they have not been marketed in any significant quantities. This is possibly due to the relatively small production and high cost of fatty amines as compared with fatty acids.

Other amino-bearing hydrophobic substances suitable for conversion to sulfoacetamides have been made by inserting other intermediate linkages.

The amino alkyl esters of fatty acids, $RCOOC_2H_4NH_2$, have been reacted with chloroacetyl chloride and subsequently sulfited. The same final product:

$$RCOOC_2H_4NHCOCH_2SO_3Na$$

has also been made by condensing[105] a fatty acid chloride with sulfoacetyl-monoethanolamine:

$$RCOCl + HOC_2H_4NHCOCH_2SO_3Na \rightarrow RCOOC_2H_4NHCOCH_2SO_3Na$$

As in the case of the Igepon T series, sulfoacetamide type compounds have been made[106] in which a urethan linkage is used to bind the hydrophobic radical. Typical of these products is $ROOCN(CH_3)COCH_2SO_3Na$. The dialkyl amides of sulfosuccinic acid have been described but have never attained the importance of the corresponding esters. The mixed amide esters of formula:

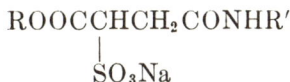

$$ROOCCHCH_2CONHR'$$
$$|$$
$$SO_3Na$$

[103] Ger. Pat. 692,925; U. S. Pat. 2,009,346, Schirm to *I. G. Farbenindustrie*.
[104] Ger. Pat. 677,013, to *I. G. Farbenindustrie*.
[105] U. S. Pat. 2,184,770; 2,245,593; 2,236,541; Katzman to *Emulsol Corp*.
[106] Brit. Pat. 497,572, to *J. R. Geigy, A.-G.*

sulfosuccinimides:

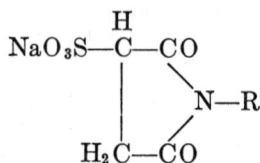

$$\begin{matrix} & \text{H} & \\ \text{NaO}_3\text{S}-\text{C}-\text{CO} & & \\ & | & \diagdown \\ & & \text{N}-\text{R} \\ & | & \diagup \\ \text{H}_2\text{C}-\text{CO} & & \end{matrix}$$

and monoamides of sulfosuccinic acid:

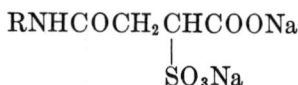

$$\text{RNHCOCH}_2\,\underset{\underset{\text{SO}_3\text{Na}}{|}}{\text{CHCOONa}}$$

have all been disclosed as surface active agents.[107] Of these the monoamides have been introduced commercially. Aerosol 18 of *American
Cyanamid Co.* is the stearyl derivative:

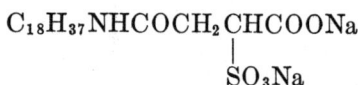

$$\text{C}_{18}\text{H}_{37}\text{NHCOCH}_2\,\underset{\underset{\text{SO}_3\text{Na}}{|}}{\text{CHCOONa}}$$

Aerosol 22, a similar product, is said to have the formula:

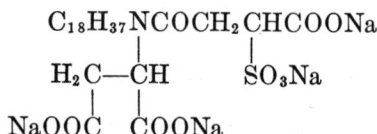

$$\begin{matrix} \text{C}_{18}\text{H}_{37}\text{NCOCH}_2\,\text{CHCOONa} \\ | \qquad\qquad\quad | \\ \text{H}_2\text{C}-\text{CH} \qquad \text{SO}_3\text{Na} \\ | \qquad | \\ \text{NaOOC} \quad \text{COONa} \end{matrix}$$

The amino alkyl esters of fatty acids have also been converted[108] to sulfo-
succinamates of general formula:

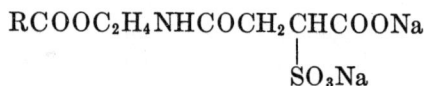

$$\text{RCOOC}_2\text{H}_4\text{NHCOCH}_2\,\underset{\underset{\text{SO}_3\text{Na}}{|}}{\text{CHCOONa}}$$

More complex sulfosuccinamides containing multiple intermediate linkages have been described,[109] a typical example being:

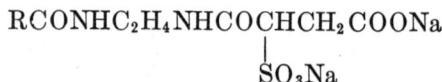

$$\text{RCONHC}_2\text{H}_4\text{NHCOCHCH}_2\,\text{COONa} \atop \qquad\qquad\qquad\qquad\underset{\text{SO}_3\text{Na}}{|}$$

made from lauric acid chloride, ethylenediamine, maleic acid, and bisulfite.
N-Substituted-2-aminopropionic esters have also been reacted with maleic
anhydride and sulfited, the resulting compounds[110] having generic formulas

[107] U. S. Pat. 2,252,401, Jaeger to *American Cyanamid Co.*
[108] U. S. Pat. 2,251,940, Katzman to *Emulsol Corp.*
[109] U. S. Pat. 2,239,720, Katzman to *Emulsol Corp.*
[110] U. S. Pat. 2,368,067, Lynch to *American Cyanamid Co.*

such as:

$$\underset{\overset{|}{SO_3H}}{HOOCCH}CH_2\,CONCH_2\,CH_2\,COOCH_3$$

$$\overset{|}{C_6H_{13}}$$

Dialkylamides of sulfomalonic acid have been described[111] as surface active agents:

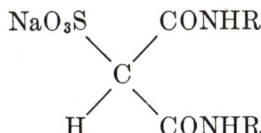

$$\begin{array}{c} NaO_3S \qquad CONHR \\ \diagdown \qquad \diagup \\ C \\ \diagup \qquad \diagdown \\ H \qquad CONHR \end{array}$$

As in the case of the dialkylsulfosuccinates, the alkyl groups are in the range of six to eight carbon atoms. It is interesting that they may be made by the direct sulfonation of the malonic diamides, using energetic sulfonating agents such as oleum. The malonic diamides are made by reacting malonic ester with the appropriate amine:

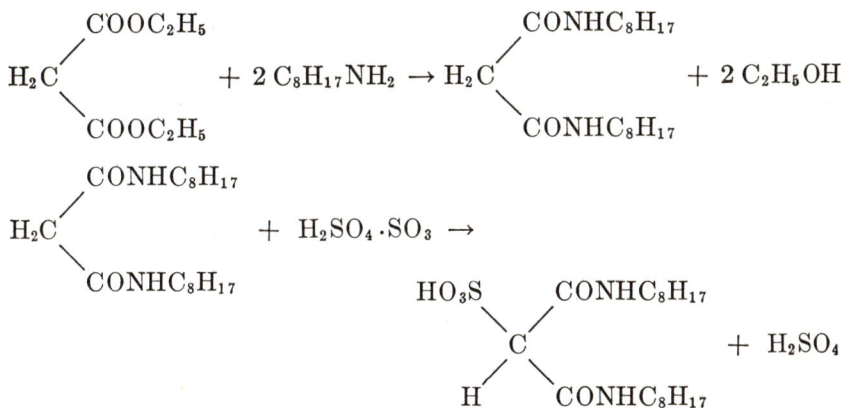

$$H_2C\begin{array}{c} \diagup COOC_2H_5 \\ \diagdown COOC_2H_5 \end{array} + 2\,C_8H_{17}NH_2 \rightarrow H_2C\begin{array}{c} \diagup CONHC_8H_{17} \\ \diagdown CONHC_8H_{17} \end{array} + 2\,C_2H_5OH$$

$$H_2C\begin{array}{c} \diagup CONHC_8H_{17} \\ \diagdown CONHC_8H_{17} \end{array} + H_2SO_4 \cdot SO_3 \rightarrow$$

$$\begin{array}{c} HO_3S \qquad CONHC_8H_{17} \\ \diagdown \qquad \diagup \\ C \qquad\qquad\qquad + H_2SO_4 \\ \diagup \qquad \diagdown \\ H \qquad CONHC_8H_{17} \end{array}$$

(c) *Ether and Other Miscellaneous Intermediate Linkages*

The ether-linked alkane sulfonic acids may be divided into two classes: those in which the ether oxygen is joined to two aliphatic radicals, and those in which it is joined on one side to an aromatic nucleus and on the other side to the sulfo-bearing aliphatic group. The non-aromatic ethers appear to have attained little if any commercial importance, probably because they are inherently expensive to manufacture. Products of the type of 1-lauryl-oxyethane-2-sulfo acid ($ROC_2H_4SO_3H$) have been made[112] by reacting lauryl

[111] Brit. Pat. 499,130; French Pat. 834,205; to *I. G. Farbenindustrie.*
[112] U. S. Pat. 1,985,747, Steindorff, Daimler, and Platz to *General Aniline Works.* Brit. Pat. 377,258, to *I. G. Farbenindustrie.*

alcohol with carbyl sulfate, isethionic acid or β-chloroethane sulfonates. Other higher fatty alcohols may be used in this reaction, and similar products have been described[113] in which R is the residue of hydroabietyl alcohol. Analogous surface active agents made by fusing the sodium derivative of a higher fatty alcohol with the sodium salt of 2-chloro-2'-sulfo ethyl ether under anhydrous conditions have been described recently:[114]

$$RONa + ClC_2H_4\text{—}O\text{—}C_2H_4SO_3Na \rightarrow R\text{—}O\text{—}C_2H_4\text{—}O\text{—}C_2H_4SO_3Na$$

The aromatic ethers have the general formula:

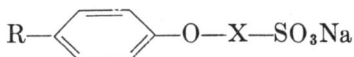

$$R\text{—}\langle\bigcirc\rangle\text{—}O\text{—}X\text{—}SO_3Na$$

where R is a typical hydrocarbon hydrophobic group and X is an alkylene or interrupted alkylene link. (The benzene ring may be nuclearly substituted or may be replaced by other aromatic rings such as naphthalene.) The essential steps in making these products are: (1) the alkylation of phenol with an appropriate olefin, alcohol, or alkyl halide, thereby introducing the hydrophobic group; (2) etherification of the alkyl phenol with one side of an alkylene dihalide; (3) replacement of the remaining halogen with a sulfonate group, using sodium sulfite as the reagent. The first and second steps can be interchanged and the halogenated alkyl phenyl ether formed initially can be nuclearly alkylated to introduce the hydrophobic group. The phenols and the phenyl alkyl ethers are both readily alkylated in the nucleus. The alkylation of phenols will be considered in greater detail in the section of this book dealing with alkyl aromatic sulfonates. The alkylation of phenyl haloalkyl ethers requires certain precautions, since the use of the usual Friedel-Crafts type of catalysts will cause the molecule to condense with itself. Nuclear alkylation of phenyl haloalkyl ethers can be accomplished satisfactorily by using an olefin or alcohol as the alkylating agent and silica or acid clay as a contact catalyst, and carrying out the reaction at about 200°C. The olefins or alcohols which are used may be cycloaliphatic, terpenoid, or aliphatic in nature.[115] The basis for the —X— group in the above generic formula may be ethylene dichloride, but more frequently is 2,2'-dichlorodiethyl ether, or 2(2-chloroethoxy)-2'-chlorodiethyl ether. The latter compound is more commonly known as triglycol dichloride and has the formula:

$$ClC_2H_4\text{—}O\text{—}C_2H_4\text{—}O\text{—}C_2H_4Cl$$

[113] U. S. Pat. 2,001, 275, Henke and Prahl to *E. I. du Pont de Nemours & Co.*
[114] U. S. Pat. 2,394,834, Young and Spitzmueller to *Industrial Patent Corp.*
[115] U. S. Pats. 2,176,834; 2,249,111; 2,273,622; Bruson to *Rohm and Haas Co.*

The only widely used surface active agents of this class are Triton 720 and Triton X-200 of *Rohm and Hass Co.*[116] They are probably made by condensing di- or triisobutene with phenol, using sulfuric acid as the alkylation catalyst, etherifying with dichlorodiethyl ether, and finally reacting with sodium sulfite.

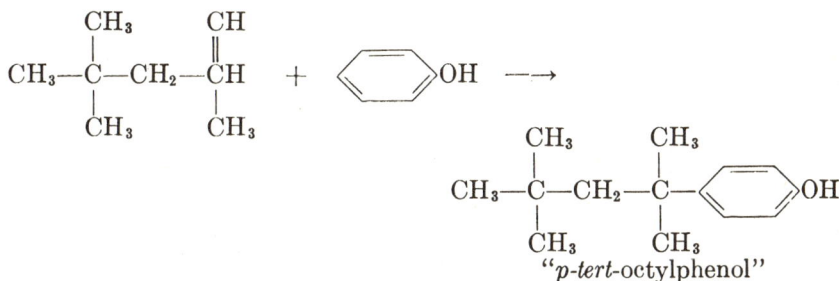

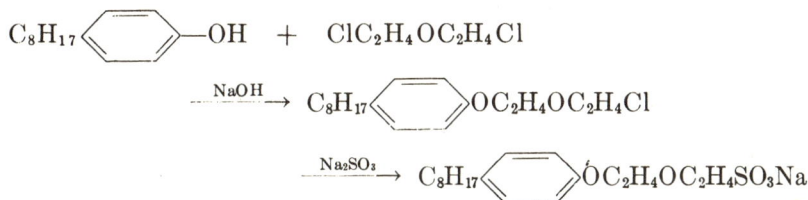

$$
\begin{array}{ccc}
CH_3 & & CH \\
| & & \| \\
CH_3-C-CH_2-CH & + & \bigcirc OH \longrightarrow \\
| & & \\
CH_3 & CH_3 &
\end{array}
$$

$$
\begin{array}{cc}
CH_3 & CH_3 \\
| & | \\
CH_3-C-CH_2-C-\bigcirc OH \\
| & | \\
CH_3 & CH_3
\end{array}
$$

"*p-tert*-octylphenol"

$$C_8H_{17}\bigcirc-OH \; + \; ClC_2H_4OC_2H_4Cl$$

$$\xrightarrow{NaOH} \; C_8H_{17}\bigcirc OC_2H_4OC_2H_4Cl$$

$$\xrightarrow{Na_2SO_3} \; C_8H_{17}\bigcirc OC_2H_4OC_2H_4SO_3Na$$

These products are highly effective in most of the usual applications of surface active agents. They have the advantage[117] of great stability toward both acid and alkali since there is no readily hydrolyzable linkage in the molecule. The aromatic nucleus is in itself an effective part of the hydrophobic portion of the compound, being about equivalent to a straight chain of four aliphatic carbon atoms.

A number of other interesting methods for linking an alkane sulfonate solubilizing group with a hydrophobic group have been disclosed, none of which appear to have been commercially exploited to any great extent. Higher alkyl vinyl sulfides, sulfoxides, and sulfones have been treated with sodium bisulfite to yield surface active alkane sulfonates:[118]

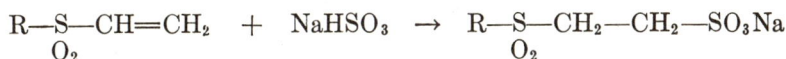

$$
\begin{array}{ccc}
R-S-CH=CH_2 & + & NaHSO_3 \rightarrow R-S-CH_2-CH_2-SO_3Na \\
O_2 & & O_2
\end{array}
$$

[116] U. S. Pats. 2,098,203; 2,106,716; 2,115,192; Brit. Pat. 494,767–9; French Pat. 830,851; U. S. Pat. 2,148,432; Bruson to *Rohm and Haas Co.*

[117] Van Antwerpen, *Ind. Eng. Chem.* **31,** 64 (1939).

[118] Ger. Pat. 681,338; Brit. Pat. 446,992; U. S. Pat. 2,140,569, Ufer and Hecht to *I. G. Farbenindustrie.*

The intermediates may be made by "vinylation" of the higher alkyl mercaptans with acetylene:

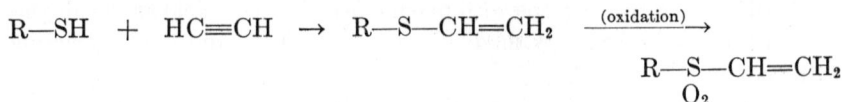

$$R—SH + HC≡CH → R—S—CH=CH_2 \xrightarrow{\text{(oxidation)}}$$

$$\begin{array}{c} R—S—CH=CH_2 \\ O_2 \end{array}$$

The sulfide link has also been used to join hydrophobe groups with an alkane sulfonic acid solubilizing group. Thus products of the general formula:

$$ROOC—CH_2—S—C_2H_4—SO_3Na$$

may be prepared[119] from the fatty alkyl ester of mercaptoacetic acid by reacting with chloroethane sodium sulfonate, or from the alkyl chloroacetate and 2-mercaptoethane sodium sulfonate.

The sulfide link has also been used in products[120] of the type:

$$RNHCO—CH_2—S—C_2H_4—SO_3Na$$

which is made by condensing the chloroacetyl derivative of a higher amine with 2-mercaptoethane sodium sulfonate ($HSC_2H_4SO_3Na$).

The sulfonamide link is represented by compounds of the type:

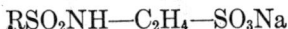

$$RSO_2NH—C_2H_4—SO_3Na$$

which may be made by condensing a Mersol type sulfonyl chloride with taurine or its analogues.

Products of the type $RSO_2NH—CH_2—SO_3Na$ have also been made[121] by condensing RSO_2NH_2 with formaldehyde–bisulfite.

Similar products containing the thioamide linkage have been disclosed[122] and are introduced in an ingenious way by adding hydrogen sulfide to a higher fatty nitrile. The resulting thioamide is then reacted with amino sulfo acids such as taurine. The final products have the general formula:

$$R—CS—NHC_2H_4SO_3H$$

[119] Ger. Pat. 644,275, to *Henkel et Cie.*
[120] Brit. Pat. 436,862; Ger. Pat. 619,299 to *Henkel et Cie.*
[121] U. S. Pat. 2,207,603, Wesche to *I. G. Farbenindustrie.*
[122] U. S. Pat. 2,201,171, Hanford to *E. I. du Pont de Nemours & Co.*

CHAPTER 5

Alkyl Aromatic Sulfonates

I. General Considerations

The aromatic sulfonic acids in which the aromatic nucleus is an integral part of the hydrophobic group are currently the most extensively used of all synthetic detergents in this country. Until very recently this was due largely to the great popularity of two products, Nacconol NR (National Aniline Division, *Allied Chemical and Dye Corp.*) and Santomerse (*Monsanto Chemical Co.*), which accounted for about half the total poundage of synthetic detergents produced in the United States in 1943. Within the past few years, however, a number of other manufacturers have entered the field and the tonnage of alkyl aryl sulfonates is increasing at an amazing rate. As a class the aromatic sulfonic acids are potentially among the least expensive anionic surface active agents. They also offer many other advantages such as great stability toward hydrolysis and the possibility of wide variation in chemical structure (and, therefore in physical properties).

The sulfonates of unsubstituted aromatic hydrocarbons such as benzene, naphthalene, anthracene, diphenyl, and even the terphenyls have little if any surface active character. The substitution of an aliphatic, cycloaliphatic, or aralkyl side chain for one or more of the nuclear hydrogen atoms confers surface activity on these sulfonates, provided the substituent is sufficiently large. The number of substituents present as well as the size of each individual substituent is important. Furthermore, the size of the aromatic nucleus itself is important since the aromatic nucleus functions as an integral part of the hydrophobic group. It is apparent from these considerations that by varying the aromatic nucleus, the number of side chain substituents, and the size or chemical nature of each substituent it is possible to arrive at almost any degree of surface activity in the final product. Another factor which may be varied is the number of sulfonic acid groups present on the nucleus. In the vast majority of actual cases only one sulfonic acid group is used. Unless the side chain is exceptionally long and heavy two or more sulfonic acid groups make the molecule as a whole too hydrophilic, thus reducing the surface activity.

In actual practice the three important aromatic nuclei used are benzene, diphenyl, and naphthalene, although several of the others have been mentioned in the patent literature and enjoy a limited use. Rosin, which consists largely of abietic acid, and various alkylated and condensed rosin derivatives have been sulfonated to form surface active agents.[1] These may be regarded as derivatives of partially hydrogenated phenanthrene nuclei. Retene, which is methylisopropylphenanthrene, and can be obtained by the oxidative pyrolysis of rosin, also yields a monosulfonic acid of mild surface active properties. Crude coal tar and high-boiling coal tar fractions, which probably contain lower alkylated condensed ring systems, have been sulfonated to yield surface active agents.[2] In general, however, the alkyl aromatic sulfonates are obtained by pure synthetic methods in two stages, an alkylation stage and a sulfonation stage. Alkylation may precede or follow sulfonation, or the two steps may be carried out simultaneously. In most instances alkylation is the first and most critical step. Once the alkylated aromatic hydrocarbon (or substituted hydrocarbon) is obtained, it is usually a simple matter to carry out the sulfonation and obtain a satisfactory final product. Accordingly, the art of making surface active agents of this class is largely the art of alkylating aromatic nuclei.

Probably the oldest and best-known procedure for alkylating an aromatic nucleus is the classical Friedel-Crafts synthesis. In this reaction an alkyl halide, RX, reacts with an aromatic hydrocarbon, ArH, in the presence of anhydrous aluminum chloride:

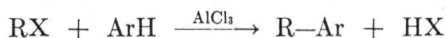

$$RX + ArH \xrightarrow{AlCl_3} R—Ar + HX$$

This reaction has been studied exhaustively from many points of view and has formed the subject of many excellent reviews.[3] It may be regarded as the prototype for practically all of the various alkylations of aromatic nuclei used in the commercial syntheses of surface active agents. There is a vast number of variations in specific reactants and reaction conditions which may be employed, but the fundamental reaction mechanism appears to be similar in every case. In place of the alkyl halides there may be used alcohols, olefins, ethers, dialkyl sulfates, alkyl sulfuric esters, alkyl sulfonates and phosphates, and various other derivatives all of which function in this reaction as the equivalent of an alkyl halide. In place of aluminum chloride a wide variety of other catalysts have been used in alkylations of the aromatic nucleus. In some cases gaseous HCl alone will bring about

[1] See, for example, U. S. Pats. 1,931,257, Henke and Charlton to *E. I. du Pont de Nemours & Co.*; 2,348,200, Fronmuller and Thomas to *Institute of Paper Chemistry*; 2,376,381, Price and May to *National Oil Products Co.*

[2] Ger. Pats. 545,968; 546,943, to *Chemische Fabrik Pott and Co.* French Pat. 789,993, to *I. G. Farbenindustrie.* Canadian Pat. 426,101, to *Colgate-Palmolive-Peet, Co.*

[3] See for example C. A. Thomas, *Anhydrous Aluminum Chloride in Organic Chemistry*, Reinhold, New York, 1941.

alkylation. Among the more commonly used catalysts are sulfuric acid, boron fluoride, hydrofluoric acid, phosphoric acid, phosphoric anhydride, and the halides of zinc, ferric iron, stannic tin, or pentavalent antimony. Acid-treated clays, silica, and phosphotungstic and phosphomolybdic acids have also been used as alkylation catalysts.

The ease with which alkylation takes place depends among other factors on the nature of the aromatic nucleus and the nuclear substituents already present. Benzene itself is fairly readily alkylated and may be taken as a convenient reference compound with which to compare other aromatic compounds. Naphthalene is somewhat more easily alkylated than benzene, i.e., milder catalysts are required and these will act at lower temperatures. Toluene, the xylenes, and the polymethyl benzenes are also more easily alkylated than benzene itself. Halogenated benzenes and phenyl-substituted benzenes (e.g., diphenyl) react more slowly than benzene but offer no great practical difficulties. Benzene derivatives which already bear a meta-directing substituent such as nitro, sulfonic acid, carbonyl, or carboxyl are extremely difficult to alkylate and will not enter into the Friedel-Crafts type of reaction under normal circumstances. Phenols and naphthols as well as their ethers, on the other hand, are much more easily alkylated than benzene, and are among the most readily alkylated of all aromatic compounds. It is a general rule, therefore, that the meta-directing substituents inhibit alkylation, whereas the ortho- and para-directing substituents greatly facilitate this reaction.

The ease of alkylation also depends upon the configuration of the entering alkyl group. Methyl and ethyl halides require powerful catalysts. Secondary and tertiary alkyl halides are more reactive in alkylation than the primary halides. The primary halides undergo rearrangement during alkylation and the product is a·secondary or tertiary alkyl aromatic compound. The more reactive halides such as benzyl and allyl chlorides react very readily with aromatic nuclei. Under certain conditions naphthenes such as cyclopropane can be used to introduce alkyl groups into benzene.[4] Naphthenes and hydrogen in the presence of zinc and HCl at 200–450°C. can also act to alkylate aromatic hydrocarbons.[5]

The variety of experimental facts concerning the alkylation reaction can be well correlated theoretically on the basis of one mechanism, which is generally called the carbonium ion mechanism.[6] According to this theory

[4] Ipatieff et al., J. Org. Chem. 5, 253 (1940).

[5] U. S. Pat. 2,361,065, Schmerling and Ipatieff to Universal Oil Products Co.

[6] For detailed treatments of the theory of Friedel-Crafts type reactions see Hammett, Physical Organic Chemistry, McGraw-Hill, New York 1940, Chapter 10; Nightingale, Chem. Revs. 25, 329 (1939).

the alkylating agent actually reacts in the form of a cation or positively charged configuration called the carbonium ion. This is formed from, *e.g.*, alkyl halides according to the equation:

$$RX \rightarrow R^+ + X^-$$

The carbonium ion may also be formed in an analogous manner from alcohols. Olefins or ethers can form carbonium ions in the presence of a proton source (such as sulfuric acid). Considering propene as an example:

$$CH_3-CH=CH_2 + H^+ \rightarrow [CH-CH-CH_3]^+$$

In this formula it appears that the positive charge resides on the number 2 carbon atom. Actually this is the usual case since *n*-propyl halides, as well as propene, react in alkylations to give isopropylated aromatic compounds.

Carbonium ions are formed from their parent compounds by reaction with strongly electrophilic reagents, that is, electron acceptors or "acids" in the generalized sense. All the alkylation catalysts fall into this class. The extent to which they are formed, *i.e.*, the effective concentration of carbonium ions in a given reaction mixture, depends on the "strength" or electrophilic power of the catalyst and on the ease with which the parent compound forms the carbonium ion. These are intrinsic properties of the molecules themselves and are functions of their configurations.

The carbonium ion is itself an electron-attracting substance, and it attacks the aromatic nucleus by displacing a proton. Thus the over-all reaction between *n*-propyl chloride and benzene, for example, may be considered stepwise as follows:

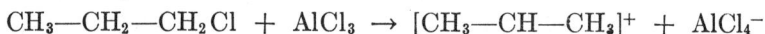

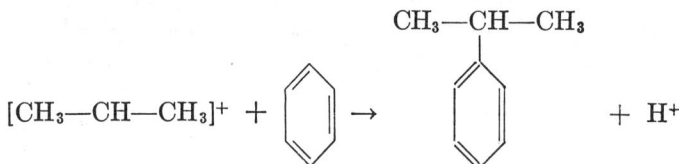

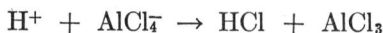

$$CH_3-CH_2-CH_2Cl + AlCl_3 \rightarrow [CH_3-CH-CH_3]^+ + AlCl_4^-$$

$$H^+ + AlCl_4^- \rightarrow HCl + AlCl_3$$

The point of attack on the aromatic nucleus by the carbonium ion will be at that position where the electron density is greatest, *i.e.* the position of maximum effective negative charge. This is the point from which a nuclear hydrogen is most easily displaced in the form of a proton.[7] Meta-

[7] See Pauling, *Nature of the Chemical Bond* Cornell Univ. Press, Ithaca, 1940, p. 147 ff.

directing substituents on the benzene ring reduce the electron density by virtue of their inductive effect, and thus slow down the rate of alkylation. The ortho, para-directing substituents, either by an inductive effect (as in the case of methyl) or by a resonance effect, increase the electron density and facilitate alkylation.

Although a large number of specific procedures and catalysts for alkylation have been described, relatively few are used commercially. Aluminum chloride, sulfuric acid, and more recently anhydrous hydrofluoric acid are used more than any of the other condensing agents, although zinc chloride, boron trifluoride, and phosphoric acid are also quite widely used. The contact catalysts such as acidic clays are seldom used for introducing the relatively large alkyl groups which go into surface active agents. For alkyl halides, aluminum chloride and HF are the favored catalysts. For olefins H_2SO_4, HF, and $AlCl_3$ are most widely used. It is interesting that H_2SO_4 has a marked selective action with reagents containing both an olefinic double bond and a halide radical. Thus benzene reacts with $CH_3CH{=}CHCH_2Cl$ in the presence of 100% sulfuric acid at 20°C. to form:[8]

$$CH_3CH{-}CH_2CH_2Cl$$

For alcohols, H_2SO_4 and BF_3 are preferred. Mixtures of BF_3 with P_2O_5 or B_2O_3 are said to be particularly effective.[9] $ZnCl_2$ is also an effective agent for condensing alcohols with phenols. It is particularly easy to use in the laboratory since no special precautions need be taken against moisture contamination.[10] One of the interesting variations which have been described consists in using $AlCl_3$ dissolved in a nitroparaffin to form a homogeneous reaction mixture.[11] Hydrogen chloride alone has also been used to effect the alkylation of aromatic hydrocarbons with olefins.[12] Phosphotungstic acid,[13] B_2O_3, and oxalic acid[14], and glycerol-H_2SO_4 mixtures[15] have

[8] U. S. Pat. 2,349,779, Van Zoeren to *Wm. S. Merrell Co.* See also Archer and Suter, *J. Am. Chem. Soc.* **67**, 43 (1945).

[9] See U. S. Pats. 2,390,835–6, Hennion *et al.* to *E. I. du Pont de Nemours & Co.*

[10] See, for example, French Pat. 778,890, to *British Industrial Solvents Ltd.*

[11] U. S. Pat. 2,385,303, Schmerling to *Universal Oil Products Co.*; Brit. Pat. 559,779; to *Universal Oil Products Co.*

[12] U. S. Pat. 2,357,978, Schmerling and Durinski to *Universal Oil Products Co.*

[13] U. S. Pat. 2,103,736, Skraup to *Schering-Kahlbaum A.-G.*

[14] U. S. Pat. 2,353,282, Turkington, Whiting, and Rankin to *Bakelite Corp.*

[15] U. S. Pat. 2,290,604, Stevens and Nickels to *Gulf Research and Development Co.*

all been used to effect the alkylation of phenol with olefins, a reaction which proceeds under mild conditions.

II. Hydrophobic Group Linked Directly to Sulfonated Aromatic Nucleus

(a) Lower Alkyl Aromatic Sulfonates

ALKYL NAPHTHALENE SULFONATES

The simplest and longest known alkyl aromatic surface active agents are the propylated naphthalene sulfonates. These were invented in Germany during World War I in an attempt to develop soap substitutes from non-fatty raw materials. Isopropyl alcohol, naphthalene, and sulfuric acid are heated together, according to the original process, forming a mixture of mono-, di-, and polypropyl naphthalenesulfonic acids.[16] This process is still probably the most widely used method of making this type of surface active agent. The molar ratio of isopropanol to naphthalene controls the relative quantities of each homologue present in the final mixture. The dipropyl compound is far more active than the monopropyl. The tri-propyl compound shows proportionately less increase in activity. The dipropylnaphthalene sulfonate, or a mixture in which this product predominates, is the usual commercial product. The propyl groups are attached to the nucleus through the number 2 carbon atom and are, therefore, strictly speaking, isopropyl groups. The positions which they occupy on the naphthalene ring are not established with certainty, and a mixture of position isomers as well as homologues doubtless exists in the commercial products. Beside the simultaneous one-step reaction of isopropanol, naphthalene, and sulfuric acid a number of other procedures may be used for making these products. They are mostly two-stage processes in which the naphthalene is first alkylated, the alkyl naphthalene isolated, and optionally purified, and the alkyl naphthalene finally sulfonated. In the alkylation stage the readily available sources of isopropylcarbonium ion such as isopropanol, n-propanol, n-propyl, or isopropyl halides, diisopropyl ether, or propylene itself may be used. As catalysts the usual strongly electrophilic alkylation catalysts have all been described. These include sulfuric acid of various concentrations, AlCl$_3$, ZnCl$_2$, BF$_3$, HF, etc.[17] The mixed propylated naphthalenes resulting from the condensation are

[16] Ger. Pat. 336,558, to *Badische Anilin- & Soda-Fabrik.*

[17] For various modifications in the preparation of lower alkyl naphthalene sulfonates see, for example, U. S. Pat. 2,133,282, Cook and Valjavec to *Arkansas Co.*; Ger. Pat. 544,889, Pospiech to *Chemische Fabrik Pott and Co.*; U. S. Pats. 2,020,385; Todd to *Imperial Chemical Industries Ltd.*; 1,980,543, to Eli Lurie; 2,143,493, to Stanley *et al.*

sometimes purified by washing and distillation before being sulfonated. Most often, however, they are simply washed free of excess catalyst and subjected to sulfonation without any attempt at separation. The propylated naphthalene sulfonates are made by a great many different manufacturers. They are a particular favorite with smaller manufacturing companies since their preparation is simple and the raw materials readily available. They are usually sold in dry powder form which may run over 90% active ingredient or as an aqueous paste of about 70% active ingredient. The products are fairly good wetting-out agents for fabrics and are used largely for general wetting-out purposes in textile mills. They do not exhibit any appreciable detergent power for soiled fabrics until relatively high concentrations are reached (1–2% or more). Their strong wetting power for glass, porcelain, and other solid surfaces makes them a useful ingredient in dishwashing powders, household scouring compositions, and other similar items which are generally known commercially as "janitor's supplies." They are also used in making dispersions and emulsions, having been used to a considerable extent in the emulsion polymerization of synthetic rubbers in Germany during World War II. Other important uses are in metal cleaning, preparatory to plating or finishing, and as a spreading ingredient in insecticides. Among the better-known trade names for the propylated naphthalene sulfonates are Alkanol B (*E. I. du Pont de Nemours & Co.*), Nekal A (*General Dyestuffs Corp.* and *I. G. Farbenindustrie*) Aerosol OS (*American Cyanamid Corp.*), and Naccosol A (National Aniline Division, *Allied Chemical and Dye Corp.*). These products are often used in admixture with solvents or with other synthetic surface active agents, usually of higher molecular weight. They act as solubilizing agents for these latter compounds, probably reducing the average micelle size, and forming a mixture which is more effective at lower temperatures. This dispersing and solubilizing action is general for the lower alkyl aromatic sulfonates and in fact for most of the surface active agents of lower molecular weight and smaller micelle size. It makes them particularly useful in increasing the penentration of dye baths, textile emulsion treatments, etc. They thus find commercial application in a wide variety of fields, and continue to hold their position on account of their low cost and general versatility. This is true in spite of the fact that in many individual applications their effectiveness could be surpassed by other surface active agents specifically "tailored" to perform the job in hand.

The butylated[18] napththalene sulfonates are today more widely used than the propylated products. They resemble the latter very closely and are

[18] U. S. Pats. 1,737,792; 1,836,588; 1,901,507; Günther to *I. G. Farbenindustrie*.

used for the same purposes and under the same conditions. They have a more pronounced wetting, detergent, and emulsifying effect at lower concentrations due to their slightly higher molecular weight.[19] They are prepared in the same manner as the propyl derivatives usually from butanol, naphthalene, and sulfuric acid. Any one of the butyl alcohols may be used, although n-butanol appears to be the starting material preferred by most manufacturers. The butyl groups of the final product are either secondary (if n-butanol or butanol-2 is used) or tertiary (if isobutanol or tertiary butanol is used). As in the case of the propylated compounds the butyl naphthalene sulfonates are made by a large number of manufacturers including many smaller companies. They are marketed in the same forms and used for the same applications as the propyl compounds. The best-known trade name is Nekal B or BX, that of the original manufacturer *I. G. Farbenindustrie*. Current German practice in manufacturing this product is given by Hoyt as follows:

1000 kg. pure anhydrous butanol is mixed with 865 kg. naphthalene in a jacketed, agitated lead-lined kettle. 1810 kg. of 98% H_2SO_4 and 2400 kg. oleum are added slowly, maintaining the temperature at 25°C. The mixture is agitated for a half hour after all the acid is added and the temperature is then gradually raised to 45–55°C. over a period of two hours. At the end of this time the mixture will separate into an upper layer of "Nekal" sulfonic acid and a lower layer of spent acid. After a further four-hour period of agitation the lower layer is drawn off. The upper layer of sulfonic acid is then diluted, neutralized with NaOH solution, bleached with NaClO solution,[20] and filtered. An appropriate amount of Glauber's salt is added for standardization and the solution is finally dried on a double drum drier and milled.

The butylation of naphthalene has recently been studied from the academic point of view by Rowe and co-workers.[21] They have separated a number of pure components from the alkylation mixtures and identified their structures. Although the propylated and butylated derivatives are the most widely used of the lower alkyl naphthalenes, a great number of others have been described and produced. Mono- and diamyl naphthalene sulfonates have been introduced recently. These products show the expected increase in surface activity as compared with the butyl derivatives. The tri- and polyamyl derivatives begin to exhibit the insolubility characteristic of the higher alkylated naphthalene sulfonates. There is no sharp line of demarcation between "higher" and "lower" derivatives in this series but rather a gradual decrease in solubility (increase in micellar size) and consequent decrease in general applicability at ordinary tempera-

[19] See Hetzer, *Chem. Ztg.* **64**, 325 (1940).
[20] The bleaching of Nekal type surface active agents with hypochlorite is disclosed in Ger. Pat. 729,189, to *Rudolf and Co.*
[21] Bromby, Peters, and Rowe, *J. Chem. Soc.* **1943**, 144.

tures. As the molecular weight of the alkyl group increases it becomes necessary to limit the final product to one alkyl substituent rather than having two or more present. With the lowest members of the series two substituents are necessary in order to attain a reasonable degree of surface activity. Dibutylnaphthalene sulfonate, for example, is a more effective product under most circumstances than the monobutyl compound. Dicapryl napthalene sulfonate, on the other hand, is too insoluble to be highly effective under ordinary circumstances. The monocapryl derivative is quite useful, however, as a detergent. It is superior to the dibutyl derivative in this respect. This point will be considered in more detail in Part II of this book.

Mixed dialkyl dervatives in which one alkyl group is propyl or butyl and the other alkyl group is hexyl or octyl have recently been described and are said to be highly effective.[22]

Another class of lower alkyl naphthalene sulfonates includes those products in which two or more naphthalene nuclei are joined by methylene groups. The prototype of this class would be dinaphthylmethanemono- or disulfonic acid:

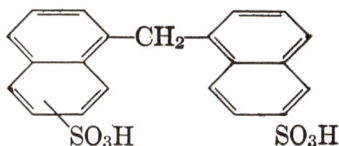

Products of this class are actually of indefinite composition and are manufactured by heating naphthalene, formaldehyde, and sulfuric acid together or by treating naphthalenesulfonic acids with formaldehyde. In this way three or more naphthalene nuclei may be joined together by methylene groups in what amounts to a low-degree condensation polymer. Lower alkylated naphthalenes such as monoisopropylnaphthalene may also be used in this reaction. Products of this type in which phenolic hydroxyl groups are present, obtained from phenols or naphthols, are widely used as synthetic tanning agents. They have an unusually high power to coagulate protein dispersions irreversibly. The unhydroxylated products are used, however, as dispersing agents for pigments[23] and may therefore be classed as surface active agents. Many commercial dyes and pigments, as produced, may contain 50–80% water and yet appear to be only slightly

[22] U. S. Pat. 2,268,140, Hengstenberg *et al.* to *General Aniline and Film Corp.* See also Ger. Pats. 704,353; 713,515; 729,787; to *I. G. Farbenindustrie.*

[23] See, for example, Canadian Pat., 347,865, Tucker to *Dewey and Almy Chemical Co.*

moist cakes. This is a serious disadvantage in handling. A small amount
of the naphthaleneformaldehyde sulfonate added to such a cake will often
cause it to liquefy into the thin watery suspension normally expected of a
system with so high a water content. The sulfonated naphthalene formaldehyde condensates are known under a variety of trade names, some prominent ones being Tamol, Leukanol, and Daxad.

Among other lower alkylated naphthalene sulfonates which have been
described but which seem to have had little if any commercial use are:
cyclohexyl and methylcyclohexyl derivatives,[24] benzyl derivatives,[25] products made by condensing glycols or alkylene dihalides with naphthalene,[26]
naphthalene-furfural condensates,[27] and napthalene alkylated with various
terpenes,[28] and terpene alcohols. The latter products appear to fall within
the border territory between "higher" and "lower" alkylated derivatives.
Recently described products in which abietene or rosin itself is used as
alkylating agent are certainly higher alkylated products and behave as
such. They have not, however, been clearly defined or characterized and
doubtless contain a variety of products other than the desired condensate
in their make-up. Despite the vast number of alkyl naphthalenes which
have been described, mentioned, or claimed in the patent literature, the
simple propyl, butyl, and amyl derivatives together with the formaldehyde
condensates remain the most important commercially.

OTHER LOWER ALKYL AROMATIC SULFONATES

It is a very rough empirical rule that in order for a lower alkylated
aromatic sulfonate to be an effective surface active agent, the aromatic
nucleus itself must be fairly large. Thus dipropylnaphthalene sulfonate is
appreciably surface active whereas dipropylbenzene sulfonate is not.
To define where appreciable surface activity begins, however, one must
specify temperatures and concentrations. At high concentrations even
dipropylbenzene sulfonate exhibits dispersing and solubilizing properties
which would class it as a mild surface active agent. The di- and trialkyl
benzene and alkyl toluene sulfonates, in which the alkyl group is amyl or
butyl have good wetting and dispersing properties in many systems. They

[24] Ger. Pat. 625,638; French Pat. 678,684, to *Organienburger Chemische Fabrik
A.-G.*
[25] Ger. Pats. 663,983; 678,134, to *Chemische Fabrik Baumheier.* U. S. Pats. 1,833,-
245, Felix to *Ciba Co.;* 1,853,414, Günther et al. to *I. G. Farbenindustrie;* 1,912,260,
Dachlauer and Thomsen to *I. G. Farbenindustrie;* 1,781,981, Kahn and Thauss to *General Aniline Works.*
[26] Brit. Pat., 302,258; U. S. Pat. 1,872,736; Günther et al. to *I. G. Farbenindustrie.*
[27] Brit. Pat. 322,737, Ellis et al. to *British Celanese Ltd.*
[28] French Pat., 763,990, to *Ciba Co.*

have not, however, achieved wide commercial acceptance. The same is true of the analogous alkylated phenols and naphthols. This is possibly because they have shown no particular advantages to offset their higher cost. The butylated and amylated phenols are widely used as such, but their sulfonates (which have appreciable surface activity) find little use.

Sodium tetrahydronaphthalene sulfonate (which may be regarded as a lower alkylated benzene sulfonate) is marketed as a wetting and dispersing agent under the trade name Alkanol S (*E. I. du Pont de Nemours & Co.*) and has enjoyed considerable success.

Sulfonated rosin and rosin transformation products such as abietene may also be classified as alkyl benzene derivatives since there is usually only one aromatic ring present if the final product is a true aromatic sulfonic acid. The extent to which these products are used commercially is difficult to ascertain, but is probably not very great. The same considerations apply to the so-called lignin sulfonates which are more-or-less purified by-products of the sulfite process for making cellulose pulp from wood. These products are surface active and have been proposed for a variety of wetting, dispersing, and emulsifying functions. Their lack of uniformity and undesirable color and odor have doubtless militated against their extensive use.

Abietic acid itself has been sulfonated with oleum of 20–60% free SO_3 content in a liquid SO_2 solvent to produce surface active agents.[29]

The butylated diphenyl sulfonates and phenylphenol sulfonates are excellent wetting agents. These products are made by *Monsanto Chemical Co.* under the trade names Aresket (monobutyldiphenyl sodium monosulfonate), Areskap (monobutylphenylphenol sodium monosulfonate), and Aresklene (dibutylphenylphenol sodium disulfonate). It is noteworthy that the last-named product contains two sulfonic acid groups. The excess of hydrophilic character thus imparted to the molecule is balanced by the presence of two butyl groups.

Propylated, butylated, and amylated diphenyl ether sulfonates as well as the sulfonated terpene-diphenyl ether condensates have also been described as wetting agents.[30] They are effective and attractive products but have not been introduced commercially so far as the writers are aware. Alkylated sulfonates of α- and β-naphthols as well as their lower alkyl ethers have been mentioned in the patent literature and have possibly been put to limited use at some time. They are not, however, standard established products.

[29] U. S. Pat. 2,348,200, Fronmuller and Thomas to *The Institute of Paper Chemistry.*
[30] U. S. Pat. 2,081,876, Prahl to *E. I. du Pont de Nemours & Co.* See also U. S. Pats. 2,202,686; 2,320,846; Borglin to *Hercules Powder Co.*; 2,315,951, Fox and Bare to *E. I. du Pont de Nemours & Co.*

(b) Higher Alkyl Aromatic Sulfonates

The higher alkyl aromatic sulfonates currently account for the largest tonnage of synthetic surface active agents produced in this country. This is due to their demonstrated versatility, effectiveness and low cost. Nacconol NR (National Aniline Division, *Allied Chemical and Dye Corp.*) and its various modifications were the first products of this class to be introduced, and are typical examples of the class.

According to the extensive patent literature, Nacconol is made essentially by starting with a carefully selected petroleum fraction in the kerosene range. This is chlorinated in the liquid phase to produce a mixture of alkyl chlorides, predominantly the monochlorinated products. This alkyl chloride is condensed with benzene in a typical Friedel-Crafts reaction using $AlCl_3$ as the catalyst. The alkyl benzene thus formed is freed of excess reactants and by-products and is then sulfonated. The sulfonated product is neutralized and dried. The nature of the kerosene fraction constituting the starting material has, of course, a profound influence on the properties of the final product. The kerosene which is used commercially is highly saturated and paraffinic in nature. It is refined from Pennsylvania or Pennsylvania type crudes, has a high aniline point (indicative of low olefinic and aromatic content) and boils in the range of 200 to 300°C. The average molecular weight corresponds to a saturated aliphatic hydrocarbon of about fourteen carbon atoms. The degree of chlorination is also an important variable in the production of Nacconol type detergents. The chlorinaton is carried out at about 60–70°C. in the presence of iodine or actinic light and is continued until a weight increase of about 20% has been attained. This corresponds to somewhat more than 1 mole chlorine absorbed per mole hydrocarbon. The condensation of this chlorinated product (which is usually referred to as keryl chloride) with benzene is carried out at about 35–50°C. using about 5 moles benzene and about 0.1 mole $AlCl_3$ per mole keryl chloride. The total react'on product consists of a heavy lower sludge layer (which is separated and discarded) and an upper layer of crude keryl-benzene. The kerylbenzene is washed and then "stripped" by distillation of excess benzene, unreacted kerosene, and keryl chlorides. It may be further purified if desired by vacuum distillation. In this case a high-boiling residue consisting of di- and polykeryl benzenes and compounds in which two or more benzene nuclei are attached to the same aliphatic chain is obtained. The amount of this residue together with the amount of forerun is an inverse measure of the yield or efficiency of the over-all conversion.

The most desirable single type of kerylbenzene appears to be that in which one aliphatic chain is attached through a single linkage to one benzene nucleus. In actual practice, however, some of the polyalkylated and polyarylated products are almost always present. The sulfonation is usually carried out with monohydrate or oleum at temperatures under 60°C. Higher temperatures tend to give undesirable dark color to the product and also to produce a certain amount of disulfonated material. The disulfonates tend to lower the detergent properties. The sulfonated product may be neutralized without removing the spent acid, or the spent acid may be separated by first adding small amounts of water. By carefully adjusting the water content of the system a layer of kerylbenzene-sulfonic acid may be separated which contains 10% or less of free sulfuric acid. This product, when neutralized and dried, is probably the basis of the "concentrated" or "low-salt content" grade called Nacconol NRSF. The most widely used grade, Nacconol NR, contains about 60% inorganic salt which is mainly sodium sulfate. Surface active agents derived from kerosene as a raw material tend to have a residual odor suggestive of their origin. Furthermore, they tend to retain some of the dark color which is developed during the Friedel-Crafts condensation. Accordingly a large amount of research has been devoted to eliminating these undesirable characteristics. Color and odor may both be improved by hypochlorite treatment, by refining the kerylbenzene with sulfuric acid washes before sulfonating and by treatment with phosphates. Numerous other procedures have also been described in the patent disclosures.[31]

Santomerse No. 1 (*Monsanto Chemical Co.*) is very similar in composition and properties to Nacconol NR. Santomerse D is a sulfonated keryl benzene in which the keryl group has an average molecular weight equivalent to a decyl group. It is marketed as a low-salt-content flake (90% or more active detergent) and also as a substantially salt-free aqueous solution.

Beside benzene and diphenyl, the lower homologues of benzene such as toluene, xylene, and cumene have also been used for alkylation. Ultrawet E (*Atlantic Refining Co.*) and Oronite Detergent (*Oronite Chemical Co.*) are probably products of this nature. The large petroleum refineries have had an obvious economic advantage in the production of alkyl aromatic detergents due to their supply of alkylhydrocarbons. This advantage has

[31] Among the important patents dealing with higher alkyl aromatic sulfonates in which the alkyl group is derived from kerosene and the aromatic nucleus is benzene, are the following: Brit. Pat. 416,379, to *I. G. Farbenindustrie;* French Pat. 766,903; U. S. Pat. 2,220,099, Günther *et al.* to *I. G. Farbenindustrie;* U. S. Pat. 2,161,173, Kyrides to *Monsanto Chem. Co.* U. S. Pats. 2,223,364; 2,233,408; 2,247,365; 2,283,199; 2,314,929; 2,340,654; 2,387,572; 2,390,295; 2,267,725; 2,393,526; 2,397,133; 2,394,851; Flett to *Allied Chemical and Dye Corp.* U. S. Pats. 2,314,255; 2,333,830; Toone to *Allied Chemical and Dye Corp.* U. S. Pat. 2,330,922, Riegler to *Allied Chemical and Dye Corp.* U. S. Pat. 2,364,767, Zizinia *et al.* to *Allied Chemical and Dye Corp.* U. S. Pat. 2,210,962, Thomas to *Sharples Solvents Corp.*

been increased in recent years with the introduction of aromatization processes designed to produce toluene from petroleum hydrocarbons. It is not surprising, therefore, that as this is being written several of the larger petroleum companies are planning or building large-capacity installations for producing alkyl aromatic hydrocarbons suitable for sulfonation to detergents of the Nacconol type.

Naphthalene has also been converted to higher alkyl derivatives, and subsequently sulfonated to give detergent products. It has previously been noted that increasing the weight of an alkyl substituent on the naphthalene nucleus above about ten carbon atoms brings about a sharp decrease in solubility of the monosulfonated product. This effect, however, is dependent on the configuration of the alkyl group. Furthermore, mixtures of monoalkylated products, such as are necessarily present in the case of petroleum-derived detergents, are considerably more soluble than individual components. A monoalkylnaphthalene with the alkyl group averaging nine carbon atoms (nonylnaphthalene) is currently being offered to detergent manufacturers. When monosulfonated this product gives a surface active agent remarkably similar to the higher alkyl benzene sulfonates in its working characteristics.

Alkylated phenols, cresols, anisoles, and phenetols, in which the alkyl group is derived from kerosene and averages twelve to sixteen carbon atoms, have been thoroughly investigated.[32] When monosulfonated, they make excellent surface active agents of all around applicability. They have, however, no outstanding advantages over the less expensive benzene analogues and accordingly have had very limited commercial use. This is rather surprising in view of the fact that the Triton 720 (*Rohm and Haas Co.*) type, mentioned above among the alkane sulfonates and alkyl sulfuric esters, is also based on an alkylated phenol. The solubilizing group in the Triton class is introduced by syntheses which are more laborious and costly than straight sulfonation. Nevertheless, they have continued to enjoy a modest success in the market whereas the straight sulfonated alkyl phenols have not. Alkylated phenols also form the basis of the important Igepal group of non-ionic detergents which will be discussed later.

So far we have considered the more important aromatic nuclei on which the higher alkyl aromatic sulfonates are based. One extremely important alkyl source, the chlorinated kerosene fractions, has also been considered. There are several other alkyl sources which may be used. Probably the

[32] Among patents disclosing alkyl phenol and phenol ether sulfonates in which the alkyl group is keryl are the following: U. S. Pat. 2,337,924, Platz *et al.* to *General Aniline & Film Corp.*; Brit. Pat. 495,414 to *I. G. Farbenindustrie.* U. S. Pats. 2,178,571; 2,196,985; 2,205,946–7; 2,223,363; 2,249,757; Flett to *Allied Chemical and Dye Corp.*

most important and potentially one of the cheapest is the group of low-polymerized simple alkenes. Propene, isobutene, isopentenes, and various isomeric hexenes and heptenes can be readily converted to their dimers, trimers, or tetramers. These low polymers are in themselves monoalkenes and, therefore, suitable alkylating agents. The monomers are obtained indirectly from petroleum or gas-refining operations by large-scale methods which are well known and inexpensive. Isobutene in particular has been produced in tremendous tonnages for use in making aviation fuel. The di-, tri-, and even tetraisobutenes can be used to alkylate benzene etc., and the resulting products when sulfonated, from an excellent series of surface active agents. The higher non-polymeric olefines obtainable by the cracking of paraffin wax also constitute an important potential source of alkyl groups for the higher alkyl aromatic sulfonates. It appears that these types, together with the chlorinated kerosene derived products are becoming the big-tonnage detergents which will compete with, supplement, and in some instances even replace soap in this country, just as the Mersols did in Germany. It is interesting that the Germans with their enormous potential for coal-derived aromatics should have emphasized the Mersols rather than the alkyl aromatics, whereas in the United States the alkyl aromatic type appears to be increasing its lead in tonnage. This is due in some measure at least to the patent situations and reflects the difference in the degree of state control over economic life in the two nations. In the United States the *du Pont Co.* has pioneered in the development of the Reed reaction products, and has a well-established patent position in this field. In the alkyl aromatic field, where the patent situation is by no means so clean-cut, numerous other large companies of high producing capacity feel freer to enter. Their combined production and marketing facilities can readily outstrip those of a single company no matter how large. In Germany, on the other hand, where the *I. G. Farbenindustrie* was so large as to be practically a branch of the government, completely dominating the organic chemical field, those products which they chose for development inevitably led in tonnage. Furthermore, patent restrictions were of secondary importance.

The Germans did, in fact, manufacture at least three products of the higher alkyl aromatic sulfonate series. One of these, called Igepal NA[33] although totally unrelated to the other members of the Igepal series, was made by starting with a hydrogenated C_{12-14} fraction of Fischer-Tropsch hydrocarbon called Mepasin. This was chlorinated, condensed with benzene using $AlCl_3$ as the catalyst, and sulfonated. A similar product was

[33] Hoyt, *PB 3868*, Offices of Technical Services, Dept. of Commerce, Washington, D. C.

made by condensing tetrapropene (the C_{12} hydrocarbon that is the tetramer of propene) with benzene and subsequently sulfonating. The condensation was carried out at 5–10°C. using anhydrous HF as a catalyst. The product, made at Hochst, was known simply as Ho 1/181. The cyclohexylamine salt of Igepal NA was used as an emulsifying agent under the name of Emulphor STL. Propene tetramer was also condensed with xylene and sulfonated. The potassium salt of this product was called Emulphor STX.

TABLE I

Alkyl Aromatic Sulfonates

Aromatic nucleus	Alkyl group	
A. Benzene	1. Methyl and ethyl	16. Mixed alkyl from polymers of C_3-C_7 monoölefins
B. Toluene	2. Propyl	
C. Xylenes, ethyl, benzene, mesitylene, cymene, etc.	3. Butyl	17. Mixed alkyl from naphthenes
	4. Amyl	
	5. Hexyl	18. Terpenoid, from terpene olefins or alcohols
D. Phenol	6. Heptyl	
E. Cresols, xylenols, and lower alkylated phenols	7. Straight-chain octyl	
	8. Octyl, 2-ethylhexyl	19. Oleic acid derivative condensates, condensed through the double bond of the oleic chain
	9. Octyl, diisobutyl	
F. Phenol ethers, anisole, phenetole, etc.	10. Nonyls	
	11. Decyls	
G. Diaryl ethers, diphenyl ether, etc.	12. Keryls	
	13. Straight-chain alkyl C_{9-18}	20. Acyl groups, i.e., alkyl aromatic ketones usually made by a Friedel-Crafts acylation reaction
H. Naphthalene		
I. Naphthols	14. Straight chain hydroxylated or unsaturated alkyl, oleyl, ricinoleyl. These may be attached to one or more aromatic nuclei.	
J. Naphthol ethers		
K. Diphenyl		
L. Phenyl phenols		21. Branched alkyl group derived from a ketone or aldehyde
M. Di- and triphenyl methanes		
N. Benzoins and desoxybenzoins	15. Mixed alkyl from cracked paraffin wax olefins	22. Olefins from misc. synthetic processes
O. Rosin and modified rosins		23. Steroid and complex alkyl-aromatic

The products discussed above include most of the commercially important members of the alkyl aromatic sulfonate series. A great number of others have been described, some of which have also been used to a greater or lesser extent. Table I lists various aromatic nuclei, coded in letters, and alkyl source groups, coded in numerals, which have been combined into surface active intermediates. For the sake of simplicity, these will be referred to in the discussion by the appropriate code combination. Thus A-12 refers to a kerylbenzene sulfonate of the Nacconol type; H-3 to a butylated naphthalene sulfonate of the Nekal B type, etc. In every case, unless otherwise specified, the final surface active product contains one sulfonic acid group attached to the aromatic nucleus. This may be combined with alkali metal or organic base cations.

In connection with products of the A-12 series, an interesting method for sulfonation of the alkyl benzene has been described, in which liquid sulfur dioxide is used as a solvent.[34] Another modification of the A-12 type of product consists of having two aromatic nuclei attached to one alkyl chain, only one of the aromatic nuclei being sulfonated.[35]

Products of the A-13 type have been described both in patents and in the scientific literature. The ultimate source of the type 13 (straight-chain C_{9-18}) alkyl group is practically limited to naturally occurring fats.[36]

The A-16 and H-16 classes include some of the most important potential large-scale detergents, since the polypropenes and butenes can be produced on a large scale as petroleum refinery by-products. The patent literature in this field is comparatively limited in view of its apparent potential. The di-, tri-, and tetraisobutenes have received more attention than the polypropenes, pentenes, or hexenes.[37] Di-, tri-, and tetraisobutene have also been used as the source of alkyl groups in the higher alkyl naphthalene sulfonates. Alkylation is effected with 98% H_2SO_4 as the condensing agent at temperatures below those required for sulfonation. The product is subsequently sulfonated.[38]

Aromatic petroleum extracts, separated by extraction with liquid sulfur dioxide, have been alkylated with polymeric isobutenes and sulfonated to give surface active agents.[39]

The sulfonation of various higher alkyl benzenes has been studied from the academic point of view by Nametkin and Robinzon.[40]

Detergents of the A-19, D-19, and H-19 types are of interest in that they use oleic acid or its esters or amides as the alkyl or hydrophobic portion of the molecule. The double bond condenses rather readily with aromatic nuclei forming, in the case of benzene, 9- or 10-phenylstearic acid:

$$CH_3(CH_2)_7-CH-(CH_2)_8COOH$$

[34] U. S. Pat. 2,244,512; Brit. Pat. 539,281, Brandt to *Colgate-Palmolive-Peet Co.*

[35] U. S. Pat. 2,180,314, Thomas to *Sharples Solvents Corp.*

[36] See, for example, Swiss Pat. 166,491, to *Ciba Co.* Gilman and Turck, *J. Am. Chem. Soc.* **61,** 478 (1939).

[37] See U. S. Pat. 2,232,117, Kyrides to *Monsanto Chemical Co.* Also French Pat. 766,903, to *I. G. Farbenindustrie.*

[38] U. S. Pat. 2,072,153, Bruson and Stein to *Rohm and Haas Co.*

[39] U. S. Pat. 2,214,051, Gaylor to *Standard Oil Development Co.*

[40] Nametkin and Robinzon, *J. Applied Chem. U. S. S. R.* **12,** 1378 (1939).

The benzene ring can be sulfonated and the resulting product is a typical lime-resistant wetting agent and detergent. Naphthalene, phenol, phenol ethers, as well as toluene, xylene, etc., may be used in place of benzene.[41]

The alkyl phenol sulfonates, particularly those in classes D-11, D-12, D-13, and D-16, have been the subject of considerable technical development and many patents. This is a natural consequence of the ease with which phenol is both alkylated and sulfonated. The products resemble the better-known and less costly alkyl benzene sulfonates. The capryl, diisobutyl, and decylphenol sulfonates exhibit foaming and mild wetting properties but are a little too low in molecular weight to function effectively in economical concentrations.[42] When straight-chain alkyl groups derived from the fatty alcohols of coconut oil, cetyl alcohol, etc. are used, excellent detergents are obtained. The keryl phenol sulfonates in which the keryl group has about the same molecular weight as that in the Nacconol type of detergent (i.e., about 13–14 carbon atoms) have also been described, and are fully as effective as their benzene analogues.[43] 1,12-Octadecanediol has been condensed with phenol and sulfonated to produce a detergent. The octadecanediol is made by hydrogenating castor oil.[44] An interesting detergent of the D-14 class is made by sulfonating cashew shell liquid. This product contains anacardol (sometimes called "cardonol"), which is a phenol substituted in the nucleus by a normal pentadecenyl chain.[45] Analogous detergents of the F-14 series have been described in which cardanol is first etherified with a lower alkyl group and subsequently sulfonated.[46]

The D-16 series is represented largely by the sulfonates of phenol alkylated with tri- or tetraisobutene, or tetra- or pentapropene. Most of the patents in this field refer to methods of alkylating phenol with these

[41] U. S. Pats. 2,081,075, to Vobach; 2,302,070, to Groggins et al. Ger. Pats. 583,686, Lindner and Zickerman to Oranienburger Chemische Fabrik A.-G.; 611,443, Lindner and Russe to Oranienburger Chemische Fabrik A.-G. French Pats. 809,342, to A. Ofner; 782,612, to A. Beyer. Zaganiaris, Ber. B71, 2002 (1938). See also Stirton and Peterson, Ind. Eng. Chem. 31, 856 (1939); Stirton and Groggnis, ibid., 32, 1136 (1940); cf. Ger. Pat. 546,914, Petroff, and U. S. Pat. 2,381,115, de Groote and Keiser to Petrolite Corp.

[42] See, for example, U. S. Pats. 2,008,017, Hester to Rohm and Haas Co.; 2,332,555, Buc to Standard Oil Development Co.; 2,189,805, Kyrides to Monsanto Chemical Co.; 2,205,949, Schwartz to Allied Chemical and Dye Corp.

[43] U. S. Pats. 2, 134,711–2; 2,178,571; Brit. Pat. 447,898; French Pat. 790,447; U. S. Pats. 2,133,287; 2,166,136; 2,196,985; Flett to Allied Chem. and Dye Corp. See also Brit. Pat. 530,415, Watanabe and Kawamura.

[44] U. S. Pat. 2,321,620, Pratt to E. I. du Pont de Nemours & Co.

[45] U. S. Pat. 2,324,300, Harvey to Harvel Corp. See also U. S. Pats. 2,098,824; 2,317,607, Harvey to Harvel Corp.

[46] U. S. Pat. 2,377,552, Harvey to Harvel Corp.

olefins. Among the unusual alkylating agents which have been described are stannic chloride, phosphotungstic and phosphomolybdic acids, silica activated with phosphoric acid, and acid-activated clays. Sulfuric acid can also be used, but considerable depolymerization takes place resulting in lower yields of the desired product unless special precautions are taken.[47]

Numerous examples have been described of phenols condensed with terpenes and subsequently sulfonated. Among the terpenes which have been used are α- and β-pinene and the triply unsaturated acyclic terpenes.[48]

In place of phenol itself lower alkylated phenols such as p-tert-butylphenol and p-tert-amylphenol have been used for condensation with higher alkyl groups. The latter may be derived from diisobutylene, fatty alcohols, etc.[49]

The lower alkyl phenyl ethers such as phenetole, anisole, and phenoxyethyl alcohol are about as readily alkylated as phenol itself. Accordingly, most of the alkyl phenol sulfonates which have been described as surface active agents have their etherified counterparts.[50] Among the more unusual derivatives of this nature are the alkyl phenyl ethers in which the short alkyl ether-forming group bears a free hydroxyl group. These compounds may form sulfuric esters on the hydroxyl as well as nuclear sulfonic acids. The resulting products are said to be good cotton-washing agents.[51] Terpenoid phenyl alkyl ether sulfonates have also been described.[52]

Alkylated naphthalene and naphthol sulfonates have been explored almost as thoroughly as their phenol analogues. The higher straight-chain fatty alcohols condense with naphthalene in the presence of H_2SO_4, BF_3, and other acidic condensing agents. The products are sulfonated, usually with oleum or chlorosulfonic acid. It is probable that the products are disulfonic acids, as most monosulfonic acids of higher alkyl naphthalenes are only slightly soluble. It has been claimed, however, that dilaurylnaphthalenemonosulfonic acid and dimyristylnaphthalenemonosulfonic acid are soluble and are useful surface active agents at ordinary tempera-

[47] U. S. Pat. 2,162,269; 2,192,689, Mikeska to Standard Oil Development Co. U. S. Pats. 2,332,555, Buc to Standard Oil Development Co.; 2,301,966, Michel and Müller vested in Alien Property Custodian; 2,256,610, Buc to Standard Oil Development Co. French Pat. 795,429, to I. G. Farbenindustrie.

[48] U. S. Pats. 2,340,901; 2,378,436; Rummelsberg to Hercules Powder Co. U. S. Pat. 2,186,132; Brit. Pat. 504,417; Zink to American Cyanamid and Chemical Co.

[49] U. S. Pats. 2,237,066, Weisberg and Slachman to Alrose Chemical Co.; 2,267,687, Kyrides to Monsanto Chemical Co.

[50] U. S. Pat. 2,337,924, Platz et al. to General Aniline and Film Corp.

[51] Brit. Pat. 492,905, to I. G. Farbenindustrie. U. S. Pat. 2,184,935, Bruson and MacMullen to Rohm and Haas Co.; Brit. Pat. 505,769; to Rohm and Haas Co.

[52] U. S. Pats. 2,202,686; 2,320,846; Borglin to Hercules Powder Co.

tures.[53] Oleyl alcohol has been condensed with naphthalene and the product sulfonated. In this case the oleyl alcohol may condense through the hydroxyl group or the double bond. Furthermore, sulfonation may take place either in the ring or at that sulfatable location on the side chain which has not already condensed with the aromatic ring.[54] Lower alkylated and terpene-alkylated naphthols and naphthol ethers have also been described. These are similar in properties to the lower alkyl naphthalene sulfonates.[55] The lower alkylated diphenyls, phenylphenols, and dihydroxydiphenyls also yield sulfonates which resemble the Nekals in their general properties.[56] In the same class may also be included the lower alkylated hydroxydiphenylmethane and hydroxytriphenylmethane sulfonates.[57]

An unusual series of higher alkyl aromatic sulfonates consists of the straight-chain alkyl desoxybenzoin sulfonates developed and patented by *J. R. Geigy Co.* They have the formula:

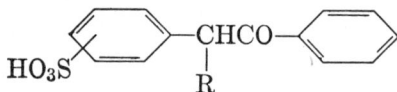

$$HO_3S \overset{}{\underset{}{\bigcirc}} - \underset{\underset{R}{|}}{CH}CO - \bigcirc$$

where R represents a lauryl, cetyl, or octadecyl, etc. group.[58] The closely related lower alkyl hydroxydiphenylmethane- and triphenylmethanesulfonic acids have also been described as wett ng agents.[59]

The higher acyl benzenes and naphthalenes have been made in at least two different ways and have been sulfonated to give surface active agents. The first way is to condense a fatty acid chloride such as stearic or lauric chloride with benzene in a Friedel-Crafts reaction. A typical product would be stearophenone:[60]

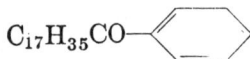

$$C_{17}H_{35}CO - \bigcirc$$

[53] Brit. Pat. 559,265, to *National Oil Products Co.* See also Brit. Pat. 539,281; Ger. Pat. 609,456; to *Oranienburger Chemische Fabrik A.-G.* Brit. Pat. 502,964, to R. Kimbara. U. S. Pat. 2,017,995, Stanley, Olpin, and Ellis to *Celanese Corp.*

[54] French Pat. 780,798, to E. A. Mauersberger. Swiss Pats. 166,491; 168,344; to *Ciba Co.*

[55] French Pat. 791,974, to *Imperial Chemical Industries Ltd.* U. S. Pats. 2,145,369, Osterhof to *Hercules Powder Co.;* 2,148,087, Rummelsberg to *Hercules Powder Co.*

[56] See U. S. Pat. 2,135,978, Magoun to *Monsanto Chemical Co.* Brit. Pat. 409,773; Ger. Pat. 605,445; Brit. Pat. 523,479; U. S. Pat. 1,994,927, Sibley to *Rubber Service Lab.* U. S. Pat. 2,285,563, Britton and Livak to *Dow Chemical Co.*

[57] Ger. Pat. 737,762; Brit. Pats. 477,196; 509,096; to *J. R. Geigy A.-G.*

[58] Swiss Pat. 198,705; 200,241–2, French Pat. 838,538; Brit. Pat. 506,610; to *J. R. Geigy A.-G.*

[59] Ger. Pat. 737,762, to *J. R. Geigy A.-G.*

[60] U. S. Pat. 2,089,154, Ralston to *Armour and Co.*

Another and more unusual way is the condensation of an aromatic carboxy acid chloride with a kerosene hydrocarbon using a Friedel-Crafts catalyst. A typical sulfonated product has the formula:[61]

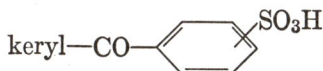

$$\text{keryl—CO—}\bigcirc\text{SO}_3\text{H}$$

Rosin and modified rosins have been used as aromatic nuclei in the formation of alkyl-aromatic-type detergents. They have been condensed with fatty alcohols, alkyl halides, and olefins, and the products sulfonated.[62]

III. Hydrophobic Group Joined to Sulfonated Aromatic Group through an Intermediate Linkage

Numerous surface active compounds have been described in which the hydrophilic group is joined by an intermediate linkage to a sulfonated aromatic nucleus. These compounds may be systematized on the same basis as has been used for the aliphatic sulfates and sulfonates, *i.e.*, on the nature of the intermediate linkage. Relatively few of these have attained great commercial importance.

(a) Ester Intermediate Linkage

The mono, di, and mixed esters of 3- or 4-sulfophthalic acid may be prepared by reacting the corresponding sulfophthalic anhydrides with appropriate alcohols. These products have the generic formula:

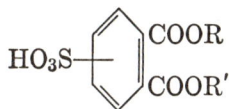

$$\text{HO}_3\text{S}\text{—}\bigcirc\begin{matrix}\text{COOR}\\\text{COOR}'\end{matrix}$$

A wide variety of alcohols may be used for the esterification.[63] Among these are the aliphatic alcohols of four or more carbon atoms, phenols such as o-chlorophenol, and various di- or polyhydric alcohols.[64] More complicated hydrophobic combinations bearing esterifiable hydroxyl groups have also been esterified with the sulfophthalic acids. These include ricinoleic acid and such products as coconut fatty hydroxyethylamide, hydroxypropyl dodecyl ether, and similar products containing multiple intermediate links.[65]

[61] U. S. Pat. 2,291,778, Wakeman to *Allied Chemical and Dye Corp.*

[62] U. S. Pat. 2,392,945, Price to *National Oil Products Co.;* 2,397,692, Price and May to *National Oil Products Co.*

[63] U. S. Pats. 1,935,264; 1,957,155 and 2,168,660; Albrecht *et al.* to *Ciba Co.* Also Brit. Pat. 370,845; 371,144; to *Ciba Co.*

[64] Ger. Pat. 705,356, to *Ciba Co.*

[65] Ger. Pat. 705,529. Swiss Pats. 167,797; 199,085-8; Brit. Pat. 504,031; Ger. Pat. 548,799; French Pats. 705,071; 767,736; to *Ciba Co.*

Beside sulfophthalic acid itself, the substituted sulfophthalic acids as well as sulfonated isophthalic, terephthalic, and naphthalic acids have been used. The disulfonated analogues have likewise been described. The member of this series which appears to be preferred for general use by investigators in the field is di-n-hexyl sulfophthalate.[66]

The alkyl esters of the sulfobenzoic acids and sulfosalicylic acids:[67]

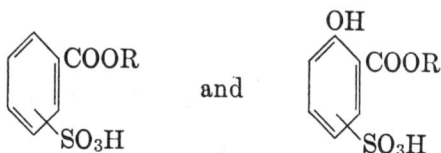

have also been described as surface active agents.

The esters of sulfonated phenoxyacetic acid, having the general formula:

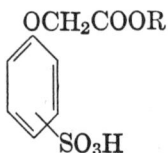

where R is a dodecyl, cetyl, or oleyl radical, have received considerable attention in the patent literature but have probably not been used commercially to any great extent.[68] These products in common with others which are based on the higher fatty alcohols are inherently more expensive than the simple fatty alcohol sulfates. It is, therefore, to be expected that they could not compete commercially unless they possessed some unusual and outstanding advantages in certain specialized applications.

Ester-linked products of analogous structure have been made by esterifying a fatty acid with an aromatic sulfonic acid containing hydroxyl groups. The hydroxyl groups may be phenolic, as in formula I,[69] or aliphatic as in formula II:[70]

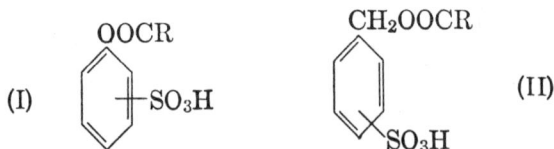

[66] U. S. Pat. 2,355,592, Kosmin to *Monsanto Chemical Co.* See also Brit. Pat. 418,334, to *Ciba Co.*

[67] U. S. Pat. 2,359,291, Gluesenkamp, and Kosmin to *Monsanto Chemical Co.*

[68] Brit. Pat. 408,749; Ger. Pats. 623,919; 616,847; U. S. Pat. 2,007,869; Lübbert to *Henkel et Cie.* Swiss Pats. 162,733; 165,401–2; to *J. R. Geigy A.-G.*

[69] French Pat. 782,280, to Lindner.

[70] Ger. Pat. 704,948; French Pat. 837,744; to *I. G. Farbenindustrie.*

As an example, cresol may be esterified by heating with coconut fatty acid chloride, and the resultant ester sulfonated with $ClSO_3H$ at 25°C.[71]

A product that is probably an ester-linked aromatic sulfonate has been made by condensing a fatty acid, glycerol, naphthalene, and sulfuric acid. It may be assumed that the glycerol condenses with the naphthalene and sulfuric acid forming a hydroxypropylated naphthalenesulfonic acid. A free hydroxy group of this product can then esterify the fatty acid.[72]

(b) Amide Intermediate Linkage

Compounds of the general formula:

$$\text{NHCOR}$$

$$\text{SO}_3\text{H}$$

where the aromatic ring or the amido nitrogen may be variously substituted are closely related to the Igepon T type of detergent. They may be made either by sulfonating a previously prepared aryl amide of a fatty acid, or by condensing a fatty acid chloride with an aryl aminosulfonic acid.[73] A wide variety of individual compounds in this group has been studied. It is reported[74] that oleylsulfanilic acid was in pilot plant production in Germany during the recent war as a possible substitute for Igepon T. The fatty acyl metanilic and N-monoethylmetanilic acids are also stated to be excellent general-purpose surface active agents. The sulfonated fatty acyl o-toluidides are stated to be more resistant to hydrolysis at the amide link due to the steric hindrance effect of the ortho substituent.[75]

Naphthenic acids have been reacted with primary aromatic amines to form arylides, which on sulfonation yield products said to be well suited for textile processing operations.[76]

Among the less usual examples of this general structure are the acylated

[71] Ger. Pat. 666,626, to *Oranienburger Chemische Fabrik. A.-G.*

[72] U. S. Pat. 2,195,512, Dreger and Ross to *Colgate-Palmolive-Peet Co.* Ger. Pat. 644,131, Brodersen and Quaedvlieg, to *I. G. Farbenindustrie.* U. S. Pat. 2,014,502, Marx *et al.* to *I. G. Farbenindustrie.*

[73] Ger. Pat. 673,730-1; French Pat. 720,529; to *I. G. Farbenindustrie.* Brit. Pats. 341,053; 343,524; to *I. G. Farbenindustrie.* See also Ger. Pat. 642,885; French Pat. 806,372; to *I. G. Farbenindustrie.*

[74] Hoyt, *PB 3868,* Office of Publication Board, Dept. of Commerce, Washington, D. C.

[75] U. S. Pat. 2,166,949, Flett to *Allied Chemical and Dye Corp.*

[76] Blum, *Bull. Soc. Ind. Mulhouse* **105**, 116 (1939); *Chem. Abstracts* **33**, 4698.

carbazolesulfonic acids[77] and the acylated aminodiphenyl ether sulfonic acids of formula:[78]

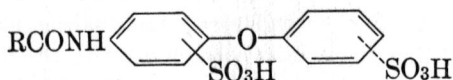

Venkataraman and co-workers[79] have made a study of this class and have found that the fatty acyl derivatives of anisidine, phenetidine, and toludine sulfo acids in which the sulfonic acid group is meta to the amino group are excellent detergents. They compare very favorably with the Igepons and Gardinols. The preferred products, said to be scheduled for commercial production in India, are:

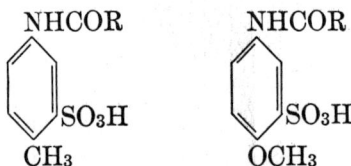

where R is the coconut fatty radical.[80] They are made by condensing coconut fatty chloride with the aminosulfo acid in 10% NaOH solution at 5–10°C.

Amide-linked aromatic sulfonic acids have also been made in which the hydrophobic group is derived from an amine and this is acylated with the carboxy group of an aromatic carboxysulfo acid. A typical example is the sulfonated phenoxyacetyl derivative of octylcyclohexylamine:[81]

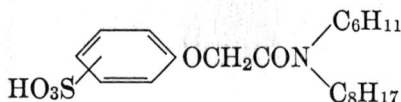

Another example of this type is the 4-sulfophthalamide of hexadecylamine.[82]

[77] French Pat. 827,185, to *Ciba Co.*

[78] Swiss Pats. 213,745; 214,900; to *J. R. Geigy A.-G.*

[79] Venkataramen, *J. Indian Chem. Soc., Ind. & News Ed.,* **7**, 24, (1944).

[80] Brit. Pat. 545,496, to Venkataraman *et al.* Ger. Pat. 667,794; French Pat. 797,631, to *Imperial Chemical Industries Ltd.*

[81] U. S. Pat. 2,253,179, Hentrich *et al.* to *Procter and Gamble Co.* Brit. Pat. 428,153, to *Henkel et Cie.*

[82] Swiss Pat. 185,942; Brit. Pat. 461,054; to *Ciba Co.*

4-Sulfophthalic acid has also been condensed with such amines as the *p*-aminoanilide of lauric acid. The final product has the formula:[83]

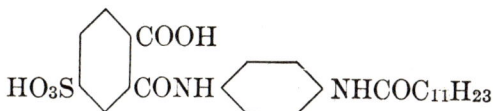

$$HO_3S \underset{CONH}{\overset{COOH}{\bigcirc}} - \bigcirc - NHCOC_{11}H_{23}$$

A more complex type of surface active agent in this class has the generic formula:

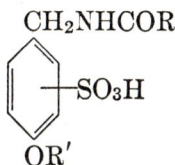

$$\overset{CH_2NHCOR}{\underset{OR'}{\bigcirc}} - SO_3H$$

where R' is hydrogen or lower alkyl. It is made by starting with the fatty acid amide, $RCONH_2$. This is condensed with formaldehyde, forming the hydroxymethylamide, $RCONHCH_2OH$. This is further condensed with phenol or anisole, and the product is finally sulfonated. Naphthalene may be used instead of phenol or phenol ethers, and the chloromethyl acid amides may be used instead of the hydroxymethyl derivatives.[84]

A similar type of compound is made by condensing the hydroxyethylamide of a fatty acid with a phenol, and subsequently or simultaneously sulfonating. The generic formula is:[85]

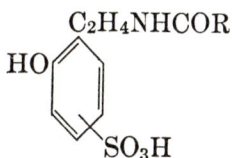

$$HO \underset{SO_3H}{\overset{C_2H_4NHCOR}{\bigcirc}}$$

Still another type in which the amide-forming amino group is not linked directly to the aromatic nucleus is represented by the compound:[86]

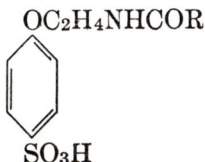

$$\overset{OC_2H_4NHCOR}{\underset{SO_3H}{\bigcirc}}$$

[83] French Pat. 808,270; Swiss Pats. 188,763–4; to *Ciba Co.*

[84] U. S. Pats. 2,237,296; 2,259,602; Brit. Pat. 508,477; to *Imperial Chemical Industries Ltd.* Ger. Pat. 581,955; Brit. Pat. 554,717; to *Ciba Co.*

[85] U. S. Pat. 2,277,805, Weisberg *et al.* to *Alrose Chemical Co.*

[86] Brit. Pat. 443,902, to *Imperial Chemical Industries Ltd.*

This may be made by sulfonating the fatty acyl phenoxyethylamide or by acylating the sulfo acid of phenoxyethylamine.

Fatty acyl hydrazides such as phenyl hydrazide of lauric acid have been sulfonated to form surface active agents. In this case the product has the formula:[87]

$$NHNHCOC_{11}H_{23}$$

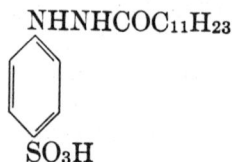

$$SO_3H$$

(c) Ether and Miscellaneous Intermediate Linkages

The simplest type of ether-linked product in this class would be a long-chain alkyl ether of a phenolsulfonic acid having the generic formula:

$$OR$$

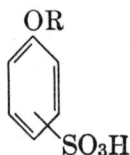

$$SO_3H$$

Although there is surprisingly little reference to this type of surface active agent in the patent literature, Hartley[88] has prepared numerous examples in pure form and has studied their properties. They were made by reacting phenol with the appropriate alkyl bromide in alcoholic potassium hydroxide solution and subsequently sulfonating the isolated purified alkyl phenyl ether. He also prepared and sulfonated the higher alkyl ethers of o-, m-, and p-cresols, as well as the mono, di, and mixed ethers of resorcinol.

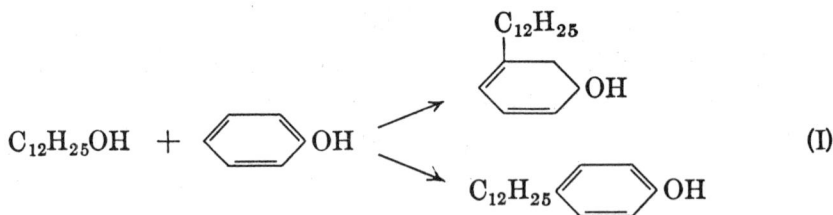

(I)

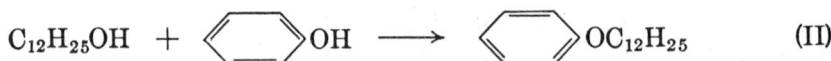

(II)

[87] Ger. Pat. 743,226, Stocker to J. R. Geigy A.-G.   Brit. Pat. 547,569, to J. R Geigy A-G.

[88] Hartley, J. Chem. Soc. **1939**, 1828.

It is probable that a certain amount of phenyl alkyl ether is always found as a by-product during the acid-catalyzed nuclear alkylation of phenols. Considering, *e.g.*, lauryl alcohol and phenol in the presence of zinc chloride, different courses of reaction are possible (I and II, p. 136).   Both reactions take place through a carbonium ion mechanism, the reacting molecular configuration being the cation $C_{12}H_{25}^+$.   This will react with phenol at the region of maximum electron density.   In acidic media phenol has a high statistical concentration of negative charge at the *o*- and *p*-carbon atoms, and nuclear alkylation is the predominant reaction.   In alkaline media, however, phenol ionizes to form the phenolate anion:

which has a high concentration of negative charge on the oxygen.   Hence, the predominant reaction is ether formation.   Even in acid media the electron density in the neighborhood of the oxygen is sufficient to result in the formation of some ether.   Conversely, there is evidence that some nuclear-alkylated material is formed during the normal ether-forming reaction in the presence of strong alkali.[89]

Surface active products of the type represented by the formula:

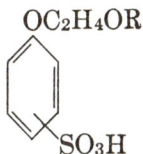

have been made by reacting higher fatty alcohols with, for example:

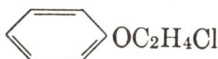

in the presence of alkali and sulfonating the resulting ether.[90]   In a similar manner by reacting, for example, *p-tert*-octylphenol with phenoxyethyl chloride products of the type:

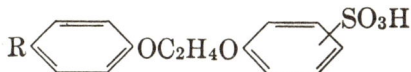

have been made.[91]

[89] See Cornforth, Cornforth, and Robinson, *J. Chem. Soc.* **1942,** 682.
[90] U. S. Pats. 2,178,830–1, Bruson to *Rohm and Haas Co.*
[91] U. S. Pat. 2,171,498; Bruson to *Rohm and Haas Co.*

Sulfonated benzyl ethers of the type:

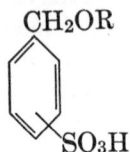

$$CH_2OR$$

$$SO_3H$$

have also been described as useful surface active agents.[92]   The intermediate unsulfonated ethers may be made by heating a higher fatty alcohol with benzyl chloride in the presence of alkaline condensing agents.   Similar products have been described in which the hydrophobic group is formed by an alkylated phenol.   This is etherified with benzyl chloride sulfonic acid.   A typical product has the formula:[93]

$$C_{18}H_{37} \langle \rangle OCH_2 \langle \rangle SO_3H$$

The use of a sulfide or thio ether group as an intermediate group is exemplified in the following synthesis:

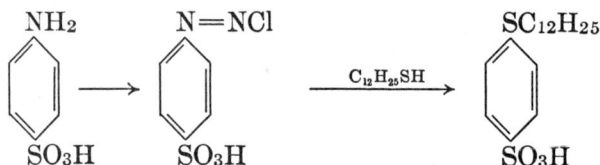

A variety of diazotized aromatic aminosulfonic acids and alkyl mercaptans may be used.[94]

Sulfone groups have also been described as intermediate linkages.   They may be made by oxidation of the above-mentioned thioethers.   More directly they can be made by reacting on aromatic sulfinate with a higher alkyl halide.[95]

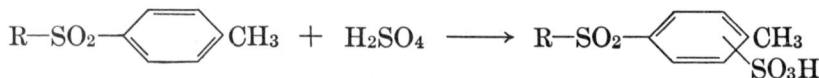

$$R-Br + NaSO_2 \langle \rangle CH_3 \longrightarrow R-SO_2 \langle \rangle CH_3 + NaBr$$

$$R-SO_2 \langle \rangle CH_3 + H_2SO_4 \longrightarrow R-SO_2 \langle \rangle \begin{array}{c} CH_3 \\ SO_3H \end{array}$$

[92] Brit. Pat. 378,454; U. S. Pat. 1,999,315; to *Imperial Chemical Industries Ltd.*
[93] Brit. Pat. 516,188, to *Deutsche Hydrierwerke A.-G.*
[94] Ger. Pat. 614,311, to *Henkel et Cie.*
[95] Brit. Pat. 415,877; Ger. Pat. 646,630; Fr. Pat. 758,078; Swiss Pat. 163,536; to *Ciba Co.*; U. S. Pat. 2,010,754, Felix and Albrecht to *Ciba Co.*

Use of the acetal group as an intermediate linkage is described in a process whereby benzaldehydesulfo acid is condensed with cetyl alcohol.[96]

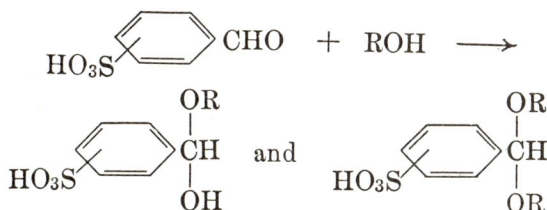

$$HO_3S \langle \rangle CHO \ + \ ROH \ \longrightarrow$$

$$HO_3S \langle \rangle \overset{OR}{\underset{OH}{CH}} \quad and \quad HO_3S \langle \rangle \overset{OR}{\underset{OR}{CH}}$$

The sulfonic ester linkage has been utilized by condensing Reed reaction sulfonyl halides with phenols and subsequently sulfonating.[97] The products presumably have the formula:

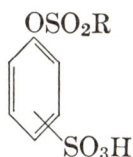

$$\overset{OSO_2R}{\underset{SO_3H}{\bigcirc}}$$

Beside the simple intermediate linkages such as have been considered, use has been made of more complicated structures in which the hydrophobic group is linked with a sulfonated aromatic ring through an intermediate condensed heterocyclic ring. Probably the most interesting products of this type are the alkyl benzimidazolesulfonic acids of formula:

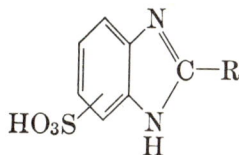

$$HO_3S \langle \rangle \overset{N}{\underset{\underset{H}{N}}{\bigg\rangle}} C{-}R$$

These are made by sulfonating the intermediate alkyl benzimidazole, which in turn may be prepared by condensing o-phenylenediamine with a fatty acid:

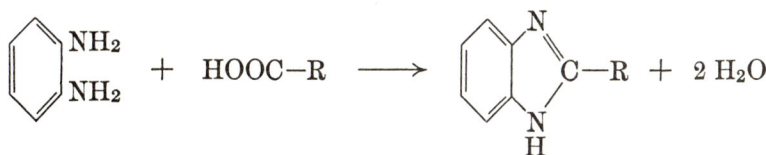

$$\overset{NH_2}{\underset{NH_2}{\bigcirc}} + \ HOOC{-}R \ \longrightarrow \ \overset{N}{\underset{\underset{H}{N}}{\bigcirc}} C{-}R \ + \ 2 \ H_2O$$

---

[96] Ger. Pat. 642,829, to *Böhme Fettchemie G.m.b.H.*
[97] Ger. Pat. 730,465, Herold *et al.* to *I. G. Farbenindustrie.*

Detergents of this nature have been developed and elaborated by the *Ciba Co.*, and are sold under the names Ultravon and Tergavon.[98] *o*-Phenylene diamine is generally considered a costly "fine chemical." The fact that the Ultravons can be made and marketed at a price competitive with the Igepons Gardinols, and other similar detergents is accordingly rather surprising. It is probable that the particular operations and conditions obtaining within this particular manufacturer's organization enable the production of *o*-phenylenediamine at an unusually low cost. The detergents themselves are of good color and odor and excellently adapted for general application in textile processing.

Indoles and dihydroindoles, substituted in the alpha position by a long-chain alkyl radical, have also been sulfonated to form surface active agents. The indoles have the formula:[99]

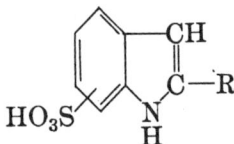

The unsulfonated intermediates may be made from the corresponding fatty acyl *o*-toluidides.

[98] Swiss Pats. 163,005; 164,730–6; French Pat. 754,626; Swiss Pats. 163,274; 196,533; 208,534; 210,980; 214,093; 213,253, to *Ciba Co.*; U. S. Pats. 2,036,525, Gränacher to *Ciba Co.*; 2,170,474; 2,297,760, Gränacher and Meyer to *Ciba Co.*
[99] U. S. Pat. 2,180,801; Ger. Pat. 675,616; Engel to *J. R. Geigy A.-G.*

CHAPTER 6

Miscellaneous Anionic Hydrophilic Groups

Although the carboxylic, sulfuric ester, and sulfonic acid groups are by far the most widely used solubilizing radicals for anionic detergents, several others have been described. None of these with the possible exception of the phosphoric esters have attained any commercial importance except for highly specialized applications.

Sulfinic ac d salts of generic formula:

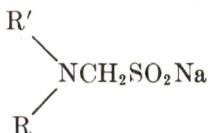

$$\begin{array}{c} R' \\ \diagdown \\ NCH_2SO_2Na \\ \diagup \\ R \end{array}$$

in which R is a hydrophobic radical and R' is either alkyl or hydrogen, are said[1] to have surface active properties. They are made by reacting the corresponding amine, R—NH—R', with sodium formaldehyde sulfoxylate, a substance widely used for reducing vat dyestuffs and in dye stripping.

Surface active sulfinic acids have been made[2] by reducing the sulfonyl chlorides which are primary products of the Reed reaction. Alkaline reducing agents such as sulfites or sulfides are used:

$$RSO_2Cl + H_2 \xrightarrow{\quad NaOH \quad} NaCl + RSO_2Na$$

Thiosulfate esters, made by reacting an alkyl halide with sodium thiosulfate, form the subject of numerous patents. As an example, dodecyl chloroacetate may be reacted with $Na_2S_2O_3$ to yield:

$$C_{12}H_{25}OOCCH_2—S_2O_3Na$$

This method of synthesis is analogous to the preparation of a sulfonic acid from sodium sulfite and an alkyl halide.

A wide variety of other intermediates containing a hydrophobic group and a replaceable halogen may be reacted[3] with $Na_2S_2O_3$ to give surface active thiosulfates. Among these are the chloroacetamides, $RNHCOCH_2Cl$, and the alkyl chloromethyl ethers, $ROCH_2Cl$. Thiosulfates of this type

---

[1] U. S. Pat. 2,146,280, Adams to *E. I. du Pont de Nemours & Co.*

[2] U. S. Pat. 2,315,514, Fox to *E. I. du Pont de Nemours & Co.*

[3] U. S. Pats. 2,004,873, to Kirstahler and Kaiser; 2,012,073, Schirm to *Unichem.*

have also been made[4] by treating the corresponding mercaptans with the
SO_3-pyridine complex. Thus lauryl mercaptan yields the pyridine salt
of lauryl thiosulfate:

$$C_{12}H_{25}SH + SO_3—(C_5H_5N) \rightarrow C_{12}H_{25}S_2O_3 \cdot C_5H_5N$$

Surface active products strictly analogous to the thiosulfates but of uncertain constitution have been made[5] by reacting halides with sodium hy-
drosulfites. For example:

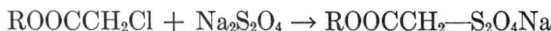

$$ROOCCH_2Cl + Na_2S_2O_4 \rightarrow ROOCCH_2—S_2O_4Na$$

Higher alkyl persulfates have been made[6] by treating a higher alcohol
or olefin with persulfuric acid, $H_2S_2O_8$. The resulting products are bleaching agents as well as surface active agents.

Surface active sulfamic acids have been made[7] by treating a higher secondary amine, such as dioctylamine, with chlorosulfonic acid.

$$C_8H_{17}—NH—C_8H_{17} + ClSO_3H \rightarrow C_8H_{17}—\overset{\displaystyle SO_3H}{\underset{\displaystyle |}{N}}—C_8H_{17}$$

Acylated diamines such as laurylethylenediamine have similarly been converted[8] to sulfamic acids:

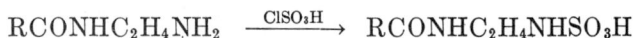

$$RCONHC_2H_4NH_2 \xrightarrow{ClSO_3H} RCONHC_2H_4NHSO_3H$$

Among other products with sulfur-containing solubilizing groups should
be mentioned the sulfonamides, disulfonimides, and acyl sulfamides. The
long-chain sulfonamides, RSO_2NH_2, are only soluble at very high pH
values corresponding to free caustic soda solutions, and have very limited
use. The acyl sulfonamides, however, are soluble in sodium carbonate
solutions. They have the general formula:

$$RSO_2—NH—CO—R'$$

where either R or R' represents a hydrophobic group and R' or R a smaller
alkyl or aryl group. Both radicals may be of about the same size if desired.
Typical examples are:

$$C_{12}H_{25}SO_2—NH—CO—C_6H_5$$

[4] Brit. Pat. 417,965; Fr. Pat. 758,756; to *Henkel et Cie*. See also French Pat. 770,539,
to *Henkel et Cie*.
[5] Swiss Pat. 163,000–3; 165,403–4; to *J. R. Geigy A.-G*.
[6] Ger. Pat. 589,778, Bertsch to *H. Th. Böhme A.-G*.
[7] U. S. Pat. 2,247,921, Orthner et al. to *I. G. Farbenindustrie*.
[8] Ger. Pat. 572,283, Günther and Holsten to *I. G. Farbenindustrie*.

made by treating dodecanesulfonamide with benzoyl chloride, and:

$$C_6H_5SO_2\text{---}NH\text{---}CO\text{---}C_{11}H_{23}$$

made by treating benzenesulfonamide with lauroyl chloride. The disulfon-imides:

$$R\text{---}SO_2\text{---}NH\text{---}SO_2\text{---}R'$$

are made[9] in analogous manner. The dibutyl and dihexyl compounds are said to be exceptionally good foaming agents.

Numerous combinations of linked and unlinked hydrophobic groups have been used in structures of this type. Alkane sulfonyl chlorides from the Reed reaction as well as alkyl aromatic sulfonyl chlorides are readily available. These can be reacted with acyl amides or they can be converted to sulfonamides and subsequently be reacted with acyl halides. Among the carboxylic halides the aryl oxyacetyl and higher fatty acyl are most frequently mentioned.[10]

The higher sulfonyl ureas which are closely analogous to the acyl sulfon-amides have also been described[11] as surface active agents. They have the formula:

$$RSO_2\text{---}NHCONH_2$$

Surface active dithiocarbamates, suitable as mercerizing wetting agents, have been made[12] by reacting a secondary aliphatic amine of suitable molecular weight with carbon bisulfide and caustic soda:

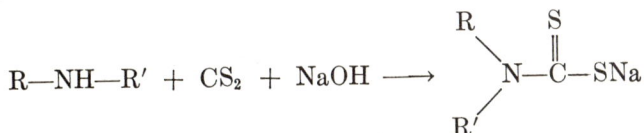

$$R\text{---}NH\text{---}R' + CS_2 + NaOH \longrightarrow \begin{array}{c} R \\ \diagdown \\ N\text{---}\overset{\overset{\textstyle S}{\|}}{C}\text{---}SNa \\ \diagup \\ R' \end{array}$$

Numerous anionic surface active agents have been described in which the solubilizing group contains phosphorus. These may be grouped into two general classes, the partial esters of various phosphoric acids, especially orthophosphoric acid, and the true phosphonic acids. In the latter compounds, the phosphorus and carbon atoms are directly joined to each other.

[9] U. S. Pats. 2,383,859; 2,394,307; Hentrich and Schirm Vested in Alien Property Custodian.

[10] U. S. Pats. 2,292,997, Hentrich and Schirm to *Unichem*. U. S. Pat. 2,341,614; 2,374,934; Hentrich *et al.* vested in Alien Property Custodian. See also Helferich and Flechsig, *Ber.* **B75**, 532 (1942).

[11] U. S. Pat. 2,390,253, Henke to *E. I. du Pont de Nemours & Co.*

[12] French Pat. 756,158, to *I. G. Farbenindustrie.*

The lecithins are salts of naturally occurring phosphoric esters with choline. They may be typified by the formula:

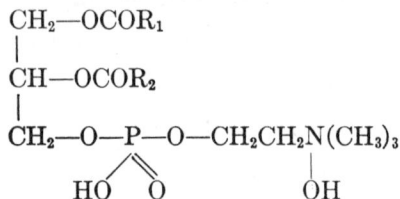

$$CH_2—OCOR_1$$
$$|$$
$$CH—OCOR_2$$
$$|$$
$$CH_2—O—P—O—CH_2CH_2N(CH_3)_3$$
$$HO \quad O \qquad\qquad OH$$

where $—COR_1$ and $—COR_2$ are radicals of fatty acids, such as stearic, palmitic, oleic, linoleic, etc. The lecithins are widely used as emulsifying acids.

The simple synthetic higher alkyl acid phosphates have been known about as long as the corresponding sulfates. They are made by treating fatty alcohols with phosphorus pentoxide or a lower hydrate of the same. They may also be made[13] by treating the alcohol with one equivalent of phosphorus oxychloride and subsequently removing the remaining chlorine atoms by hydrolysis. Just as in the case of the sulfuric esters a wide variety of hydroxylated or olefinic intermediates which contain hydrophobic groups may be "phosphated" to give surface active agents.[14] Oleic and ricinoleic acids and their esters, for example, have been phosphated at the double bond or the hydroxyl group to give analogues of the sulfated fatty oil series.[15] Among other intermediates which have been phosphated are: the polyoxy alkyl ethers[16] of fatty alcohols, $R—(OC_2H_4)_xOH$, or alkyl phenols:

$$R—\langle\bigcirc\rangle(OC_2H_4)_xOH$$

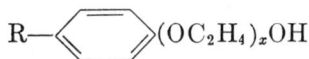

the fatty esters of polyhydric alcohols, $RCOO—X—OH$, and the higher alkyl esters[17] of hydroxycarboxylic acids, $ROOC—X—OH$.

Whether or not the above products are all partial esters of orthophosphoric acid is highly questionable. Phosphoric acid is noted for forming numerous polymeric species and mixtures of these are probably formed during esterification. Both the alkyl pyrophosphates and also the pyrophosphates of hydrophobes with intermediate linkages to the hydroxyl group

---

[13] U. S. Pat. 2,005,619, Graves to *E. I. du Pont de Nemours & Co.* Ger. Pat. 619,019 to *Böhme Fettchemie G.m.b.H.*

[14] Adler and Woodstock, *Chem. Ind.* **51**, 516–21, 557 (1942).

[15] U. S. Pat. 1,900,973, Bertsch to *H. Th. Böhme A.-G.*

[16] Ger. Pat. 696,317, Ulrich and Sauerwein to *I. G. Farbenindustrie.*

[17] U. S. Pats. 2,027,785; 2,052,029; 2,177,983; to **Benjamin R. Harris.**

have been specifically described.[18] They may be made by treating an alkyl sulfate with sodium pyrophosphate. The pyrophosphates have also been treated with hydrogen peroxide to obtain perphosphate derivatives[19] with oxidizing as well as surface active properties.

Partial esters of phosphorous acid with alcohols of eight or more carbon atoms have been made[20] by treating the alcohols with phosphorus trichloride and rapidly saponifying.

The acid salts of higher fatty alkyl phosphoric esters have been prepared and studied by Wagner-Jauregg and Wildermuth.[21] These investigators prefer to formulate them as hydrogen-bonded acid salts of orthophosphoric esters rather than as pyrophosphates.

Blown or oxidized fatty oils have been phosphated[22] with phosphoric acid to form surface active agents. Presumably the acid breaks one of the oxide rings formed during the oxidation.

One of the few commercial phosphate wetting agents is a complex product of formula $R_5(P_3O_{10})_2Na_5$, where R is either 2-ethylhexyl or octyl-2. This material is produced as Victawet 35B and 58B by the *Victor Chemical Works*. It is characterized by high wetting power combined with very low tendency toward foaming. In certain textile-treating operations this lack of foam is an advantage, although in most instances consumers associate foam with surface activity and tend to judge the efficacy of a product by its foaming power.

The phosphates, in contrast to the sulfates and sulfonic acids, have very poor hard-water resistance, *i.e.*, their calcium and magnesium salts are quite insoluble. This is a very serious disadvantage so far as general applicability is concerned. One of the principal advantages of most successful synthetic surface active agents over soap is their lime resistance. The phosphates, therefore, are commercially advantageous only in specialized applications where neither soap nor other synthetics perform well and where lime resistance is unimportant. They have been used, for example, in commercial bakeries to impart a desirable fluffy character to certain types of cake.

The phosphonic acids have not been nearly so thoroughly investigated

[18] French Pat. 772,787, to *H. Th. Böhme*. See also U. S. Pat. 2,345,388, Ericks and Williams to *American Cyanamid Co.*; Brit. Pat. 419,868, to *H. Th. Böhme A.-G.*; U. S. Pats. 2,053,653, Butz to *American Hyalsol Corp.*; 2,177,650, to Benjamin R. Harris.

[19] Brit. Pat. 425,804, to *H. Th. Böhme A.-G.*

[20] Ger. Pat. 646,480, to *I. G. Farbenindustrie*.

[21] Wagner-Jauregg and Wildermuth, *Ber.* **B77**, 481 (1944); *Chem. Abstracts* **40**, 4658[3].

[22] U. S. Pat. 2,386,250, McNally and Dickey to *Eastman Kodak Co.*

as the phosphoric esters.   The monoalkyl phosphonic acids have the formula:

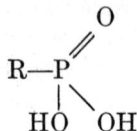

$$R-P{\overset{\displaystyle O}{\underset{HO\quad OH}{\big\|}}}$$

while the dialkyl phosphonic acids have two carbon atoms attached to phosphorus and have the formula:

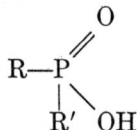

$$R-P{\overset{\displaystyle O}{\underset{R'\quad OH}{\big\|}}}$$

Closely related to these are the phosphine oxides (which do not form salts) and have the formula:

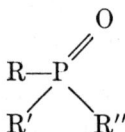

$$R-P{\overset{\displaystyle O}{\underset{R'\quad R''}{\big\|}}}$$

Long-chain monoalkyl phosphonic acids whose salts are surface active have been made by reacting phosphorus trichloride with a kerosene or similar aliphatic hydrocarbon in the presence of aluminum chloride to form a dichloromonoalkylphosphine. This is then hydrolyzed to the phosphinic acid and finally oxidized to the phosphonic acid.[23]

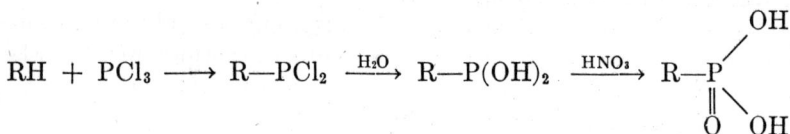

$$RH + PCl_3 \longrightarrow R-PCl_2 \xrightarrow{H_2O} R-P(OH)_2 \xrightarrow{HNO_3} R-P{\overset{\displaystyle OH}{\underset{O\quad OH}{\Big\|}}}$$

Similarly alkyl aryl phosphinic acids have been made[24] from phosphorus trichloride and alkyl aromatic hydrocarbons such as kerylbenzene. The alkylolamine salts of these phosphinic acids are said to be good dispersing agents.

Hydroxylated phosphonic acids may also be made by the reaction of phosphorus trichloride on an aldehyde or ketone, followed by hydrolysis. The mechanism of the reaction possibly involves the addition of the phos-

[23] U. S. Pat. 2,137,792, Woodstock to *Victor Chemical Works*.
[24] U. S. Pat. 2,347,633, Kosolapoff to *Monsanto Chemical Co*.

phorus atom to the carbon atom of the carbonyl group and one chlorine to the oxygen atom. On hydrolysis this chlorine has the same oxidizing capacity as an equivalent amount of hypochlorite, and oxidizes the phosphinic acid to a phosphonic acid. Thus formaldehyde and phosphorus trichloride give hydroxymethanephosphonic acid, possibly by the following reaction chain:

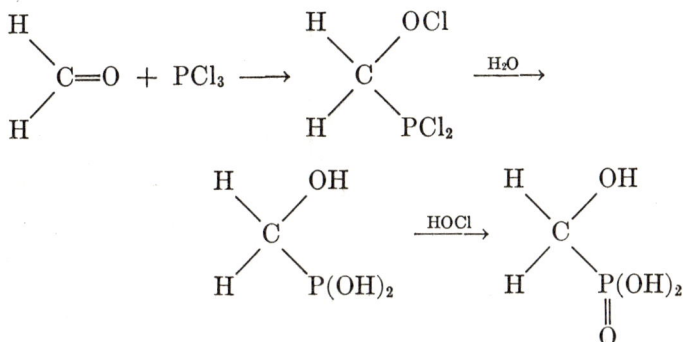

$$
\begin{array}{c}
H \\
\diagdown \\
\quad C{=}O \ + \ PCl_3 \ \longrightarrow \\
\diagup \\
H
\end{array}
\qquad
\begin{array}{cc}
H & OCl \\
\diagdown \diagup \\
C \\
\diagup \diagdown \\
H & PCl_2
\end{array}
\xrightarrow{\ H_2O\ }
$$

$$
\begin{array}{cc}
H & OH \\
\diagdown \diagup \\
C \\
\diagup \diagdown \\
H & P(OH)_2
\end{array}
\xrightarrow{\ HOCl\ }
\begin{array}{cc}
H & OH \\
\diagdown \diagup \\
C \\
\diagup \diagdown \\
H & P(OH)_2 \\
& \| \\
& O
\end{array}
$$

Hydroxymethanephosphonic acid has been used to condense with fatty acid chlorides to form surface active agents:[25]

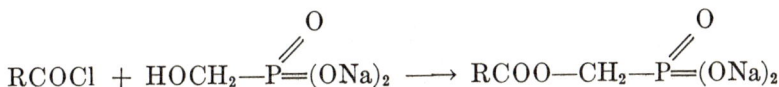

$$
RCOCl \ + \ HOCH_2{-}P{\overset{O}{=}}(ONa)_2 \ \longrightarrow \ RCOO{-}CH_2{-}P{\overset{O}{=}}(ONa)_2
$$

Pentadecane-8-one and similar higher ketones have been reacted with phosphorus trichloride and subsequently hydrolyzed to give hydroxylated phosphonic acids which are surface active.[26]

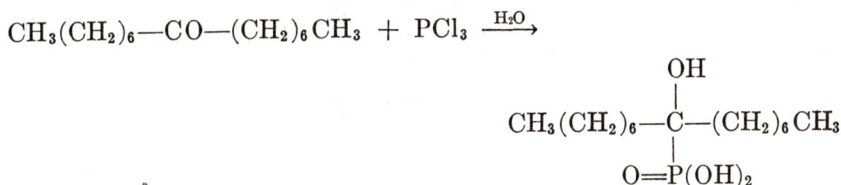

$$
CH_3(CH_2)_6{-}CO{-}(CH_2)_6\,CH_3 \ + \ PCl_3 \ \xrightarrow{\ H_2O\ }
$$

$$
\begin{array}{c}
OH \\
| \\
CH_3(CH_2)_6{-}C{-}(CH_2)_6\,CH_3 \\
| \\
O{=}P(OH)_2
\end{array}
$$

When phosphorus trichloride is reacted in a similar manner with α-unsaturated aldehydes or ketones, the product is a β-ketophosphonic acid. Presumably in this case 1,4 addition of phosphorus trichloride takes place. Thus, starting with 5-ethylheptene-3-one-2, a surface active keto-

---

[25] French Pat. 767,793, to *I. G. Farbenindustrie;* Ger. Pat. 646,290; to *Böhme Fett-chemie G.m.b.H.*

[26] U. S. Pat. 2,254,124, Stevens and Turner to *E. I. du Pont de Nemours & Co.*

phosphonic acid is obtained,[27] possibly by the following mechanism:

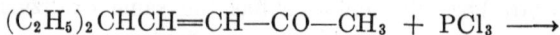

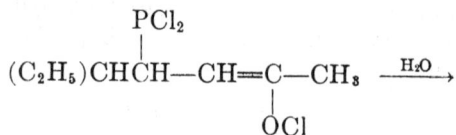

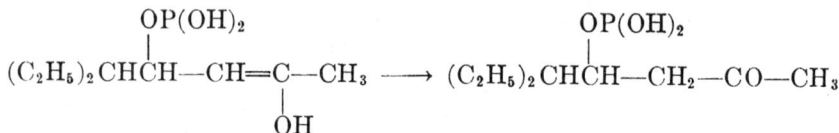

$$(C_2H_5)_2CHCH=CH-CO-CH_3 + PCl_3 \longrightarrow$$

$$\overset{\displaystyle PCl_2}{\underset{\displaystyle OCl}{(C_2H_5)CHCH-CH=C-CH_3}} \xrightarrow{H_2O}$$

$$\overset{\displaystyle OP(OH)_2}{(C_2H_5)_2CHCH-CH=C-CH_3} \longrightarrow (C_2H_5)_2CHCH-CH_2-CO-CH_3$$
$$\underset{\displaystyle OH}{}$$

Naphthenyl ketones have also been used[28] to produce surface active hydroxymethanephosphonic acids.

A surface active agent has recently been made[29] by sulfating the free hydroxyl groups on a dialkyl phosphonic acid of formula:

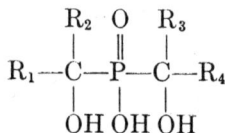

$$\overset{\displaystyle R_2 \quad O \quad R_3}{R_1-\underset{\displaystyle OH}{C}-\underset{\displaystyle OH}{P}-\underset{\displaystyle OH}{C}-R_4}$$

where the R may be alkyl or aryl.

To the writers' knowledge none of the phosphonic acid series has been marketed commercially as a surface active agent.

[27] U. S. Pat. 2,268,157, Marvel to *E. I. du Pont de Nemours & Co.*

[28] U. S. Pat. 2,376,130, Dickey and McNally to *Eastman Kodak Co.; cf.* U. S. Pat. 2,364,348, Dickey to *Eastman Kodak Co.*

[29] U. S. Pat. 2,369,443, Dickey to *Eastman Kodak Co.* See also U. S. Pat. 2,336,230, Dickey and Loria to *Eastman Kodak Co.*

PART I

PROCESSES FOR SYNTHESIZING AND MANUFACTURING SURFACE ACTIVE AGENTS

2. Cationic, Non-ionic, and Miscellaneous Surface Active Agents

CHAPTER 7

Cationic Surface Active Agents

I. Introduction

The cation-active or cationic surface active agents are characterized by the fact that the hydrophobic group forms part of a cation when the compound is dissolved in water. A typical example of this class is octadecyl ammonium chloride which dissociates in water according to the equation:

$$C_{18}H_{37}NH_3Cl \rightarrow [C_{18}H_{37}NH_3]^+ + Cl^-$$

The class may be regarded as consisting broadly of those bases which contain a typical hydrophobic group, and may be sub-classified according to the essential nature of the functional basic group. The amines and quaternary ammonium salts constitute by far the largest groups of cationic surface active agents. Since the amines are readily converted to quarternary ammonium compounds by very simple syntheses the two groups will be discussed together. They differ from each other in one very important physical property, namely, solubility. The primary, secondary, and tertiary amines containing a hydrophobic group sufficiently large to be in the surface active range are usually insoluble in water or aqueous alkaline solutions. They are soluble in acidic solutions where the pH is low enough to convert the bases to their conjugate soluble cations. The quaternary ammonium compounds by contrast are soluble in basic as well as in acidic aqueous media. This is in accordance with the strong basicity of the quaternary ammonium hydroxides. Aside from the amines and quaternary ammonium compounds there is a group of nitrogenous bases including guanidines, hydrazines, amine oxides, basic nitrogen heterocyclic substances, etc. around which surface active agents have been synthesized. Finally, there is the group of non-nitrogenous bases, of which the most noteworthy are the sulfonium compounds.

In general the cationic and anionic surface active agents will mutually precipitate when brought together in aqueous solution. This is due to the formation of the high-molecular-weight, poorly ionizable salt of the hydrophobic anion with the hydrophobic cation. In some cases, however, where both anion and cation are at the lower end of the hydrophobic range, the salts are soluble and have been used as surface active agents. In these cases it cannot be truly said that the product is anion-active or cation-active since both ions are active. Typical examples of such products include: (1) triamylbutylammonium cymenesulfonate, made by heating triamylamine with the butyl ester of cymenesulfonic acid:[1]

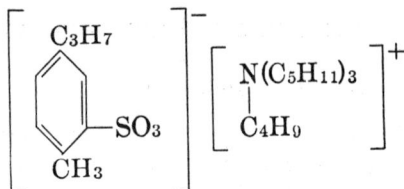

$$\left[\begin{array}{c} C_3H_7 \\ \\ \\ CH_3 \end{array} SO_3 \right]^- \left[\begin{array}{c} N(C_5H_{11})_3 \\ | \\ C_4H_9 \end{array} \right]^+$$

(2) the dimethylphenylbenzylammonium salt of dibutylnaphthalene sulfonic acid:[2]

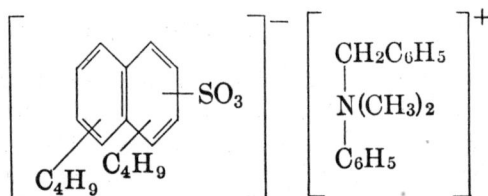

$$\left[\begin{array}{c} SO_3 \\ C_4H_9 \quad C_4H_9 \end{array} \right]^- \left[\begin{array}{c} CH_2C_6H_5 \\ | \\ N(CH_3)_2 \\ | \\ C_6H_5 \end{array} \right]^+$$

(3) the trimethylheptylammonium salt of sulfated butyl oleate:[3]

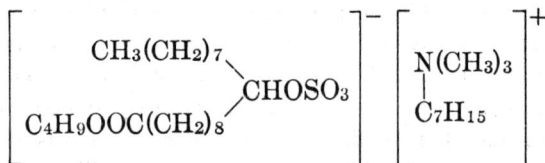

$$\left[\begin{array}{c} CH_3(CH_2)_7 \\ \qquad\qquad CHOSO_3 \\ C_4H_9OOC(CH_2)_8 \end{array} \right]^- \left[\begin{array}{c} N(CH_3)_3 \\ | \\ C_7H_{15} \end{array} \right]^+$$

(4) it has been claimed that the mixtures of rosin acids such as abietic acid with such high-molecular-weight cations as dodecylpyridinium are sufficiently soluble to serve as detergents and emulsifying agents.[4] If the actual salts are formed in this case their solubility is an anomaly.

Many of the simpler cationic compounds which exhibit surface activity have been known for a long time. The higher fatty amines, their salts, and quaternary compounds were prepared and studied before the present century. The fact that they gave soapy, foaming solutions with wetting, detergent, and emulsifying properties was also recognized. The name "in-

[1] Brit. Pat. 502,517, to *Ciba Co.*
[2] Brit. Pat. 316,090, to *I. G. Farbenindustrie.*
[3] Brit. Pat. 499,203, to *I. G. Farbenindustrie.*
[4] Brit. Pat. 501,020, to *Courtaulds, Ltd.*

vert soaps" was applied to them in recognition of the fact that they differed
from ordinary soap in the sign of the charge carried by the long-chain ion.
The intensive development of the cationic detergents during the past
two decades has paralleled that of the anionic group. There are, however,
salient differences in the commercialization and technological use of the
two series. The anionic series has been applied mainly for wetting, detergency, emulsification, and the other functions which are usually associated
with the term "surface activity." The cationics have also been used
for these purposes, and it is these applications which will be emphasized
in this book. Their development has been conditioned, however, by three
or four major uses in which their gross surface active properties are only
remotely important. The first of these uses is as germicides, fungicides
and disinfectants. Following the discovery, a few years ago, that quaternary ammonium compounds containing a primary normal aliphatic chain
in the range of twelve to eighteen carbon atoms were potent bactericides,
a tremendous number of new quaternary ammonium compounds containing various hydrophobic structures was synthesized and tried out for this
purpose. The mechanism by which these reagents kill microorganisms
has not been clearly demonstrated, but it bears no obvious relation to their
effectiveness as emulsifiers, wetting agents, or foaming agents. Nor does it
appear to be connected with their ability to lower surface tension. The
presence of a typical hydrophobic group is necessary for the compounds to
be germicides, and of course this will invariably make the compound a
typical surface active agent.[5] The long-chain amine salts and other long-
chain nitrogenous cationic compounds, as well as the quaternary ammonium salts, have germicidal powers. They appear to be adsorbed by the
bacteria quite strongly and interfere with the respiratory functions. It
may be argued that this is a "surface effect" since it involves preferential
adsorption at the interface between the organism and the surrounding
aqueous medium, but other factors which may be equally important or
more important doubtless enter into the killing effect.

Another extremely important technological process which involves the
use of substances which happen to be cationic surface active agents is the
treatment of fabrics to make them water-repellent. Fabrics have been
rendered water-repellent by a number of different processes, most of which
produce a finish that is not resistant to laundering. The "Velan" and "Zelan" processes, originated and developed within recent years, are representative of a type of process which produces a lasting, wash-resistant,

[5] See, for example, U. S. Re 20,869, Bousquet, Graves, and Salzberg to *E. I. du Pont
de Nemours & Co.* U. S. Pats. 2,295,504–5, Shelton to *Wm. S. Merrell Co.* Green
and Birkeland, *Proc. Soc. Exptl. Biol. Med.* **51**, 55 (1942).

water-repellent finish. They make use of a special type of cationic long-chain compound in which a quaternary nitrogen is linked through a methylene group to an amide or ester group. The amide or ester group is in turn linked to a long straight-chain aliphatic radical, usually the octadecyl radical. Zelan (*E. I. du Pont de Nemours & Co.*) has the structure:

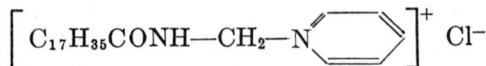

$$\left[C_{17}H_{35}CONH-CH_2-N \diagdown \diagdown \diagup \right]^+ \quad Cl^-$$

This product is soluble in water, giving the foaming soap-like solution typical of surface active agents. It is applied by padding onto the fabric, drying, and finally heating at an elevated temperature. The Zelan breaks down during this treatment, pyridine is evolved, and the fabric acquires a water-repellent finish. There is some question as to the exact chemical reactions involved. Some investigators maintain that the hydroxyl groups of the cellulosic fabric are etherified, forming a surface structure represented by the formula:

$$C_{17}H_{35}CONH-CH_2-O-cellulose$$

Others maintain that methylenedistearamide:

$$(C_{17}H_{35}CONH-)_2CH_2$$

is produced and that this forms a thin, firmly adherent coating on the fibers. Being resinous and waxy in nature it serves as a macroscopic water-repellent coating. The preponderence of evidence seems to be in favor of this latter view. In either event the original cation active compound used does not serve as a surface active agent, in the usual sense. It serves as a raw material which is broken down by a chemical reaction to form the desired coating on the fibers. The hydrophobic group present in the original molecule is separated from its solubilizing group and forms a truly hydrophobic substance on the fiber surfaces. The cationic nature of the original molecule is significant in at least two respects. The special cationics used are capable of undergoing the desired breakdown reaction on heating. Secondly, the original molecule is strongly adsorbed by the cellulose and accordingly the water-repellent reaction products are formed in intimate contact with the cellulosic surface.

A third important use to which cationic surface active agents are put, and in which the surface activity is of no particular significance, is in the finishing of rayon and cotton piece goods. Many of the cationic surface active agents, particularly those which contain stearyl groups, have the property of imparting a soft rich hand when applied to these fabrics. They are effective in much lower concentrations than the sulfonated oils and tallows which are also used for this purpose. Furthermore, they resist washing to a significant extent and are thus semi-permanent. The softening effect is due essentially to a lubricating action between the individual fibers of the fabric. The fibers adsorb a thin layer of the cationic compound, which reduces the friction between their surfaces. The surface active agent functions as a lubricant in much the same way that any other lubricant would. The fact that it is surface active in aqueous solution is immaterial in this process. The fact that it is cationic renders it firmly and rapidly adsorbed.

Still another use of these products which does not involve their surface active properties is in the fixation of direct dyestuffs. The direct dyes are for the most part soluble salts of aromatic sulfonic acids in which the colored anion is of rather high molecular weight. They are adsorbed by the fabrics during the dyeing process, the adsorption equilibrium favoring attachment of the dye molecule to the fabric. The dyed fabrics, as would be expected, lose some of their adsorbed dye on immersion in water. This effect is called bleeding. Because they bleed more or less easily, direct dyes are not used for good washable fabrics. They are less expensive than the wash-resistant azoic dyes or vat dyes, however, and much effort has been spent in developing new direct-dye structures and after-treating processes to improve their resistance to bleeding. One such process is the treatment of the dyed fabric with an aqueous solution of a typical long-chain cationic surface active agent. This treatment results in the formation on the fabric of an insoluble salt of the long-chain cation with the large dyestuff anion. The bleeding effect is thereby either greatly reduced or eliminated. Although this treatment will not stop bleeding in soap solutions, the fact that it stops bleeding in water makes it quite valuable. It is doubtful if this particular use of cationic surface active agents is of great economic importance, but it is the subject of a large number of patents, and many new cationic agents have been developed with their dye-fixing properties as the primary objective.

The general methods of synthesis used for producing the cationic surface active agents are quite similar to those used for the anionics. In particular, the same raw material sources of hydrophobic groups are used. The same types of intermediate linkages, i.e., the amide, ester, and ether linkages, are also used to unite the hydrophobic radical to the cation-forming radical. One salient difference between the two series is in the ionogenic radicals themselves. In the anionic series there are three ion-forming groups of major importance, the carboxy, sulfonic acid, and sulfuric ester. In the cationic series the basic nitrogen atom is the only ionogen of any practical importance. This basic nitrogen atom, however, is attached by four covalent linkages to other organic radicals. In most cases only one of these is a hydrophobic group, the others being of lower molecular weight. The nature of these shorter radicals may have a pronounced effect on the properties of molecule as a whole. For example we may consider n-octadecylamine ($C_{18}H_{37}NH_2$) as a primary structure for a cationic detergent. By substituting the two hydrogens on the nitrogen atom, and finally by quaternizing with still another radical, it is possible to produce an enormous number of different surface active agents. Some of these will exhibit markedly different properties from others. It is a fact that many of the patented cationic surface active agents differ among themselves solely in the nature of the non-hydrophobic substituents on the nitrogen atom.

An important parallel between the anionic and cationic series is in the effect of the non-surface-active ion on the properties of the ion pair. The non-surface-active ions commonly used in the anionic series are sodium, potassium, and the simple basic nitrogen radicals. These contribute very significantly to the properties of the molecule as a whole, the different salts exhibiting different properties. Similarly different salts of a typical surface active cation will show different solubilities, surface properties, etc. The halides, sulfate, aryl sulfonate, and lower alkyl sulfuric ester are anions most usually encountered in combination with quaternary ammonium cations. Non-quaternary basic nitrogen compounds are usually produced in the form of free bases and may be solubilized by combining with any one of a wide variety of appropriate acids. Organic as well as inorganic acids may be used. The salts generally differ quite remarkably among themselves with respect to solubility and surface active properties.

In some instances both ions of the product contribute equally to the action. An example is in the use of long-chain quaternary fluosilicates as mothproofing agents. The cation helps the spreading properties of the product and the anion is the active pesticide.[6] These facts are seldom emphasized to the degree their importance warrants, since attention is usually focused on the structure of the active cation. It is well to bear them in mind, however, whenever actual development or application work in this field is undertaken.

II. Basic Nitrogen Joined Directly to the Hydrophobic Group

Probably the largest and most important group of cationic detergents is built around the straight-chain fatty amines of eight to eighteen carbon atoms. These are analogous to the naturally occurring fatty acids, from which they are in fact synthesized directly or indirectly. In order to convert a fatty acid or its derivative to a fatty amine two essential steps

[6] U. S. Pat. 2,150,601, Flint to *E. I. du Pont de Nemours & Co.*

must be taken.   Nitrogen must be introduced into the molecule at the functional group, and a reduction of the carboxyl group must be effected.   The carboxyl group or a derivative in the same oxidation state such as the amide or nitrile must be reduced to the oxidation state of the final amine, *i.e.*, the oxidation state corresponding to an alcohol.   The order in which these steps are taken is immaterial.   Thus nitrogen may be introduced first by converting the fatty acid to its nitrile, which is then reduced to the amine.   Conversely the fatty acid or an appropriate ester may be first reduced to the fatty alcohol.   The alcohol can then be converted by ammonolysis to the amine.

The most widely used commercial method for preparing the higher fatty amines is by the catalytic hydrogenation of the corresponding nitriles. The nitriles may be made by heating the fatty acid ammonium salts or amides under dehydrating conditions:[7]

$$RCOONH_4 \rightarrow H_2O + RCONH_2 \rightarrow H_2O + RCN$$

The hydrogenation may be carried out in the liquid or gas phase using Ni or the other well-known hydrogenation catalysts.   Under normal conditions a considerable quantity of the dialkyl amine is obtained in addition to the desired monoalkyl amine.   The presence of excess ammonia in the reaction mixture minimizes the formation of dialkyl amine.[8]   This is to be expected in view of the mechanism by which the dialkyl amines are formed.   The primary addition product of hydrogen to the nitrile is an aldimine.   This may then add a molecule of previously formed amine. The addition product splits off ammonia, and the resulting substituted imine is hydrogenated to the secondary amine:

$$RC\equiv N + H_2 \longrightarrow RCH=NH$$

$$RCH=NH + RCH_2NH_2 \longrightarrow RCH-NH_2 \longrightarrow RCH=NCH_2R + NH_3$$
$$\overset{|}{NH-CH_2R}$$

$$RCH=N-CH_2R + H_2 \longrightarrow RCH_2-NH-CH_2R$$

It is apparent that in the presence of excess ammonia the addition of ammonia (rather than of RCH_2NH_2) to the imine will be favored.   This results in a higher yield of primary amine:

$$RCH=NH + NH_3 \longrightarrow RCH-NH_2 \xrightarrow{H_2} RCH_2NH_2 + NH_3$$
$$\overset{|}{NH_2}$$

[7] French Pat. 785,622, to *I. G. Farbenindustrie*.
[8] Brit. Pat. 438,793, to *I. G. Farbenindustrie*.

If hydrogenation of the nitrile is carried out in the presence of an excess of a primary or secondary amine, N-substituted fatty amines are obtained. For example, if stearonitrile is hydrogenated in the presence of methylamine the products $C_{18}H_{37}NHCH_3$ and $C_{18}H_{37}N(CH_3)_2$ are obtained. If piperidine is used:

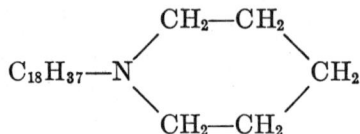

$$C_{18}H_{37}-N \underset{CH_2-CH_2}{\overset{CH_2-CH_2}{\diagdown}} CH_2$$

is obtained.[9] Secondary and tertiary amines of this type are more usually prepared, however, by alkylation of a previously prepared primary amine or by reacting a long-chain alkyl halide with the low-molecular-weight primary or secondary amine.

Current German industrial practice in the conversion of fatty acids to fatty amines has been described in detail by Sheely.[10] Nitrile formation and hydrogenation are carried out in one stage in the vapor phase.

Fatty acid amides and fatty acid ammonium salts, as well as the nitriles, may be hydrogenated to produce the primary fatty amines. Using these materials more energetic conditions are required to effect the reduction and water is split off during the reaction. Copper, chromium, and cobalt catalysts, promoted by alkaline-earth oxides, have been described as being effective.[11] Large yields of dialkyl amines have been reported in the hydrogenation of higher fatty amides and anilides using a copper – chromium – barium oxide catalyst.[12]

Fatty amines have also been made by the ammonolysis of fatty alcohols. For example, cetyl alcohol and ammonia are reported to give 95% yields of hexadecylamine when reacted at 380–400°C. and 120–130 atmospheres pressure in the presence of alumina.[13] The lower secondary or primary amines may be used in place of ammonia, in which case mono- or di-N-substituted fatty amines are formed.[14] The reaction of an alkyl halide with ammonia or with primary or secondary amines is a classical method for

[9] Brit. Pat. 439,274; French Pat. 773,367; U. S. Pat. 2,160,578; Schmidt to *I. G. Farbenindustrie*.

[10] Sheely, *PB 2424*, Office of Technical Services, Dept. of Commerce, Washington, D. C.

[11] U. S. Pat. 2,166,971, Schmidt and Huttner to *I. G. Farbenindustrie*; Ger. Pat. 667,627; to *I. G. Farbenindustrie*. Brit. Pat. 425,927, to *Rohm and Haas Co.* U. S. Pat. 2,143,751, Adkins to *Rohm and Haas Co.*

[12] Ueno and Takase, *J. Soc. Chem. Ind. Japan* **44**, 29, 58 (1941); *Chem. Abstracts* **34**, 2787; **35**, 3963.

[13] Ger. Pat. 611,924, to *Deutsche Hydrierwerke A.-G.*

[14] French Pats. 779,913; 780,028; Brit. Pat. 436,414; to *I. G. Farbenindustrie*.

preparing amines. When a tertiary amine is used a quaternary ammonium halide is formed. In forming higher alkyl secondary and tertiary amines of the general formula:

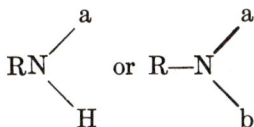

$$RN\Big\langle\begin{matrix} a \\ H \end{matrix} \quad \text{or} \quad R-N\Big\langle\begin{matrix} a \\ b \end{matrix}$$

where R represents a higher alkyl radical and a and b represent low-molecular-weight radicals, it is more usual to start with RNH_2 and replace the hydrogen atoms by a or b radicals. It is also possible, however, to react RCl or RBr with aNH_2 or abNH, or with ammonia. The reactions of higher alkyl halides with liquid ammonia to form the higher alkyl amines have been studied. The method is not applicable to halides above C_{20}, due to their poor solubility in liquid ammonia.[15] The reaction of higher alkyl halides with aqueous ammonia or lower amines has also been described as a commercial preparative method.[16] It is particularly applicable where the amino group is part of a heterocycle. For example, N-cetyl-piperidine is best prepared from cetyl bromide and piperidine.[17]

Beside the higher alkyl halides the sulfates, aryl sulfonates, or monosulfuric esters may be used. Higher alkyl chlorosulfonates have also been used to react with ammonia or amines.[18]

Although it is not usual commercial practice to use the higher alkyl halides, sulfates, etc. in the preparation of primary secondary or tertiary amines, they are used in the preparation of quaternary ammonium salts. This method is particularly useful where the tertiary amine is readily available. The higher alkyl pyridinium halides such as cetylpyridinium bromide are important cationic surface active agents. They are made by reacting pyridine with the alkyl halides.[19] Quinolinium salts are also made in this manner by simply heating quinoline together with the higher alkyl halide. The higher alkyl pyridinium sulfates can be made directly by sulfating the alcohol in the usual manner with chlorosulfonic acid and treating the sulfation mixture with pyridine. The alcohol may also be treated with the

---

[15] Von Braun and Klar, *Ber.* **B73**, 1417 (1940).

[16] French Pat. 784,599, to *I. G. Farbenindustrie*.

[17] Brit. Pat. 433,356, to *Deutsche Hydrierwerke A.-G.*; French Pat. 784,599, to *I. G. Farbenindustrie*.

[18] Brit. Pats. 410,956; 414,299; to *H. Th. Böhme, A.-G.* Brit. Pat. 435,863, to *I. G. Farbenindustrie*.

[19] Brit. Pat. 379,396, Baldwin and Hailwood to *Imperial Chemical Industries Ltd.* See also Fawcett and Gibson, *J. Chem. Soc.* **1934**, 396.

complex sulfating agent made by treating chlorosulfonic acid with pyridine. After bringing the reactants together the mixture is heated in the range of 150°C. and the alkyl pyridinium sulfate is formed.[20]   Salts of the higher alkyl pyridinium ions with phenols or mononuclear aromatic sulfonates are said to be highly surface active but non-foaming.[21] Nicotinium salts of the higher alkyl halides have been described as insecticides.[22] Tertiary benzylamines such as N,N-dimethyl-2-chlorobenzylamine have also been added to long-chain halides to form the quaternary compounds.[23] In general it is a matter of convenience and availability of reagents whether the long chain is introduced in the form of an amine or a halide. The higher alkyl halides are appreciably less reactive than the lower members of the series and, therefore, less convenient to use when the reverse method is applicable. This difference in reactivity is less than an order of magnitude, however, and is usually not a serious factor in the choice of methods.

The higher normal primary amines are excellent detergents and foaming agents in the form of their hydrochlorides, acetates, etc.[24]   Their physical and surface active properties have been thoroughly studied, particularly by Ralston and co-workers.[25]   Although they are widely used as such for their surface activity, they are mainly used as raw materials in the preparation of other cationic surface active agents.

Branched-chain higher alkyl amines have been made by the reaction of the higher ketones with formamide. The reaction takes place at about 180°C. and the formyl radical of the formamide is oxidized to carbon dioxide:

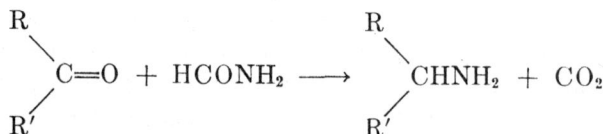

$$\begin{array}{c} R \\ \diagdown \\ \diagup \\ R' \end{array} C{=}O \ + \ HCONH_2 \ \longrightarrow \ \begin{array}{c} R \\ \diagdown \\ \diagup \\ R' \end{array} CHNH_2 \ + \ CO_2$$

The ketones may be derived from the oxidation products of paraffin wax[26] or may be synthesized by various other methods previously discussed.

[20] U. S. Pat. 2,104,728, Bertsch and Stober to *Böhme Fettchemie G.m.b.H.* Ger. Pat. 724,991 to *Böhme Fettchemie G.m.b.H.*

[21] Ger. Pat. 719,633, to *Böhme Fettchemie G.m.b.H.*

[22] U. S. Pat. 2,048,885; Brit. Pat. 401,707; Oakeshott to *Imperial Chemical Industries Ltd.*

[23] Ger. Pats. 681,850; 658,321; to *I. G. Farbenindustrie.*

[24] U. S. Pat. 2,274,807, Rawlins and Sweet to *Parke, Davis and Co.*

[25] Ralston *et al.*, *J. Am. Chem. Soc.* **63**, 2576 (1941); **64**, 498,2067 (1942); **65**, 328 (1943); etc.

[26] U. S. Pat. 2,378,880, Burwell and Camelford to *Alox Chemical Co.* Brit. Pat. 502,102 to *I. G. Farbenindustrie.*

Higher aldehydes or ketones may also be hydrogenated in the presence of ammonia to give the corresponding higher amines. The intermediate product is an aldimine or ketimine, which is then hydrogenated at the double bond.

$$RCHO + NH_3 \rightarrow RCH{=}NH \rightarrow RCH_2NH_2$$

2-Ethylhexaldehyde has been converted to the corresponding amine by this process.[27] This aldehyde and other higher-branched chain aldehydes have been developed on a commercial scale by *Carbide and Carbon Chemicals Corp.* They are made by extensions of the aldol synthesis. If ammonia is replaced by primary or secondary amines in this reaction the resulting products are secondary or tertiary amines.[28]

Keryl amines and polyamines have been described as obtainable from keryl chloride and ammonia or polyethylene polyamines. It is doubtful if high yields are obtainable and no commercial products of this nature are known to the writers.[29]

A product made by the action of keryl chloride or chlorinated paraffin oil on pyridine has also been described as a useful surface active agent.[30] Keryl diethanolamines, prepared by the reaction of keryl chloride with diethanolamine, have been studied and found to be less effective in reducing surface tension than the corresponding lauryl derivatives.[31]

Amines have also been made by adding ammonia across the double bond of an olefin in the surface active range. Dodecene and ammonia, for example, are passed over a cobalt–silica catalyst at 450–750°F. and 500–3000 psi. Nitriles are also produced in the reaction.[32]

Cholanylamines and their alkylated and quaternized derivatives have been used as germicides and they show the surface activity typical of the quaternary ammonium germicides.[33]

Dicaprylamine and its *N*-alkyl derivatives have been prepared by reacting dicapryl alcohol or its corresponding ketone, 9-methyl-7-pentadecanone,

---

[27] Canadian Pat. 409,004; U. S. Pat. 2,319,848; Clark and Wilson to *Carbide and Carbon Chemicals Corp.*

[28] Brit. Pat. 414,712 to *I. G. Farbenindustrie.*

[29] See Profft, *Fette u. Seifen* **49**, 868 (1942); *Chem. Abstracts* **37**, 6482. U. S. Pats. 2,272,489, Ulrich to *General Aniline and Film Corp.*; 2,305,830, Profft vested in Alien Property Custodian; 2,361,457, Clark to *Mathieson Alkali Works.*

[30] Brit. Pat. 506,083, to *Vereinigte Glanzstoff- Fabriken A.-G.*

[31] Padgett and Degering, *Ind. Eng.Chem.* **32**, 486 (1940).

[32] U. S. Pats. 2,381,470-2, Teter to *Sinclair Refining Co.*; 2,381,709, Apgar and Teter to *Sinclair Refining Co.*

[33] Caldwell, *J. Am. Chem. Soc.* **60**, 991 (1938). U. S. Pat. 2,252,863, Raymond and Dillon to *G. D. Searle and Co.*

with hydrogen and ammonia oxamines in the presence of hydrogenation catalysts.[34]

Following the discovery some years ago of the germicidal effects of the long-chain cations a large number of products were developed having the general formula:

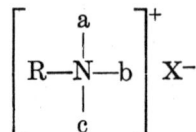

$$\left[\begin{array}{c} a \\ | \\ R{-}N{-}b \\ | \\ c \end{array} \right]^{+} X^{-}$$

where a, b, and c represent relatively small hydrocarbon radicals (or form a ring structure) and X^- represents an anion such as Cl^-, Br^-, $SO_4CH_3^-$, etc.

The cetyltrimethylammonium salts and the corresponding coconut fatty and octadecyl analogues have enjoyed considerable popularity. They are made by exhaustive methylation of the primary fatty amines with methyl bromide, chloride, or sulfate. If the latter is used the anion associated with the final product is methosulfate, $-O{-}SO_3{-}CH_3^-$.

One of the best-known cationic surface active agents is the coconut fatty alkyl dimethylbenzylammonium chloride.[35] This is made by methylating the primary amine to give the dimethyl alkyl amine, which is then quaternized by the addition of benzyl chloride. The product is marketed as Zephyran, Roccal, Ammonyx, Triton K-60, and under various other trade names. It is widely used as a disinfectant and antiseptic. The pure product is a white crystalline solid. It is usually furnished, however, as an aqueous solution of about 10% strength or as an aqueous paste of 20–30% strength. Halogenated benzyl groups have been introduced into the molecule, an example being cetyldimethyl-o-chlorobenzylammonium bromide.[36]

The higher alkyl dimethylmethallylammonium halides are made in an entirely similar manner, using methallyl chloride rather than benzyl chloride as the quaternizing agent. They are stated to be particularly active germicides.[37]

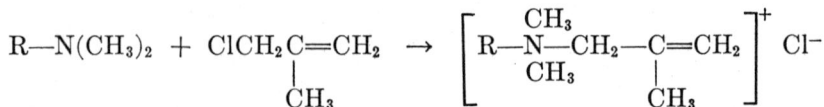

$$R{-}N(CH_3)_2 + ClCH_2C{=}CH_2 \rightarrow \left[\begin{array}{c} CH_3 \\ | \\ R{-}N{-}CH_2{-}C{=}CH_2 \\ | \qquad\qquad | \\ CH_3 \qquad CH_3 \end{array} \right]^{+} Cl^-$$

---

[34] U. S. Pat. 2,160,058, Covert to *Rohm and Haas Co.*

[35] U. S. Pat. 2,108,765, Domagk to *Alba Pharmaceutical Co.* See also Ger. Pat. 680,599; 708,076, to Domagk.

[36] French Pat. 806,662, to *I. G. Farbenindustrie.*

[37] U. S. Pat. 2,191,922, Bruson to *Rohm and Haas Co.*

The usual procedure in making quaternary ammonium compounds is to react a tertiary amine with an alkyl halide to form the salt. It is possible however, to make the quaternary ammonium hydroxide directly by reacting a tertiary amine with ethylene oxide or other active epoxides such as propylene oxide or glycide in the presence of water. A typical reaction using ethylene oxide and an alkyl dimethylamine is:

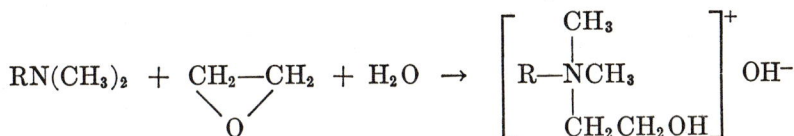

$$RN(CH_3)_2 \ + \ CH_2\text{—}CH_2 \ + \ H_2O \ \rightarrow \ \begin{bmatrix} CH_3 \\ | \\ R\text{—}NCH_3 \\ | \\ CH_2CH_2OH \end{bmatrix}^+ \ OH^-$$

By starting with the primary amine and using excess ethylene oxide three β-hydroxyethyl groups are introduced:[38]

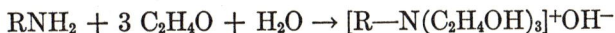

$$RNH_2 + 3 \ C_2H_4O + H_2O \rightarrow [R\text{—}N(C_2H_4OH)_3]^+OH^-$$

The reaction of ethylene oxide with RNH_2 may be stopped at the secondary or tertiary amine stage, giving:

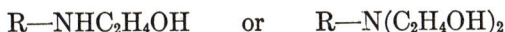

$$R\text{—}NHC_2H_4OH \qquad or \qquad R\text{—}N(C_2H_4OH)_2$$

These compounds are highly surface active.[39]   Among other surface active amines in which the non-hydrophobic substituents contain hydroxyl groups are the higher alkyl glucamines. These may be made by reacting the alkyl halide with glucamine or methyl glucamine or by hydrogenating a mixture of glucose and the primary alkyl amine. The products have the formula:[40]

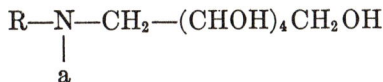

$$\underset{\underset{a}{|}}{R\text{—}N}\text{—}CH_2\text{—}(CHOH)_4CH_2OH$$

They may be quaternized in the usual manner or used in the form of their salts.

The unreduced intermediate Schiff's bases formed from higher fatty amines with aldoses such as glucose have also been described as surface active agents.

$$RN\text{=}C\text{—}(CHOH)_4CH_2OH$$

[38] U. S. Pats. 2,127,476; 2,137,314; 2,173,069, Ulrich and Ploetz to *I. G. Farbenindustrie*; 2,143,388, Schlack to *I. G. Farbenindustrie*; Brit. Pats. 380,851; 419,588; to *I. G. Farbenindustrie*.

[39] Brit. Pat. 372,325 to *I. G. Farbenindustrie*. French Pat. 809,360, to *I. G. Farbenindustrie*. See also Rumpf and Kwass, *Bull. soc. chim.* **10**, 347 (1943).

[40] Brit. Pat. 428,142; U. S. Pat. 2,016,956; Calcott and Clarkson to *E. I. du Pont de Nemours & Co.* U. S. Pats. 2,060,850; 2,181,929; Werntz to *E. I. du Pont de Nemours & Co.*

These products are probably too unstable for practical use unless they are hydrogenated.[41]

Highly hydroxylated bases may be made by reacting fatty amines with the chloride of polyglycerol,[42] or by the action of excess ethylene oxide on such compounds as:

$$R-NH-C(CH_2OH)_3$$

This product is made by reacting RCl with trimethylolaminomethane, which is in turn derived from nitromethane and formaldehyde:[43]

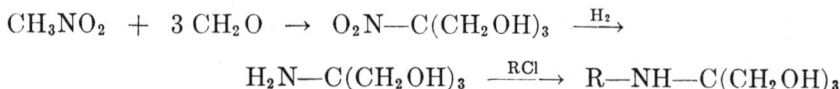

$$CH_3NO_2 + 3 CH_2O \rightarrow O_2N-C(CH_2OH)_3 \xrightarrow{H_2}$$

$$H_2N-C(CH_2OH)_3 \xrightarrow{RCl} R-NH-C(CH_2OH)_3$$

Fatty alkyl halides have been condensed with alkylene polyamines such as ethylenediamine and diethylenetriamine to give products containing two or more basic amino groups:[44]

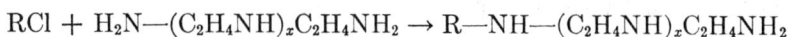

$$RCl + H_2N-(C_2H_4NH)_xC_2H_4NH_2 \rightarrow R-NH-(C_2H_4NH)_xC_2H_4NH_2$$

Similarly, the alkyl halides may be condensed with N,N-diethylethylene-diamine to yield:[45]

$$R-NH-C_2H_4N(C_2H_5)_2$$

Among other products containing a plurality of basic amino groups are the polyamines made by reducing polymerized unsaturated fatty nitriles (e.g., polymerized nitrile of linseed oil fatty acids),[46] and the quaternary compounds formed from alkyl halides and hexamethylenetetramine.[47]

Chlorinated paraffin wax has been condensed with triethylenetetramine to give a polybasic surface active agent.[48] Alkyl guanidines, made by the addition of cyanamide to a long-chain fatty primary amine, have also

---

[41] French Pat. 791,372, to *Henkel et Cie.*

[42] Ger. Pat. 606,236, to *I. G. Farbenindustrie.*

[43] Brit. Pat. 420,066; French Pat. 768,732; to *Imperial Chemical Industries Ltd.*

[44] U. S. Pats. 2,267,204–5; 2,246,524; Kyrides to *Monsanto Chem. Co.* See also U. S. Pat. 2,243,329, De Groote and Blair to *Petrolite Corp.* Linsker and Evans, *J. Am. Chem. Soc.* **67**, 1581 (1945).

[45] Ger. Pat. 547,987, to *I. G. Farbenindustrie.*

[46] U, S. Pat. 2,178,522, Ralston and Vander Wal to *Armour and Co.*

[47] Brit. Pat. 424,717, to Hunsdiecker and Vogt.

[48] Ger. Pat. 668,744, to *I. G. Farbenindustrie.*

been described as surface active agents.[49] The alkyl guanidines and bi-guanides (made in a similar manner by the addition of dicyandiamide to a fatty amine) have been used primarily as softening agents for fabrics but are said to be effective germicides also.[50]

Long-chain guanamines have been made by heating a biguanide with a fatty acid ester in the presence of a metal alkoxide.[51]

A number of surface active long-chain acyl biguanides have been described. On heating they are converted to guanamines. These products can be exhaustively alkylated to increase the solubility and range of effectiveness.[52] Still another group of polybasic compounds are made by reacting higher alkyl halides with amidines such as dimethylphenylbenz-amidine. These are quaternary salts.[53]

The higher unsubstituted amidines such as stearoamidine are basic in nature and their salts are said to be particularly useful in flotation. They are made from the corresponding nitriles by first reacting with alcoholic hydrochloric acid and then with ammonia:[54]

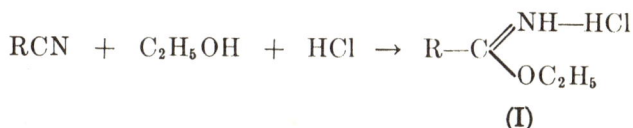

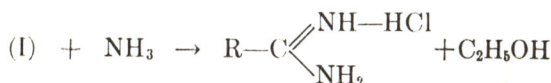

$$RCN \ + \ C_2H_5OH \ + \ HCl \ \rightarrow \ R\!-\!C\!\!\begin{array}{c} \diagup NH\!-\!HCl \\ \diagdown OC_2H_5 \end{array}$$

$$(I)$$

$$(I) \ + \ NH_3 \ \rightarrow \ R\!-\!C\!\!\begin{array}{c} \diagup NH\!-\!HCl \\ \diagdown NH_2 \end{array} +C_2H_5OH$$

Bisamidines have been made by similar procedures from the α,ω-dini-triles of six to ten carbon atoms. These amidines can be substituted on the NH_2 groups by alkyl radicals. They are described as effective germicides.[55]

The higher alkyl halides have been quaternized with a wide variety of tertiary bases. The products are all surface active; yet with few exceptions

[49] Brit. Pat. 421,862, to Hunsdiecker and Vogt.

[50] U. S. Pat. 2,213,474, Puetzer to *Winthrop Chemical Co.* French Pat. 805,768, to *I. G. Farbenindustrie.*

[51] U. S. Pats. 2,309,664, to Oldham; 2,309,679, Thurston to *American Cyanamid Co.*

[52] Ger. Pat. 735,596; Brit. Pat. 561,548; U. S. Pat. 2,315,765; Bindler to *J. R. Geigy A.-G.*

[53] Brit. Pat. 498,090, to *J. R. Geigy A.-G.*

[54] U. S. Pat. 2,389,681, Mikeska to *Standard Oil Development Co.*

[55] U. S. Pats. 2,364,074–5, Hunt, Kirby, and Lontz to *E. I. du Pont de Nemours & Co.*

their particular properties and advantages have been incompletely described. Among the bases used are:

(1) N-Ethyl- and N-hydroxyethylmorpholine:[56]

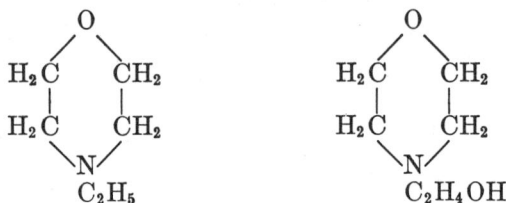

$$
\begin{array}{cc}
\underset{\substack{\diagup\;\diagdown\\ H_2C\quad CH_2\\ |\qquad |\\ H_2C\quad CH_2\\ \diagdown N \diagup\\ C_2H_5}}{O}
&
\underset{\substack{\diagup\;\diagdown\\ H_2C\quad CH_2\\ |\qquad |\\ H_2C\quad CH_2\\ \diagdown N \diagup\\ C_2H_4OH}}{O}
\end{array}
$$

One group of products in this case has the formula:

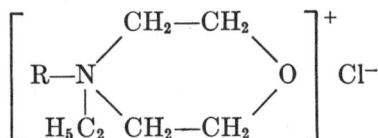

$$
\left[\begin{array}{c}
\text{CH}_2\text{—CH}_2 \\
R\text{—N}\diagup\qquad\diagdown O \\
\underset{H_5C_2}{|\diagdown}\;\;\text{CH}_2\text{—CH}_2\diagup
\end{array}\right]^{+} \text{Cl}^{-}
$$

Closely related products have also been made by condensing RNH_2 with 2,2′-dichloroethyl ether to give the intermediate amine:

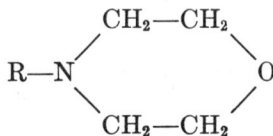

$$
R\text{—N}\underset{\diagdown\;\text{CH}_2\text{—CH}_2\diagup}{\overset{\diagup\;\text{CH}_2\text{—CH}_2\;\diagdown}{}}O
$$

which is then quaternized with, e.g., ethyl chloride.[57]

(2) 2-Methylbenzimidazole:[58]

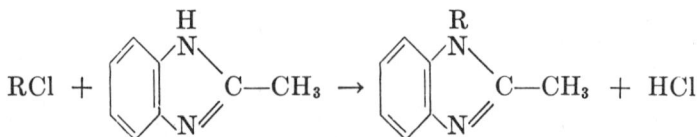

$$
RCl + \underset{}{\overset{H}{\underset{N}{\bigcirc}}}C\text{—}CH_3 \rightarrow \underset{}{\overset{R}{\underset{N}{\bigcirc}}}C\text{—}CH_3 + HCl
$$

This product is a tertiary amine and can be quaternized with another mole of higher alkyl halide or with a mole of a lower alkyl halide or sulfate.

[56] McGreal and Niederl, J. Am. Chem. Soc. 63, 1476 (1941).
[57] U. S. Pat. 2,380,325, to Niederl, Niederl, and McGreal.
[58] Brit. Pat. 439,261, to Ciba Co.

(3) Dimethyl- and diethylcyclohexylamines.[59] The product has the formula:

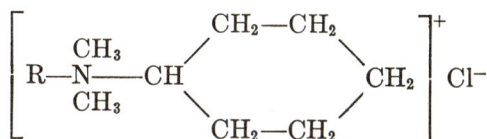

$$\left[\begin{array}{c} CH_3 \\ R-N----CH \\ CH_3 \end{array} \underset{CH_2-CH_2}{\overset{CH_2-CH_2}{\diagdown}} CH_2 \right]^+ Cl^-$$

(4)

$$\underset{C_2H_5}{\overset{C_2H_5}{\diagdown}} N^*-C_2H_4NH-CO-NH_2$$

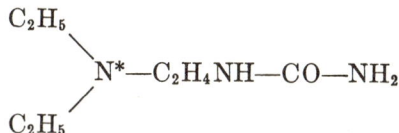

and homologues of this basic urea derivative. The alkyl halide adds at the starred nitrogen atom.[60]

(5) Bis-(morpholinomethyl) polyhydric phenols, made by the reaction of formaldehyde, morpholine, and a di- or trihydric phenol.[61]

Still more complex non-hydrophobic groups have been introduced by using a long-chain primary, secondary, or tertiary amine and reacting with a complex lower halide. For example the following series of reactions yields an effective germicide:[62]

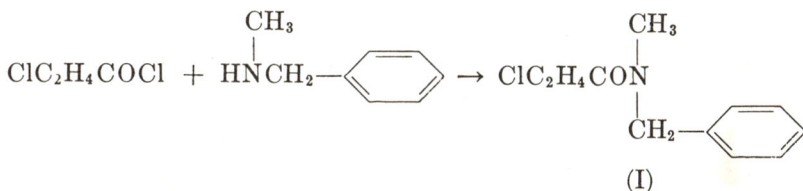

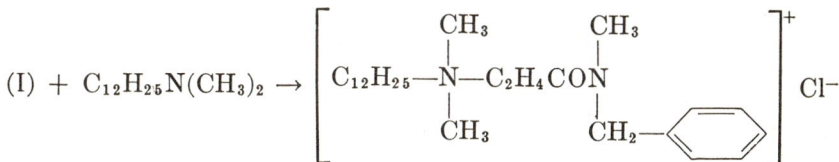

$$ClC_2H_4COCl + \underset{}{\overset{CH_3}{\overset{|}{H\,N}}}CH_2- \bigcirc \rightarrow ClC_2H_4CO\underset{CH_2-\bigcirc}{\overset{CH_3}{\overset{|}{N}}}$$

(I)

$$(I) + C_{12}H_{25}N(CH_3)_2 \rightarrow \left[C_{12}H_{25}-\underset{CH_3}{\overset{CH_3}{\overset{|}{N}}}-C_2H_4CO\underset{CH_2-\bigcirc}{\overset{CH_3}{\overset{|}{N}}} \right]^+ Cl^-$$

Similarly the long-chain fatty amines RNH_2 have been exhaustively al-

[59] U. S. Pat. 2,080,143, Johnson and Lubs to *E. I. du Pont de Nemours & Co.*
[60] U. S. Pat. 2,203,505; French Pat. 843,081; to *Imperial Chemical Industries Ltd.*
[61] U. S. Pat. 2,040,040, Bruson to *Rohm and Haas Co.*
[62] U. S. Pat. 2,336,179, Leuchs to *Winthrop Chemical Co.*

kylated with the halide $ClCH_2CONHC_2H_4OH$, which is made from chloroacetyl chloride and monoethanol amine. The product has the formula:[63]

$$[R-N(CH_2CONHC_2H_4OH)_3]^+Cl^-$$

According to another disclosure cetyldimethylamine has been quaternized
with chloroacetylbenzylmethylamine to give the compound:[64]

$$\left[\begin{array}{c} CH_3 \\ | \\ C_{16}H_{33}-N-CH_3 \quad CH_3 \\ | \quad\quad\quad / \\ CH_2CON \\ \quad\quad\quad\backslash \\ \quad\quad\quad CH_2C_6H_5 \end{array} \right]^+ \quad Cl^-$$

(6) Bisammonium compounds are represented by such products as:

$$\begin{array}{ccc} CH_3 & & CH_3 \\ | & & | \\ C_{12}H_{25}-N-C_2H_4-N-C_{12}H_{25} \\ | & & | \\ Cl \quad CH_3 & & Cl \\ & CH_3 \end{array}$$

from dodecyl chloride and tetramethylethylenediamine. They are stated
to be excellent germicides and fungicides.[65]

A group of cationic surface active agents has been made by reacting α-
halogenated fatty acid derivatives with tertiary amines. α-Chlorostearic
anilide or α-bromostearic ethyl ester, for example, may be reacted with
pyridine, trimethylamine, etc.[66] Halogenated long-chain aldehydes and
ketones may be used instead of the carboxylic acid derivatives. For example, α-chloromethyl nonyl ketone may be reacted with pyridine to give a
quaternary salt.[67]

The application of these products in textile processing has been studied
by Mehta and Trivedi, who prepared and carefully purified several members of the series.[68]

Numerous long-chain amines have been described in which one of the

[63] U. S. Pat. 2,328,021, Katzman and Epstein to *Emulsol Corp.*

[64] U. S. Pat. 2,336,179, Leuchs to *Winthrop Chemical Co.*

[65] Brit. Pat. 441,473, to *I. G. Farbenindustrie*. U. S. Pat. 2,375,853, Kirby and
Lontz to *E. I. du Pont de Nemours & Co.*

[66] U. S. Pat. 2,202,328, Albrecht to *Ciba Co.* French Pat. 788,898, to *Ciba Co.*
Swiss Pats. 179,654; 177,574, to *Ciba Co.* U. S. Pats. 2,087,565, Balle and Eisfeld to
General Aniline Works; 2,168,253, Balle and Schulz to *I. G. Farbenindustrie*. U. S.
Pat. 2,359,884, Tinker and Linch to *E. I. du Pont de Nemours & Co.*

[67] Brit. Pat. 493,592, to *N. V. Onderzoekingsinstituut Research.*

[68] Mehta and Trivedi, *J. Soc. Dyers Colourists* **56**, 343 (1940).

substituents on the nitrogen atom is an aromatic ring.[69]   Dodecyl-, hexa-
decyl- and octadecylanilines have been made by reacting aniline with the
corresponding alkyl halide.   These compounds may be further alkylated
and finally quaternized with such reagents as methyl sulfate or lower alkyl
halides:[70]

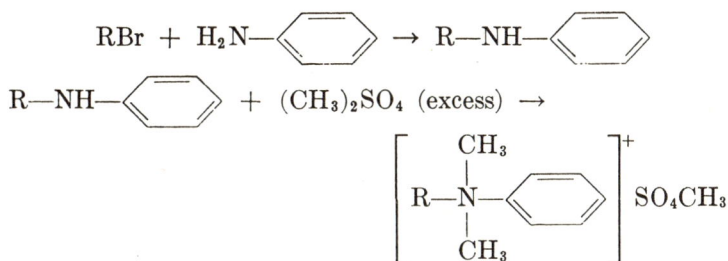

$$RBr + H_2N-\left\langle\!\!\bigcirc\!\!\right\rangle \rightarrow R-NH-\left\langle\!\!\bigcirc\!\!\right\rangle$$

$$R-NH-\left\langle\!\!\bigcirc\!\!\right\rangle + (CH_3)_2SO_4 \ (excess) \rightarrow$$

$$\left[\begin{array}{c} CH_3 \\ | \\ R-N-\left\langle\!\!\bigcirc\!\!\right\rangle \\ | \\ CH_3 \end{array}\right]^+ SO_4CH_3$$

The higher alkyl anilines may also be made by decarboxylating the α-
anilino fatty acids.   This is accomplished by heating in an inert diluent:[71]

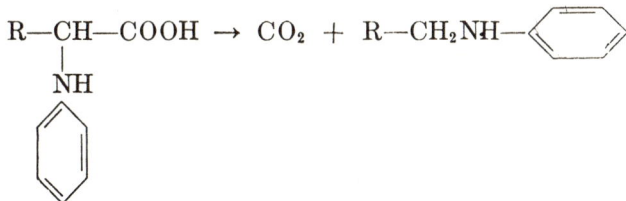

$$R-\underset{\underset{\displaystyle \bigcirc}{\overset{|}{NH}}}{CH}-COOH \rightarrow CO_2 + R-CH_2NH-\left\langle\!\!\bigcirc\!\!\right\rangle$$

Long-chain alkylated and acylated amines have also been made by a
variety of procedures.   By condensing stearic acid chloride with dimethyl-
aniline, for example, the ketone is obtained:

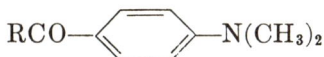

$$RCO-\left\langle\!\!\bigcirc\!\!\right\rangle-N(CH_3)_2$$

This may be quaternized with methyl sulfate.   The same compound may
be made by acylating chlorobenzene with stearoyl chloride and reacting the
product with dimethylamine:[72]

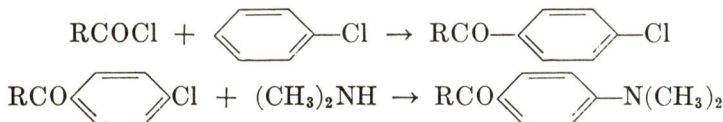

$$RCOCl + \left\langle\!\!\bigcirc\!\!\right\rangle-Cl \rightarrow RCO-\left\langle\!\!\bigcirc\!\!\right\rangle-Cl$$

$$RCO\left\langle\!\!\bigcirc\!\!\right\rangle Cl + (CH_3)_2NH \rightarrow RCO\left\langle\!\!\bigcirc\!\!\right\rangle-N(CH_3)_2$$

[69] See, for example, U. S. Pat. 2,161,322, Balle *et al.* to *I. G. Farbenindustrie.*
[70] Swiss Pats. 200,669; 211,790–1; French Pat. 842,299; Ger. Pat. 734,991; Brit. Pat.
509,542; to *J. R. Geigy A.-G.*
[71] French Pat. 785,004, to *H. Th. Böhme A.-G.*
[72] French Pat. 840,734, to *J. R. Geigy A.-G.*; U. S. Pat. 2,097,640, Piggott to *Imperial
Chemical Industries Ltd.*

p-Aminodimethylaniline has been alkylated with cetyl bromide to give a diamino aromatic surface active agent:[73]

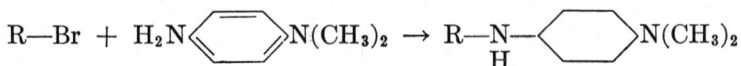

$$R—Br + H_2N⟨\ ⟩N(CH_3)_2 → R—\underset{H}{N}—⟨\ ⟩N(CH_3)_2$$

p-tert-Octylanisole and similar alkylated phenol ethers and phenols have been nitrated and reduced to give long-chain substituted aromatic amines:

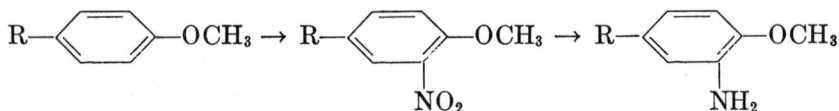

$$R—⟨\ ⟩—OCH_3 → R—\underset{NO_2}{⟨\ ⟩}—OCH_3 → R—\underset{NH_2}{⟨\ ⟩}—OCH_3$$

These amines may be used as such in the form of their salts, or they may be further alkylated or quaternized.[74]

Substituted benzylamines of the type:

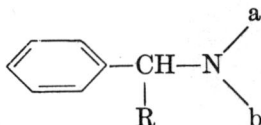

$$⟨\ ⟩—\underset{R}{CH}—N\overset{a}{\underset{b}{⟨}}$$

and their quaternized products have been described as surface active

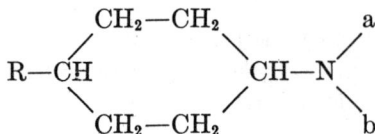

$$R—CH\overset{CH_2—CH_2}{\underset{CH_2—CH_2}{⟨}}CH—N\overset{a}{\underset{b}{⟨}}$$

agents.[75] The hydro aromatic series is represented by products of the type:[76]

The long-chain alkyl phenols are relatively easy and inexpensive to produce, as has been pointed out in a previous section. This is particularly true of such products as di- and triisobutyl phenols. These intermediates may be converted by the Mannich reaction into an interesting series of cationic surface active agents. *p-tert*-Octylphenol, for example, may be reacted with formaldehyde and dimethylamine as follows:

[73] U. S. Pat. 2,097,640, Piggott to *Imperial Chemical Industries Ltd.*

[74] Brit. Pats. 493,294; 493,339; to *I. G. Farbenindustrie.* See also Niederl and Dexter, *J. Am. Chem. Soc.* **63,** 1475 (1941).

[75] Brit. Pat. 536,881, to *J. R. Geigy A.-G.*

[76] U. S. Pat. 2,265,838, Hentrich *et al.* to *Procter and Gamble Co.*

$$R\text{---}\langle\bigcirc\rangle\text{---OH} + CH_2O + (CH_3)_2NH \rightarrow R\text{---}\langle\bigcirc\rangle\text{---OH}$$

$$\begin{array}{c} | \\ CH_2 \\ | \\ N(CH_3)_2 \end{array}$$

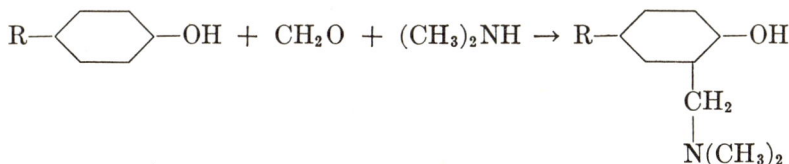

The tertiary amine thus obtained may be quaternized with any of the usual reagents.[77]   Morpholine or other lower secondary amines may be used in place of dimethylamine.[78]   In an alternative process phenol is first condensed with formaldehyde and the lower secondary amine, and the product condensed with a higher fatty alcohol such as lauryl alcohol:[79]

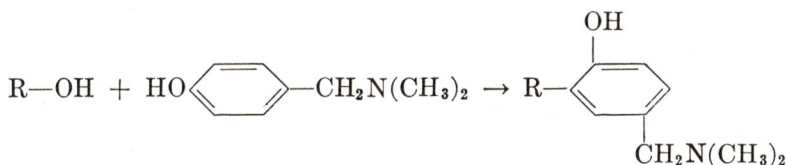

$$R\text{---OH} + HO\langle\bigcirc\rangle\text{---}CH_2N(CH_3)_2 \rightarrow R\text{---}\langle\bigcirc\rangle$$

with substituents OH and $CH_2N(CH_3)_2$

Alkylated phenols may also be chloromethylated, using formaldehyde and hydrogen chloride.   The resulting substituted benzyl chlorides are very reactive and can be quaternized with lower tertiary amines or they can be converted to non-quaternary amines:[80]

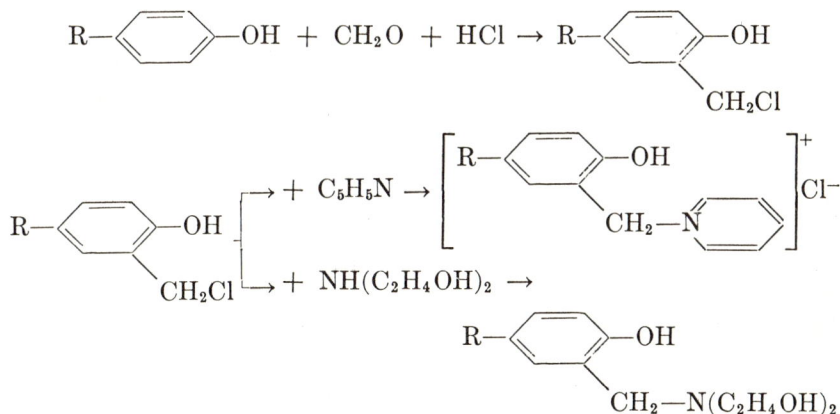

$$R\text{---}\langle\bigcirc\rangle\text{---OH} + CH_2O + HCl \rightarrow R\text{---}\langle\bigcirc\rangle\text{---OH}$$

with CH_2Cl substituent

$$R\text{---}\langle\bigcirc\rangle\text{---OH} \underset{CH_2Cl}{\overset{}{\Big|}}$$

$$\xrightarrow{+ C_5H_5N} \left[R\text{---}\langle\bigcirc\rangle\text{---OH} \atop CH_2\text{---}N\langle\bigcirc\rangle \right]^+ Cl^-$$

$$\xrightarrow{+ NH(C_2H_4OH)_2} R\text{---}\langle\bigcirc\rangle\text{---OH}$$

with $CH_2\text{---}N(C_2H_4OH)_2$ substituent

[77] U. S. Pat. 2,033,092; Brit. Pàt. 444,351; Bruson to *Rohm and Haas Co.*
[78] Niederl and Abbruscato, *J. Am. Chem. Soc.* **63,** 2024 (1941).
[79] U. S. Pat. 2,045,517, Bruson and Stein to *Rohm and Haas Co.*
[80] U. S. Pats. 2,165,956; 2,180,791, Brunner to *I. G. Farbenindustrie.* Brit. Pat. 500,550, to *I. G. Farbenindustrie.*

III. Hydrophobic Group Joined to Cationic Group through an Intermediate Linkage

(a) Amide Intermediate Linkage

Among the best-known long-chain cationic compounds to be used specifically for their surface active properties are the products manufactured by *Ciba Co.* under the name of Sapamine. They are made by acylating an unsymmetrical dialkyl ethylenediamine with a fatty acid chloride, usually oleic acid chloride:

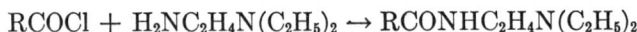

$$RCOCl + H_2NC_2H_4N(C_2H_5)_2 \rightarrow RCONHC_2H_4N(C_2H_5)_2$$

These products are used in the form of their acetates or hydrochlorides. They are also quaternized by the addition of alkyl halide or sulfate.[81] Both the tertiary amine salts and quaternary salts are sold as Sapamines, being differentiated by their suffix letter designation.

In place of the fatty acid chloride the free fatty acid may be reacted with the unsymmetrical dialkyl ethylenediamines. The ammonium salt is the initial reaction product, but on heating to temperatures in the range of 150°C. water is eliminated and the amide is formed. Unsymmetrical fatty acyl ethylenediamine itself may be prepared by the action of a fatty acid ester on an excess of ethylenediamine.[82] Some of the diacyl compound is formed at the same time:

$$RCOOCH_3 + H_2NC_2H_4NH_2 \rightarrow RCONHC_2H_4NH_2$$

By methods similar to those used in preparing the Sapamines, products containing hydroxy alkyl substituents on the basic nitrogen atom have been synthesized, such as:

$$RCONH-C_2H_4N(C_2H_4OH)_2 \quad or \quad RCONH-C_2H_4NHC_2H_4OH, \quad etc.$$

These may be further alkylated and/or quaternized.[83]

A compound closely analogous to the Sapamines but based on 1,3-propylenediamine has been introduced recently under the name Surface Active Agent M by *American Cyanamid Corp.* It has the formula:

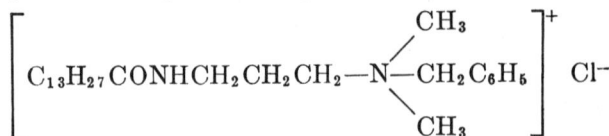

$$\left[C_{13}H_{27}CONHCH_2CH_2CH_2-N \begin{matrix} CH_3 \\ \diagup \\ \diagdown \\ CH_3 \end{matrix} CH_2C_6H_5 \right]^+ Cl^-$$

---

[81] U. S. Pat. 1,737,458, Hartmann and Kägi to *Ciba Co.* See also French Pat. 716,500 to *Ciba Co.*

[82] U. S. Pat. 2,387,201, Weiner to *Bonneville, Inc.*

[83] U. S. Pats. 2,186,464; 2,329,406; 2,357,598; Mauersberger to *Alframine Corp.*

where $C_{13}H_{27}$ is the myristyl radical.

Diethylenetriamine and higher polyethylene polyamines have been acylated with one mole or more of fatty acid to give cationic surface active agents. The number of moles of fatty acid condensed onto the amine must be at least one less than the number of free basic amino groups in the starting material. These reactions are normally carried out by heating a mixture of the amine and the free fatty acid until the equivalent quantity of water has been eliminated. The reaction products are mixtures, and are not usually separated into their components or purified. Esters, glycerides, or fatty acid chlorides may be used in place of the free fatty acids, but this is not normally done in commercial practice. The products are for the most part viscous liquids of strong odor and are used in the form of their salts.[84] Substances of this nature have also been made by reacting a fatty acid or its amide with ethyleneimine in excess.[85]

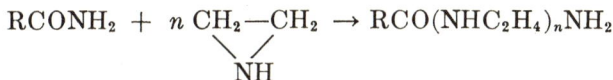

$$RCONH_2 + n\ CH_2{-}CH_2\ (NH) \rightarrow RCO(NHC_2H_4)_nNH_2$$

The fatty acyl derivatives of hydroxylated diamines such as N-(aminoethyl)-ethanolamine are useful surface active agents,[86] although they are usually further condensed to give the imidazolines (see Section IV of this chapter).

$$RCOOH + H_2NC_2H_4NHC_2H_4OH \rightarrow RCONHC_2H_4NHC_2H_4OH$$

The acyl polyethylene polyamines have been modified by further acylation or by treatment with urea.[87] They remain cationic as long as one amino group is unacylated. The acylated polyethylene polyamines have also been quaternized by the addition of excess methyl or ethyl sulfate.[88]

[84] French Pat. 716,238; U. S. Pats. 1,947,951; Neelmeier et al. to General Aniline and Film Corp. U. S. Pats. 2,304,369; Morgan and McLeod to Arnold, Hoffman and Co.; 2,347,178, Fritz and Robinson to National Oil Products Co.

[85] U. S. Pat. 2,163,807, Piggott and Statham to Imperial Chemical Industries Ltd.

[86] U. S. Pats. 2,340,881, Kelley and Robinson to National Oil Products Co.; 2,391,-830, Jayne and Day to American Cyanamid Co.

[87] U. S. Pats. 2,345,632, Kelley and Robinson to National Oil Products Co.; 2,344,-259, Morgan and McLeod to Arnold Hoffman Co.; 2,201,041, Katz to Warwick Chemical Co. Ger. Pat. 731,981, to I. G. Farbenindustrie.

[88] U. S. Pat. 2,329,406, Mauersberger to Alframine Corp.

Aminoethylmorpholine can also be condensed with fatty acids to give products of the Sapamine type:

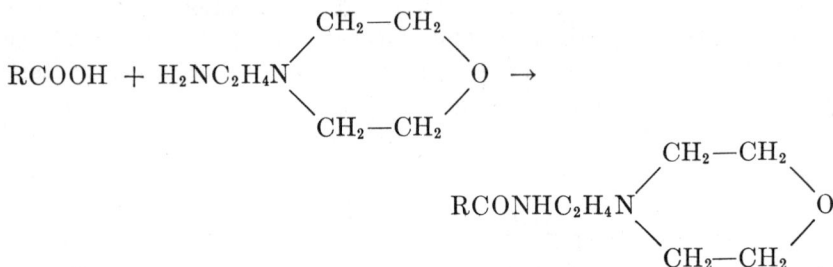

$$RCOOH + H_2NC_2H_4N \underset{CH_2-CH_2}{\overset{CH_2-CH_2}{\diagdown \diagup}} O \rightarrow$$

$$RCONHC_2H_4N \underset{CH_2-CH_2}{\overset{CH_2-CH_2}{\diagdown \diagup}} O$$

When fatty acylated monoethanolamine is heated with pyridine hydrochloride to temperatures in the range of 120°C., water is eliminated and a pyridinium salt is formed.[89]   The yields in this reaction are excellent, and the raw materials are relatively inexpensive, but the products have apparently had little if any commercial exploitation, although they are said to be useful in the flotation of acidic minerals.[90]

Higher fatty acids can be condensed with aminoguanidine or with alkylolcyanamides to give surface active bases.[91]   These may be alkylated further, if desired, with the usual alkylating agents.

Several interesting surface active basic compounds which contain two or more intermediate linkages of the amide and/or ester type have been described by *Emulsol Corp.*,[92] one of the pioneering laboratories in the field of surface active agents.   The following synthesis exemplifies a general method for obtaining products with two amide linkages:

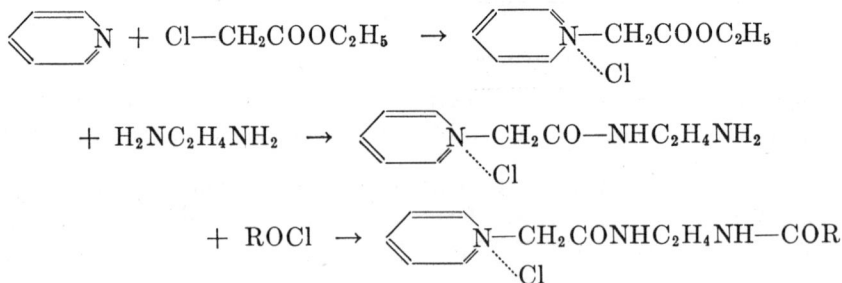

$$\bigcirc N + Cl-CH_2COOC_2H_5 \rightarrow \bigcirc N-CH_2COOC_2H_5$$
$$\cdots Cl$$

$$+ H_2NC_2H_4NH_2 \rightarrow \bigcirc N-CH_2CO-NHC_2H_4NH_2$$
$$\cdots Cl$$

$$+ ROCl \rightarrow \bigcirc N-CH_2CONHC_2H_4NH-COR$$
$$\cdots Cl$$

[89] U. S. Pat. 2,242,211, Haack to *Heyden A.-G.*   See also U. S. Pat. 2,273,181, De Groote to *Petrolite Co.*

[90] U. S. Pat. 2,336,015, Jayne *et al.* to *American Cyanamid Co.*

[91] Brit. Pat. 520,394, to *I. G. Farbenindustrie.*   Also U. S. Pats. 2,258,320–1, Ericks to *American Cyanamid Co.*

[92] U. S. Pat. 2,176,896, Epstein and Katzman to *Emulsol Corp.*

Compounds with one amide and one ester linkage may be prepared as follows:

$$ClCH_2COOC_2H_5 + H_2NC_2H_4OH \rightarrow ClCH_2CONC_2H_4OH$$
$$H$$
$$(I)$$

$$(I) + RCOCl \rightarrow ClCH_2CONHC_2H_4OOCR$$
$$(II)$$

$$(II) + (CH_3)_3N \rightarrow [(CH_3)_3N-CH_2CONHC_2H_4OOCR]^+Cl^-$$

This end product, where RCO represents the mixed coconut fatty acyl residues, foams strongly in aqueous solutions.[93] It is stated to be an extremely powerful germicide and is marketed by the manufacturers under the name Emulsept.

Compounds in which the basic nitrogen is joined to an aromatic nucleus have also been prepared. Thus, for example:

$$RCONH-CH_2-\underset{}{\bigcirc}-N(CH_3)_2$$

may be prepared by heating the reactive p-(dimethylamino)-benzylsulfanilic acid with a fatty acid amide. The product may be quaternized or used in the form of its salts.[94] A fatty acyl urea such as the ureide of stearic acid may be used in place of the fatty amide. The product[95] then has the formula:

$$RCONH-CO-NHCH_2-\underset{}{\bigcirc}-N(CH_3)_2$$

Alkyl (rather than acyl) ureas of the type:

$$RNH-CO-NH\underset{}{\bigcirc}-N(CH_3)_2 \quad \text{and}$$

$$R-\underset{\underset{C_6H_5}{|}}{N}-CO-NH\underset{}{\bigcirc}-N\underset{\underset{CH_2C_6H_5}{\diagdown}}{\overset{\diagup CH_3}{}}$$

[93] U. S. Pats. 2,245,593; 2,189,664; Katzman to *Emulsol Corp.* For more complicated structures with multiple ester and amide linkages, see U. S. Pats. 2,299,756; 2,388,154; Katzman and Harris to *Emulsol Corp.*

[94] U. S. Pat. 2,229,803, Engel and Pfaehler to *J. R. Geigy A.-G.*; Brit. Pat. 500,412; Swiss Pat. 199,452; to *J. R. Geigy A.-G.*

[95] Swiss Pats. 207,299; 207,300-3; to *J. R. Geigy A.-G.* U. S. Pat. 2,312,395, Engel and Pfaehler to *J. R. Geigy A.-G.*

have also been used as surface active agents, mostly for softening cellulosic fabrics.[96]

Products of the type:

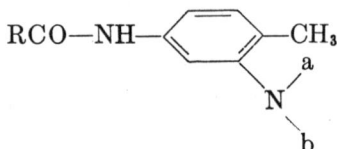

$$RCO—NH—\underset{\underset{b}{\overset{|}{\underset{N}{}}}}{\bigcirc}\underset{a}{—CH_3}$$

can be prepared in the usual manner by acylating the free amine with a fatty acid chloride. These relatively simple products are good wetting agents when quaternized.[97] Their hydroaromatic analogues such as:

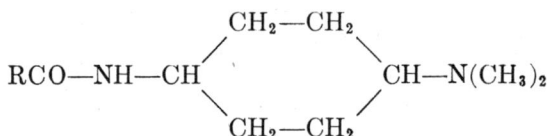

$$RCO—NH—CH\underset{CH_2—CH_2}{\overset{CH_2—CH_2}{}}CH—N(CH_3)_2$$

can be made by heating the corresponding cyclo aliphatic base with a free fatty acid.[98]

In the compounds considered above the hydrophobic group is contained in the fatty acid which forms the amide linkage. The essential structural elements are a hydrophobic carboxy acid and a diamine. One of the amino groups of the diamine forms the amide link and the other remains free to form the essential hydrophilic residue. Amide-linked cationic detergents can also be formed by combining a long-chain hydrophobic amine with an aminocarboxy acid or an equivalent structure. The starting materials are inherently more costly in preparing this type of compound. Nevertheless numerous examples have been described in the patent literature.

Probably the best-known members of this series are made by acylating a fatty amine with chloroacetic acid, usually in the form of its halide or ester, and subsequently replacing the chlorine by an amino group or quaternizing with a tertiary amine.[99]

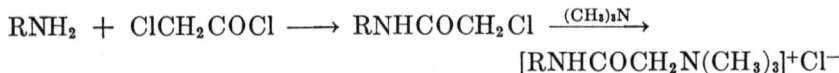

$$RNH_2 + ClCH_2COCl \longrightarrow RNHCOCH_2Cl \xrightarrow{(CH_3)_3N}$$
$$[RNHCOCH_2N(CH_3)_3]^+Cl^-$$

[96] Swiss Pats. 198,680; 206,594. U. S. Pat. 2,253,773; Engel and Pfaehler to *J. R. Geigy A.-G.*

[97] Swiss Pat. 206,718, to *J. R. Geigy A.-G.*

[98] See French Pat. 838,597, to *Ciba Co.*, which deals mainly with the conversion of these products to their corresponding amine oxides.

[99] See Brit. Pat. 505,429 and Ger. Pat. 710,480, to *Deutsche Hydrierwerke A.-G.*, for examples of this type and germicidal mixtures with other substances.

The products can be varied by using different hydrophobic amines, halocarboxylic acids, and hydrophilic amines. The following examples are typical of numerous others in the literature:

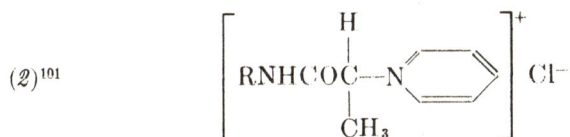

$(1)^{100}$
$$\left[RNHCOCH_2\overset{\overset{\displaystyle CH_3}{|}}{\underset{\underset{\displaystyle CH_3}{|}}{N}}CH_2C_6H_5 \right]^+ \quad Cl^-$$

$(2)^{101}$
$$\left[RNHCOC\overset{\overset{\displaystyle H}{|}}{\underset{\underset{\displaystyle CH_3}{|}}{---N}}\langle \rangle \right]^+ \quad Cl^-$$

This product is made by condensing ethyl lactate with coconut fatty amine, and heating the resultant lactic amide with pyridine hydrochloride.

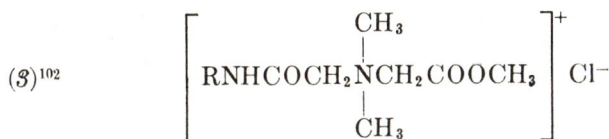

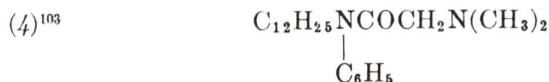

$(3)^{102}$
$$\left[RNHCOCH_2\overset{\overset{\displaystyle CH_3}{|}}{\underset{\underset{\displaystyle CH_3}{|}}{N}}CH_2COOCH_3 \right]^+ \quad Cl^-$$

$(4)^{103}$
$$C_{12}H_{25}\overset{}{\underset{\underset{\displaystyle C_6H_5}{|}}{N}}COCH_2N(CH_3)_2$$

$(5)^{104}$
$$RNHCO\langle \overset{=N}{\quad} \rangle$$

This product is made by heating nicotinic acid with laurylamine, for example, until the amide is formed with the elimination of water.[104]

The urea linkage has also been used to combine hydrophobic and hydrophilic amines into a surface active molecule:

$$R—NHCONH—CH_2C_6H_4N(CH_3)_2$$

has been prepared from the carbamate $RNHCOOCH_3$.[105]   It is used in the form of a quaternized derivative as a softening agent.

[100] U. S. Pats. 2,295,655, to Reuss *et al.*; 2,317,999, Leuchs to *Winthrop Chemical Co.*
[101] U. S. Pats. 2,242,211; 2,329,703; 2,348,613; De Groote to *Petrolite Corp.*
[102] U. S. Pat. 2,191,978, Balle and Horst to *I. G. Farbenindustrie.*
[103] U. S. Pat. 2,153,707, Becherer and Iselin to *J. R. Geigy A.-G.*; French Pat. 815,634; to *J. R. Geigy A.-G.*
[104] U. S. Pat. 2,304,830, Katzman to *Emulsol Corp.*
[105] Swiss Pats. 213,837–40; 199,452; to *J. R. Geigy A.-G.*

(b) Cationic Surface Active Agents Used Primarily for Water-Repellent Treatments

There is a large class of cationic surface active agents the members of which are used almost exclusively for producing water-repellent finishes on fabrics. Since these products are very closely related to those used for other purposes, and since most of the patent disclosures state that they may also be used for foaming, wetting, emulsifying, etc. (although they very seldom are used for these purposes), it might be well to consider them at this point. The product which is currently of greatest commercial importance in this field has an amide linkage between the hydrophobic group and the solubilizing quaternary ammonium group. Many of the others do not have this linkage, but will nevertheless be included in the discussion. Practically all of the products use the n-octadecyl or n-heptadecyl group (derived directly or indirectly from stearic acid) as the essential hydrophobic radical. These substances are by no means the only ones which are used in rendering fabrics water-repellent, but they are widely and successfully applied for this purpose. The finishes which they afford are much more resistant to laundering and cleaning than the simple wax treatments and hence have become very popular.

The best-known product of this class in this country is Zelan (*E. I. du Pont de Nemours & Co.*). It has the formula:

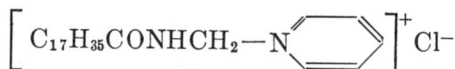

$$\left[C_{17}H_{35}CONHCH_2 - N \bigcirc \right]^+ Cl^-$$

and may be made by treating stearic amide with formaldehyde, HCl, and pyridine. It is soluble in water giving a soapy solution. It is applied to the fabric in aqueous solution, usually buffered with sodium acetate. The fabric is then dried and subsequently "baked," or heated to temperatures in the range of 120–150°C. for a few minutes. During this baking operation pyridine is given off. The fabric is finally rinsed and dried. It is generally accepted today that the quaternary compound is decomposed during baking and forms methylenedistearamide, $(C_{17}H_{35}CONH)_2CH_2$, which deposits on the fiber surfaces as a tough, thin, adherent waxy coating which is responsible for the water-repellent properties. Much of the earlier literature explains the permanent repellent effect by postulating an actual etherification of the hydroxyl groups on the cellulose. This may possibly be an important factor but is currently not considered to be so by practitioners in this field.

The essential structural feature of these products appears to be a methylene group between a quaternary nitrogen and an amide nitrogen. In some cases the quaternary nitrogen may be replaced by a tertiary nitrogen, and the amide group by an ether or ester oxygen. Velan (*Imperial Chemical Industries*), the original product in this field, has the structure:

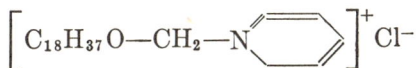

$$\left[C_{18}H_{37}O-CH_2-N \right]^+ Cl^-$$

and is made from octadecyl chloromethyl ether and pyridine. Norane (*Warwick Chemical Co.*), a very successful product, is said to have the structure:

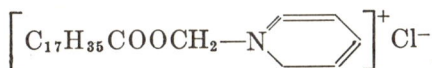

$$\left[C_{17}H_{35}COOCH_2-N \right]^+ Cl^-$$

and to be made by the combination of stearic acid chloride, formaldehyde and pyridine. These products are applied in the same general manner as Zelan. A voluminous patent literature has been built up in this field, but the above essential features are present in practically all the described products.[106] Sulfonamides have been used in place of the more usual fatty acyl amides:[107]

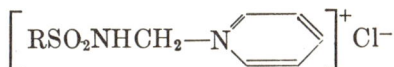

$$RSO_2NH_2 + CH_2O + HCl + C_5H_5N \longrightarrow$$

$$\left[RSO_2NHCH_2-N \right]^+ Cl^-$$

[106] Typical examples: U. S. Pats. 2,146,392, Baldwin and Walker to *Imperial Chemical Industries Ltd.*; 2,146,406, Piggott to *Imperial Chemical Industries Ltd.*; 2,147,811, 2,303,191, Baldwin and Piggott to *Imperial Chemical Industries Ltd.*; 2,369,776, Cusa, Salkeld, and Walker to *Imperial Chemical Industries Ltd.* Brit. Pat. 497,856, to *Imperial Chemical Industries Ltd.* U. S. Pats. 1,952,008, Bruson to *Rohm and Haas Co.*; 2,204,653, Bock to *Rohm and Haas Co.*; 2,282,701, Bock and Bruson to *Rohm and Haas Co.* Brit. Pat. 418,247, to *Rohm and Haas Co.* U. S. Pats. 2,146,408, Shipp to *E. I. du Pont de Nemours & Co.*; 2,268,395, Henke and Pikl to *E. I. du Pont de Nemours & Co.*; 2,327,160, Bacon to *E. I. du Pont de Nemours & Co.*; 2,352,152, Kaplan to *Richards Chemical Co.*; 2,333,452, Thurston and Nagy to *American Cyanamid Co.*; 2,344,934, West to *American Cyanamid Co.*; 2,242,565, Wolf to *Heberlein Patent Corp.* French Pat. 834,949; Brit. Pats. 501,480; 492,669; Ger. Pat. 703,501, to *Farberei-Gesellschaft Flores und Co. vormals Stolte-Missy.* U. S. Pat. 2,301,676, Balle and Orthner to *I. G. Farbenindustrie.* French Pat. 847,599, to *I. G. Farbenindustrie.* U. S. Pats. 2,279,497, Sallmann and Albrecht to *Ciba Co.*; 2,338,177, Gränacher, Sallmann, and Albrecht to *Ciba Co.*; 2,338,178, Gränacher and Sallmann to *Ciba Co.*

[107] U. S. Pat. 2,344,321, Orthner, Balle, and Schild to *General Aniline and Film Co.*

Thiourea may be used as the solubilizing basic group in this type of product in place of pyridine or other tertiary bases:[108]

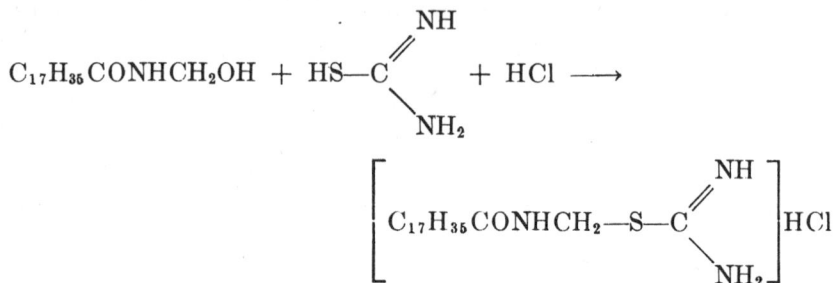

$$C_{17}H_{35}CONHCH_2OH + HS—C \overset{\displaystyle \nearrow NH}{\underset{\displaystyle \searrow NH_2}{}} + HCl \longrightarrow$$

$$\left[C_{17}H_{35}CONHCH_2—S—C \overset{\displaystyle \nearrow NH}{\underset{\displaystyle \searrow NH_2}{}} \right] HCl$$

Many of these products are fairly stable in aqueous solution and have been described as softening agents as well as emulsifiers, etc. Their success as water repellents, however, depends to a large degree on their instability. If they are too stable they will not decompose with sufficient rapidity during the baking process. In most applications of surface active compounds stability is a highly desirable property. Accordingly, the types listed above are not used to any extent except as water-repellent finishes.

(c) Ester Intermediate Linkage

Cationic surface active compounds with intermediate ester linkages may be synthesized by linking a hydrophobic carboxy acid with an amino alcohol or by linking a hydrophobic alcohol with an amino acid. The general reactions are:

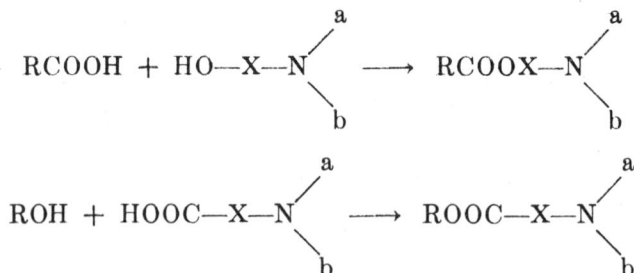

$$RCOOH + HO—X—N \overset{\displaystyle \nearrow a}{\underset{\displaystyle \searrow b}{}} \longrightarrow RCOOX—N \overset{\displaystyle \nearrow a}{\underset{\displaystyle \searrow b}{}}$$

$$ROH + HOOC—X—N \overset{\displaystyle \nearrow a}{\underset{\displaystyle \searrow b}{}} \longrightarrow ROOC—X—N \overset{\displaystyle \nearrow a}{\underset{\displaystyle \searrow b}{}}$$

Indirect synthetic methods may be used to attain the final structures, or the reactions may be carried out in stages, but the essential structural features remain as shown.

The higher fatty acids such as stearic, oleic, ricinoleic, mixed coconut

[108] U. S. Pats. 2,331,387; 2,345,109; Gränecher, Sallmann, and Albrecht to *Ciba Co.*

fatty, etc. may be esterified with tertiary amino alcohols by direct heating since there is no chance for amide formation. They may alternatively be converted to the corresponding acid chlorides and reacted with the tertiary amino alcohols. Triethanolamine is a preferred amino alcohol in this reaction, and one, two, or all three of its hydroxyl groups may be esterified. The esters are used in the form of their salts.[109]

In place of triethanolamine other tertiary amino alcohols such as diethyl-ethanolamine, methyldiethanolamine, N-(hydroxyethyl)-morpholine, etc. have been used. More complex amino alcohols such as those formed by reacting the ethanolamines with glycerol chlorohydrin or epichlorohydrin have also been used[110] to esterify with fatty acids.

Triethanolamine has also been "polymerized" by heating before being esterified[111] with the higher fatty acids. The polymerization is presumably an ether formation between hydroxyl groups on different molecules and is analogous to the formation of polyglycerols from glycerol.

Amino alcohols in which the amino group is substituted by amide-bearing substituents have also been used,[112] an example being the compound:

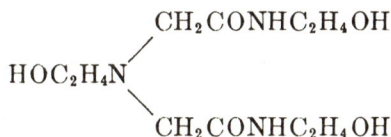

$$HOC_2H_4N \begin{array}{c} CH_2CONHC_2H_4OH \\ \diagdown \\ CH_2CONHC_2H_4OH \end{array}$$

Long-chain amino alcohols have also been used,[113] a good example being the product:

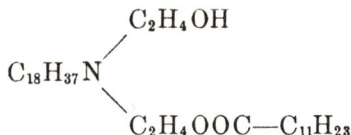

$$C_{18}H_{37}N \begin{array}{c} C_2H_4OH \\ \diagdown \\ C_2H_4OOC—C_{11}H_{23} \end{array}$$

which can be used in the form of its salts as a softener.

The fatty acid esters of these various tertiary amino alcohols have been

[109] U. S. Pats. 1,946,079–80, Kern and Sala to *E. I. du Pont de Nemours & Co.*; 2,173,058, to Wolf Kritchevsky; 2,022,678, to Wolf Kritchevsky *et al.*; 2,228,985, De Groote *et al.* to *Petrolite Corp.*

[110] U. S. Pat. 2,149,527, Kartaschoff to *Chemische Fabrik Sandoz.* Ger. Pat. 737-738; French Pat. 866,855; to *Sandoz Ltd.*

[111] U. S. Pat. 2,248,089, Katzman and Epstein to *Emulsol Corp.*

[112] U. S. Pats. 2,368,208; 2,390,942; Katzman and Epstein to *Emulsol Corp.* U. S. Pat. 2,290,881, Katzman to *Emulsol Corp.*

[113] U. S. Pat. 2,359,043, Link and Jaccard to *Sandoz Ltd.*

used to a considerable extent in Germany as finishing agents.[114]   Soromine
DB is the stearic ester of dibutylaminoethanol.   Soromine A Base is
stearyltriethanolamine and Emulphor FM is oleyltriethanolamine.

If diethanolamine or monoethanolamine is heated with a fatty acid the
amide rather than the ester is formed.   In order to form the ester it is
necessary to block the amino group.   This is usually done by converting
the amino alcohol to its hydrochloride and esterifying with the fatty acid
chloride:

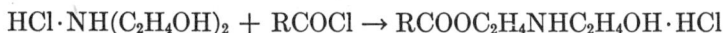

$$HCl \cdot NH_2C_2H_4OH + RCOCl \rightarrow RCOOC_2H_4NH_2 \cdot HCl$$

$$HCl \cdot NH(C_2H_4OH)_2 + RCOCl \rightarrow RCOOC_2H_4NHC_2H_4OH \cdot HCl$$

Both of these products[115] are effective cationic surface active agents.
They are readily hydrolyzed in alkaline solution but are normally used in
acid solution where they exist as salts.

Quaternary ammonium compounds of similar structure may be made by
heating a halogenated ester with a tertiary amine.   As an example, lauric[116]
acid may be esterified with ethylenebromohydrin, and the resulting β-
bromoethyl laurate heated with pyridine:

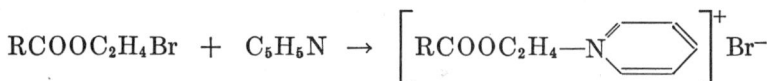

$$RCOOC_2H_4Br + C_5H_5N \rightarrow \left[RCOOC_2H_4\text{—N} \bigcirc \right]^+ Br^-$$

Hydroxyethylcyanamide,[117] aminoethylethanolamine, and acylated am-
inoethylethanolamines are among other basic alcohols which have been
esterified with fatty acids.

Among the amino acid esters of hydrophobic alcohols the simplest are
the glycine esters.   The glycine ester of lauryl alcohol:

$$H_2N\text{—}CH_2COOC_{12}H_{25}$$

has been made[118] in the form of its p-toluenesulfonic acid salt by heating
glycine, p-toluenesulfonic acid, and lauryl alcohol together to about 200°C.

[114] Ger. Pat. 546,406; U. S. Pat. 2,187,823; Ulrich and Nüsslein to I. G. Farbenindustrie.

[115] U. S. Pats. 2,355,442; 2,305,083; 2,322,202; Jayne and Day to American Cyanamid
Co.   U. S. Pat. 2,354,320, Johnson to United Shoe Machinery Co.

[116] Ger. Pats. 626,718; 677,698; to Ciba Co.   U. S. Pat. 2,391,831, Jayne and Day to
American Cyanamid Co.

[117] U. S. Pat. 2,258,320, Ericks to American Cyanamid Co.

[118] U. S. Pat 2,293,026, Day and Jayne to American Cyanamid Co.   See also U. S.
Pat. 2,382,360, Weiner to Bonneville Ltd.

The products may be used as such or may be converted to salts of other acids.

Analogous products are made[119] indirectly by esterifying a fatty alcohol (or more generally any hydroxyl-bearing hydrophobic group) with chloroacetic acid. The chlorine atom of the resultant chloroacetic ester is then replaced by reacting with an amine. If a tertiary amine is used the final product is a quaternary ammonium salt:

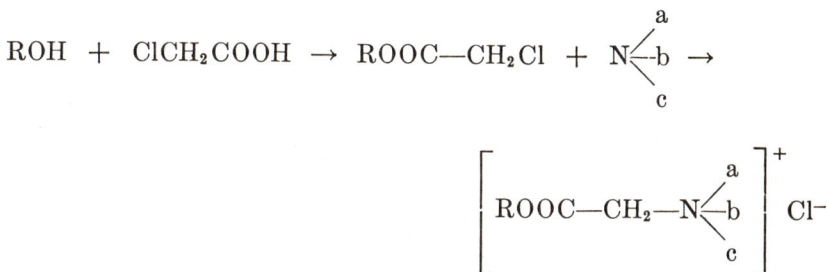

$$\text{ROH} + \text{ClCH}_2\text{COOH} \rightarrow \text{ROOC—CH}_2\text{Cl} + \text{N}\begin{smallmatrix}a\\b\\c\end{smallmatrix} \rightarrow$$

$$\left[\text{ROOC—CH}_2\text{—N}\begin{smallmatrix}a\\b\\c\end{smallmatrix}\right]^+ \text{Cl}^-$$

The flexibility afforded by the sequence of reactions is obviously very great. The "ROH" component may be a simple fatty alcohol such as lauryl alcohol, a more complex one such as cholesterol, or one which in itself contains intermediate linkages, examples being:

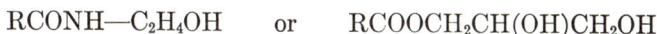

$$\text{RCONH—C}_2\text{H}_4\text{OH} \quad \text{or} \quad \text{RCOOCH}_2\text{CH(OH)CH}_2\text{OH}$$

Among amines which have been reacted[120] with the ROOC—CH$_2$Cl intermediate are pyridine, aniline, mono- and dihydroxyethylanilines, piperidine, morpholine, benzyldimethylamine, triethylamine, and others.

The simple betaine esters:

$$[\text{R—OOC—CH}_2\text{—N (CH}_3)_3]^+ \text{Cl}^-$$

are stated to be valuable antiseptics and effective surface active agents. Ether esters of formula:

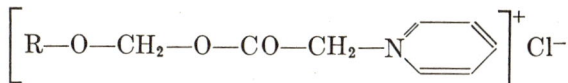

$$\left[\text{R—O—CH}_2\text{—O—CO—CH}_2\text{—N}\bigcirc\right]^+ \text{Cl}^-$$

[119] U. S. Pats. 2,023,075, Harris to *Emulsol Corp.*; 2,190,133 and 2,213,979, Epstein and Katzman to *Emulsol Corp.* See also U. S. Pats. 2,317,378; 2,321,594, to Benjamin R. Harris.

[120] Brit. Pats. 403,883; 408,754; to *Henkel et Cie.* Swiss Pat. 206,717 to *J. R. Geigy A.-G.* U. S. Pats. 2,356,587, Hentrich *et al.* vested in Alien Property Custodian; 2,367,878, Lee to *Hoffman-LaRoche Co.*; 2,087,565, Balle and Eisfeld to *General Aniline Works.*

and diesters of formula:

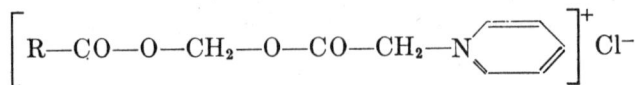

$$\left[R—CO—O—CH_2—O—CO—CH_2—N\underset{=}{\bigcirc} \right]^+ Cl^-$$

may be made[121] by reacting the chloromethyl ether of a fatty alcohol or the chloromethyl ester of a fatty acid with sodium chloroacetate, and subsequently quaternizing with pyridine.

Dialkyl maleates such as dioctyl maleate have been converted[122] to cationic surface active agents by adding monoethanolamine across the double bond. The product is a dialkyl ester of N-β-hydroxyethylaspartic acid:

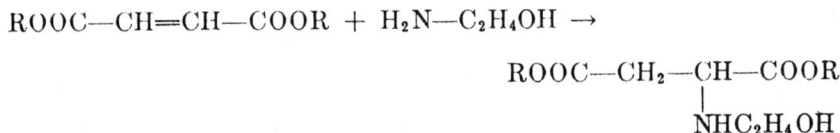

$$ROOC—CH{=}CH—COOR + H_2N—C_2H_4OH \rightarrow$$

$$
\begin{array}{c}
ROOC—CH_2—CH—COOR \\
| \\
NHC_2H_4OH
\end{array}
$$

Surface active esters of aromatic amino acids are represented by such compounds as cetyl p-aminobenzoate or dodecyl anthranilate. The alkyl dimethylamino benzoates can be quaternized[123] readily with reagents such as dimethyl sulfate or the lower alkyl halides:

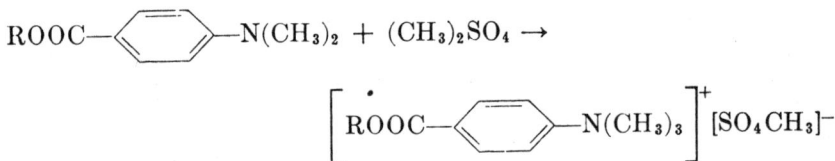

$$ROOC—\bigcirc—N(CH_3)_2 + (CH_3)_2SO_4 \rightarrow$$

$$\left[ROOC—\bigcirc—N(CH_3)_3 \right]^+ [SO_4CH_3]^-$$

More complex aromatic amino esters have also been described.[124] A typical example is the diphenylmethane derivative:

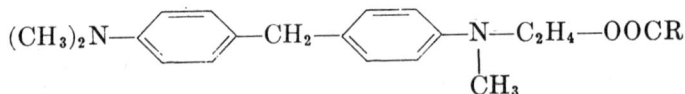

$$
\begin{array}{c}
(CH_3)_2N—\bigcirc—CH_2—\bigcirc—N—C_2H_4—OOCR \\
| \\
CH_3
\end{array}
$$

which can be used as such or can be quaternized.

[121] Swiss Pats. 211,243; 211,244; 209,972; to *Ciba Co.*

[122] U. S. Pat. 2,324,712; Brit. Pat. 553,166; Lynch to *American Cyanamid Co.*

[123] U. S. Pat. 2,202,864, Piggott and Woolvin to *Imperial Chemical Industries Ltd.* French Pat. 759,821, to *Henkel et Cie.*

[124] Swiss Pats. 208,532; 210,986–7; to *J. R. Geigy A.-G.*

Fatty acid esters of basic heterocyclic alcohols have been made,[125] a typical reaction being:

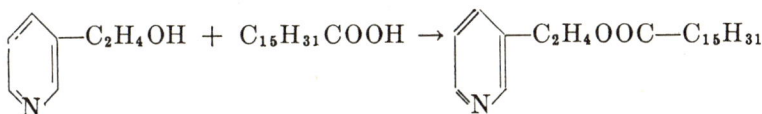

$$\text{[pyridine]}-C_2H_4OH + C_{15}H_{31}COOH \rightarrow \text{[pyridine]}-C_2H_4OOC-C_{15}H_{31}$$

In place of straight carboxylic acids there may be used sulfonic acids, and in place of hydrophobic alcohols there may be used the hydrophobic alkylated phenols. Starting with such reagents as *m*-nitrobenzenesulfonyl chloride and *p-tert*-octylphenol, the following reactions may be carried out:

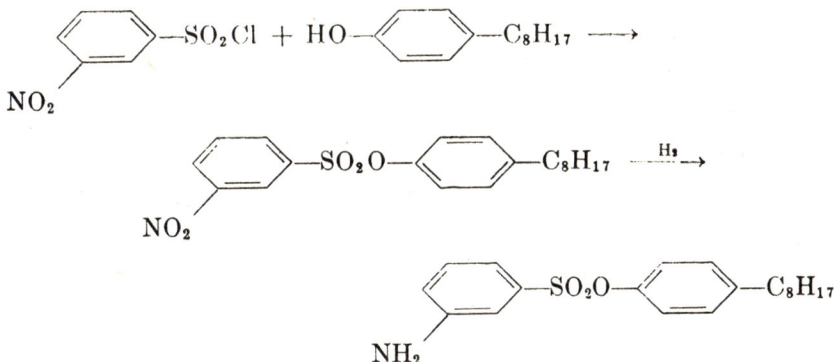

$$\text{[benzene ring, } NO_2]-SO_2Cl + HO-\text{[benzene ring]}-C_8H_{17} \longrightarrow$$

$$\text{[benzene ring, } NO_2]-SO_2O-\text{[benzene ring]}-C_8H_{17} \xrightarrow{H_2}$$

$$\text{[benzene ring, } NH_2]-SO_2O-\text{[benzene ring]}-C_8H_{17}$$

This product may then be alkylated and quaternized.[126]

Few of the ester-linked products described in this section appear to have been produced on a large scale under well-known trade names, but some of them such as the fatty acid esters of the ethanolamines and the aminated chloroacetic esters have been made and used commercially.

(d) Ether and Related Intermediate Linkages

The simplest ether-linked cationic surface active agents have the general formula:

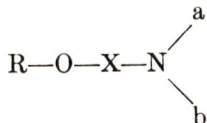

$$R-O-X-N\begin{matrix} a \\ b \end{matrix}$$

[125] Ger. Pat. 702,583, to *Chemische Fabrik Heyden A.-G.*
[126] Brit. Pat. 517,339, to *Deutsche Hydrierwerke A.-G.*

where X represents an alkylene group.   Mention has already been made of the products where X is a methylene group and the basic nitrogen is contained in a pyridine ring.   These are the Velan type products used as water-repellent finishes.   When pyridine or quinoline is not used as the basic component, more stable products can be prepared.   Thus the product made[127] by treating a higher alkoxymethyl chloride with triethylamine is said to be sufficiently stable for use as a wetting or emulsifying agent under usual conditions:

$$R\!-\!O\!-\!CH_2Cl + (C_2H_5)_3N \rightarrow [ROCH_2N(C_2H_5)_3]^+Cl^-$$

When the alkylene group —X— in the above generic formula contains two or more carbon atoms, the ethers are fully stable under all normal conditions of use.   They may be made by several different synthetic methods.

Epichlorohydrin has been reacted with fatty alcohols and also with such hydrophobic alcohols as hydroabietenol to form γ-chloro-α-glyceryl ethers. Boron trifluoride catalysts are effective in this condensation.

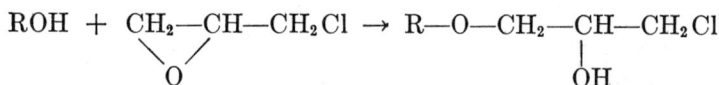

$$ROH + CH_2\!-\!CH\!-\!CH_2Cl \rightarrow R\!-\!O\!-\!CH_2\!-\!\underset{\underset{OH}{|}}{CH}\!-\!CH_2Cl$$

This halide can then be reacted[128] with amines to give the desired surface active agents.

Ethers of tertiary amino alcohols such as triethanolamine have been made directly by converting the amino alcohol to its sodium derivative and then reacting with a long-chain alkyl halide such as cetyl bromide.   Metallic sodium is used to make the sodium derivative.   Sodamide and sodium hydroxide are also said to be effective in this conversion.   If triethanolamine is used one or more of the hydroxyl groups may be etherified.[129] If a product such as N,N-diethylethanolamine is used, the reaction is more clean-cut and only one product is obtained:[130]

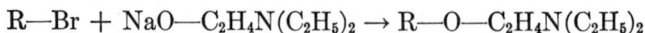

$$R\!-\!Br + NaO\!-\!C_2H_4N(C_2H_5)_2 \rightarrow R\!-\!O\!-\!C_2H_4N(C_2H_5)_2$$

These products may be used as salts or they may be quaternized.   They

[127] Ger. Pat. 704,338, to *Deutsche Hydrierwerke A.-G.*

[128] U. S. Pats. 2,231,502, Krzikalla *et al.* to *General Aniline and Film Corp.*; 2,334,-517, Tucker to *Procter and Gamble Co.*

[129] U. S. Pat. 2,290,880, Katzman and Epstein to *Emulsol Corp.*

[130] U. S. Pats. 2,297,221, Huttenlocher, vested in Alien Property Custodian; 2,312,135, Ulrich and Küspert to *General Aniline and Film Corp.*   Brit. Pat. 517,-114–5; French Pat. 832,288; to *Zschimmer and Schwarz Chemische Fabrik.*   See also French Pats. 837,604; 771,746; to *I. G. Farbenindustrie.*

are said to be highly effective surface active agents, but are essentially expensive to produce.

Long-chain ethers of N-β-hydroxyethylmorpholine have been prepared and studied by Niederl and co-workers.[131] The Williamson synthesis was used:

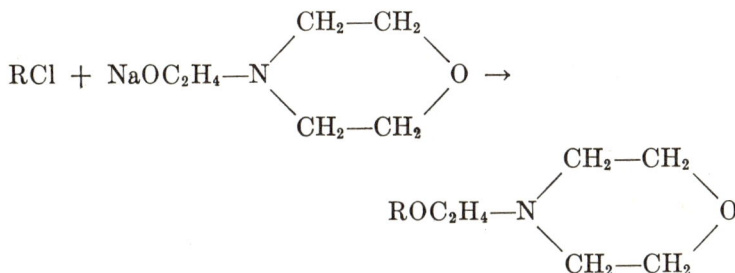

$$RCl + NaOC_2H_4{-}N\underset{CH_2-CH_2}{\overset{CH_2-CH_2}{<\qquad>}}O \rightarrow$$

$$ROC_2H_4{-}N\underset{CH_2-CH_2}{\overset{CH_2-CH_2}{<\qquad>}}O$$

An ingenious method for producing an ether-linked primary amine has recently been described. A fatty alcohol is added to the double bond of acrylonitrile, using sodium ethylate as a catalyst. The resulting ether nitrile is then hydrogenated[132] to the amine:

$$ROH + CH_2{=}CHCN \rightarrow RO{-}CH_2CH_2CN \xrightarrow{H_2} RO{-}CH_2CH_2CH_2NH_2$$

Dihydric phenols may also serve as intermediate links between a hydrophobic chain and a basic solubilizing group. Starting with resorcinol, for example, one of the hydroxyl groups may be etherified with cetyl bromide, and the other hydroxyl group with an amino alkyl halide. A typical product has the formula:

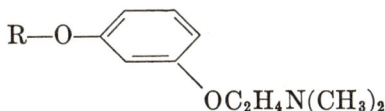

$$R{-}O{-}\langle\!\!\bigcirc\!\!\rangle{-}OC_2H_4N(CH_3)_2$$

This may be quaternized with benzyl chloride or other suitable quaternizing agents. The resulting products are said to be very effective germicides.[133]

Hydrophobic monohydric phenols in which the hydrophobic chain is

---

[131] Niederl, Wolf, and Slobodiansky, *J. Am. Chem. Soc.* **67**, 1227 (1945).

[132] See disclosure in U. S. Pats. 2,371,892 and 2,372,624, Carpenter to *American Cyanamid Co.*

[133] Ger. Pat. 697,297 to Domagk. Ger. Pat. 627,880; French Pat. 769,444; to *I. G. Farbenindustrie.* U. S. Pats. 2,087,131-2, Taub and Leuchs to *Alba Pharmaceutical Co.*

directly attached to the benzene nucleus can also be etherified with amino alkyl halides. Alternatively they may be etherified with one side of an alkylene dihalide, the other halogen then being replaced by an amine, or quaternized with a tertiary amine. For example p-sec-dodecylphenol, made by condensing lauryl alcohol with phenol, is etherified[134] with N-β-chloroethylpiperidine:

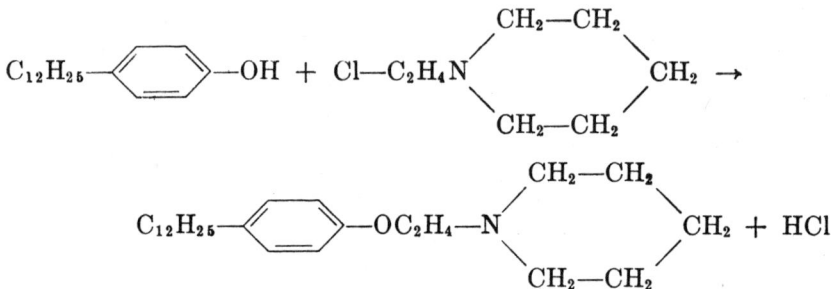

$$C_{12}H_{25} - \langle \rangle - OH \ + \ Cl - C_2H_4N \begin{array}{c} CH_2 - CH_2 \\ \diagup \qquad \diagdown \\ \qquad \qquad CH_2 \rightarrow \\ \diagdown \qquad \diagup \\ CH_2 - CH_2 \end{array}$$

$$C_{12}H_{25} - \langle \rangle - OC_2H_4 - N \begin{array}{c} CH_2 - CH_2 \\ \diagup \qquad \diagdown \\ \qquad \qquad CH_2 \ + \ HCl \\ \diagdown \qquad \diagup \\ CH_2 - CH_2 \end{array}$$

Acyl phenols such as p-lauroylphenol can also be converted to basic ethers in a similar manner.[135]

One of the commercially important cationic surface active agents, used mainly as a disinfectant and germicide, is of this general type. It is marketed under the name Hyamine by the manufacturer, *Rohm and Haas Co.*, and under the name Phemerol by *Parke, Davis, and Co.* It may be made by condensing p-*tert*-octylphenol (from diisobutene and phenol) with dichlorodiethyl ether. The resulting ether monohalide is then quaternized with benzyldimethylamine. The final product therefore has the formula:

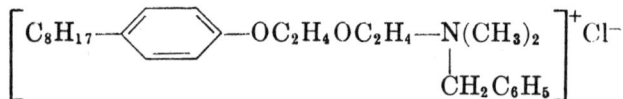

$$\left[C_8H_{17} - \langle \rangle - OC_2H_4OC_2H_4 - \underset{\underset{CH_2C_6H_5}{|}}{N}(CH_3)_2 \right]^+ Cl^-$$

A large number of other specific examples are given[136] in the patents covering this compound. These products, like their anionic analogues, which have already been discussed, have a high degree of surface activity, being excellent surface tension depressants and good foaming and wetting agents. Straight aliphatic ether amines have been prepared by similar procedures.[137]

Heterocyclic bases may in some cases be linked by an ether oxygen to a hydrophobic group, and the resulting compounds used as tertiary or quater-

[134] French Pat. 829,228; Brit. Pat. 493,865; to *Henkel et Cie.*
[135] Brit. Pats. 488,700; 484,906; to *I. G. Farbenindustrie.*
[136] Brit. Pat. 494,766; U. S. Pats. 2,170,111; 2,229,024; Bruson to *Rohm and Haas Co.*
[137] U. S. Pat. 2,355,337, Spence to *Rohm and Haas Co.*

nary salts. The hydroxyquinolines are adaptable starting materials for this type of synthesis:[138]

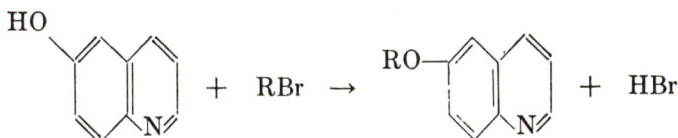

(e) *Sulfide and Sulfone Intermediate Linkages*

By starting with higher alkyl mercaptans rather than alcohols, compounds with thioether linkages may be prepared. Dodecyl mercaptan, for example, may be reacted with diethylaminoethylchloride, in the presence of caustic soda, to give the product:[139]

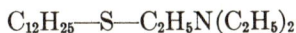

$$C_{12}H_{25}\text{—}S\text{—}C_2H_5N(C_2H_5)_2$$

This same mercaptan has also been treated with formaldehyde and hydrochloric acid and subsequently quaternized:[140]

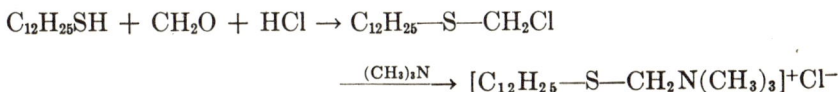

$$C_{12}H_{25}SH + CH_2O + HCl \rightarrow C_{12}H_{25}\text{—}S\text{—}CH_2Cl$$

$$\xrightarrow{(CH_3)_3N} [C_{12}H_{25}\text{—}S\text{—}CH_2N(CH_3)_3]^+Cl^-$$

The higher mercaptans may also be etherified with dichloroethyl ether or chloroethyl chloromethyl ether and the resulting products quaternized:

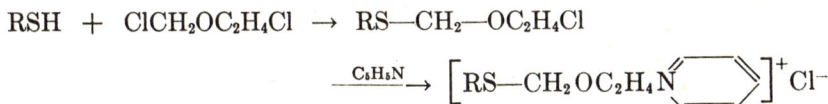

$$RSH + ClCH_2OC_2H_4Cl \rightarrow RS\text{—}CH_2\text{—}OC_2H_4Cl$$

$$\xrightarrow{C_5H_5N} \left[RS\text{—}CH_2OC_2H_4N \right]^+Cl^-$$

Mixed-ester, sulfide-linked products of this nature have also been described:[141]

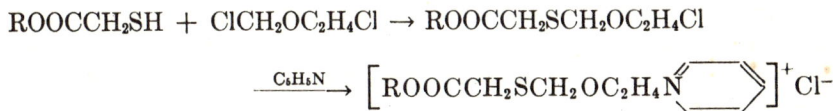

$$ROOCCH_2SH + ClCH_2OC_2H_4Cl \rightarrow ROOCCH_2SCH_2OC_2H_4Cl$$

$$\xrightarrow{C_5H_5N} \left[ROOCCH_2SCH_2OC_2H_4N \right]^+Cl^-$$

A more complex sulfide-linked compound has been described[142] in which

[138] U. S. Pat. 2,152,047, Hahl and Leuchs to *Alba Pharmaceutical Co.*
[139] Brit. Pat. 436,725, to *I. G. Farbenindustrie.*
[140] U. S. Pat., 2,086,585, Taub and Leuchs to *Alba Pharmaceutical Co.*
[141] Swiss Pats. 211,245; 209,972; 211,246; to *Ciba Co.*
[142] Brit. Pat. 437,285, to *Imperial Chemical Industries Ltd.*

mercaptobenzothiazole is etherified with octadecyl bromide and subsequently quaternized with dimethyl sulfate:

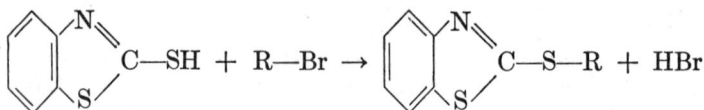

It is not clear whether the quaternized compound made from this intermediate is a sulfonium or an ammonium compound.

Surface active amino aryl sulfones have also been described. They may be made by the following series of reactions:

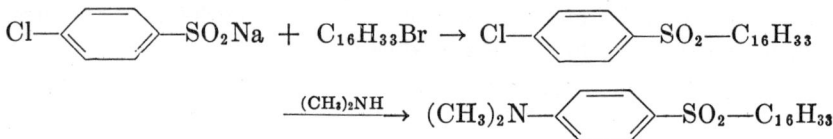

The replacement of the chlorine atom by the dimethylamino group proceeds readily because of the activating effect of the sulfone group in the para position. In place of cetyl bromide such products as ethyl α-bromo-laurate may be used. The products may be quaternized.[143]

Cationic surface active agents containing a sulfonamide linkage have been made by reacting those sulfonyl chlorides which are primary products of the Reed reaction with ethylenediamine or other di- or polyamines. If at least one amino is left unacylated the products[144] are acid soluble and surface active.

(f) Imino Intermediate Linkage

Compounds with a basic imino linkage between the hydrophobic group and the solubilizing amino group may be regarded as directly linked poly-basic compounds. At least one patent,[145] however, implies a point of view in which the imino group is regarded primarily as a linking group. It describes the reaction of heptadecylamine with 1-diethylamino-2-chloro-ethane and states that the product is an effective foaming agent:

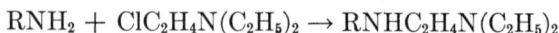

$$RNH_2 + ClC_2H_4N(C_2H_5)_2 \rightarrow RNHC_2H_4N(C_2H_5)_2$$

[143] Swiss Pats. 200,667; 203,590–92; French Pat. 841,506; to *J. R. Geigy A.-G.*
[144] U. S. Pat. 2,361,188, Fox to *E. I. du Pont de Nemours & Co.*
[145] French Pat. 716,560, to *I. G. Farbenindustrie.*

The same product can, of course, be made from the long-chain halide and
N,N-diethylethylenediamine.

Imino-linked compounds have also been made[146] by alkylating 2-amino-
pyridine with long-chain alkyl halides:

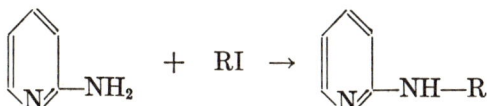

The aminoquinolines may similarly be alkylated to produce[147] imino-linked
surface active compounds.

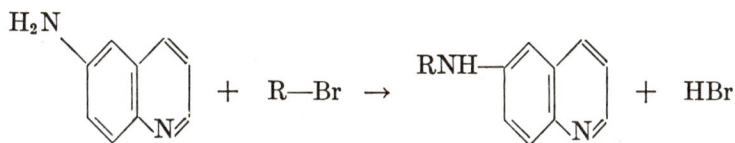

IV. Basic Hydrophilic Group Linked in Heterocyclic Ring

For classification purposes a distinction may be made between those
heterocyclic compounds in which the long chain is attached directly to the
basic nitrogen atom and those in which it is attached to another atom of a
heterocycle which includes the nitrogen atom. The former class has already been considered. The latter class may be represented by a generic
formula:

where the ring symbol represents a cyclic system including the basic nitrogen.

The C-alkyl pyridines in which the alkyl group is in the surface active
range (seven or more carbon atoms) have been described as good wetting
and emulsifying agents as well as bactericides.[148]

2-Alkyl quinolines have also been disclosed as effective surface active

---

[146] Sharp, *J. Chem. Soc.* **1939,** 1855.

[147] U. S. Pat. 2,152,047, Hahl and Leuchs to *Alba Pharmaceutical Co.*

[148] French Pat. 849,399; U. S. Pat. 2,247,266; Canadian Pat. 399,050; Wibaut and
Overhoff to *Shell Development Co.*

agents provided the alkyl group contains six or more carbon atoms. These products may be quaternized or used as their salts.[149]

The 2-alkyl benzimidazoles have already been described as intermediates for a series of anionic detergents which may be formed by sulfonating the benzene nucleus. In themselves, however, they are basic and can be alkylated to form acid-soluble amines. These can be further alkylated to form quaternary salts:[150]

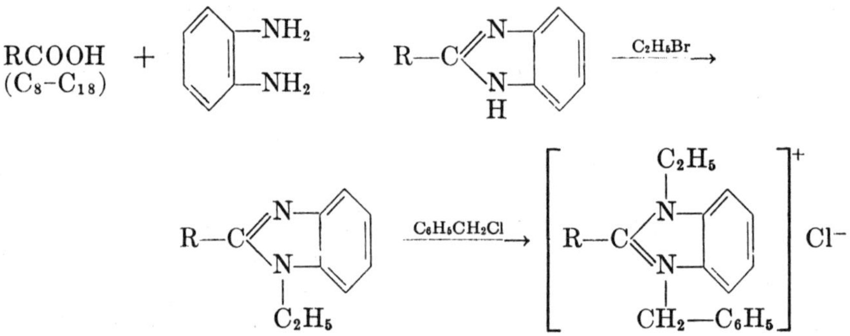

The location of the alkyl groups on the nitrogen atoms is uncertain. They may be symmetrically or unsymmetrically disposed. The products are stable and form typical foaming solutions in water.

Closely related to these compounds are the 2,3-dihydroindoles in which carbon atom 2 bears a long-chain alkyl substituent. These may be made by starting with a higher fatty acyl o-toluidide. By treating with sodium amide or sodium ethylate the ring is closed, forming the indole. This is then hydrogenated to the dihydroindole:

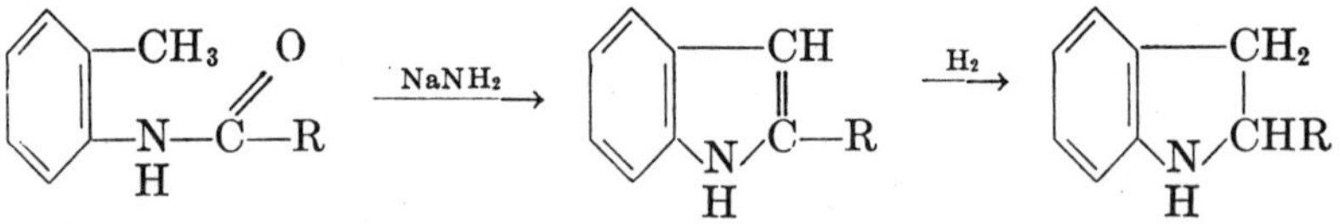

These products are quite strongly basic and may be used in the form of their salts. They may also be alkylated and peralkylated to the quaternary salts.[151]

---

[149] French Pat. 781,812, to I. G. Farbenindustrie.

[150] Swiss Pats. 163,274; 170,762; 172,953–62; Brit. Pats. 419,010; U. S. Pat. 2,043,164; Gränacher to Ciba Co.

[151] U. S. Pats. 2,211,771; 2,320,580; Engel and Pfaehler to J. R. Geigy A.-G. Ger. Pat. 676,852, to J. R. Geigy A.-G.

Long-chain alkyl aminotriazoles have been made by condensing a fatty acid with aminoguanidine. The reaction is effected by heating to about 140°C. until two molecules of water have been split off:[152]

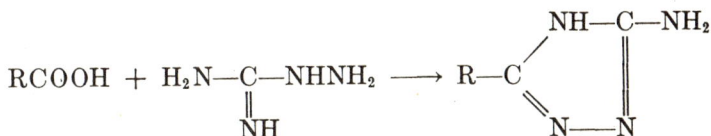

$$RCOOH + H_2N-\overset{\displaystyle \underset{\|}{NH}}{C}-NHNH_2 \longrightarrow R-C\overset{\displaystyle NH-C-NH_2}{\underset{\displaystyle N-N}{\big\langle}}$$

Higher alkyl oxazolines are made by condensing fatty acids with β-amino alcohols. The reaction does not go well with ethanolamine itself but goes almost quantitatively with certain analogues such as 2-methyl-2-amino-1,3-propanediol or trimethylolaminomethane.[153]

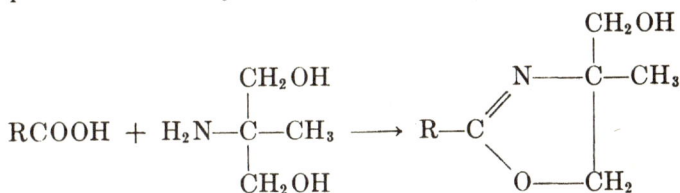

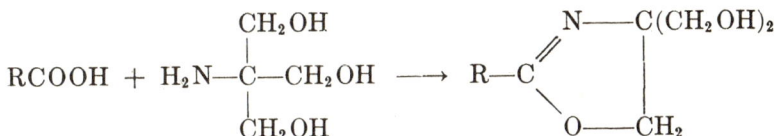

$$RCOOH + H_2N-\overset{\displaystyle \overset{CH_2OH}{|}}{\underset{\displaystyle \underset{CH_2OH}{|}}{C}}-CH_3 \longrightarrow R-C\overset{\displaystyle N-C-CH_3}{\underset{\displaystyle O-CH_2}{\big\langle}} \; \overset{CH_2OH}{|}$$

$$RCOOH + H_2N-\overset{\displaystyle \overset{CH_2OH}{|}}{\underset{\displaystyle \underset{CH_2OH}{|}}{C}}-CH_2OH \longrightarrow R-C\overset{\displaystyle N-C(CH_2OH)_2}{\underset{\displaystyle O-CH_2}{\big\langle}}$$

Both these amino alcohols are made commercially from the nitroparaffins. The alkyl oxazolines dissolve readily in aqueous acids to form powerfully surface active solutions. The acid solutions slowly decompose on storage, probably by a hydrolytic mechanism.

Long-chain oxazoline salts of organic acids such as citric or lactic acid are stable for longer periods than the mineral acid salts.[154] The alkyl oxazolines can be converted to quaternary salts but these also decompose on long standing in aqueous solution. The free bases are quite stable. They are oil soluble and have marked surface activity in these non-aqueous solutions. The oxazoline from oleic acid and 2-amino-2-methyl-1,3-pentanediol is made and sold under the name Alkaterge O by *Commercial*

[152] U. S. Pat. 2,233,805, Broderson and Quaedvlieg to *General Aniline and Film Corp.*

[153] French Pat. 830,125, to *Ciba Co.*

[154] U. S. Pat. 2,402,791, Wampner to *Commercial Solvents Corp.*

Solvents Corp. The corresponding coconut fatty acid derivative has also been produced.

The six-membered ring analogues of the oxazolines have been described and may be made by ring closure from the higher acyl 1,3-amino alcohols:[155]

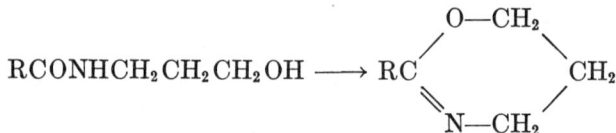

$$RCONHCH_2CH_2CH_2OH \longrightarrow RC \begin{array}{c} O\text{---}CH_2 \\ \diagup \qquad \diagdown \\ \qquad\qquad CH_2 \\ \diagdown \qquad \diagup \\ N\text{---}CH_2 \end{array}$$

A series of cationic surface active agents has been made by condensing ketones or aldehydes in the surface active range with amino-1,3-diols such as 2-amino-2-methyl-1,3-propanediol. The products are 1,3-dioxanes or cyclic aldols and ketols. A typical product has the formula:[156]

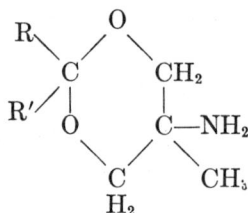

$$\begin{array}{c} R \diagdown \qquad O \\ \qquad\diagdown\;\diagup \qquad\diagdown \\ \qquad C \qquad\qquad CH_2 \\ R' \diagup \; | \qquad\qquad | \\ \qquad O \qquad\qquad C\text{---}NH_2 \\ \qquad\diagdown \qquad \diagup \qquad\diagup \\ \qquad\quad C \qquad CH_3 \\ \qquad\quad H_2 \end{array}$$

Another group of surface active heterocyclic compounds which has been intensively developed is made up of the long-chain imidazolines (or glyox-alidines). The simpler members of the series may be made by condensing a fatty acid or its ester with the monohydrochloride of ethylenediamine. The yield of the desired product is not high, since a large proportion of the diacyl ethylenediamine is formed at the same time:[157]

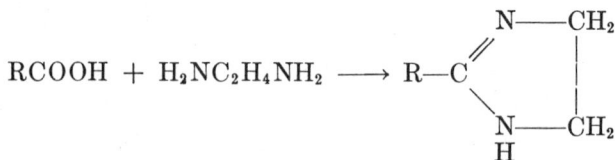

$$RCOOH + H_2NC_2H_4NH_2 \longrightarrow R\text{---}C \begin{array}{c} N\text{------}CH_2 \\ \diagup\diagup \qquad\quad | \\ \qquad\qquad | \\ \diagdown \qquad\qquad | \\ N\text{------}CH_2 \\ H \end{array}$$

They may also be made from the iminoether of the fatty acid and ethyl-

---

[155] U. S. Pat. 2,329,619, Jayne and Day to *American Cyanamid Co.*

[156] U. S. Pats. 2,317,555; 2,320,707; Robinette to *Commercial Solvents Corp.*

[157] U. S. Pats. 2,155,877–8; 2,215,861–4; Chwala and Waldmann to *General Aniline and Film Corp.* Brit. Pats. 479,491; 501,727; to Waldmann and Chwala.

enediamine. The imino ether is made from the nitrile by the action of alcohol and HCl:[158]

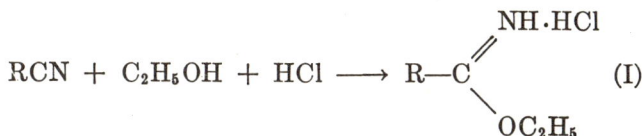

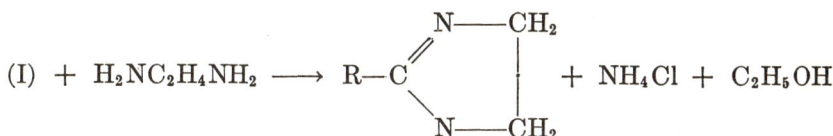

$$RCN + C_2H_5OH + HCl \longrightarrow R-C\underset{OC_2H_5}{\overset{NH \cdot HCl}{<}} \quad (I)$$

$$(I) + H_2NC_2H_4NH_2 \longrightarrow R-C\begin{array}{c} N-\!\!\!-CH_2 \\ < \quad | \\ N-\!\!\!-CH_2 \end{array} + NH_4Cl + C_2H_5OH$$

Still another method of preparation involves heating the fatty acid with ethyleneurea to about 250°C.:[159]

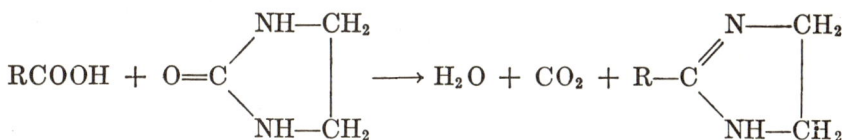

$$RCOOH + O=C\begin{array}{c} NH-CH_2 \\ < \quad | \\ NH-CH_2 \end{array} \longrightarrow H_2O + CO_2 + R-C\begin{array}{c} N-\!\!\!-CH_2 \\ < \quad | \\ NH-CH_2 \end{array}$$

Bisimidazolines have been prepared by similar procedures using dibasic long-chain acids such as sebacic acid and condensing with two molecules of ethylenediamine or its equivalent.[160]

The imidazolines may be alkylated and further converted to quaternary ammonium salts. They can be condensed with ethylene or propylene oxide to give the N-hydroxyl alkyl derivatives.[161] The hydroxyethylimidazolines may also be made by condensing a fatty acid with 2-aminoethylethanolamine.

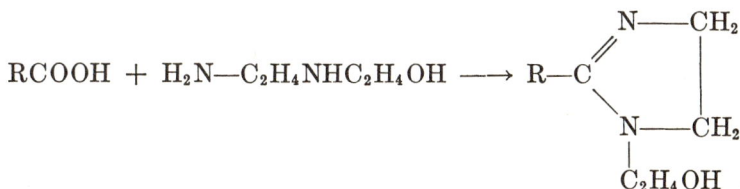

$$RCOOH + H_2N-C_2H_4NHC_2H_4OH \longrightarrow R-C\begin{array}{c} N-\!\!\!-CH_2 \\ < \quad | \\ N-\!\!\!-CH_2 \\ | \\ C_2H_4OH \end{array}$$

[158] U. S. Pat. 1,958,529, Bockmühl and Knoll to *Winthrop Chemical Co.*; French Pat. 836,873; Brit. Pat. 492,812; to *I. G. Farbenindustrie.* U. S. Pat. 2,161,938, Sonn to *Ciba Co.*

[159] U. S. Pat. 2,176,843, Kränzlein and Bestian to *I. G. Farbenindustrie.*

[160] U. S. Pat. 2,210,588, Kränzlein and Bestian to *General Aniline & Film Corp.*; Ger. Pat. 704,410; Brit. Pat. 501,522; to *I. G. Farbenindustrie.*

[161] U. S. Pat. 2,211,001, Chwala to *General Aniline and Film Corp.*

This reaction proceeds very readily at temperatures above 150°C. particularly if a water-immiscible solvent such as xylene is used to remove the water of reaction as it is formed.[162]   The hydroxyethylimidazoline made in this way from oleic acid is manufactured by *Carbide and Carbon Chemicals Corp.* under the name Cationic Amine 220.   It is a brownish-colored oily liquid, readily soluble in aqueous acids to give foamy solutions of high surface activity.   The product may be quaternized readily with the lower alkyl-halides, sulfates, or sulfonates.   The quaternary salts are quite stable, highly surface active, and show high germicidal power.

When fatty acids are condensed with diethylenetriamine, β-amino-ethylimidazolines are formed.   Polyethylene polyamines may be used in an analogous manner:

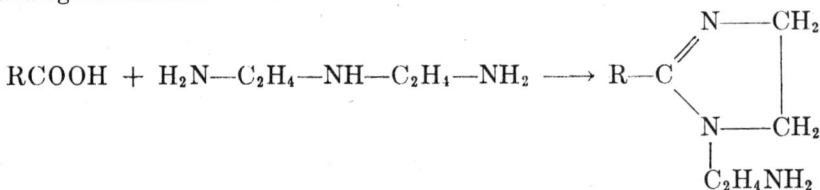

$$RCOOH \;+\; H_2N-C_2H_4-NH-C_2H_4-NH_2 \;\longrightarrow\; R-C \overset{\displaystyle N----CH_2}{\underset{\displaystyle \underset{\displaystyle C_2H_4NH_2}{N----CH_2}}{\diagup\diagdown}}$$

Derivatives of these products in which the primary amino group is acetylated have been used as softening agents for textiles.[163]   They are marketed under the trade name Onyxsan by *Onyx Oil and Chemical Co.*

2-Alkyl tetrahydroindoles have been described as surface active agents. A typical formula is:[164]

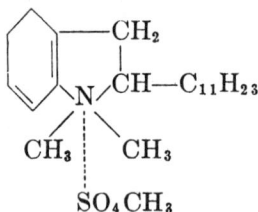

V. Miscellaneous Surface Active Nitrogen Bases

Numerous basic nitrogen compounds which are surface active and which do not fit into simplified classification schemes have been reported.

[162] U. S. Pats. 2,267,965, Wilson to *Carbide and Carbon Chemicals Corp.*; 2,268,273, Wilson and Wilkes to *Carbide and Carbon Chemicals Corp.*

[163] U. S. Pat. 2,355,837, Wilson to *Carbide and Carbon Chemicals Corps.*   U. S. Pat. 2,200,815; Brit. Pat. 549,328; Ackley to *Richards Chemical Works.*

[164] U. S. Pat. 2,211,771, Engel and Pfaehler to *J. R. Geigy A.-G.*

Some of these have been made in the course of scientific studies and some for their industrial potentialities. Richard Kuhn and co-workers,[165] in the course of an extensive study of the physiological effects of "invert soaps," synthesized and purified a large number of unusual products in this class. Among others they describe long-chain substituted tetrazolium and benzo-triazolium salts and azinium salts from N,N-dialkyl hydrazines.

Dialkylguanidines having the generic formula:

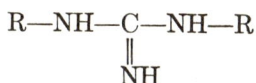

$$R—NH—\underset{\underset{NH}{\|}}{C}—NH—R$$

have been described as surface active agents. These products[166] have also been reacted with ethylene oxide to increase their solubility. Monosub-stituted guanidines and biguanidines have also been described. They may be made by reacting a higher alkyl halide with guanidine or a higher alkyl amine with cyanamide.[167]

Higher fatty acyl biguanides are made[168] by acylating dicyandiamide, and reacting the product with a primary or secondary amine:

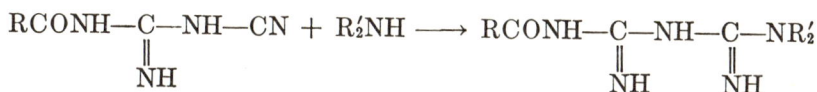

$$RCONH—\underset{\underset{NH}{\|}}{C}—NH—CN + R_2'NH \longrightarrow RCONH—\underset{\underset{NH}{\|}}{C}—NH—\underset{\underset{NH}{\|}}{C}—NR_2'$$

Isothioureas are formed readily by the action of a fatty alkyl halide or sulfate on thiourea. The products are basic and surface active.

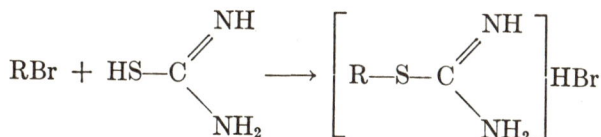

$$RBr + HS—C\overset{\nearrow NH}{\underset{\searrow NH_2}{}} \longrightarrow \left[R—S—C\overset{\nearrow NH}{\underset{\searrow NH_2}{}} \right]HBr$$

They are hydrolyzed readily in alkaline solution to the corresponding mer-captans, which limits their usefulness. They have been described[169] as useful in the concentration of acidic minerals, being used in the same manner as numerous other cationic surfaces active agents. They have been used in the decomposable waterproofing type of composition to a greater extent.

[165] Kuhn and co-workers, *Ber.* **B73,** 1080 (1940); **B74,** 941, 1365 (1941); **B75,** 75 (1942).

[166] U. S. Pats. 2,299,012; 2,338,627; Ericks to *American Cyanamid Co.*

[167] French Pat. 788,429, to *I. G. Farbenindustrie.*

[168] U. S. Pat. 2,324,354, Bindler and Schläpfer to *J. R. Geigy A.-G.*

[169] U. S. Pat. 2,336,868, Jayne *et al.* to *American Cyanamid Co.*

A product of the type

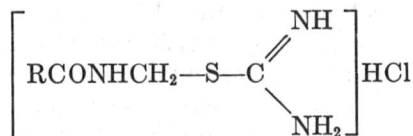

$$\left[RCONHCH_2-S-C\begin{smallmatrix} \\ \nearrow NH \\ \searrow NH_2 \end{smallmatrix} \right] HCl$$

can be made[170] by reacting, for example, chloromethylstearamide with thiourea.

O-Alkyl isoureas have also been described[171] as surface active agents. They have the formula:

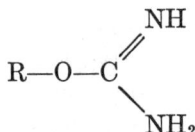

$$R-O-C\begin{smallmatrix} \\ \nearrow NH \\ \searrow NH_2 \end{smallmatrix}$$

and may be regarded as ammono analogues of the carbamic esters or urethans.

Fatty acyl hydrazides have been condensed with reducing sugars to give cationic surface active products[172] of uncertain constitution.

Numerous patents have been issued relating to the use of long-chain amine oxides as surface active agents. The secondary amine oxides are identical with disubstituted hydroxylamines. They are prepared by oxidizing a secondary amine with hydrogen peroxide, and may be used in the form of their salts. From dodecylmethylamine, for example, the following product is formed:[173]

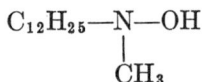

$$C_{12}H_{25}-\underset{\underset{CH_3}{|}}{N}-OH$$

The tertiary amine oxides are formed by oxidizing tertiary amines with hydrogen peroxide, benzoyl peroxide, Caro's acid, or hypochlorites. A wide variety of hydrophobic tertiary amines have been converted to their oxides. Representative of the types which have been used are:

(1)[174] $RCONH-C_2H_4-N(CH_3)_2$

[170] U. S. Pat. 2,051,947, to Hunsdiecker and Vogt.
[171] Ger. Pat. 705,355, to *Böhme Fettchemie G.m.b.H.*
[172] U. S. Pat. 2,355,911, Gränacher and Sallman to *Ciba Co.*
[173] French Pat. 786,334, to *I. G. Farbenindustrie.*
[174] U. S. Pat. 2,169,976, Guenther and Saftien to *I. G. Farbenindustrie.*

(2)[175]

$$R-\overset{\displaystyle H}{\underset{\displaystyle N(CH_3)_2}{C}}-COOH$$

(3)[176]

R—⟨benzene ring⟩—OH, with CH$_2$N(CH$_3$)$_2$

$$RCONH\langle\ \rangle-N(CH_3)_2$$

$$RCO\langle\ \rangle-N(CH_3)_2$$

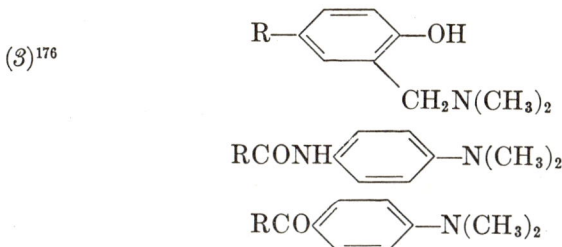

and others.[177] The product formed by oxidizing dodecyldimethyl amine has the formula:

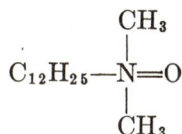

$$C_{12}H_{25}-\overset{\displaystyle CH_3}{\underset{\displaystyle CH_3}{N}}=O$$

or in the hydrated form:

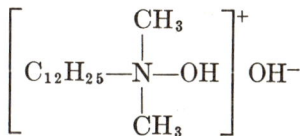

$$\left[C_{12}H_{25}-\overset{\displaystyle CH_3}{\underset{\displaystyle CH_3}{N}}-OH \right]^+ OH^-$$

Such products are soluble in water and show surface activity dependent on the nature of the hydrophobic chain. Furthermore, these products are basic in nature and form salts with acids. They can be oxyethylated and polyoxyethylated by treatment with ethylene oxide under pressure. Presumably the hydroxyethyl groups add to the hydroxyl which is covalently bound to the nitrogen atom.[178]

A water-dispersible oil-soluble cationic surface active agent containing both phosphorus and nitrogen has recently been introduced by *Victor*

[175] U. S. Pat. 2,159,967, Engelmann to *E. I. du Pont de Nemours & Co.*
[176] U. S. Pat. 2,220,835, Bruson and McCleary to *Rohm and Haas Co.*
[177] Swiss Pats. 175,351; 181,444; 182,592; 183,587; 199,451; French Pats. 786,911; 792,822; Brit. Pats. 437,566; 462,881; to *Ciba Co.*
[178] Ger. Pat. 675,411; Brit. Pat. 460,710; U. S. Pat. 2,185,163; Ulrich to *I. G. Farbenindustrie.*

Chemical Works under the name Victamine D. It is said to have the formula:

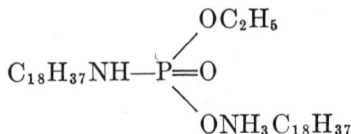

$$C_{18}H_{37}NH-P\overset{\displaystyle OC_2H_5}{\underset{\displaystyle ONH_3C_{18}H_{37}}{=}}O$$

and is made[179] from stearylamine and ethyl metaphosphate.

VI. Non-nitrogenous Cationic Surface Active Agents

The only non-nitrogenous cationic surface active agents which have received appreciable attention in the patent literature are the sulfonium compounds. They are made by adding lower alkyl halides or sulfates to a thio ether containing at least one group in the hydrophobic range. A typical product may be made from cetyl methyl sulfide and ethyl bromide, for example:

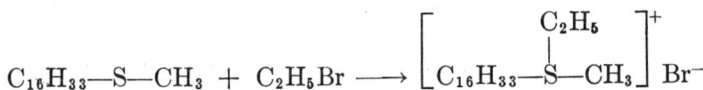

$$C_{16}H_{33}-S-CH_3 + C_2H_5Br \longrightarrow \left[C_{16}H_{33}-\overset{\displaystyle C_2H_5}{\underset{\displaystyle }{S}}-CH_3 \right]^+ Br^-$$

The thio ethers themselves are made by alkylating the mercaptans in alkaline solution with an alkyl halide or sulfate. Bost and Everett[180] have studied the rates of formation and yields of the higher sulfonium salts in various solvents, and find methanol to be an optimum reaction medium.

Aromatic amide-linked sulfonium compounds of formula:

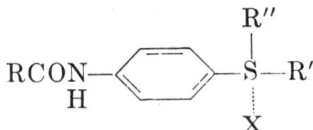

$$RCON-\!\!\!\left\langle\ \right\rangle\!\!\!-\overset{\displaystyle R''}{\underset{\displaystyle X}{S}}-R'$$
$$\ \ \ \ H$$

have been disclosed[181] as surface active agents.

Higher sulfonium sulfates have been made[182] by heating a thio ether with sulfuric acid and a lower alcohol such as methanol or ethanol:

[179] U. S. Pat. 2,406,423, Woodstock to *Victor Chemical Works.*

[180] Bost and Everett, *J. Am. Chem. Soc.* **62**, 1752 (1940).

[181] U. S. Pat. 2,077,831, Felix to *Ciba Co.*

[182] U. S. Pats. 2,185,654, van Peski and Hoeffelmann to *Shell Development Co.*; 2,335,119, Hoeffelmann to *Shell Development Co.*; French Pat. 840,778; Ger. Pat. 705,224; Canadian Pat. 388,120; to *Shell Development Co.*   U. S. Pat. 2,316,152, Bearse and Shutt to *Battelle Memorial Institute.*

$$R\!-\!S\!-\!CH_3 + CH_3OH + H_2SO_4 \longrightarrow \left[\begin{array}{c} R\!-\!S\!-\!CH_3 \\ | \\ CH_3 \end{array}\right]^+ HSO_4^-$$

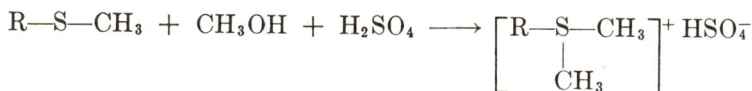

Other complex sulfonium salts containing intermediate linkages have also been described.[183] None of the surface active sulfonium salts, however, appear to have had any large-scale commercial use and none are being manufactured currently to the writers' knowledge.

Jerchel[184] has synthesized a large number of phosphonium and arsonium bases containing hydrophobic groups, and studied their bactericidal and hemolytic properties.

---

[183] U. S. Pats. 2,193,963, to Benjamin R. Harris; 2,230,587, Chwala and Waldmann to *General Aniline and Film Corp.* Ger. Pat. 671,882; French Pat. 807,213; to *I. G. Farbenindustrie.*

[184] Jerchel, *Ber.* **B76**, 600 (1943); see *Chem. Abstracts* **38**, 60 (1944).

CHAPTER 8

Non-ionic Surface Active Agents

I. Water-Soluble Types

The most strongly hydrophilic groups encountered in surface active agents are the ionogenic groups, and the majority of water-soluble surface active agents are in fact ionogens. There are, however, certain configurations which are hydrophilic, but non-ionogenic. The two best-known radicals in this class are ether oxygen and hydroxyl. Other groups containing oxygen, such as carboxylic ester, and those containing non-basic nitrogen, such as the amide group, —CONH—, are also distinctly hydrophilic. It is very unusual for groups other than hydroxyl or ether to occur in sufficient number in a single large molecule to render the molecule water soluble. This is particularly true when the molecule contains a typical hydrophobic group. Modern synthetic methods allow the introduction of a controlled number of ether or hydroxyl groups into a hydrophobic molecule, thereby confering on it any desired degree of compatibility with water. Thus it is possible to start with a substance such as octadecyl alcohol (which has more hydrophilic character than octadecane but is still water insoluble) and progressively introduce ethylene ether groups by means of ethylene oxide. The products become more and more hydrophilic, and finally completely soluble in water when about twelve to fifteen ethylene ether groups have been introduced.

The non-ionic surface active agents are usually considered to include only those compounds which are soluble in water. There is a large number of compounds which are dispersible in water and which have surface active properties but do not have enough hydrophilic character to be fully soluble in water. These products are widely used as emulsifying agents and include such well-known commercial products as glycerol monostearate, glycol oleate, etc. The dividing line between these substances and the soluble non-ionic detergents is not sharp. They are used, however, for different purposes and should be considered in different categories. The water-insoluble compounds are for the most part oil soluble and are sometimes grouped under the name oil-soluble emulsifying agents.

The best-known and best-studied series of non-ionic detergents is made by reacting a hydrophobic hydroxy compound (either a phenol or an alcohol) with several moles of ethylene (or propylene) oxide. Ethylene oxide adds to hydroxyl groups under moderate conditions of temperature and pressure in the presence of alkaline catalysts.

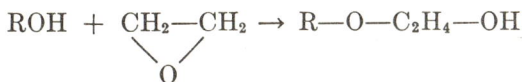

$$ROH + CH_2—CH_2 \rightarrow R—O—C_2H_4—OH$$
$$\underset{O}{\diagdown \diagup}$$

(I)

$$(I) + CH_2—CH_2 \rightarrow R—O—C_2H_4—O—C_2H_4—OH$$
$$\underset{O}{\diagdown \diagup}$$

$$(I) + n\ CH_2—CH_2 \rightarrow R—O—C_2H_4(—OC_2H_4—)_nOH$$
$$\underset{O}{\diagdown \diagup}$$

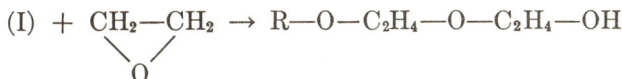

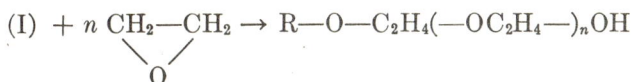

Propylene oxide reacts in the same manner. The number of ethylene ether groups necessary to effect complete water solubility depends, of course, on the molecular weight and structure of the hydrophobic portion of the molecule. When, for example, ROH in the above equation is hexylphenol fewer moles of ethylene oxide will be required than when ROH is octadecyl alcohol. Conversely, with any given ROH starting material the solubility of the product will vary with the number of ethylene oxide molecules added. It is apparent that in this series alone there is the possibility for hundreds of different combinations. These polyoxyethylene ethers of the higher fatty alcohols and alkyl phenols have been developed largely in Germany. The trade names Igepal, Peregal, Leonil, and Emulphor include numerous members of this series.

Starting with oleyl alcohol and adding increasing amounts of ethylene oxide a useful degree of water solubility is attained with about 6 to 8 moles of ethylene oxide. These products are suitable for wool processing and for emulsifying neutral oils or mineral oils. When 10–15 moles of ethylene oxide have been added the products are good detergents for cotton and rayon. With 20–30 oxyethylene groups present excellent emulsifying agents are obtained. The manufacture of Leonil OX, a detergent particularly suited for washing wool, is described by Hoyt as follows:[1]

. 800 kg. of oleyl alcohol are charged to a glass-lined kettle of 3-m.^{3} capacity and about 80 g. (0.01%) powdered caustic soda are added. The charge is heated to 160–180°C. and the kettle evacuated to 10 mm. pressure. Ethylene oxide from a storage tank holding 2000 kg. is allowed to flow in under nitrogen pressure of about 3 atmospheres. The flow is regulated by this pressure. The reaction is highly exothermic

[1] Hoyt, *PB 3868*, Office of Technical Services, Dept. of Commerce, Washington, D. C.

and requires about 12–14 hours for completion. The product is a soft waxy paste. It gives clear solutions in cold water and these become turbid on heating to about 40°C. The 30% aqueous solution of this product is called Leonil O. A 20% solution is sold as Peregal O. This product contains 15 moles of ethylene oxide per mole of oleyl alcohol. A similar product containing 25 moles of ethylene oxide is made under the name of Diazopon A. It is used as a dyeing assistant and is said to be particularly effective in minimizing crocking. Aqueous solutions of this product remain clear up to 70–80°C. It is a less effective detergent than Leonil O.

Leonil FFO is made in a similar manner by condensing 8 moles of ethylene oxide with an alkylated β-naphthol. The latter substance is prepared by alkylating β-naphthol with a C_6–C_8 olefin fraction (obtained in synthetic gasoline manufacture) in the presence of aluminum chloride. Leonil FFO has the same performance characteristics as Leonil O in equal concentrations. It was used to a considerable extent as a household laundering agent in Germany during the war. For this purpose it was diluted to 10% with water and thickened with sodium polyacrylate.

Non-ionic detergents of the Leonil O and Leonil FFO type are quite effective in washing both wool and cotton. In washing wool they have the particular advantage of being non-substantive. They are therefore not depleted or exhausted by adsorption. They are poor foamers in comparison with ionic detergents of similar effectiveness. It is noteworthy that these non-ionic detergents are frequently sulfated (sulfation taking place at the terminal hydroxyl group). This converts them into anionic detergents but preserves their excellent wool-washing properties and even improves their cotton-washing properties. It also improves their foam. Sulfation enables the saving of a considerable amount of the ethylene oxide otherwise required. A sulfated product made from oleyl alcohol and about 5 moles of ethylene oxide is similar to an unsulfated product of the Leonil O type with 15 moles of ethylene oxide.

The Igepal types are similar to Leonil O and FFO except that they are based on an alkyl phenol. They are highly recommended for washing wool, cotton, rayon, feathers, etc. and also as ingredients of spotting compositions. One of the important advantages of these non-ionic polyethers is their complete stability in hard water or salt water and their stability toward hydrolytic agents. They are unaffected by changes in pH and wash equally well in acid, alkaline, or neutral liquors.

The polyoxyethyl ethers have also been widely used in Germany as emulsifying agents. Emulphor A Extra is made from 6 moles of ethylene oxide and 1 mole of a dodecyl phenol. It has negative temperature coefficient of solubility in water, becoming less soluble at higher temperatures. It is used in preparing textile-treating emulsions such as spinning oils etc. Emulphor STS is made from the same alkyl phenol and 7.5 to 8 moles of ethylene oxide. Emulphor O is oleyl alcohol condensed with 20 moles of ethylene oxide. Emulphor ELN is dodecyl phenol with 20 moles of ethylene oxide. Emulphor EL is castor oil with 40 moles of ethylene oxide. In this case the ethylene oxide forms an ether link with the hydroxyl group at the 12 position in the ricinoleic acid residue. Dismulgan III is dodecyl phenol with 30 moles of ethylene oxide. All of these substances are designed for special emulsification processes.

The Peregals, Igepals, Leonils, and Emulphors have been marketed in this country by *General Dystuff Corp.* At least one similar product has also been manufactured in this country by *Rohm and Haas Co.* under the name of Triton NE. This is a polyoxyethylene ether of an alkyl phenol. Richardson, Kern, Murray, and Sudhoff have reported in full detail on the manufacture of the polyoxyethylene ether detergents in Germany. The patent literature on these products includes numerous German, British, and French patents but relatively few United States patents.[2] Their behavior in practical applications has been reported by Nüsslein and other workers.[3]

A considerable number of modifications of the above products and procedures have been described. Successive treatments with ethylene and propylene oxides for example have been used to attain increased solubility.[4] Most of the modifications, however, are concerned with the type of hydroxy compound used as a base material for reaction with ethylene oxide. Among these are: (*1*) Sod oil (degras, moellon) and similar hydroxylated oils recovered in the making of chamois leather.[5] (*2*) Higher alcohols made by the reaction of olefins, carbon monoxide, and hydrogen.[6] (*3*) Ether alcohols of the type:

$$R—O—CH_2—CHOH—R'$$

made by etherifying an alcohol or phenol, ROH, with a halo ketone such as chloroacetophenone:

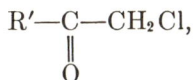

$$R'—\underset{\underset{O}{\|}}{C}—CH_2Cl,$$

[2] Richardson *et al.*, *PB 6684*, Office of Technical Services, Department of Commerce, Washington, D. C. French Pats. 713,426; 717,427; 727,202; 752,831; 770,804; Ger. Pats. 548,201; 670,419; 680,245; Brit. Pats. 346,550; 367,420; 380,431; 409,336; 443,559; 443,631; U. S. Pats. 1,970,578; 2,213,417; 2,213,477; Steindorff *et al.* to *General Aniline and Film Co.* U. S. Pat. 2,214,352, Schoeller, Ulrich, and Ploetz to *General Aniline and Film Corp.*

[3] Nüsslein, *Melliand Textilber.* **18**, 248 (1937). Krüger, *Klepzig's Textil.-Z.* **41**, 558 (1938).

[4] U. S. Pat. 2,174,761, Schuette and Wittwer to *I. G. Farbenindustrie.*

[5] Brit. Pat. 434,424; U. S. Pat. 2,048,512; Piggott and White to *Imperial Chemical Industries Ltd.* French Pat. 785,800, to *Imperial Chemical Industries Ltd.*

[6] Belgian Pat. 444,625, to *Th. Goldschmidt A.-G.*

and reducing the resulting ketone ether.[7] (4) Complex phenols made by condensing a simple phenol with an aldehyde or ketone, for example:

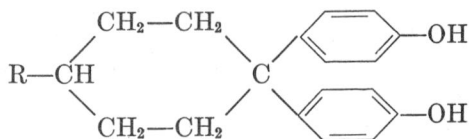

$$R-CH \begin{array}{c} CH_2-CH_2 \\ \\ CH_2-CH_2 \end{array} C \begin{array}{c} \langle \rangle-OH \\ \\ \langle \rangle-OH \end{array}$$

from alkyl cyclohexanone and phenol.[8] (5) Secondary alcohols made by reducing the ketones obtained by pyrolysis of synthetic paraffin wax fatty acids.[9] (6) Polyalkyl cyclohexanols made by the reduction of the corresponding isophorones. The isophorones are made by the trimeric condensation of ketones.[10]

The higher alkyl mercaptans add ethylene oxide in the same way that the higher alcohols do:[11]

$$RSH + n\,C_2H_4O \rightarrow RS(-C_2H_4O-)_nH$$

Polyethylene glycols of controlled molecular weight may be made by reacting ethylene oxide with water or with ethylene glycol:

$$HOH + n\,C_2H_4O \rightarrow HO(-C_2H_4O-)_nH$$

These glycols have been added across the double bond of dioctyl maleate to give a non-ionic analogue of the well-known dialkyl sulfosuccinates:

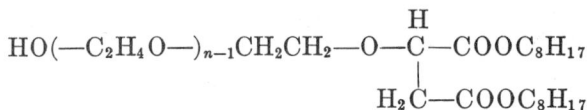

$$HO(-C_2H_4O-)_{n-1}CH_2CH_2-O-\overset{\displaystyle H}{\underset{\displaystyle H_2C-COOC_8H_{17}}{C}}-COOC_8H_{17}$$

The same product may be made by adding ethylene oxide to the dialkyl malate.[12]

In most of the products discussed so far the hydrophobic portion of the molecule is linked directly to the polyethylene oxide chain which acts as the hydrophilic group. The hydrophobic residue is accordingly either an alcohol or a phenol. Non-ionic surface active products of this type have also been made in which the hydrophobic group is derived from the less expensive higher fatty acids.

[7] Brit. Pats. 516,879; 516,978, to *Deutsche Hydrierwerke A.-G.*
[8] Brit. Pat. 515,907, to *Deutsche Hydrierwerke A.-G.*
[9] U. S. Pat. 2,355,823, Schlegel to *Procter and Gamble Co.*
[10] U. S. Pat. 2,148,103, Bruson to *Rohm and Haas Co.*
[11] French Pat. 780,144, to *I. G. Farbenindustrie.*
[12] U. S. Pat. 2,341,846, Meincke to *American Cyanamid Co.*

The simplest of these are the straight fatty acid esters of polyethylene glycol:

$$RCOO(-C_2H_4O-)_nH$$

These may be made by reacting the fatty acid with ethylene oxide under pressure, in the same manner as the polyoxyethylene ethers are made. Beside ethylene oxide there may be used propylene oxide or mixtures of the two oxides.[13] They are more usually made, however, by esterifying the previously prepared polyethylene glycol with a fatty acid. Polyethylene glycols of definite molecular weight are produced and marketed in this country by *Carbide and Carbon Chemicals Corp.*, and other producers are reported to be entering this field. Polyethylene glycols are readily esterified with fatty acids by the usual procedures.[14] A certain proportion of the diacylated polyglycol is formed at the same time as the desired monoacyl compound. When the ethylene oxide – fatty acid process is used only the monoacyl compound is formed. Fatty acid esters of the polyglycols behave similarly to the corresponding fatty alcohol polyglycol ethers. Their solubility relationships and their effectiveness as detergents are approximately the same. Like the polyglycol ethers they are relatively poor foaming agents but good emulsifying agents.[15] The main disadvantage of the esters is their lack of stability in hot alkaline solutions. As would be expected they are readily saponified to the fatty acid soap and polyethylene glycol. They are sufficiently resistant to mild alkalis, however, to be useful under ordinary laundering conditions.

Several manufacturers have recently started to produce the fatty esters of polyethylene glycols. With the availability of the polyglycols themselves it is a relatively simple operation to produce the esters. Neutronyx (*Onyx Oil and Chemical Co.*) and Glycaid (*Glyco Products Co.*) are representative of this class. Products of this type are stated to be excellent antistatic agents in the processing of cellulose acetate fibers.[16]

Numerous products have been described in which a hydroxy ester or hydroxyamide of a fatty acid is treated with ethylene oxide to form the soluble polyethylene ether. Urea, for example, or dimethylolurea may be condensed with about 20 – 30 times its weight of ethylene oxide, and the resulting product acylated with stearic acid.[17]

[13] U. S. Pat. 2,174,760, Schuette and Wittwer to *I. G. Farbenindustrie.*
[14] U. S. Pat. 2,275,494, to Harry Bennett.
[15] See Ger. Pats. 623,482; 626,491; Marx *et al.* to *I. G. Farbenindustrie.*
[16] Brit. Pat. 563,725, to *British Nylon Spinners Co.*
[17] Brit. Pat. 432,356, Piggott; French Pat. 768,732; Supplement 46,129; to *Imperial Chemical Industries Ltd.*

Fatty acyl methylglucamine:

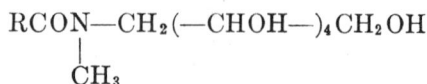

$$\text{RCON--CH}_2(\text{--CHOH--})_4\text{CH}_2\text{OH}$$
$$|$$
$$\text{CH}_3$$

which by itself is dispersible but not soluble in water, has been treated with ethylene oxide to give water-soluble products. These are much more resistant to hydrolysis than the fatty esters of the polyglycols.[18]

Fatty amides and hydrophobic sulfonamides have also been reacted with ethylene oxide to give surface active agents:[19]

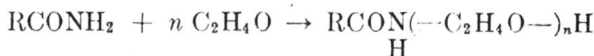

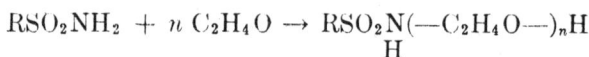

$$\text{RCONH}_2 + n\,\text{C}_2\text{H}_4\text{O} \rightarrow \text{RCON(--C}_2\text{H}_4\text{O--)}_n\text{H}$$
$$\text{H}$$

$$\text{RSO}_2\text{NH}_2 + n\,\text{C}_2\text{H}_4\text{O} \rightarrow \text{RSO}_2\text{N(--C}_2\text{H}_4\text{O--)}_n\text{H}$$
$$\text{H}$$

In these products it is probable that the second hydrogen on the amido nitrogen also reacts with the ethylene oxide forming disubstituted amides.

Fatty acyl derivatives of trimethylolaminomethane have been reacted with ethylene oxide:[20]

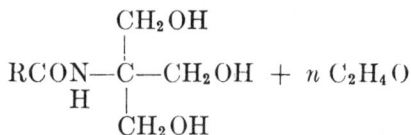

$$\text{CH}_2\text{OH}$$
$$|$$
$$\text{RCON--C--CH}_2\text{OH} + n\,\text{C}_2\text{H}_4\text{O}$$
$$\text{H}\ \ |$$
$$\text{CH}_2\text{OH}$$

Fatty acyl monoethanolamides have also been treated with ethylene oxide to make them water soluble:[21]

$$\text{RCONH--C}_2\text{H}_4\text{OH} + n\,\text{C}_2\text{H}_4\text{O} \rightarrow \text{RCONH(--C}_2\text{H}_4\text{O--)}_{n+1}\text{H}$$

Stearic hydrazide has been treated first with glycidol and subsequently with ethylene oxide to give surface active agents:[22]

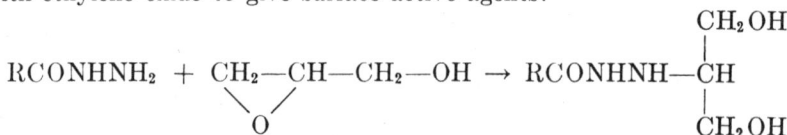

$$\text{CH}_2\text{OH}$$
$$|$$
$$\text{RCONHNH}_2 + \text{CH}_2\text{--CH--CH}_2\text{--OH} \rightarrow \text{RCONHNH--CH}$$
$$\diagdown\!\diagup$$
$$\text{O} \qquad\qquad\qquad\qquad \text{CH}_2\text{OH}$$

Then, $n\,\text{C}_2\text{H}_4\text{O}$ is used.

[18] Brit. Pat. 420,518, to *Imperial Chemical Industries Ltd.* and Piggott.
[19] Brit. Pat. 415,718, to *I. G. Farbenindustrie.*
[20] Brit. Pat. 420,066, *Imperial Chemical Industries Ltd.* and H. A. Piggott.
[21] French Pat. 789,522, to *I. G. Farbenindustrie.*
[22] U. S. Pat. 2,371,133, Gränacher *et al.* to *Ciba Co.*

Fatty alcohol glucosides, which in themselves are surface active but are not very water soluble, have been solubilized by addition of ethylene oxide.[23] Higher fatty aldehydes have been condensed with urea and subsequently with ethylene oxide to give surface active agents.[24]

One of the best-known and widely used series of non-ionic water-soluble surface active agents is made by *Atlas Powder Co.* under the tradename Tween. The Tweens are fatty acid esters of anhydrosorbitols which have been solubilized by etherifying the free hydroxyl groups with ethylene oxide. Sorbitol itself is produced by the hydrogenation of glucose. The three anhydrosorbitols have the formulas:

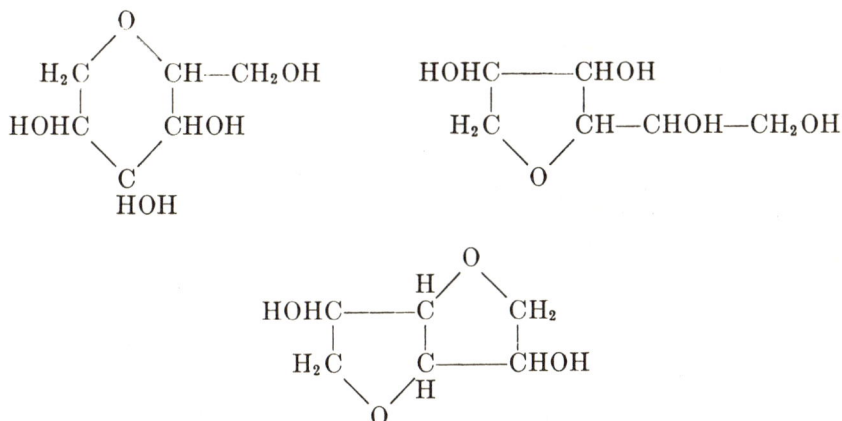

These products are obtained commercially as a mixture. Since they all contain at least two hydroxyl groups one of these groups is available for esterification with a fatty acid, and the others for etherification with ethylene oxide. The anhydrosorbitol esters which have not been etherified are sold under the name Span. The Tweens are made so that one fatty acyl residue is present for each mole of sorbitol or anhydrosorbitol. Lauric, palmitic, oleic, and stearic acids are used in preparing various members of the series. The Tweens are used primarily as emulsifying agents, but they are also useful as wetting and solubilizing agents. They are soluble in most oxygenated solvents and in many hydrocarbon and halogenated solvents.[25]

A water-soluble non-ionic surface active agent based on a phosphoric

[23] French Pat. 838,863, to *I. G. Farbenindustrie.*
[24] U. S. Pat. 2,172,747, Bowles and Kaplan to *Richards Chemical Works.*
[25] U. S. Pat. 1,959,930; Ger. Pats. 544,921; 628,715; Schmidt and Meyer to *I. G. Farbenindustrie.* U. S. Pat. 2,322,820–1, Brown to *Atlas Powder Co.*; 2,380,166, Griffin to *Atlas Powder Co.*

ester is made commercially by *Victor Chemical Works* under the name Victawet 12.   It is an ester of H_3PO_4 in which one of the hydrogen atoms is esterified with polyethylene glycol and the other two hydrogen atoms are esterified with hydrophobic alcohols of medium chain length.

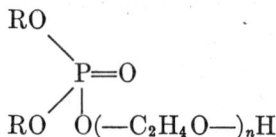

$$\begin{array}{c} RO \\ \diagdown \\ P{=}O \\ \diagup \quad | \\ RO \quad O(-C_2H_4O-)_nH \end{array}$$

Natural phosphatides such as soybean lecithin have been oxyethylated or reacted with propylene oxide to produce water-soluble surface active agents.[26]

Disulfimides of the type:

$$RSO_2-NH-SO_2R$$

have been reacted with ethylene oxide to give water-soluble surface active agents.   The ethylene oxide alkylates the imido nitrogen atom.[27]

Although the polyethylene glycol chain is the most important hydrophilic configuration used in the preparation of water-soluble non-ionic detergents, it is not the only one which has been described.   Katzman, for example, has disclosed numerous non-ionic detergents, solubilized by a multiplicity of amide and hydroxy groups.   The product:

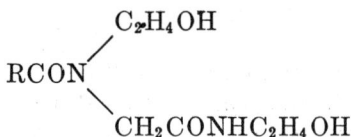

$$\begin{array}{c} C_2H_4OH \\ \diagup \\ RCON \\ \diagdown \\ CH_2CONHC_2H_4OH \end{array}$$

is typical of these.   It is made by condensing chloroacetylethanolamine with another mole of ethanolamine, and acylating the product with coconut fatty acid chloride.[28]

The low-order condensation polymers made by heating polyhydric alcohols with polybasic water-soluble acids have also been used as hydrophilic groups.   Glycol tartrate or glycerol citrate, for example, may be esterified (on one of its free hydroxyl groups) with stearic acid to give a water-soluble non-ionic detergent.[29]

[26] U. S. Pat. 2,310,679, DeGroote and Keiser to *Petrolite Corp.*
[27] U. S. Pat. 2,345,121, Hentrich *et al.* vested in the Alien Property Custodian.
[28] U. S. Pats. 2,303,366; 2,388,154; Katzman to *Emulsol Corp.*
[29] U. S. Pat. 2,329,166 to N. B. Tucker.

Mixed methylene–ethylene ether chains have been attached to hydrophobes containing active hydrogen just as the polyethylene ethers are attached. Formaldehyde and ethylene oxide, or formaldehyde and ethylene glycol are reacted with an ROH compound (for example) to give:[30]

$$R—OC_2H_4(—OCH_2—OC_2H_4—)_nOH$$

In some cases where the hydrophobic group is of lower molecular weight a glucose or sorbitol radical is sufficient to confer a useful degree of water solubility. Amyl- and hexylphenols, for example, can be hydroxymethylated and then etherified with sorbitol to give water-soluble nonionic wetting agents:[31]

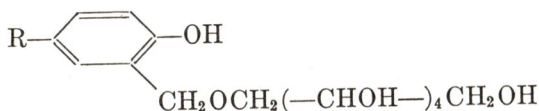

$$R—\langle\!\!\!\bigcirc\!\!\!\rangle—OH$$
$$CH_2OCH_2(—CHOH—)_4CH_2OH$$

The glucosides of the medium-molecular-weight alkyl phenols (C_4–C_8 side chains) are fairly soluble in water and have surface active properties. Those of the higher fatty alcohols (C_{12} and over) are not soluble enough to be classed as water-soluble surface-active agents. These substances, however, have been sulfated or reacted with ethylene oxide to give typical detergents and wetting agents.[32]

Glucosides of hydrophobic molecules which already contain some solubilizing hydroxyl groups are generally water soluble and useful as detergents or wetting agents. They may be made directly or indirectly. β-Aminoethylglucoside, for example, may be made by first making the glucoside of ethylenechlorohydrin and then replacing the chlorine atom by an amino group. This product can then be condensed with coconut fatty acid chloride to give a water-soluble surface-active agent.[33] Similarly, long-chain polyhydric alcohols such as dodecyl glyceryl ether or dodecylthiosorbitol may be converted to glucosides directly by heating with glucose or methyl glucoside. The products are water soluble.[34]

[30] U. S. Pats. 2,366,738, Loder and Gresham to *E. I. du Pont de Nemours & Co.*; 2,395,265, Gresham to *E. I. du Pont de Nemours & Co.*

[31] U. S. Pat. 2,268,126, Orthner and Sönke; Brit. Pat. 502,080, to *I. G. Farbenindustrie.*

[32] Ger. Pat. 593,422; French Pat. 838,863; U. S. Pats. 1,951,784–5; 2,049,758; Bertsch *et al.* to *H. Th. Böhme A.-G.*

[33] U. S. Pat. 2,356,565, Chwala, vested in the Alien Property Custodian.

[34] U. S. Pat. 2,374,236, Salzberg and Werntz to *E. I. du Pont de Nemours & Co.*

The sulfone group has also been used as a hydrophilic group in non-ionic detergents. *tert*-Octyl-β-hydroxyethyl sulfone:

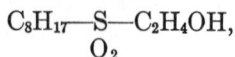

$$C_8H_{17}—S—C_2H_4OH,$$
$$O_2$$

is a typical example. It is prepared by condensing octylene with β-thio-ethanol and oxidizing the thioether thus formed.[35]

One of the very important and widely used water-soluble detergents which may be classed as non-ionic is made by heating together 1 mole of a higher fatty acid with approximately 2 moles of diethanolamine. Coconut fatty acid is usually used in making this product. It is made by several different manufacturers and sold under their various trade names (N-100, *E. F. Houghton and Co.;* Dianol G, *Quaker Chemical Co.;* Alrosol, *Alrose Chemical Co.;* Detergent 1000, *E. C. Drew Co.*, etc.), but is most often referred to as the Ninol type detergent from the name of the company where it was originated.[36]

The chemistry involved in the formation of this detergent has not as yet been explained in detail. The process itself is extremely simple. One mole of coconut fatty acid and 2 moles of diethanolamine are mixed in an open kettle or a still. The mixture is heated to about 150–170°C., at which temperature water is slowly distilled off. As the water is removed the acid number (a measure of free fatty acid or fatty acid amine salt) falls, indicating formation of the amide according to the equation:

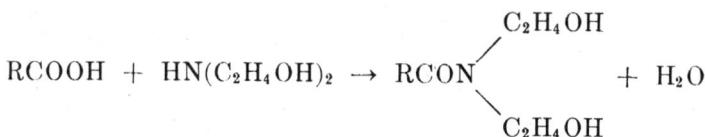

$$RCOOH + HN(C_2H_4OH)_2 \rightarrow RCON \begin{array}{c} C_2H_4OH \\ \diagup \\ \diagdown \\ C_2H_4OH \end{array} + H_2O$$

When the amount of water removed has almost reached the theoretical quantity required by the above equation, and the free fatty acid content has fallen to the range of 5% or less, the reaction is stopped. The product is a clear yellowish-brown liquid of high viscosity. It is readily soluble to a clear solution in water and has a pH of about 9. This is about the same pH as an aqueous solution of diethanolamine itself. It is a powerful detergent and wetting agent. In acidified aqueous solutions it is much less effective and becomes cloudy and gelled if sufficiently concentrated. From the method of formation it would appear that the detergent is a

[35] Swiss Pats. 204,845; 207,645–6; to *Ciba Co.*

[36] U. S. Pat. 2,089,212, to Wolf Kritchevsky. See also U. S. Pats. 2,094,608–9, to Wolf Kritchevsky.

simple mixture of fatty acyl diethanolamine with excess diethanolamine. It is in fact possible to distill pure diethanolamine from the product with no difficulty. A mixture of these two components, however, does not behave at all like the Ninol detergent. It has nowhere near the detergent power, gives turbid solutions, is not a good wetting agent, etc. If this mixture is heated to 120–140°C. for about an hour it becomes fully equivalent to the Ninol detergent. This transformation takes place with no loss of water or change in weight. This precludes any chemical condensation. It is possible that a partial rearrangement of hydroxyamide to amino ester takes place, but this is contraindicated by the fact that the product does not behave like a cationic detergent:

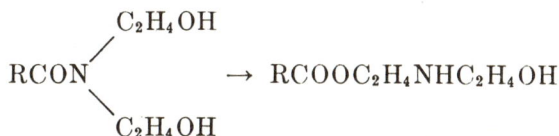

$$RCON\begin{array}{c} C_2H_4OH \\ \\ C_2H_4OH \end{array} \rightarrow RCOOC_2H_4NHC_2H_4OH$$

One possible explanation of this behavior is that a physicochemical change takes place on heating and the molecules are combined into oriented groupings or micelles. These micelles would presumably have to be of such a nature as to maintain their identity even on dilution. Another point of interest in this connection is that a small amount of uncondensed fatty acid must be present in the Ninol type detergent for it to be fully effective. This indicates that a mixed micelle of free amine, amine soap, and fatty acid may be the active material. A mixture of these three components heated together actually gives an effective detergent.[37]

Diethanolamine is not the only amine which can be used in preparing the Ninol type of detergent. Trimethylolaminomethane (I) can also be used. Bis(dihydroxypropyl)amine (diglycerylamine) (II) likewise produces an effective detergent. With trimethylolaminomethane the product is a high melting waxy solid.

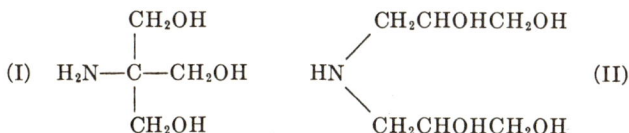

$$(I) \quad H_2N-\overset{\displaystyle CH_2OH}{\underset{\displaystyle CH_2OH}{\overset{|}{\underset{|}{C}}}}-CH_2OH \qquad HN\begin{array}{c} CH_2CHOHCH_2OH \\ \\ CH_2CHOHCH_2OH \end{array} \quad (II)$$

The Ninol type of detergent has been further reacted with such substances as thionyl chloride, chloroacetyl chloride, or chloroacetic acid to produce detergents which are effective in acid solution.[38]

[37] U. S. Pat. 2,404,297, Kroll to *Alrose Chemical Co.*
[38] U. S. Pats. 2,192,664; 2,325,062; Kritchevsky to *Ninol, Inc.*

II. Oil-Soluble Emulsifying Agents

A class of compounds which is very widely used commercially consists of those non-ionic substances containing too few hydrophilic groups to render them water soluble, but enough such groups to give them a distinct polar character. These substances are for the most part soluble in oxygenated, aromatic, and halogenated solvents and in some cases in mineral oils. They have in general a powerful emulsifying effect, usually forming W/O (water-in-oil) emulsions when introduced into an oil–water system. In themselves they are often dispersible to form O/W (oil-in-water) systems in water. It has been pointed out above that the borderline between these products and the water-soluble non-ionic substances is not sharp. Their water solubility increases progressively with the number of hydrophilic groups present and inversely with the relative size of the hydrophobic group.

The simplest of these oil-soluble emulsifying agents are the higher fatty alcohols themselves. The sterols in particular have a strong tendency to emulsify water in oils. Sterols and fatty alcohols, however, are seldom used by themselves. More often they are used in conjunction with more powerful emulsifying agents.

The best-known oil-soluble non-ionic emulsifying agents are the fatty acid esters of polyhydric alcohols or of ether alcohols. Glycerol monostearate is a typical example. It is usually made by heating tristearin with an excess of glycerol.

The polyglycerols have also been used to esterify with fatty acids. In some cases the products are quite soluble in water. Polyglycerols are made by heating glycerol in the presence of alkaline catalysts. Water is eliminated and glyceryl ethers of glycerol are formed. The fatty esters are made by heating the polyglycerol with one or more equivalents of fatty acid. Other hydrophobic acids such as resin acids or naphthenic acids have also been used.[39]

The esters of ethylene glycol, di-, and triethylene glycol, and polyethylene glycols are also widely used. These may be made either by esterifying the glycols or by treating the fatty acids with ethylene oxide.[40] The ester made by condensing oleic acid with 6 moles of ethylene oxide was used in Germany during the war under the name Emulphor A.

The fatty esters of the sugar alcohols have already been mentioned. They are made and sold under the trade name Span by *Atlas Powder Co.*

[39] U. S. Pats. 2,023,388; 2,022,766; to Benjamin R. Harris. Ger. Pat. 623,482, Marx *et al.* to *I. G. Farbenindustrie*; French Pat. 767,788; to *Farb- u. Gerbstoff Werke Flesch.* Brit. Pat. 494,639, to Herbert Schou. U. S. Pat. 2,154,977, Furness and Fairbourne to *Lever Bros.*

[40] See, for example, U. S. Pat. 1,914,100, to Harry Bennett; French Pat. 842,943, to *I. G. Farbenindustrie.*

Sugar alcohols other than the anhydrosorbitols, and also sugars themselves have been used to esterify with the higher fatty acids.[41]

The higher fatty esters of pentaerythritol:

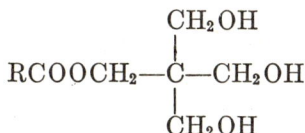

$$RCOOCH_2-\underset{\underset{CH_2OH}{|}}{\overset{\overset{CH_2OH}{|}}{C}}-CH_2OH$$

and corresponding esters of dipentaerythritol are produced by *Heyden Chemical Co.* under the name Pentamul. Esters of pentaerythritol containing a plurality of fatty acid radicals are said to be particularly useful as additives in lubricating oils.[42]

The chemistry of the fatty acid esters of polyhydric alcohols has been reviewed very thoroughly by Goldsmith.[43] The extent to which this field has been studied is indicated by the fact that almost 600 references to the original literature are cited by this author.

Esters of lower molecular weight such as hexyl lactate, amyl tartrate, octyl citrate, etc. are said to have pronounced surface activity. It is probable that their behavior resembles that of pine oil and terpineol which are better-known synergists or promotors for surface active agents.[44]

An unusual emulsifying type of ester recently described has the formula:

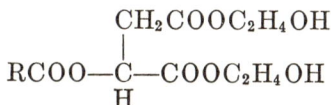

$$RCOO-\underset{\underset{H}{|}}{\overset{\overset{CH_2COOC_2H_4OH}{|}}{C}}-COOC_2H_4OH$$

It may be made by esterifying a fatty acid with bishydroxyethyl malate.[45] Higher alkyl urethans with surface-active properties are made by reacting ammonia or amines with the chlorocarbonates of fatty alcohols as:

$$ROOC-Cl + H_2NC_2H_4OH \rightarrow ROOC-NHC_2H_4OH$$

They may be rendered more soluble in water by treatment with ethylene oxide.[46]

[41] U. S. Pat. 2,147,241, Cantor to *Corn Products Refining Co.* French Pat. 716,458; Brit. Pat. 375,842; to *I. G. Farbenindustrie.*

[42] U. S. Pat. 2,371,333, Johnston to *Shell Development Co.*

[43] Goldsmith, *Chem. Revs.* **33**, 257 (1943).

[44] Brit. Pat. 460,140; French Pat. 790,041; to *Alexander Wacker Gesellschaft für Elekt. Ind.* U. S. Pat. 2,375,905, Engelhard and Petri, vested in the Alien Property Custodian.

[45] U. S. Pat. 2,329,166, Tucker to *Procter and Gamble Co.*

[46] Ger. Pats. 619,500, Ulrich and Körding to *I. G. Farbenindustrie*; 696,042, Gündel to *Deutsche Hydrierwerke A.-G.*; French Pat., 849,020, to *I. G. Farbenindustrie.* See also Ger. Pat. 697,730, Steindorff *et al.* to *I. G. Farbenindustrie.*

Several ether-linked emulsifying agents have been described. The ethers of oleyl alcohol or cetyl alcohol with di- or triethylene glycol are excellent emulsifying agents.[47] The mono higher alkyl ethers of glycerol and polyglycerol have also been prepared and used as emulsifiers. They are said to be particularly useful in preparing cosmetic creams.[48]

Monoethers of 1,2-dihydroxycyclohexanol have been made by reacting fatty alcohols with cyclohexene oxide:[49]

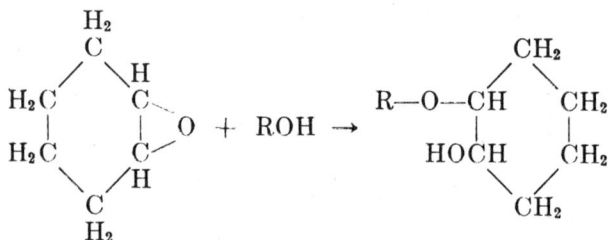

$$
\begin{array}{c}
H_2 \\
C \\
\diagup \quad \diagdown H \\
H_2C \qquad C \\
| \qquad\qquad | \quad O \; + \; ROH \; \rightarrow \\
H_2C \qquad C \\
\diagdown \quad \diagup H \\
C \\
H_2
\end{array}
\qquad
\begin{array}{c}
CH_2 \\
\diagup \quad \diagdown \\
R\text{---}O\text{---}CH \qquad CH_2 \\
| \qquad\qquad | \\
HOCH \qquad CH_2 \\
\diagdown \quad \diagup \\
CH_2
\end{array}
$$

The simplest amide-linked non-ionic oil-soluble emulsifying agents are the fatty amides of monoethanolamine. The C_8 to C_{12} members of the series are readily dispersible in water but the higher members are much more inert and waxy in behavior.[50] A wide variety of other alkylol amines have been reacted with fatty acids to form amides which may be used in preparing emulsions.[51] Among them are glucamine and methyl glucamine,[52] alkoxyl alkylol amines[53] such as:

$$H_2N\text{---}C_2H_4OC_2H_4OH$$

diethanolamine, the alkylol amines derived from the nitroparaffins, and dihydroxypropylamine ("glycerylamine").

More or less closely related amines have also been condensed with long-chain alkane sulfonyl chlorides such as those produced by the Reed reaction.[54]

[47] U. S. Pat. 2,164,431, Schöller and Nüsslein and Brit. Pat. 404,931, to *I. G. Farbenindustrie.*

[48] U. S. Pats. 2,258,892, to Benjamin R. Harris; 2,038,705 Baldwin *et al.* to *Imperial Chemical Industries Ltd.*; French Pat. 788,663, to *Henkel et Cie.*

[49] Brit. Pat, 524,470, Holt to *E. I. du Pont de Nemours & Co.*

[50] Ger. Pats. 612,686; 546,406, to Ulrich and Nüsslein; French. Pat. 852,792, to *I. G. Farbenindustrie.*

[51] U. S. Pat. 2,173,448, Katzman and Epstein to *Emusol Corp.*

[52] Brit. Pats. 414,403; 420,066; 420,518, to *Imperial Chemical Industries Ltd.* French Pat. 771,614, to *Ciba Co.*

[53] U. S. Pat. 2,187,823, Ulrich and Nüsslein to *I. G. Farbenindustrie.*

[54] Brit. Pat. 413,457, to *I. G. Farbenindustrie*; U. S. Pat. 2,334,186, Fox to *E. I. du Pont de Nemours & Co.*

Acid amides and sulfonamides have been treated with glycidol to introduce hydrophilic groups. The reaction is similar to that which takes place with ethylene oxide:[55]

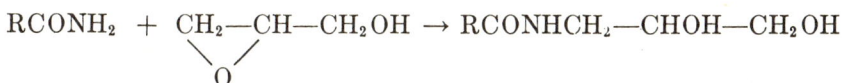

$$RCONH_2 \; + \; CH_2\!\!\underset{\underset{O}{\diagdown\diagup}}{-\!\!CH\!\!-}CH_2OH \; \rightarrow \; RCONHCH_2\!-\!CHOH\!-\!CH_2OH$$

Glycidol also reacts with hydroxyethylamides in the same manner as ethylene oxide:[56]

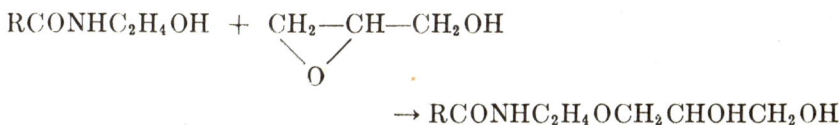

$$RCONHC_2H_4OH \; + \; CH_2\!\!\underset{\underset{O}{\diagdown\diagup}}{-\!\!CH\!\!-}CH_2OH$$

$$\rightarrow RCONHC_2H_4OCH_2CHOHCH_2OH$$

Fatty acid ethanolamides have been treated with formaldehyde to give emulsifying agents. It is probable that the $-CH_2OH$ group is introduced on the hydroxyl in this reaction.[57]

The fatty acid methylolamides, $RCONHCH_2OH$, are water dispersible and are good oil-soluble emulsifying agents. They are, however, relatively unstable chemically. They have been treated with urea[58] or with alkylating agents such as methanol and sulfuric acid to give more stable products with similar properties.[59]

Unsymmetrically substituted urea derivatives which are surface active have been made by starting with a hydrophobic amine such as dodecyl amine, treating with phosgene to give the carbamyl chloride, and finally treating with a hydrophilic amine such as methyl glucamine, or diethanolamine:[60]

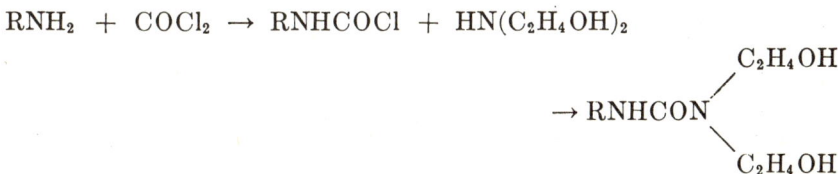

$$RNH_2 \; + \; COCl_2 \; \rightarrow \; RNHCOCl \; + \; HN(C_2H_4OH)_2$$

$$\rightarrow RNHCON\!\!\underset{\diagdown C_2H_4OH}{\overset{\diagup C_2H_4OH}{}}$$

[55] Brit. Pats. 420,883–4, to *I. G. Farbenindustrie*.
[56] Brit. Pat. 420,903, to *I. G. Farbenindustrie*.
[57] U. S. Pat. 2,186,464, Mauersberger to *Alframine Corp*.
[58] Swiss Pats. 210,976; 209,637, to *Ciba Co*.
[59] U. S. Pat. 2,361,185, Engelmann and Pikl to *E. I. du Pont de Nemours & Co*.
[60] U. S. Pat. 2,251,892, Orthner *et al*. to *I. G. Farbenindustrie*.

CHAPTER 9

Ampholytic Surface Active Agents

I. Introduction

Of the many patents which describe surface active agents containing both basic and acidic groups in the same molecule relatively few are explicit in pointing out the essential ampholytic nature of the compounds. This feature has been emphasized, however, in at least one important case. Rumpf prepared the higher N-alkyl taurines by heating the higher aliphatic amines with sodium bromoethane sulfonate. He also prepared the dialkylated derivatives and the N-alkyl N-phenyltaurines by similar reactions.[1] In his studies on these products he specifically points out their ampholytic properties.

$$RNH_2 + BrC_2H_4SO_3Na \rightarrow RNH—C_2H_4SO_3H$$

None of the ampholytic surface active agents have, to the writer's knowledge, achieved any great commercial importance. It might be expected that at least some members of the group should exhibit unusual properties that could not readily be achieved by anionic, cationic, or non-ionic agents. The small amount of study which has been devoted to them has evidently not brought such properties to light as yet.

Ampholytic detergents may be grouped into classes corresponding to the nature of the anion-forming group, which is usually carboxyl, sulfonic acid, or sulfuric ester. They may also be classified according as the basic nitrogen atom is quaternary or non-quaternary.

II. Carboxy Acid Ampholytes

The simplest products of this series are the higher alkyl amino acids. These may be made from the higher alkyl amines such as dodecyl, cetyl, or oleyl amine by reacting with a halogenated carboxy acid, such a chloroacetic acid:

$$RNH_2 + ClCH_2COOH \rightarrow RNHCH_2COOH$$

[1] Rumpf, *Compt. rend.* **212**, 83 (1941).

By further reaction with chloroacetic acid the disubstituted amines, and finally the betaines are formed:

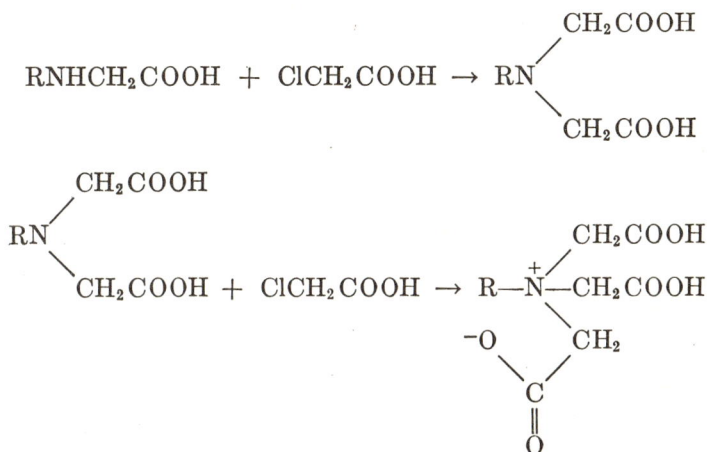

$$RNHCH_2COOH + ClCH_2COOH \rightarrow RN\begin{smallmatrix}CH_2COOH \\ \\ CH_2COOH\end{smallmatrix}$$

$$RN\begin{smallmatrix}CH_2COOH \\ \\ CH_2COOH\end{smallmatrix} + ClCH_2COOH \rightarrow R-\overset{+}{\underset{-O}{N}}\begin{smallmatrix}CH_2COOH \\ \\ CH_2\end{smallmatrix}$$

The latter products are stated to be excellent wetting and washing agents in either acid or alkaline solutions. They are also good sequestering agents for the lime and magnesium salts in hard water.[2] The mono- and dicarboxymethyl fatty amines can also be made from the amines by the formaldehyde–cyanide synthesis.[3]

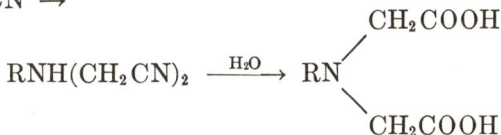

$$RNH_2 + CH_2O + HCN \rightarrow RNHCH_2CN \xrightarrow{H_2O} RNHCH_2COOH$$

$$RNH_2 + 2 CH_2O + 2 HCN \rightarrow$$

$$RNH(CH_2CN)_2 \xrightarrow{H_2O} RN\begin{smallmatrix}CH_2COOH \\ \\ CH_2COOH\end{smallmatrix}$$

Surface active aminocarboxy acids of this general type have also been made by reducing the Schiff's base formed between a higher aldehyde or ketone (such as 2-ethylhexaldehyde or methyl pentadecyl ketone) and a lower amino acid:[4]

[2] U. S. Pat. 2,206,249, Daimler et al. to General Aniline and Film Co.

[3] Brit. Pat. 460,372, to I. G. Farbenindustrie. See also French Pat. 793,473, to I. G. Farbenindustrie.

[4] Brit. Pat. 518,656, to I. G. Farbenindustrie.

$$RCHO + H_2N-CH_2COOH \rightarrow$$

$$RCH{=}N-CH_2COOH \xrightarrow{H_2} RCH_2-N-CH_2COOH$$
$$\phantom{RCH{=}N-CH_2COOH \xrightarrow{H_2} RCH_2-}H$$

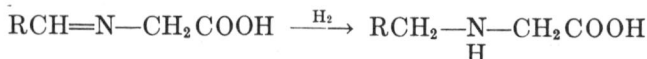

By starting with a tertiary higher alkyl amine such as dodecyldimethyl-amine and heating with halogenated carboxylic acids, of which chloroacetic acid is most commonly used, betaines are obtained directly. These products may also be made by condensing the tertiary amine with a halogenated carboxylic ester and subsequently saponifying.[5]

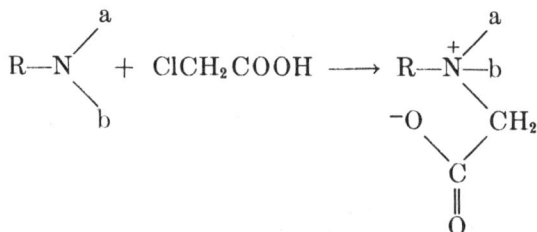

Instead of saponifying the resultant esters may be reacted further with aminocarboxy acids:

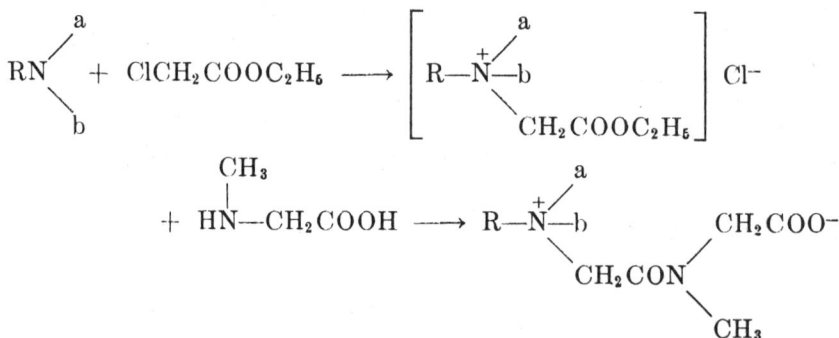

Higher alkyl halogen compounds may be reacted with lower tertiary amino acids to form surface active betaines, provided the halogen is in a reactive configuration. The chloromethyl ethers of higher alcohols, for example, will react with dimethylaminoacetic acid to produce the betaine.[6]

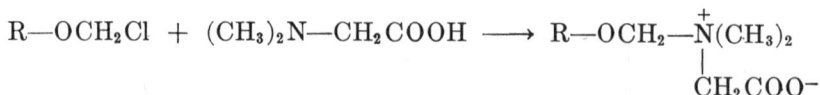

$$R-OCH_2Cl + (CH_3)_2N-CH_2COOH \longrightarrow R-OCH_2-\overset{+}{N}(CH_3)_2$$
$$CH_2COO^-$$

[5] U. S. Pat. 2,082,275, Daimler and Platz to *I. G. Farbenindustrie*. Brit. Pats. 446,416; 464,657; French Pats. 789,304; 806,790; to *I. G. Farbenindustrie*.
[6] U. S. Pat. 2,217,846, Orthner *et al.* to *General Aniline and Film Corp.*

Syntheses of the above type may be carried out using halogenated dicarboxy acids, such as monochlorosuccinic acid, as well as the halogenated monocarboxy acids:

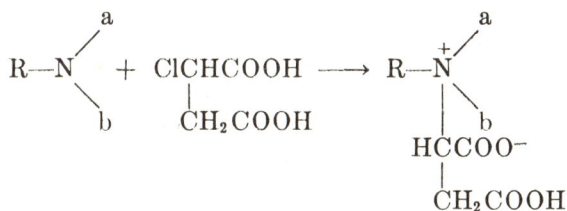

$$\underset{\substack{\diagup \text{a} \\ \text{R}-\text{N} \\ \diagdown \text{b}}}{} + \underset{\substack{\text{ClCHCOOH} \\ | \\ \text{CH}_2\text{COOH}}}{} \longrightarrow \underset{\substack{\diagup \text{a} \\ \text{R}-\overset{+}{\text{N}} \\ | \diagdown \text{b} \\ \text{HCCOO}^- \\ | \\ \text{CH}_2\text{COOH}}}{}$$

It is also stated in one interesting disclosure that quaternized aspartic acid derivatives of the above type may be made by heating the maleic or fumaric acid salts of tertiary amines:[7]

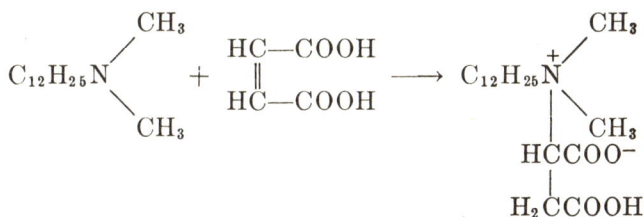

$$\underset{\substack{\diagup \text{CH}_3 \\ \text{C}_{12}\text{H}_{25}\text{N} \\ \diagdown \text{CH}_3}}{} + \underset{\substack{\text{HC}-\text{COOH} \\ \| \\ \text{HC}-\text{COOH}}}{} \longrightarrow \underset{\substack{\diagup \text{CH}_3 \\ \text{C}_{12}\text{H}_{25}\overset{+}{\text{N}} \\ | \diagdown \text{CH}_3 \\ \text{HCCOO}^- \\ | \\ \text{H}_2\text{CCOOH}}}{}$$

Surface active amino acids and betaines in which the long-chain group is attached to a carbon rather than to the nitrogen may be made from the α-halogenated long-chain fatty acids. α-Bromopalmitic acid, for example, may be aminated with ammonia, primary, secondary, or tertiary amines:[8]

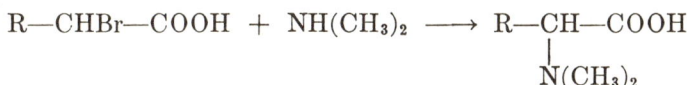

$$\text{R}-\text{CHBr}-\text{COOH} + \text{NH}(\text{CH}_3)_2 \longrightarrow \underset{\substack{\text{R}-\text{CH}-\text{COOH} \\ | \\ \text{N}(\text{CH}_3)_2}}{}$$

If tertiary amines are used, betaines are formed. Pyridine is stated to be particularly effective in this reaction:[9]

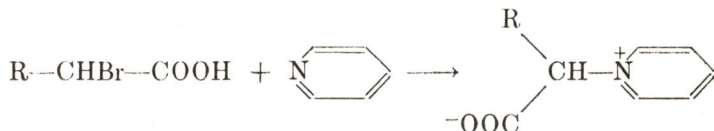

$$\text{R}-\text{CHBr}-\text{COOH} + \text{N}\langle\text{O}\rangle \longrightarrow \underset{\substack{\text{R} \\ \diagdown \\ \text{CH}-\overset{+}{\text{N}}\langle\text{O}\rangle \\ \diagup \\ {}^-\text{OOC}}}{}$$

As in other classes of surface active agents the hydrophobic group may

[7] Ger. Pat. 688,083; French Pat. 832,102; to *Deutsche Hydrierwerke A.-G.*

[8] U. S. Pats. 2,279,138, Henke and McGrew to *E. I. du Pont de Nemours & Co.;* 2,203,009, Calcott *et al.* to *E. I. du Pont de Nemours & Co.*

[9] U. S. Pat. 2,199,397, Engelmann to *E. I. du Nemours & Co.* Swiss Pat. 179,662, to *Ciba Co.* Ger. Pat. 692,417, Rieche *et al.* to *I. G. Farbenindustrie.*

be joined to either or both of the hydrophilic centers through intermediate linkages.

Higher monoalkyl esters of monochlorosuccinic acid have been converted to betaine types by reacting with tertiary amines:

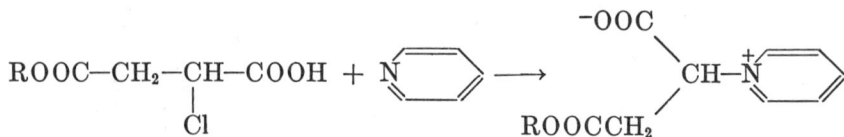

$$ROOC-CH_2-\underset{\underset{Cl}{|}}{CH}-COOH \;+\; N\hspace{-2pt}\bigcirc \longrightarrow \quad \underset{ROOCCH_2}{\overset{^-OOC}{\diagdown}}CH-\overset{+}{N}\hspace{-2pt}\bigcirc$$

More complex structures with amide and ester linkages have been built up using the same general synthetic scheme:

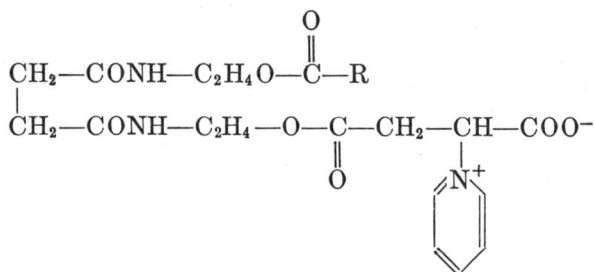

$$CH_2-CONH-C_2H_4O-\overset{\overset{\displaystyle O}{\|}}{C}-R$$
$$CH_2-CONH-C_2H_4-O-\underset{\underset{O}{\|}}{C}-CH_2-\underset{N^+}{CH}-COO^-$$

The above compound, for example, may by synthesized from succinic acid, monoethanolamine, a higher fatty acid, monochlorosuccinic acid, and pyridine.[10]

An interesting group of ampholytic surface active agents has been made by adding higher fatty amines across the double bond of acrylic acid. The products are long-chain alkyl β-aminopropionic acids:[11]

$$R-NH-CH_2CH_2COOH$$

Mono-fatty acylated polyamines have been reacted with chloroacetic acid to form ampholytic surface active agents.[12]   A typical reaction is:

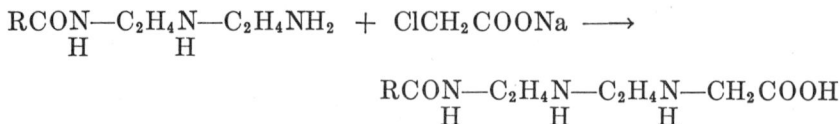

$$RCON-C_2H_4N-C_2H_4NH_2 \;+\; ClCH_2COONa \longrightarrow$$
$$\;\;\;\;H\;\;\;\;\;\;\;H$$

$$RCON-C_2H_4N-C_2H_4N-CH_2COOH$$
$$\;\;\;\;H\;\;\;\;\;\;\;\;\;H\;\;\;\;\;\;\;\;H$$

[10] U. S. Pats. 2,213,979; 2,299,756; 2,239,706; Epstein and Katzman to *Emulsol Corp.*
[11] French Pat. 843,558, to *I. G. Farbenindustrie.*   See also French Pat. 793,504, to *I. G. Farbenindustrie.*
[12] French Pat. 783,255, to *I. G. Farbenindustrie.*

A group of high-molecular-weight ampholytic softening agents has been described in which long-chain tertiary amines are reacted with chloro-methylated salicyclic acids:[13]

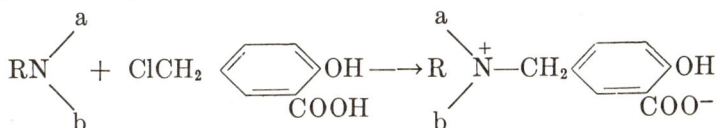

$$\underset{\text{b}}{\overset{\text{a}}{RN}} + ClCH_2 \text{—}\langle\rangle\text{—}\underset{COOH}{OH} \longrightarrow R \overset{\text{a}}{\underset{\text{b}}{\overset{+}{N}}}\text{—}CH_2\text{—}\langle\rangle\text{—}\underset{COO^-}{OH}$$

They are claimed to be exceptionally substantive to cellulosic fabrics.

The sulfonamide linkage has also been used in making ampholytic detergents. By starting with an alkane sulfonamide or alkyl aryl sulfonamide the following series of reactions may be carried out:[14]

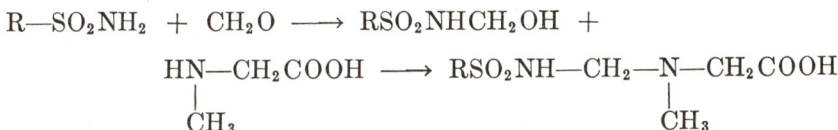

$$R\text{—}SO_2NH_2 + CH_2O \longrightarrow RSO_2NHCH_2OH +$$

$$\underset{CH_3}{\overset{|}{HN}}\text{—}CH_2COOH \longrightarrow RSO_2NH\text{—}CH_2\text{—}\underset{CH_3}{\overset{|}{N}}\text{—}CH_2COOH$$

Werntz has described an interesting series of non-nitrogenous ampholytic surface active agents which are sulfonium carboxy compounds. They are made by reacting higher alkyl thioethers with a halogenated carboxy acid. Dodecyl methyl sulfide, for example reacts as follows with chloro-acetic acid:[15]

$$C_{12}H_{25}\text{—}S\text{—}CH_3 + ClCH_2COOH \longrightarrow C_{12}H_{25}\text{—}\underset{CH_2COO^-}{\overset{CH_3}{\overset{|}{\underset{|}{S^+}}}}$$

This type of compound is called a thetine by analogy with the nitrogenous betaines.

III. Sulfuric Ester and Sulfonic Ampholytes

Ampholytic surface active agents containing sulfuric ester as the anion-forming hydrophilic group have been prepared for the most part by sulfating hydrophobic amines containing either an olfenic double bond or a hydroxyl group.

1-Amino-9,10-octadecene (oleylamine) and the corresponding 9- or

[13] Swiss Pats. 205,898; 207,448–50; to *J. R. Geigy A.-G.*

[14] U. S. Pat. 2,243,437, Orthner *et al.* to *General Aniline and Film Corp.*

[15] U. S. Pat. 2,178,353, Werntz to *E. I. du Pont de Nemours & Co.*

10-hydroxyoctadecylamine have been sulfated to give surface active ampholytes.[16]

The hydroxyethyl and dihydroxypropyl derivatives of the higher fatty amines have been sulfated in the usual manner.[17] The parent compounds may be made by the action of the corresponding epoxides or halohydrins on the amine:

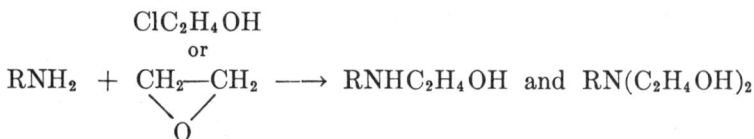

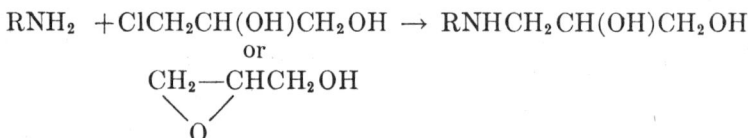

$$\text{RNH}_2 + \begin{array}{c} \text{ClC}_2\text{H}_4\text{OH} \\ \text{or} \\ \text{CH}_2\!\!-\!\!\text{CH}_2 \\ \diagdown\diagup \\ \text{O} \end{array} \longrightarrow \text{RNHC}_2\text{H}_4\text{OH} \ \text{and} \ \text{RN}(\text{C}_2\text{H}_4\text{OH})_2$$

$$\text{RNH}_2 + \begin{array}{c} \text{ClCH}_2\text{CH(OH)CH}_2\text{OH} \\ \text{or} \\ \text{CH}_2\!\!-\!\!\text{CHCH}_2\text{OH} \\ \diagdown\diagup \\ \text{O} \end{array} \rightarrow \text{RNHCH}_2\text{CH(OH)CH}_2\text{OH}$$

and

$$\text{RN[CH}_2\text{CH(OH)CH}_2\text{OH]}_2$$

More than one of the hydroxyl groups may be sulfated if desired. These products have also been converted into phosphoric acid esters, forming ampholytes which contain amino and phosphoric ester as the hydrophilic groups.[18] Similar products have been made from ricinoleic alcohol (1,12-dihydroxy-9,10-octadecene) by replacing one of the hydroxyl groups with halogen and diethylamino successively and finally sulfating the remaining hydroxyl group.[19]

Compounds such as the monooleic acid amides of ethylenediamine, diethylenetriamine, etc. may be sulfated to form ampholytes.[20]

Saturated or unsaturated fatty acyl derivatives of aminoethylethanolamine, $\text{RCONHC}_2\text{H}_4\text{NHC}_2\text{H}_4\text{OH}$, may be sulfated. Products of this type have also been made by condensing sulfated oleic acid with aminoethylethanolamine. The reaction is effected at 150°C. under substantially anhydrous conditions.[21]

The saturated and unsaturated fatty acyl derivatives of aminoethylethanolamine may be submitted to ring closure and thus converted to

[16] Brit. Pat. 353,232; Ger. Pat. 669,541; U. S. Pat. 1,951,469; Bertsch to *H. Th. Böhme A.-G.*

[17] Brit. Pat. 512,022, to *I. G. Farbenindustrie.*

[18] Brit. Pat. 377,695; U. S. Pat. 2,229,307; Ralston and Harwood to *Armour and Co.*

[19] Brit. Pat. 435,290, to *Deutsche Hydrierwerke A.-G.*

[20] U. S. Pat. 2,008,649, Ulrich and Nüsslein to *I. G. Farbenindustrie.*

[21] U. S. Pat. 2,329,086, Robinson and Webber to *National Oil Products Co.*

products of the general formula:

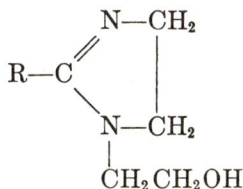

$$
\begin{array}{c}
\quad N\!-\!CH_2 \\
R\!-\!C\diagup\big| \\
\diagdown N\!-\!CH_2 \\
\big| \\
CH_2\,CH_2OH
\end{array}
$$

These imidazoline derivatives have been sulfated to form ampholytes of high surface activity. Sulfation takes place on the hydroxyl group unless the fatty acid chain contains a double bond, in which case it may occur at both positions. Unsaturated fatty acyl tetrahydropyrimidines have been made by condensing oleic acid (or other unsaturated hydrophobes such as abietic or naphthenic acids) with 1,3-aliphatic diamines:

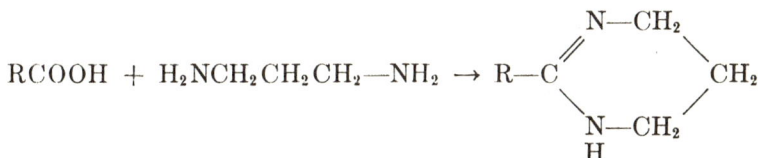

$$
RCOOH + H_2NCH_2CH_2CH_2\!-\!NH_2 \rightarrow R\!-\!C\!\!\begin{array}{c}\diagup N\!-\!CH_2\diagdown\\ \diagdown N\!-\!CH_2\diagup\\ H\end{array}\!\!CH_2
$$

These imidazoline derivatives have also been sulfated to give surface active agents.[22]

A large number of ampholytic sulfonic acids which are analogous to the ampholytic carboxylic acids considered above may be made by substituting a halogenated alkane sulfonic acid for the halogenated carboxy acid. Just as chloroacetic acid is by far the most commonly used halocarboxy acid, so 2-chloro- or 2-bromoethanesulfonic acid is the most commonly used compound in its class. There are at least three other halo alkane sulfonic acids which are frequently mentioned in the literature and which are commercially practicable under some economic conditions. These are 2-hydroxy-3-chloropropanesulfonic acid, made from glyceroldichlorohydrin:

$$ClCH_2\!-\!CHOH\!-\!CH_2SO_3H$$

2-chloroethoxyethanesulfonic acid, made from β,β'-dichlorodiethyl ether:

$$ClC_2H_4\!-\!O\!-\!C_2H_4SO_3H$$

and the p-sulfonic acid of benzyl chloride:

$$ClCH_2\!-\!\!\diagup\!\!\bigcirc\!\!\diagdown\!\!-\!SO_3H$$

[22] Ger. Pat. 700,371, Waldmann and Chwala to *I. G. Farbenindustrie.*

These halosulfonic acids, for example, have been reacted with the higher aliphatic amines to form products of which the higher alkyl taurines are typical:[23]

$$RNHC_2H_4SO_3H$$

They may also be reacted with hydrophobic tertiary amines to give quaternary inner salts.[24]

Alternatively, higher fatty halides may be reacted with amino sulfonic acids. Cetyl chloride, for example, has been reacted with sodium aniline 2,4-disulfonate to form surface active agents.[25] Alkyl benzyl chlorides have also been reacted with aminosulfo acids such as taurine to give similar compounds:[26]

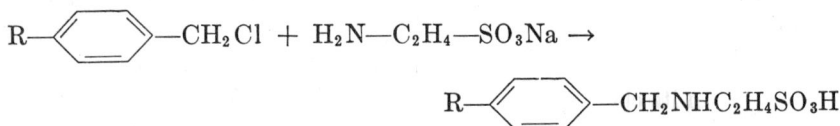

$$R-\langle\ \rangle-CH_2Cl\ +\ H_2N-C_2H_4-SO_3Na\ \rightarrow$$

$$R-\langle\ \rangle-CH_2NHC_2H_4SO_3H$$

Products analogous to the anionic sulfosuccinic acid esters have been made by reacting chlorosuccinic acid esters or maleic esters with taurine:[27]

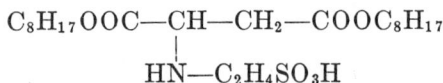

$$C_8H_{17}OOC-CH-CH_2-COOC_8H_{17}$$
$$|$$
$$HN-C_2H_4SO_3H$$

Ampholytic sulfonic acids may also be made by direct sulfonation of suitable cationic compounds containing sulfonatable aromatic rings. Zephyran itself has been sulfonated using chlorosulfonic acid at low temperatures as the reagent. The product is a true ampholyte:[28]

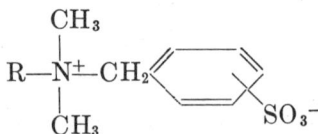

$$
\begin{array}{c}
CH_3 \\
| \\
R-N^{\pm}-CH_2-\langle\ \rangle-SO_3^- \\
| \\
CH_3
\end{array}
$$

[23] U. S. Pat. 1,944,300, Ott *et al.* to *I. G. Farbenindustrie*.

[24] Brit. Pats. 446,269; 446,337; to *I. G. Farbenindustrie*.

[25] Brit. Pat. 417,394, to *Deutsch Hydrierwerke A.-G.* See also U. S. Pat. 2,245,162, Sibley to *Monsanto Chemical Co.*

[26] U. S. Pat. 2,376,911, Gränacher, Streuli, and Meyer to *Ciba Co.*

[27] U. S. Pat. 2,379,535, Lynch and West to *American Cyanamid Co.*

[28] U. S. Pat. 2,075,958, Piggott to *Imperial Chemical Industries Ltd.*; Brit. Pat. 419,942; to *Imperial Chemical Industries Ltd.*

A number of α-substituted benzylamines of general formula:

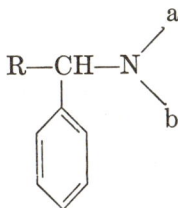

$$R-CH-N\begin{matrix} \nearrow a \\ \searrow b \end{matrix}$$

have also been sulfonated to produce ampholytic surface active agents.[29]

In the imidazoline series ampholytic products have been made by reacting the long-chain imidazolines with the usual halogenated sulfonic acids.[30] A typical product has the formula:

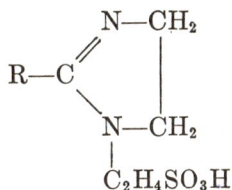

$$R-C\begin{matrix} \diagup N-CH_2 \\ \diagdown N-CH_2 \end{matrix}$$
$$C_2H_4SO_3H$$

The aminoethylimidazolines have been condensed with aldehyde-substituted sulfonic acids to form ampholytic products with a Schiff's base type of linkage:[31]

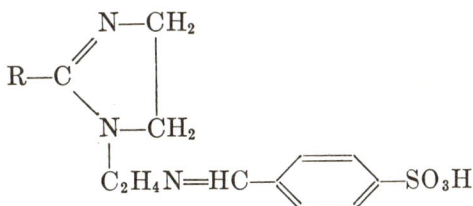

$$R-C\begin{matrix} \diagup N-CH_2 \\ \diagdown N-CH_2 \end{matrix}$$
$$C_2H_4N=HC-\langle\!=\!\rangle-SO_3H$$

The sulfonated benzimidazoles have already been considered in the anionic class. A good case might be made for classifying them with the ampholytics since the secondary nitrogen atom is sufficiently basic to

[29] Swiss Pats. 211,795–7; 213,562–4; to *J. R. Geigy A.-G.*

[30] Brit. Pats. 460,858; 507,766; French Pat. 796,917; Ger. Pat. 705,132; U. S. Pats. 2,154,922; 2,199,780; Waldmann and Chwala to *I. G. Farbenindustrie.*

[31] Brit. Pat. 499,784, to Chwala and Waldmann.

condense with chloroacetic acid.   Compounds of the type:

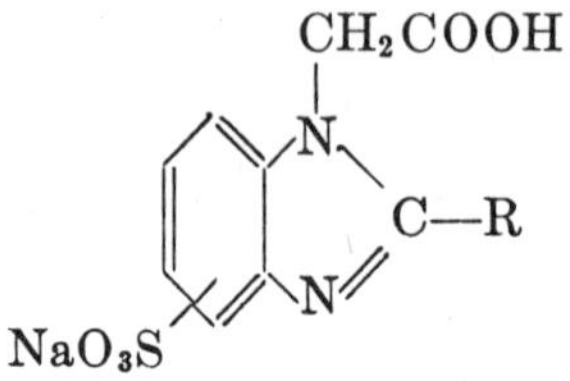

$$CH_2COOH$$

have been produced this way.[32]

Another type of detergent which may be either ampholytic or anionic has been made by using the triazine ring to link the hydrophobic and hydrophilic groups.   By starting with cyanuric chloride and reacting successively with a higher fatty amine and an aminosulfonic acid or an aminocarboxy acid, products such as:

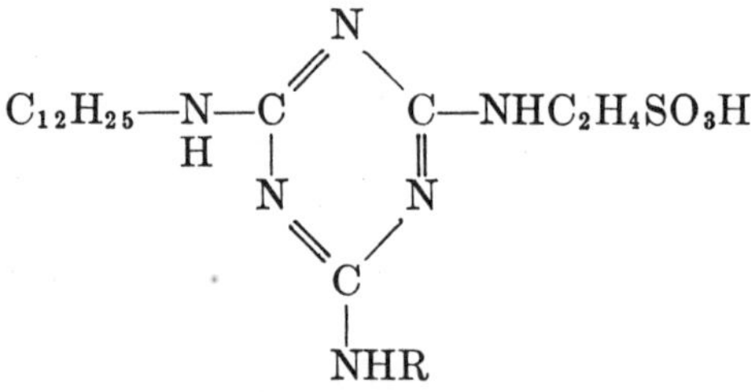

have been made.[33]   This general reaction is capable of wide variation and many interesting examples are given in the patent disclosures.

[32] U. S. Pat. 2,335,271, Gränacher, Streuli, and Frei to *Ciba Co.*   See also U. S. Pats. 2,004,864, Gränacher and Ackermann to *Ciba Co.*; 2,036,525, Gränacher to *Ciba Co.*; Ger. Pat. 735,009, to *Ciba Co.*

[33] U. S. Pat. 2,298,533, Hentrich and Schirm to *Unichem.*

CHAPTER 10

Special Compositions and Mixtures

I. Surface Active Agents in Non-aqueous Systems

The term "surface active agent" usually is applied to solutes which affect an interface involving an aqueous system. Most often it is reserved for water-soluble substances. We have already seen, however, that many substances which are relatively insoluble in water but soluble in oil can strongly affect an interface between their solvent oil and water. These are the so-called oil-soluble emulsifying agents. There are numerous technological operations in which surface phenomena are of primary importance and in which aqueous media are not involved at all or are only involved secondarily. The field of lubricants and lubrication is possibly the most noteworthy example.

The basic problem in lubrication is the maintenance of a film of relatively low shear resistance between two solid surfaces. As long as the film is maintained, the surfaces slide over each other readily. When the film is broken the solid surfaces come into intimate contact and adhere to each other or "seize." To maintain a liquid film between two solid surfaces it is necessary that the liquid have sufficient viscosity to resist being mechanically squeezed out by the pressure on the bearing interface. Fully as important, however, the liquid must wet the metal surfaces thoroughly and adhere strongly. This property of lubricants is often referred to as oiliness. One difference between good lubricating oils and poor ones is that the former contain substances which are strongly adsorbed at the oil–metal interface. Lubricants designed for extreme pressure conditions, such as obtain in modern hypoid-gear automobile differentials, have special ingredients added which are very strongly adsorbed. These ingredients are in effect surface active agents since small concentrations in the bulk of the oil have a profound effect on the interface. A vast science of lubrication and lubricant additives has grown up in recent years. It is outside our scope to consider the large numbers of different substances which have been proposed as extreme-pressure lubricant additives.[1] It

---

[1] Several review articles have appeared in connection with lubricant additives. Among them are: Schnurman, *Chem. Products* **7**, 56 (1944); Musgrave, *Iron Steel Engr.* **20**, No. 5, 40, (1943); Evans, *J. Inst. Petroleum* **29**, 333 (1943); Pritzker, *Natl. Petroleum News* **37**, R793–800 (1945).

should be pointed out that they are actually surface active within their environmental system, which consists essentially of a hydrocarbon–metal interface. Like the aqueous surface active agents we have been considering these substances have one broad feature in common. They are all electrically unbalanced molecules. One portion of the molecule might be termed "oleophilic." It is characterized by low polarity or low intensity of the stray fields associated with the interatomic bonds. The other portion of the molecule has high-intensity stray fields. It tends to be repelled by the hydrocarbon molecules and hence to be attracted to the surface or interface.

It was early discovered that the addition of fatty oils or fatty acids to hydrocarbon lubricating oil greatly improved its ability to function under high bearing pressures. One of the first synthetic substances to be deliberately added for the purpose of increasing oiliness was dichlorostearic methyl ester. Since that time a great many substances have been developed and used for this specific purpose. Typical of these are such structures as sulfurized hydrocarbons, metallic soaps such as lead soaps, long-chain alkyl esters and alkyl phenyl esters of phosphoric acid, and chlorinated alkyl aromatic hydrocarbons. These substances all have the effect of literally "oil plating" the bearing surfaces thereby preventing intimate metal-to-metal contact. The action is analogous to the wetting-out process in an aqueous system.

In the lubricating-oil field there are at least two other examples of surface effects obtained by the addition of special reagents. In the normal running of an engine the lubricating oil gradually undergoes oxidation and thermal cracking and forms a sludge consisting of suspended carbon particles. These carbon particles tend to deposit as a hard cake in the piston ring grooves and on the valves. This necessitates periodic shutdowns to scrap out the carbon deposit. If neglected too long between cleanings the engine may be seriously damaged. It has been found that the addition of small amounts of special reagents to the oil keeps the sludge deflocculated and suspended. Whatever carbon deposits are formed are soft and friable rather than compacted. These reagents exert a surface effect similar to that of a dispersing agent in an aqueous system. Products used for this purpose include mahogany soaps (which are also used for other purposes in aqueous systems) heavy-metal soaps such as calcium phenyl stearate,[2] sulfurized and chlorinated hydrocarbons, etc.

Another problem of surfaces in connection with lubricating oils is foaming. Violent agitation of the oil in an engine tends to produce foaming. where a pressure pumping system is used this may result in the delivery

[2] U. S. Pat. 2,081,075, Vobach to *Sinclair Refining Co.*

of foam rather than of a full quota of oil to vital engine parts. Antifoaming agents have been developed to overcome this tendency. Among substances which have been proposed for this purpose are dimethylsilicone,[3] various salts of dialkyl orthophosphates,[4] trialkyl thiophosphates,[5] and fluorinated oils.[6]

Non-aqueous surface active agents are used in slushing oils designed to protect metals from corrosion. Lanolin is particularly useful for this purpose, but many products have been particularly designed with this end in view. Some of them are so powerful that they will cause the oil in which they are incorporated to displace water completely from a metallic surface. It is thus possible to remove metallic objects from a pickling bath or an aqueous finishing or cleaning bath and dip them directly in the protective oil. The oil displaces the adhering aqueous liquor (which sinks to the bottom of the bath) and the metal is withdrawn dry, oil-coated, and ready to store.

Surface active agents are used in paints and enamels to deflocculate the pigment particles and keep them suspended in the oleoresinous vehicle.

It should be noted that many of the products which are surface active in aqueous systems are also oil soluble and surface active in non-aqueous systems. Some examples of this sort are considered in the last part of this book. The foregoing paragraphs are intended merely to indicate the large fields of technology where surface activity is of prime importance and yet where the better-known soap-like surface active agents are not necessarily applicable. The academic investigation of these fields has lagged behind their industrial development up to the present time.

II. Importance of Mixtures of Surface Active Agents

Surface active agents are seldom encountered in absolutely pure form. They are usually produced as mixtures with inorganic salt and may also contain considerable amounts of water. In most cases these ingredients function purely as diluents and the product is not regarded as a purposeful mixture. There are many cases, however, in which surface active agents are mixed with each other or with secondary ingredients to produce compositions for specialized applications. The number of such special compositions which has been described is far greater than the number of individual surface active agents, and it is not our purpose to consider them in detail. Many of them can be grouped into broad classes and a compre-

[3] U. S. Pat. 2,375,007, Larsen and Diamond to *Shell Development Co.*
[4] U. S. Pats. 2,397,377-81, Smith *et al.* to *Gulf Oil Corp.*
[5] U. S. Pat. 2,353,587, Rill to *Sacony Vacuum Oil Co.*
[6] U. S. Pat. 2,394,596, Davis and Zimmer to *Standard Oil Development Co.*

hensive view of the field of surface active agents should include some of these more important groups of mixtures.

III. Inorganic Builders

One of the most widely encountered types of mixture consists essentially of a surface active agent (of the water-soluble or "soapy" type) and an inorganic salt which enhances the surface activity.   It has been known for many years that certain alkaline salts such as the alkali metal carbonates, phosphates, borates, and silicates promote the detergent action of soaps. These substances are known as "builders," or more specifically as "alkaline builders."   Soap preparations with admixed alkaline builders are very widely marketed and used.   Neutral reacting inorganic salts such as sodium sulfate and sodium chloride also act as builders for soap, but this effect is nullified except at very low concentrations because they salt out the soap.   They are not regarded as useful builders for soap because of this effect.   The alkaline builders are quite useful with the sulfonic acid and sulfuric ester synthetic detergents and even with some of the non-ionic types.   Harris has made an investigation of the effect of both alkaline and neutral salts as builders for the alkyl aryl sulfonate type of detergent.[7] The alkaline builders behave with these sulfonates in much the same way as with soap.   The neutral builders such as sodium sulfate can be used in much larger proportions than is possible with soap.   They contribute to detersive efficiency even when present in amounts equal to the active detergent.   Above these proportions they may usually be regarded as diluents.

The phosphate builders have been added to akyl sulfates and alkane sulfonates to increase their detersive efficiency and their anticorrosion properties.[8]   Sodium hydrogen phosphates are claimed to be effective builders for a variety of sulfonic acid and sulfuric ester detergents.[9]   Mixtures of tetrasodium pyrophosphate and acid-reacting salts which provide a pH of less than 7 have been used for the same purpose, particularly with detergents of the Igepon type.[10]

Among the more unusual builders which have been used with the sulfuric ester and sulfonic acid detergents are the salts of sulfamic acid[11] and salts

---

[7] J. C. Harris, *Oil and Soap* **23,** 101 (1946).

[8] U. S. Pats. 2,369,278, Lind to *Procter and Gamble Co.*; 2,394,320–1, McGhie to *Colgate-Palmolive-Peet Co.*

[9] U. S. Pat. 2,383,502, Quimby to *Procter and Gamble Co.*

[10] U. S. Pat. 2,279,314, Henderson and Maxwell to *Lever Bros. Co.*

[11] Brit. Pat. 496,209, to *Imperial Chemical Industries Ltd.*

of the simple aromatic sulfonic acids such as benzene and naphthalene sulfonic acids.[12]

Sometimes inorganic salts which may or may not have a building action are added simply to convert a liquid or pasty detergent to a dry mixture. This has been done for example with Lamepon type detergents and with the non-ionic agents,[13] both of which are viscous liquids under ordinary conditions.

Mixtures of detergents with builders and solid salts which have a bleaching action are occasionally encountered, but components such as these are more usually added separately to the treating baths.[14]

Prominent among inorganic builders and fillers which are mixed with synthetic detergents is bentonite and the allied group of clays. Bentonite in itself has detergent properties by virtue of its strong emulsifying power. It is also a powerful adsorbent and forms stiff pasty products even when the water content is relatively high. Mixes of most of the various synthetic detergent types with bentonite have been described.[15]

IV. Mixtures with Oils and Solvents

Compositions comprising a surface active agent and a water-insoluble oil or solvent are made up for a wide variety of purposes and are very frequently encountered commercially. Solvents such as pine oil, terpineol, phenols, hydrogenated phenols, cellosolves, medium-boiling-range alcohols, esters and ketones, etc. are frequently added to synthetic surface active agents to improve their wetting properties.[16] The surface active agents which are used are soluble in the solvents and in many cases a large amount of water can be incorporated into the mixture without creating a second phase. The compositions are usually homogeneous viscous liquids.

Similar mixtures are often used as detergents, particularly where a heavy soil consisting largely of greasy matter is to be removed. The solvent serves to "cut" or dissolve the grease. On flushing with water the surface

[12] Ger. Pat. 685,428, to *Henkel et Cie.* Brit. Pat. 406,001, to *Imperial Chemical Industries Ltd.*

[13] Brit. Pats. 500,631, to *Chemische Fabrik Grünau*; 490,285, to *I. G. Farbenindustrie.*

[14] See, for example, U. S. Pat. 2,320,279–81, Kalusdian to *Mathieson Alkali Works.* French Pats. 770,253; 773,751; to *H. Th. Böhme A.-G.*

[15] See, for example, Brit. Pat. 461,221, to Marriott, Dugan, and *F. W. Berk and Co. Ltd.*

[16] Ger. Pat. 683,845, to *Böhme Fettchemie G.m.b.H.* Brit. Pats. 393,164; 402,295–6, to *H. Th. Böhme A.-G.* U. S. Pats. 2,243,994, Adams and MacLaren to *Standard Oil Co. of Indiana*; 2,221,933, Eitelman and Flett to *National Aniline and Chemical Co.* Ger. Pat. 617,180, to *Böhme Fettchemie G.m.b.H.*

active agent causes both solvent and greasy soil to become emulsified and easily removable.   Mixtures of this type are sometimes used without water as detergent aids in dry cleaning of fabrics and in cleaning engines or other articles where the presence of water would be undesirable.   The petroleum hydrocarbon solvents and the chlorinated solvents are most frequently used in these mixtures.   The surface active agents may be amine soaps, potash soaps, or almost any type of synthetic detergent.[17]

Surface active agents mixed with oils or solvents form the basis of numerous emulsifiable and emulsifying compositions.   "Soluble oils" of various types are usually mixtures of oil and a surface active agent together with a blending agent or mutual solvent which serves to give a homogeneous preparation.   Most sulfated vegetable oils contain an appreciable proportion of unsulfated material.   The mahogany sulfonates contain 20% or more of hydrocarbon oil and can thus be considered as mixtures of oil and surface active agent.   The soluble oils of various types are widely used as lubricants for textile fabrication and as cutting oils in metal fabrication. Many different types of surface active agents have been used in preparing these compositions.   The soaps, particularly the amine soaps, and the petroleum sulfonates are among the most popular.[18]

Specially prepared emulsifying agents for making paraffin wax emulsions are rather complex mixtures which contain soap or other surface active agents and a mixture of oils and solvents.   These substances are so designed that they can form a homogeneous mixture with the wax to be emulsified. This mixture forms a stable emulsion on dilution with water.[19]

Ternary mixtures consisting of a surface active agent, a solvent, and an alkaline builder such as a phosphate have been described and are sometimes encountered as specialty cleaners.[20]

V. Organic Builders and Mixtures

Surface active agents are often encountered in the form of mixtures with other organic substances which are not surface active but which may contribute special properties to the mixture.   Possibly the most important class of organic additives include the water-soluble high-polymeric mate-

[17] See, for example, U. S. Pats. 2,383,114, De Villiers to *Cities Service Oil Co.*; 2,388,962, Flett to *Allied Chemical and Dye Corp.*; 2,403,612, Rice to *Solvay Process Co.*; 2,403,618–9, Skinner to *National Carbon Co.*

[18] U. S. Pats. 2,388,677, Cohen to *Standard Oil Development Co.*; 2,079,803, Holtzclaw and Winning to *Standard Oil Development Co.*

[19] See, for example, U. S. Pats. 2,207,256, Kapp to *National Oil Products Co.*; 2,244,685, Fritz and Beach to *National Oil Products Co.*

[20] U. S. Pats. 1,999,628–32, Friesenhahn to *Unichem.*   Ger. Pats. 623,403; 629,646, to *Deutsche Hydrierwerke A.-G.*

rials such as gums, starches, proteins, and cellulose derivatives. The sodium salt of carboxymethylcellulose was intensively studied and widely used as a detergent builder in Germany during the war. It was made under the trade name Tylose HBR. Convincing evidence has been gathered in laboratory and field tests that this substance has a powerful building action on most of the anionic detergents.[21] Other water-soluble cellulose ethers such as methylcellulose and hydroxyethylcellulose have also been used as builders for anionic detergents of various types.[22] Complex urea-formaldehyde-hydroxyethylcellulose combinations have been described for the same purpose.[23] Sodium polyacrylate has been used as a thickening agent and a builder for synthetic detergents. Sometimes gums and starches are used in specialty products to give high viscosity and an appearance of high concentration to diluted surface active compositions.

Organic acids are often used with sulfonate or sulfate type surface active agents to form acid-cleaning compositions for special purposes. Inorganic acids may also be used but these are usually added by the user rather than by the detergent manufacturer. Citric and lactic acids have been incorporated into detergents for personal use such as shampoos.[24] Large-scale industrial uses for acid detergents are in cleaning dairy equipment and in metal cleaning prior to finishing. Acids such as $NaHSO_4$, glycolic, levulinic, etc. are used with synthetic detergents in these applications.[25]

Urea and thiourea have been used with various types of surface active agents to increase their solubility. They are also used as cheap and innocuous fillers in preparations for personal use. They have a soothing effect on the skin and modify the harsh feel normally associated with the sulfate and sulfonate detergents. The solubilizing action of urea is quite pronounced and is an example of its ability to peptize colloidal materials in general.[26] Sodium citrate and sodium lactate have similar peptizing properties and are sometimes used as additives for surface active agents.[27]

The water-soluble detergents are said to be improved by the addition

[21] Hoyt, *PB 3865*, Office of Technical Services, Dept. of Commerce, Washington, D. C.

[22] U. S. Pat. 2,347,336, Seyferth to *Allied Chemical and Dye Corp.*

[23] U. S. Pats. 2,237,240, Sponsel to *Kalle and Co.*; 2,373,863, Vitalis to *American Cyanamid Co.*

[24] French Pat. 787,819, to *I. G. Farbenindustrie.*

[25] U. S. Pats. 2,326,837, Coleman to *National Carbon Co.*; 2,338,688–9, Parker and Bonewitz to *The Rex Co.*

[26] U. S. Pats. 2,251,768, Swain to *American Cyanamid Co.*; 2,342,150, Kleinicke to *Johnson Marsh Corp.* Ger. Pat. 612,417, to *I. G. Farbenindustrie.* U. S. Pats. 2,374,187, Flett to *Allied Chemical and Dye Corp.*; 2,374,544, Hoyt to *Allied Chemical and Dye Corp.*

[27] U. S. Pat. 2,072,155, Chamberlin to *Warwick Chemical Co.*

of small amounts of the oil-soluble emulsifying type of compound. Substances such as the fatty acid amides and alkylol amides, fatty nitriles, and fatty morpholides have accordingly been added to the sulfate and sulfonate type detergents to enhance their detersive power.[28]

VI. Mixtures of Surface Active Agents with Each Other

A very widely used type of mixture consists of soap mixed with one of the lime-resistant sulfate or sulfonate detergents. A mixture of this type was used by the armed forces as an all-purpose detergent for hard water or sea water.[29] Alkyl aryl sulfonates of the Nacconol type are very effective in admixture with soap, although many other types have been used. Small amounts of alkane sulfonates made by the Reed reaction are said to confer high foaming power to tallow-rosin soaps.[30] Alkali lignin has also been mixed with soap to increase its effectiveness in hard water.[31]

Mixtures of fatty alkyl sulfates and soaps are encountered in the form of sulfonated spermaceti and other sulfonated waxes.[32] They have also been made by sulfating crude oxidation products of paraffin wax which contain fatty alcohols as well as fatty acids.[33] Numerous other surface active products are mixtures by virtue of the methods used in preparing them. Oleic and ricinoleic esters have been sulfated in the presence of other sulfatable or sulfonatable compounds such as terpene polymers,[34] aromatic hydrocarbons, etc. Fatty alcohols and fatty alkylol amides are sometimes sulfated in the presence of compounds such as dibutylnaphthalene to give mixed surface active agents.[35] Spermaceti has been sulfated in the presence of anhydrous glycerol to give a mixture of a fatty alcohol sulfate and a fatty monoglyceride sulfate.[36]

[28] U. S. Pats. 2,383,525–6; 2,383,739–40; Tucker to *Procter and Gamble Co.* U. S. Pats. 2,383,737–8; Richardson and McAllister to *Procter and Gamble Co.* Ger. Pats. 682,579, to *I. G. Farbenindustrie*; 702,242, to *Chemische Fabrik Grünau.*

[29] See U. S. Pats. 2,390,295, Flett to *Allied Chemical and Dye Corp.*; 2,407,130, Blades to *Kamen Soap Products Co.* French Pat. 718,325, to *N. V. Vereenigde Fabrieken van Stearine, Kaarsen en Chemische Producten.*

[30] U. S. Pat. 2,388,767, Safrin to *Wilson and Co.*

[31] U. S. Pat. 2,352,021, Schubert and Pierer, vested in the Alien Property Custodian.

[32] U. S. Pat. 1,957,324, Dambacher to *E. F. Houghton and Co.* Ger. Pats. 579,896; 681,225; 546,681; to *Deutsche Hydrierwerke A.-G.*

[33] Brit. Pat. 435,385, to *I. G. Farbenindustrie.*

[34] U. S. Pat. 2,003,471, Rummelsberg to *Hercules Powder Co.*

[35] Ger. Pats. 664,111; 644,131; to *I. G. Farbenindustrie.* See also for other mixtures Ger. Pats. 672,350; 614,227; Lindner and Zickermann to *Oranienburger Chemische Fabrik A.-G.*

[36] Swiss Pat. 171,359, to *Chemische Fabrik vorm. Sandoz.*

Surface active agents which have been prepared separately are often mixed to obtain special properties and numerous such compositions have been described in the patent literature.   Typical examples are mixtures of fatty alkyl sulfates with Igepons,[37] and Lamepons with Nekals.[38]

Although the cationic and anionic detergents are not usually compatible, small amounts of one type are sometimes incorporated into the other to effect thickening or special wetting properties.[39]

It should be emphasized that the mixtures we have been considering are representative of products which may be encountered as simple surface active agents.   They are not "baths" or mixtures made by the consumer for his immediate use.   The latter mixtures will be discussed at length in the section entitled "Practical Applications of Surface Active Agents" (Chapters 17 to 28).

[37] U. S. Pat. 2,247,741, Beller and Fasce to *Jasco, Ind.*
[38] Ger. Pat. 748,564, to *Chemische Fabrik Grünau.*
[39] French Pat. 839,653, to *I. G. Farbenindustrie.*   Brit. Pat. 328,675, to *I. G. Farbenindustrie.*   Also U. S. Pat. 1,867,022, Munz *et al.* to *I. G. Farbenindustrie.*

PART II

THE PHYSICAL CHEMISTRY
OF SURFACE ACTIVE AGENTS
IN THEORY AND PRACTICE

CHAPTER 11

Introduction

Surface active agents have been defined as solutes which possess the property of altering the surface or interfacial properties of their solutions to an unusual extent, even when they are present in relatively small amounts. In order to understand the behavior of surface active agents it is, therefore, necessary to understand in some detail the various phenomena which occur at phase boundaries. It should be pointed out, however, that there are many aspects of surface behavior to which a consideration of surface active agents as such is quite irrelevant. In other words, a study of the various effects of surface active agents constitutes only a specialized portion of the broad field of surface chemistry and physics. We intend to emphasize only those phases of surface chemistry which are necessary to an understanding of surface active agents.

The theory of surface active agents also involves many considerations which have nothing to do with interfacial phenomena. Differences between surface active agents and other solutes are due in large measure to certain peculiarities in the internal structure of their solutions, which might be characterized as forming a special type of colloidal solution. Surface active molecules and ions tend to form, in the interior of their solutions, aggregates known as micelles, with resultant anomalies in the physical and electrical properties of their solutions. The study of these anomalies and their relationship to micelle formation constitutes a large and important part of the physical chemistry of surface active agents. Micelle effects can be treated more or less separately, but also can be correlated closely with surface effects.

A third important part of the theory of surface active agents deals with the molecular interpretation of such secondary effects of surface active agents as wetting, detergent action, spreading, dispersion, hydrotropy, emulsification, frothing and foaming, lubrication, and other phenomena of technological importance. The primary physical effects are, of course, measurable as changes in surface tension, surface viscosity, electrical potential differences across the interface, etc. The correlation between these primary changes and practical effects such as detergency and emulsification is not always simple or easy to make. Such correlations form an important and actively investigated field.

Lastly there remains to be considered the relationships between surface activity and molecular structure. This field is still very incompletely explored. As in most other attempts to find simple relations between physicochemical properties and structural formulas, relatively few universally valid rules have been deduced. Available correlations have for the most part been empirically derived and are of limited applicability. Nevertheless they have been useful to a certain extent in guiding the efforts of the synthesist. It seems probable that empirical studies relating structure to surface activity will be more actively pursued in the future, and there is little doubt that they should contribute greatly to our understanding of the subject.

CHAPTER 12

Surface and Interfacial Relationships of Pure Liquids

I. General Considerations

Liquids are distinguished from solids by their mobility or fluidity. On a molecular scale this means that their molecules are free to migrate from point to point throughout the mass of the liquid. The molecules are accordingly changing their positions relative to each other continuously and rapidly. By contrast the individual molecules in crystalline solids can only vibrate over a very small range about their fixed centers in the crystal lattice.

All similar molecules exert an attractive force for each other. This intermolecular attraction is counteracted by the thermal agitation of the molecules. The over-all statistical effect of thermal agitation is a tendency for the molecules to separate from one another and form a gas. In a gas under ordinary conditions the intermolecular attractive forces are so weak as to be of minor importance in determining the properties of gases. Such attractive forces, known as van der Waals' forces, do, however, give rise to deviations from the ideal gas laws, and to the Joule-Thompson effect. As the intermolecular distance becomes very small, the van der Waals' forces become much more powerful. In fact, the van der Waals' forces are generally considered to vary inversely as the sixth power of the distance between the molecules. Electrostatic or coulombic forces, by contrast, vary as the inverse square of the distance, and as a consequence, are effective over much longer distances than van der Waals' forces.

At conditions of temperature and pressure appropriate for each molecular species the van der Waals' forces become strong enough to cause liquefaction, $i.e.$, to maintain the molecules in a coherent but still fluid mass. Strictly speaking the true van der Waals' forces ($i.e.$, those intermolecular forces of attraction involved in deviations from the ideal gas law) are not solely responsible for the condensed form of the liquid state. Several other factors may contribute, notably the forces of electrostatic attraction in those liquids containing ions or dipoles. We shall use the term "van der Waals' forces" loosely, however, to refer to the combination of force effects which maintains a substance in the liquid state by counteracting the vaporizing tendency of thermal molecular motion.

243

An individual molecule in the interior of a liquid at any given instant is surrounded more or less uniformly by a force field of attraction between it and its neighbors.   It is apparent that this force field will average out over a relatively short period of time to be spherically symmetrical, considering the molecule in question to be at the center of the sphere.   This means that the resultant van der Waals' force on the molecule is zero.   A molecule at the surface of a liquid, on the other hand, is subjected to a highly un-balanced force field.   On the liquid side of the molecule the attractive force of the other similar molecules is as strong as in the interior.   On the vapor side there is only the attractive force of widely scattered gas molecules. Since molecular density in the liquid phase is of the order of 1000 times greater than in the gas phase, and since van der Waals' forces are only significant at short ranges, the attraction for a surface liquid molecule exerted by the air or gas above it is inconsequential.   There exists there-fore a resultant force on every surface molecule tending to pull it perpen-dicularly inward.   From a purely mechanical point of view, this force acts in such a way that the surface strives to assume a shape having the smallest possible area.   In a mass of liquid which has temporarily been distorted so that the surface does not have the minimum equilibrium area, the molecules move away from the surface faster than other molecules come in to take their place.   This process goes on, diminishing the surface area until the surface has assumed an area and shape in equilibrium with the environment.   Mechanical factors in this environment will of course include the material and shape of the vessel in which the liquid is contained as well as gravitational effects.   If the liquid is a free-falling drop or if it is suspended in another liquid of the same density it will assume a spherical form.   This is the form which has least surface per unit volume.

The unbalanced force fields at the surface of a liquid can be equated to a definite quantity of free energy possessed by each unit area of that surface.[1] The existence of this free energy at the surface is a fundamental property of liquids.   The free surface energy is manifested by the tendency toward contraction.   Conversely a definite amount of work, equivalent to the free surface energy, must be done in order to expand the surface isothermally. The surface free energy is expressed in ergs per square centimeter.   It is a real thermodynamic property and is as easy a concept to grasp as the

[1] We shall use the term "surface" to denote the interface between a liquid and a gas.   The gas may be air, or the vapor of the liquid in question, or any foreign gas phase.   In a rigorous discussion, it is usual to consider liquid surfaces in equi-librium with their own vapor and, unless otherwise specified, this will be the implied condition in our discussion.   Strictly speaking, the term "surface" means any phase boundary, but common usage restricts it to the liquid–gas or solid–gas interface.

concept of potential energy in any other system. Just as a definite amount of work is required to move a weight up an inclined plane, thus endowing the weight with potential energy of position, so work is required to expand a liquid surface with resultant increase in the *total* surface energy. (The surface energy per unit area, of course, remains the same.)

In view of these relationships it would be logical to conclude that the free surface energy of a liquid could be calculated provided the inter-molecular forces and space relationships were known. Such calculations have been made, and good agreement with experimental values has been achieved in the case of unassociated non-polar liquids having approximately spherical molecules.[2]

The surface free energy expressed in ergs per square centimeter is mathematically and dimensionally equivalent to a *tension* expressed in units of force per linear centimeter. This quantity is called the surface tension. A certain amount of confusion has arisen with respect to the physical interpretation of this quantity. It is readily understood if the following facts are kept firmly in mind. First, the surface tension is not dimensionally a force. It is a force per unit of length, or F/d. This makes it dimensionally equivalent to energy per unit of area (the free surface energy), or Fd/d^2. The quantity in the numerator, Fd, is usually called "work" rather than energy, but it is dimensionally equivalent to energy. Surface tension is accordingly a very commonly used and convenient concept. It is expressed as dynes per centimeter, and the force (a vector quantity) is exerted tangential to the surface. Surfaces do indeed behave as if a primary force were being exerted tangential to them, and this behavior inevitably suggests the existence of a "skin" or special contractile surface layer at the boundary of the liquid. In pure liquids there is in fact no such skin, and even in systems where a pellicle or molecular layer dissimilar from the interior does exist this pellicle is not "tending to contract" after the manner of a tight sausage casing.

One difficulty involved in the concept of surface tension is its failure to fit in readily with the molecular picture. On a macroscopic scale it is a real physical effect, and mathematically it is perfectly sound. However, when we visualize a liquid as a gigantic swarm of molecules each with its own sphere of attractive force for its neighbors, it is difficult to see how a force tangential to the surface can arise. In this case it may be more satisfying to adopt the thermodynamic point of view, and regard the tendency for a surface to contract as a manifestation of the universal tendency for any system to minimize its free energy.

In discussing liquid surfaces from the molecular point of view there are

[2] Belton and Evans, *Trans. Faraday Soc.* **37**, 1 (1941). Rocard, *Compt. rend.* **197**, 122 (1933). Lennard-Jones and Corner, *Trans. Faraday Soc.* **36**, 1156 (1940).

several other features which deserve consideration. A question as to the sharpness of transition from liquid to vapor often arises. Calculations based on kinetic theory show that the molecules at the surface and in the gas phase as well as in the interior are in violent motion and that the molecular traffic across the interface is heavy. These calculations have been verified by experiments on the rate of evaporation of liquids into a vacuum. It might be logical to suppose therefore that the surface is continuously agitated to such a degree that it is hazy and ill defined. All the evidence, however, points in the other direction. Without going into the details of experimental work it is believed by almost all investigators today that the transition from liquid to vapor is very sharp, usually not more than one molecule thick. This becomes understandable and even predictable if we consider the short effective range of the van der Waals' forces. These forces vary inversely as a higher power of the distance between the molecules. Therefore a molecule only need be removed from its neighbor by a relatively short distance (about two molecular diameters) to be free enough for existence in the vapor phase. The converse is, of course, also true and the net result is a very sharp

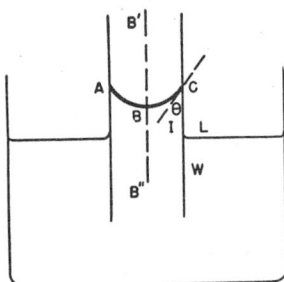

Fig. 3.   Liquid forming meniscus in a tube.

boundary between liquid and vapor even though this boundary is being traversed in both directions by enormous numbers of fast-moving molecules.

One of the consequences of the existence of free surface energy is the existence of a pressure differential across a curved liquid surface. If we consider a free droplet of liquid of non-spherical form, we see that the first effect of the free surface energy is a minimizing of the surface-to-volume ratio, causing the droplet to assume a spherical form. When this state is reached the surface still tends to decrease its area but is opposed by the resistance of the liquid to compression. In other words the interior of the droplet is under a definite pressure due to the tendency of the surface to contract. Pressure is defined as force per unit area and the force vector of this pressure points *toward the concave side of the surface* (in this case toward the interior of the liquid). Another way of visualizing this pressure is to consider a liquid forming a meniscus in a tube (Fig. 3). Neglecting for the moment the reasons that the meniscus is formed, we see that the surface ABC is a surface of revolution about the axis B'BB". The free surface energy causes a tendency for this surface to minimize. The minimum surface area would exist if ABC were planar instead of curved, *i.e.*, if point B were at the same vertical level as points A and C. The tendency for the surface to minimize is tantamount to a pressure on the convex or "B" side of the meniscus. When a liquid rises in a capillary tube it is this pressure which causes the rise. As we shall see, the energy relationships at the interface between the liquid and the walls of the capillary require the liquid surface to form with the solid wall a definite angle θ (WCI) known as the contact angle. We may conceive of capillary rise as a series of steps. The pressure at B tends to raise B to the AC level. This in turn changes the contact angle WCI, making it less acute. The wall–liquid energy relationship is thus upset and the level AC rises again until angle WCI again has the equilibrium value θ. The liquid rises until the weight of the column of liquid BL is sufficient to a balance the excess pressure at B. Adam[3] has presented an excellent simplified mathematical derivation of the pressure differential across a curved surface. This relationship, known as Laplace's equation, is:

$$p_1 - p_2 = \gamma \left(\frac{1}{R_1} + \frac{1}{R_2} \right)$$

where $p_1 - p_2$ is the pressure difference between the concave and convex sides, γ is the surface tension in dynes per centimeter, and R_1 and R_2 are the principal radii of curvature of the curved surface element under consideration. This equation is of fundamental importance.

II. Relationship of Surface Tension to Other Physical Properties

(a) Temperature Effects

The surface tension of almost all pure liquids falls with rising temperature. The only known exceptions are a few molten metals. At the critical temperature the surface tension becomes zero, and in fact the experimental criterion for critical temperature is the disappearance of the liquid–vapor meniscus. A large amount of theoretical and experimental work has been done in tracing the relationship between surface tension and temperature. That surface tension should fall with rising temperature can be seen almost intuitively. The free surface energy tends to pull the molecules inward from the surface to the interior. This is counteracted by the opposing tendency of thermal agitation to push the molecules outward through the surface and into the vapor phase. The latter tendency increases markedly with increasing temperature, whereas the former tendency (due to van der Waals' forces) would not be expected to change greatly with tempera-

[3] Adam, *Physics and Chemistry of Surfaces*, Oxford Univ. Press, London, 1942.

ture unless something equivalent to a phase change occurred. (Such a change might take place, for example, if the liquid were highly associated at one temperature and not at another.) For many liquids the relationship between surface tension and temperature is almost linear over long temperature ranges, and is described by the empirical equation of Eotvos, Ramsay, and Shields:

$$\gamma V^{2/3} = k(T_c - T - 6)$$

where γ is the surface tension, V is the molecular volume, and T_c is the critical temperature. The value of k is fairly constant within any simple homologous series. In passing, however, from one series to another, the value of k may change by several fold. Accumulated evidence has shown that this equation is probably not of fundamental theoretical significance. Kinetic theory of the liquid state appears insufficiently advanced at present to trace any accurate and fully quantitative theoretical relationships between surface tension and temperature.[4]

Since surface tension decreases with rising temperature it can be shown by the methods of thermodynamics that heat is absorbed when a surface is extended. This heat represents the work necessary to pull new molecules into the surface against the attractive forces of the interior. The amount of heat which must be added to maintain isothermal conditions during an expansion of the surface is equal to:

$$-T(d\gamma/dT)$$

and is called the latent heat of the surface. The total surface energy is accordingly equal to:

$$-T(d\gamma/dT) + \gamma$$

per square centimeter. For those liquids where the surface tension – temperature curve is linear (i.e., $d\gamma/dT$ is constant) the rate of change of total surface energy with respect to temperature is constant. Ferguson and Irons[5] have pointed out in fact that the molar total surface energy is a constant-independent of temperature (for unassociated liquids). This is also true of the molar surface entropy.

[4] Sibaiya, *Current Sci. (India)* **3**, 418, 550 (1935). Wheeler, *ibid.* **3**, 550 (1935). Lovgren, *Svensk Kem. Tid.* **53**, 359 (1941); *Chem. Abstracts* **36**, 3077. Freund, *Oel u. Kohle* **37**, 500 (1941); *Petroleum Z.* **35**, 295 (1939). Frenkel, *J. Exptl. Theoret. Phys. U.S.S.R.* **10**, 1109 (1940); *Chem. Abstracts* **35**, 4648. Teige, *Kolloid-Z.* **102**, 132 (1943).
[5] Ferguson and Irons, *Proc. Phys. Soc. London* **53**, 182 (1941).

(b) Surface Tension vs. Density

McLeod[6] derived an empirical equation connecting surface tension with density as follows:

$$\frac{\gamma}{(D - d)^4} = C$$

where C is a constant, different for each pure liquid, D is the density of the liquid and d that of the vapor. This relationship holds over wide temperature ranges. Sugden took the fourth root of the C in McLeod's equation and multiplied the result by the molecular weight to obtain a quantity he called the parachor:

$$\frac{M}{D - d} \gamma^{1/4}$$

The parachor is of course also constant over wide temperature ranges. The factor $M/(D - d)$ is called the molecular volume. The molecular volume at constant temperature is known to depend on molecular constitution and the parachor affords a means of comparing molecular volumes without taking into account their temperature dependence. By comparison of parachors a large body of data has been built up showing the contributions of various organic radicals and structures to molecular volume. The parachor has thus proved to be an extremely useful tool in deducing organic structures, in much the same way as molecular refraction, spectroscopic data, etc.[7]

A number of other relationships between free surface energy and various physical properties have been worked out, some on an empirical and some on a theoretical basis. Tripathi[8] and Pollara[9] have independently proposed equations relating surface tension to vapor pressure, following earlier work by Mokrushin[10] relating surface tension to boiling point. The surface tension – viscosity relationship has been studied by Buehler[11] and the relationship of surface tension to latent heat of vaporization by Albert and Eirich,[12] and Starobinets and Romyah.[13]

[6] McLeod, *Trans. Faraday Soc.* **19**, 38 (1923).

[7] Sugden, *The Parachor and Valency*, Routledge, London (Knopf, New York), 1930.

[8] Tripathi, *J. Indian Chem. Soc.* **20**, 197 (1943).

[9] Pollara, *J. Phys. Chem.* **46**, 1163 (1942).

[10] Mokrushin, *J. Gen. Chem. U. S. S. R.* **2**, 911 (1932).

[11] Buehler, *J. Phys. Chem.* **42**, 1207 (1938).

[12] Albert and Erich, *Z. physik. Chem.* **A183**, 9 (1938).

[13] Starobinets and Romysh, *J. Gen. Chem. U. S. S. R.* **7**, 151 (1937).

It is interesting to note that the surface tension of liquid mixtures often varies linearly with the composition even when such properties as vapor pressure and fluidity deviate widely from ideal behavior. Conrad and Hall[14] have shown this to be the case, for example, with chloroform–methanol mixtures. In studies of surface active agents these relationships involving the properties of pure or homogeneous liquids are rarely applicable, and the reader is referred to the original articles for further discussion.

III. Liquid–Liquid Interfaces

So far we have been considering the liquid–gas interface or surface. We have seen that it possesses a definite amount of free energy per unit area by virtue of the unbalanced force field acting on the surface molecules. When two mutually insoluble liquids phases are brought together the boundary between them has many characteristics in common with a liquid–gas boundary. A definite quantity of free energy is associated with each unit area of the interface and accordingly the interface tends to contract. This free energy is mathematically equivalent to a tangential force on every unit line which may be drawn on the surface, *i.e.*, to an *interfacial tension*. Due to thermal agitation there is a tremendous travel of molecules across the boundary from one phase to another but the boundary itself may be regarded as very sharp (*i.e.*, a "Maxwell's demon" would travel a distance of only two or three molecular diameters across the boundary before finding himself completely within a new phase environment). The source of interfacial free energy is the unequal attractive force exerted on the interfacial layer of molecules by molecules within the separate phases. In general the attractive force between one liquid phase and the other will be greater than that between either liquid and a gas phase. This presumption postulates a positive force of attraction between the two molecular species, and is to be expected simply because of the greater number of molecules per unit volume in a liquid phase. Accordingly the interfacial tension between two liquids is always lower than the surface tension of the liquid with higher tension. Strong forces of attraction between the two different molecular species across the interface result in low free interfacial energy (*i.e.*, low interfacial tension) and vice versa.

It is also evident that two liquid phases A and B can co-exist in contact with each other only if the attraction of the A molecules for each other and the B molecules for each other is greater than the attraction of A for B. If the A-to-B attraction were equal to or greater than either the A-to-A or B-to-B attraction, the interface would cease to exist and the liquids

14 Conrad and Hall, *J. Am. Chem. Soc.* **57**, 863 (1935).

would be mutually miscible. This would correspond to a state of zero or negative interfacial tension, and would be analogous to a hypothetical liquid–gas surface at or above the critical temperature. Indeed the criterion for miscibility of two liquid phases is that the interfacial tension be zero or negative.

A general rule proposed by Antonoff[15] states that when two liquids are mutually saturated with each other the interfacial tension between the phases is equal to the difference between the separately measured surface tensions of each phase, i.e.:

$$\gamma_{AB} = \gamma_A - \gamma_B$$

(phases A and B mutually saturated). This rule holds true in most cases.

The fact that two molecular species A and B have a positive attraction for each other across the phase boundary means that work must be done (i.e., energy added to the system) in order to separate them. This fact may be stated in another way by saying that the two liquids *adhere* to each other. The lower the interfacial energy, the stronger is the adhesion. Consider two liquids A and B in contact over an area of one square centimeter. The interfacial energy may be designated as γ_{AB}. If the liquids are now separated, the free surface energies of the two surfaces are γ_A and γ_B. The difference:

(1) $$\gamma_A + \gamma_B - \gamma_{AB} = W_{AB}$$

is equal to the amount of energy which must be put into the system to separate the two phases. This is usually called the *work of adhesion*, W_{AB}.[16] Equation 1 is called Dupré's equation after the investigator who propounded it almost eighty years ago.

We can apply Dupré's equation to measure the adhesion of a liquid to itself. If we consider A and B to be the same liquid the term γ_{AB} vanishes and the equation reduces to:

$$W = 2\gamma$$

---

[15] Antonoff, *J. chim. phys.* **5**, 372 (1907).

[16] If "work" is defined as force multiplied by distance, the above terminology may become confusing. The question will immediately arise as to how far the phases must be separated before we can consider that the "work" of adhesion has been fully expended. The writers accordingly prefer the term "adhesional energy" for W_{AB}. In much of the literature on surface physics, however, the term "work" is used synonymously with "energy" and, by definition, the two are equivalent dimensionally. This terminology will therefore be adopted in order to conform with that used by leading investigators in the field.

This may be interpreted as meaning that the adhesional energy across an imaginary plane of one-square-centimeter area within a liquid is numerically equal to twice the surface energy. This quantity, 2γ, has been termed the cohesional work by Harkins.[17]

IV. Solid–Liquid Interfaces

The surfaces of solids, just as of liquids, are possessed of free energy due to unbalanced forces of attraction acting on the surface layer of molecules. Under usual conditions most solids never approach true isotropy as closely as liquids, and the surface energy will vary from point to point. From the experimental point of view, liquid surfaces are more easily studied since they are much more readily reproducible. This is because they are in dynamic equilibrium with the interior and are continuously being renewed.

In a solid surface, however, the molecules remain in fixed positions and the properties of the true clean surface are soon obscured by dirt, by adsorbed material, and by external attrition. The free surface energy of solids may in some cases be estimated by calculations from crystallographic and other data. It may also be estimated by comparing the heats of solution of coarsely and finely divided samples of the solid, as well as by other indirect methods.[18] There are no good direct methods for measuring the free surface energies or "surface tensions" of solids. In all probability they are of the same general order of magnitude as those of liquids but somewhat higher. This would be expected because the densities are not greatly different, which indicates that the molecules are about the same distance apart. Hence the van der Waals' forces should be of the same order of magnitude. The higher enthalpy of the liquid phase would also lead us to expect it to have a lower free surface energy.

When a liquid comes into contact with an insoluble solid surface the molecules of both phases usually attract each other across the interface and the interfacial tension is accordingly lower than the sum of the separate surface tensions of the two phases. The molecular picture is somewhat different from that of a liquid–liquid interface. There is negligible molecular travel across the solid–liquid interface, and conditions on the solid side are comparatively static. The interface itself is possessed of free energy

---

[17] Harkins and Cheng, *J. Am. Chem. Soc.* **43**, 35 (1921).

[18] The total surface energy of a solid, for example, sodium chloride, appears as heat when the solid is dissolved. Hence, the smaller particles will have a higher heat of solution. These measurements require an accurate knowledge of the surface areas of the particles and are subject to large errors. For a complete discussion of this point, see Boyd, *Pub. Am. Assoc. Advancement Sci.* **21**, 128 (1943).

and therefore tends to minimize its area and it may be considered mathematically in the same manner as the liquid–liquid interface.

Dupré's equation is fully applicable, and becomes:

$$W_{SL} = \gamma_{SA} + \gamma_{LA} - \gamma_{SL}$$

where W_{SL} is the work of adhesion between solid and liquid and γ_{LA} is the liquid surface tensions. γ_{SA} and γ_{SL}, the surface tension of the solid and the solid–liquid interfacial tension respectively, are not directly measurable by simple means. The work of adhesion W_{SL} can be measured, however, by other means, which will be considered further along in the discussion.

V. Surface Relationships Involving Three Phases

So far we have been considering boundaries between two phases. These boundaries are, in the parlance of geometry, surfaces. A great many phenomena of surface physics involve the contact of three or more phases. Geometrically, three or more phases can be in contact only at a line, not over a surface. The most commonly encountered three phase systems are liquid–solid–gas, liquid–liquid–gas, and liquid–liquid–solid. Systems involving two solid phases are of great practical importance but are difficult to study on an experimental basis. Systems involving three mutually insoluble liquids are sufficiently rare to be of little importance.

A typical example of the solid–liquid–gas system is a droplet of a liquid resting on a solid plate surrounded by air. In order to simplify the discussion we may for the moment neglect the gravitational force tending to flatten out the droplet of liquid. In fact, if the droplet is small enough, the gravitational effect becomes truly negligible. This is because the surface-to-mass ratio increases as the mass decreases and surface forces become far more powerful than gravitational forces. In Figure 4, the three phases contact each other along a line of which points X are intercepts. There are three interfaces involved in the system. The liquid–air interface LA acts as it would in an isolated system and tends to minimize its area. This tendency constitutes a positive force pulling against the line of contact X in a direction tangential to LA at X and represented by the vector X_a. In the plane of the solid surface there are two interfaces, namely, the liquid–solid and the gas–solid. The liquid–solid interface tends to minimize itself and this constitutes a force parallel to the solid surface and pointed toward the interior of the droplet (vector X_l).

Although the surface force (vector X_s) exerted by the solid–air interface is too weak to effect contraction of the solid surface, nevertheless a force of such character does exist.   This force strives to minimize the area of the solid–air interface whenever that is possible.   In the system of Figure 4 this may be done by pulling the liquid over the solid surface, thus substituting solid–liquid interface for solid–air interface.   Mathematically this tendency of the solid–air interface to minimize itself may be represented as a pull at line X in the direction X_s.   The length of this vector is equivalent to the solid–air surface energy, γ_{SA}.   When this system has come to equilibrium the resultant force *in the solid plane* is zero.   Furthermore, X_a, tangential to the L–A surface at X, forms with the S–L interface an angle, θ, which is known as the contact angle.   It is a definite and real physical quantity, characteristic of the three-phase system being studied.   Contact angle is conventionally measured within the liquid, *i.e.*, in a solid–liquid-gas system it is the angle between the liquid–gas surface and the liquid-solid interface.

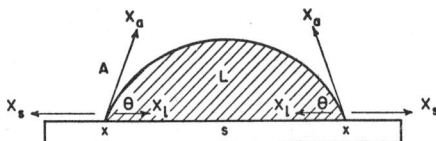

Fig. 4.   Interfacial forces acting
on liquid droplet on a solid plate.

Applying Dupré's equation to the system of Figure 4:

$$W_{SL} = \gamma_{SA} + \gamma_{LA} - \gamma_{SL}$$

and:

$$\gamma_{SA} = \gamma_{SL} + \gamma_{LA} \cos \theta$$

Combining these equations:

(1a) $$W_{SL} = \gamma_{LA}(1 + \cos \theta)$$

If the contact angle is zero, $W_{SL} = 2\gamma_{LA}$.   In this case the liquid attracts the solid as much as it attracts itself.   A negative contact angle is physically not realizable but it is perfectly possible for the liquid to attract the solid more strongly than it attracts itself, *i.e.*, $W_{SL} > 2\gamma_{LA}$.   Accordingly the non-existence of a positive contact angle does not specify a definite surface energy state of the system, but merely indicates that $W_{SL} \geqq 2\gamma_{LA}$.   Gans has pointed this out very clearly and has suggested that the term "zero contact angle" be reserved for the special case where $W_{SL} = 2\gamma_{LA}$.

He suggests[19] the term "no contact angle" for the more general case where $W_{SL} > 2\gamma_{LA}$.

These relationships may also be considered from a somewhat different point of view. Referring again to Figure 4, we see that, if θ is zero, $\cos \theta = 1$ and $X_s = X_a + X_l$ if the system is in equilibrium. This is the special case which Gans calls "zero contact angle." If, however, $X_s > X_a + X_l$ equilibrium cannot be attained and the liquid will continue to spread over the solid surface. This is the case which Gans calls "no contact angle." We shall specify these two states—"zero contact angle" and "no contact angle"—hereinafter by saying that the contact angle is *zero* or *nil*, respectively. When the contact angle is *nil*, correct numerical values of W_{SL}

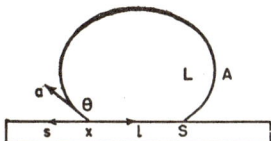

Fig. 5. Liquid droplet forming contact angle greater than 90° with a solid plate.

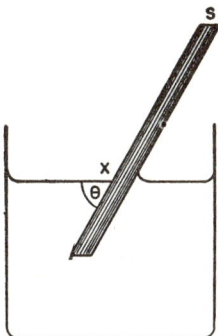

Fig. 6. Tilted-plate method of measuring contact angle.

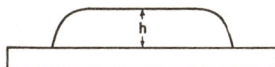

Fig. 7. Surface tension measurement from height of droplet on a solid plate.

cannot be indicated or calculated with the aid of equation 1a. The reason for this is the inability of the term, $\cos \theta$, to assume a numerical value greater than unity as would be required in order to indicate a greater attraction of the solid for the liquid than that of the liquid for itself. Such restriction in the valid scope of equations involving the term, $\cos \theta$, should be kept in mind in connection with subsequent discussions.

Contact angles of greater than 90° (Fig. 5) indicate that $X_s < X_l$ or $W_{SL} < \gamma_{LA}$. A contact angle of 180° would indicate no adhesion whatsoever between solid and liquid, *i.e.*, $W_{SL} = 0$ and $X_l \geqq X_s + X_a$. No cases of this sort are known to exist.

Liquid–liquid–gas systems may be represented by a lens of an immiscible liquid floating on the surface of another liquid, so that both liquid phases

[19] Gans, *J. Phys. Chem.* **49**, 165 (1945).

contact the air. Liquid–liquid–solid systems may be represented by a solid plate dipping vertically into a vessel containing two immiscible liquids in layers one above the other. In both these cases the force–vector pictorialization is similar to the one considered above. The contact angle exists and has the same significance as in the solid–liquid–gas system. In systems containing two liquid phases it should be specified in which phase the contact angle is being measured.

Two very important surface effects which are often discussed in terms of contact angles are wetting and spreading. The term "wetting" is applied to solid surfaces and means what it says, that the solid has been brought into intimate contact with the liquid. Most investigators agree that the degree to which a solid body is wet by a liquid may best be measured by the contact angle at the liquid–solid–air line. If the contact angle is zero or nil the solid is completely wet. If the contact angle is positive the solid is incompletely wet. Contact angles of greater than 90° indicate that the solid is difficult to wet by the particular liquid being examined. The term "unwettable" would suggest a contact angle of 180°. Actually it is loosely used in any case where the contact angle is greater than 90° and the liquid rolls freely in droplets on the solid surface.[20]

The direct measurement of contact angles is rather difficult. One of the most satisfactory methods is the plate method illustrated in Figure 6. The solid plate S is tilted until visual observation shows that the liquid is level at X right up to the point of contact with S. The contact angle is then θ. A number of other methods have been described which involve direct observation of the angle of a bubble or drop of liquid in contact with a plate or crystal face of the solid. Microscopes and mirror-lens systems are used in actual measurement of the angles. Very small angles can be detected for example by directing a beam of light at the solid and liquid surfaces near the line of contact and photographing the reflection. Contact angles can also be determined indirectly. The solid being investigated may be made into a capillary tube and immersed in the liquid. If the surface tension of the liquid is known, the contact angle can then be calculated from the equation:

$$h = \frac{2\gamma \cos \theta}{gDr}$$

where h = rise in the capillary tube, r = radius of tube, D = density of liquid, and γ is the surface tension of the liquid. Another indirect procedure involves the formation of a drop of the liquid on a plate of the solid and

[20] Compare Michaud, *J. chim. Phys.* **36**, 23 (1939).

measurement of the height, h (Fig. 7).   The drop must be sufficiently large so that the top is fairly flat.   The following formula is then applicable.[21]

$$1 - \cos \theta = \frac{gDh^2}{2\gamma}$$

One of the greatest difficulties in working with contact angle measurements is the fact the results often show poor reproducibility.   This can scarcely be regarded as surprising when we consider how difficult it is to prepare and maintain a clean reproducible solid surface.   Monomolecular layers of impurities on a solid surface can alter the contact angle considerably.   In the case of metals, oxide films are often the cause of irregular results in contact angle measurements.   Most experimental chemists have probably noted at some time that dirty or oxidized mercury will "hang" in a manometer tube, *i.e.*, its contact angle against glass has become considerably less than 90°.   Adsorbed moisture on solid surfaces is tremendously effective in altering the contact angle between these surfaces and the water–air system or the water–mineral oil system.[22]

A most striking feature concerned with experimental work on contact angles is the difference in contact angle of the same system depending on whether the liquid is advancing or receding over the surface.   The advancing angle is greater than the receding angle in practically all cases.   The difference is often noticeable even after the liquid has stopped moving relative to the solid surface.   In view of the fact that the contact angle exists as a result of free energy relationships this hysteresis should represent merely a deviation from equilibrium.   In many instances, however, there appears to be a more or less permanent difference in contact angles depending on whether the solid and liquid in question have or have not been in contact prior to the measurement.   In other words the work of adhesion is less for a dry solid surface than for one which has previously been wetted, and this points to the possibility that during the first contact of solid and liquid some of the liquid is adsorbed, thus changing the character of the

[21] For a discussion of the methods of measuring contact angles, with references to the original literature, the reader is referred to Adam, *Physics and Chemistry of Surfaces*, Oxford Univ. Press, London, 1941.   See also Bartell and Bristol, *J. Phys. Chem.* **44,** 86 (1940).   Bartell and Zuidema, *J. Am. Chem. Soc.* **58,** 1449 (1936).   Lange and Nagel, *Kolloid-Z.* **73,** 268 (1935).   Bate, *Proc. Phys. Soc. London* **53,** 403 (1941).   Irons, *Phil. Mag.* **34,** 614 (1943).   Wakeham and Skau, *J. Am. Chem. Soc.* **67,** 268 (1945).   Beament, *Trans. Faraday Soc.* **41,** 45 (1945).

[22] See, for example, Chapek and Krechun, *Colloid J. U. S. S. R.* **5,** 763 (1939); *Chem. Abstracts* **34,** 2674.

solid surface.   In general, the cleaner the solid surface the smaller is the
hysteresis between advancing and receding contact angles.   This suggests
that the larger advancing contact angle may be due to a film of impurity
on the solid surface.   This film may be adsorbed air, even in those cases
where most scrupulous care has been taken by the investigators.

Contact angles are of interest principally in calculating the energy of
adhesion between liquids and solids and in predicting the physical effects
which are related to this energy.   One important problem of this nature
concerns the differential wetting of a solid by two liquids.   The equilibrium
equations for this system can be illustrated by referring to Figure 8.   Two
liquids, A and B, are in contact with the solid S.   At equilibrium the result-

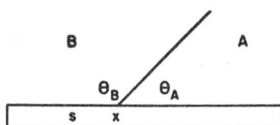

Fig. 8.   Contact an-
gle involving two
liquids and a solid.

ant force at X is zero, or:

$$\gamma_{BS} = \gamma_{AS} + \gamma_{AB} \cos \theta_A$$

From Dupré's equation:

$$W_{AS} = \gamma_A + \gamma_S - \gamma_{AS}$$

$$W_{BS} = \gamma_B + \gamma_S - \gamma_{BS}$$

$$W_{AS} - W_{BS} = \gamma_A - \gamma_B + \gamma_{AB} \cos \theta_A$$

This relationship is useful in measuring the energy of adhesion of a liquid
A for a solid even when the contact angle at the liquid–solid–air interface
is zero.   A second liquid B, immiscible with liquid A and having a finite
angle with the solid in air (and in liquid A) is used. W_{BS} can then be
measured in a separate experiment against air, and W_{AS} calculated after
θ_A has been determined.[23]   Measurements of the angles in liquid–liquid–
solid systems are more difficult than in liquid–solid–air systems.   The
hysteresis effects are in general more noticeable.   Since the equations above
apply only to equilibrium angles they often cannot be applied directly in
technological investigations.   Liquid–liquid–solid systems of technical importance include oil–water–mud mixtures in oil wells, emulsions involving
clays, bentonite, etc.

[23] Bartell and Osterhof, *Colloid Symposium Monograph* 5, 113 (1927).

The subject of contact angles has been surrounded by a certain amount of confusion, and is sometimes looked upon by many highly competent technologists in the field of surface activity as impractical of application. Although there are several possible reasons for this state of affairs, direction of attention to a few salient points may help to extend the practical application of this useful concept. In the first place some investigators convert the results of surface tension measurements into terms of contact angles and use "contact angle" almost interchangeably with "work of adhesion" in their discussions. This substitution of the *measure* of an energy effect for the effect itself is often misleading, particularly to the beginner. Similarly, the fact that contact angle effects occur only at the line boundary of three phases is often not sufficiently emphasized. The only systems in which the contact angle is of practical importance are liquid–liquid–solid, liquid–solid–gas, and liquid–liquid–gas. Finally, the importance and almost universal occurrence of the hysteresis effect is often neglected in discussions of contact angle. It seems necessary to emphasize that hysteresis more than any other single factor makes a large amount of contact angle data inapplicable in practical problems. More precisely, it is the lack of a sound theory which can account quantitatively for hysteresis. Nevertheless, an understanding of the significance of contact angles is necessary for technologists in order to follow the scientific literature, and in many instances is of great value in actual practical problems.[24]

VI. Wetting and Spreading

If a drop of liquid is placed on the surface of another immiscible liquid, or on the surface of a solid which it cannot dissolve, it may spread out to a thin film or it may remain in the form of a lens. The free-energy relationships, *i.e.*, the free surface energies of the two phases and the interfacial tension between them, determine what happens. Spreading occurs when the contact angle is zero or nil, otherwise the droplet will remain as such.

Considering a liquid A spreading on a liquid B, and referring to Figure 9 the necessary condition for spreading is that:

$$\gamma_B \gtrless \gamma_A \cos \theta + \gamma_{AB}$$

and since $\theta = 0$ when one liquid spreads on another:

$$\gamma_B \gtrless \gamma_A + \gamma_{AB}$$

[24] For discussion of the significance of contact angle, and typical methods and data, see Pease, *J. Phys. Chem.* **49**, 107 (1945); Thiessen and Schoon, *Z. Elektrochem.* **46**, 170 (1940); Bartell and Bartell, *J. Am. Chem. Soc.* **56**, 2205 (1934); Scholberg and Wetzel, *J. Chem. Phys.* **13**, 448 (1945).

The decrease in free energy of the system accompanying spreading is:

$$S = \gamma_B - \gamma_A - \gamma_{AB}$$

Dupré's equation states that:

$$W_{AB} = \gamma_B + \gamma_A - \gamma_{AB}$$

Combining:

$$S = W_{AB} - 2\gamma_A$$

and since $2\gamma_A$ is called the work of cohesion, we can say that the decrease in free energy of the total system when A spreads over B is equal to the work of adhesion of A to B minus the work of cohesion of A. This quantity, S, is called the spreading coeff cient, a term first introduced by Harkins.

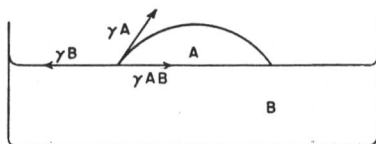

Fig. 9. Forces involved in spreading
of one liquid on another.

If the spreading coefficient is positive the liquid A will spread over B. If it is negative, liquid A will remain as a lens.

The energy relationships for the spreading of a liquid on a solid are exactly the same as for liquid on liquid.

$$W_{SL} = \gamma_S + \gamma_L - \gamma_{SL}$$

The decrease in free energy accompanying spreading of a liquid over a solid is:

$$S = \gamma_S - \gamma_L - \gamma_{SL}$$

and as before:

(2) $$S = W_{SL} - 2\gamma_L$$

The work of adhesion of liquid to solid (W_{SL}) has been shown to be (equation 1a, page 254):

$$W_{SL} = \gamma_L(1 + \cos \theta) \qquad \text{when } \theta \text{ is } positive \ or \ zero$$

and:

$$W_{SL} > 2\gamma_L \qquad \text{when } \theta \text{ is } nil$$

Substituting in equation 2:

$$S = \gamma_L (\cos \theta - 1) \qquad \text{when } \theta \text{ is } positive \text{ or } zero$$

$$S > 0 \qquad \text{when } \theta \text{ is } nil$$

From these relationships, it is apparent that the contact angle must be *zero* or *nil* if S is not to be negative and if spreading is to occur. Conversely, the condition that the contact angle is *zero* or *nil* is sufficient to ensure spreading, since the value of γ_L is always positive. It is noteworthy that θ in these relationships refers to the advancing contact angle.

The rate of spreading of a liquid over a plane solid surface is usually slow. Spreading can take place through the vapor phase if the liquid is volatile. More generally, however, it takes place by a mechanism of two-dimensional diffusion. The spreading of a liquid between two solid surfaces or over a channeled surface is usually more rapid because of a capillary effect. That is, the contact angle requirements cause a curvature of the surface which in turn causes a pressure gradient which drives the liquid mass toward the concave side of the surface.

The term "spreading pressure" has been suggested by Harkins for the quantity:

$$\gamma_B - \gamma_{AB} = \gamma_A \cos \theta$$

He has pointed out that in order for this relationship to hold γ_B should be measured using the vapor of A as the gas phase.[25] The spreading pressure is sometimes called the "adhesion tension." This terminology is confusing since it is difficult to distinguish from the free energy of adhesion W_{AB}. The term "adhesion tension" is applied in the case of liquid–solid–gas systems to the quantity $W_{AS} - \gamma_L$, which is the difference between the energy of adhesion of a liquid to a solid and the surface tension of the liquid. This is also equal to $\gamma_A \cos \theta$ where θ is the contact angle at the liquid–solid–air line. When two liquids A and B are in contact with a solid, A will displace B if its adhesion tension is the greater.[26]

Spreading of one liquid over another immiscible liquid usually takes place rapidly. If the dimensions of the vessel are large enough the spreading will continue until, at equilibrium, liquid A is present as a monomolecular layer on liquid B. On a surface of limited dimensions the equilibrium configuration will be a monomolecular film with the excess of liquid A gathered together as one or more lenses. During spreading the advancing liquid is

[25] Harkins and Livingston, *J. Chem. Phys.* **10**, 342 (1942).
[26] Bartell and Miller, *Ind. Eng. Chem.* **24**, 335 (1932).

usually of macroscopic thickness and for some time before coming to equilibrium may remain in the form of a so-called "duplex film." This term was introduced by Harkins to indicate a layer of liquid thin enough so that the gravitational forces are negligible as compared with the surface forces, but sufficiently thick so that the surface effects on its two sides are independent of each other.

The spreading coefficient is widely used as a measure of the wetting power, and the spreading coefficients of aqueous solutions of surface active agents over mineral oil are often cited in technical literature. From the energy relationships it is evident that a liquid of low surface tension may spread over one of higher surface tension but the reverse is not possible. Most organic liquids have a lower surface tension than water and many will spread on water. It should be noted that at equilibrium, where the phases are saturated with each other, the observed effects will be different from those calculated unless allowance has been made for the changes in γ due to mutual saturation. This has been mentioned in considering Antonoff's rule.

VII. Heat of Wetting

When a finely divided insoluble solid is immersed in a liquid there often occurs a marked temperature rise. This is to be expected since a solid–air interface of high free energy is disappearing and is being replaced by a solid–liquid interface of relatively low free energy. The heat of wetting per square centimeter of surface for a liquid A wetting a solid B is given by the equation:

$$-\Delta H = (\gamma_B - \gamma_{AB}) - T \frac{d(\gamma_B - \gamma_{AB})}{dT}$$

In this case γ_B is the free surface energy of the solid measured against a vacuum.[27, 28] It is often quite different from γ_B measured against the vapor of liquid A because of adsorption effects involving large energy changes.

The actual heat of wetting per square centimeter of surface is difficult to measure because the actual extent of the surface of a finely divided solid cannot be determined except by highly complicated techniques. Furthermore, the difficulties of obtaining clean solid surfaces make *absolute* data difficult to obtain. On the other hand, *reproducible* results are relatively easy to obtain since reproducible surfaces (containing adsorbed air or vapor layers) are apparently readily attained by careful experimenters. The

[27] See Bangham and Razouk, *Trans. Faraday Soc.* **1937**, 1459.
[28] Bangham and Razouk, *Proc. Roy. Soc. London* **A166,** 572 (1938).

term:

$$T \frac{d(\gamma_B - \gamma_{AB})}{dT}$$

is called the latent heat of the interface. Harkins and co-workers have developed the theory involved in connection with measurements on the heat of wetting of metal oxides.[29,30] Measurements of the heat of wetting have been used to estimate the surface area of powders.

The subject matter of preceding sections of this chapter was given without numerous references to emphasize the more important points and to avoid excessive detail. The literature is extremely voluminous and until quite recently the terminology and symbols had not been standardized, even among authoritative investigators. For a thorough presentation of the subject the reader is referred to Adam's treatise which is itself exceptionally well documented. A highly condensed presentation, in the nature of a review containing 500 references, is given by Allen, Knoll, Ryan, and Murray in Mattiello, *Protective and Decorative Coatings*, Vol. IV, Wiley, New York, 1944, chap. I. The following list is supplementary. *General and Theoretical:* Formation of liquid lenses—Miller, *J. Phys. Chem.* **45**, 1025 (1941). The solid-liquid interface—Clayton, *J. Oil Colour Chemists Assoc.* **18**, 412 (1935), a comprehensive review. Wetting and adhesion of bubbles—Frumkin, *Acta Physicochim. U. R. S. S.* **9**, 313–40 (1938) (in German and English). Wetting in capillaries—Loman, *Chem. Weekblad* **36**, 337 (1939). Spreading of liquids on filter paper—Prosad and Ghosh, *Kolloid-Z.* **79**, 19 (1937). Energy relationships of wetting—Illiin, Leontiew, and Bragin, *Phil. Mag.* **23**, 294 (1937); Harkins and Jura, *J. Chem. Phys.* **11**, 560 (1943). *Experimental Techniques and Practical Applications:* Wettability of porous surfaces—Cassie and Baxter, *Trans. Faraday Soc.* **40**, 546 (1944); *J. Textile Inst.* **36**, T67–90 (1945). The foregoing papers consider the theory involved in making water-repellent fabrics, and are recognized as of fundamental importance in this field. *Wetting and Detergency Symposium*, British Section International Society of Leather Trades Chemists, 1937. Spreading of oils on solids—Washburn and Anderson, *J. Phys. Chem.* **50**, 401 (1946). Wetting of steel surfaces—Miller, *J. Phys. Chem.* **50**, 300 (1946). Wetting of paint pigments—Bartlett, *Official Digest Federation Paint & Varnish Production Clubs,* No. 243, 76 (1945).

VIII. Measurement of Surface and Interfacial Tensions

(a) Introduction

In this section the various methods for determining surface and interfacial tensions will be considered in outline form only. Adam has devoted an entire chapter of his well-known book to this subject, and more recently Harkins has reviewed it comprehensively, going into full details of both

[29] Harkins and Boyd, *J. Am. Chem. Soc.* **64**, 1190, 1195 (1942).
[30] Harkins and Dahlstrom, *Ind. Eng. Chem.* **22**, 897 (1930).

theory and experimental methods.[31]   The reader is referred to these excellent reviews as well as to the original references for a detailed consideration.

Both Adam and Harkins cite abundant references. The following list is supplementary and includes some of the more recently published work: (1) Ring and modified ring methods: Dole and Swartout, *J. Am. Chem. Soc.* **62**, 3039 (1940). Ruyssen, *Kolloid-Z.* **101**, 170 (1942). Abribat and Dognon, *Compt. rend.* **208**, 1881 (1939). Fulmer and Andes, *J. Phys. Chem.* **36**, 467 (1932). Schultze, *Kolloid-Z.* **67**, 26 (1934). Thibaud and Lemonde, *J. Phys. Radium* **2**, 26 (1941); *Chem. Abstracts* **37**, 6163. (2) Drop weight method: Culbertson and Hedman, *J. Phys. Chem.* **41**, 485 (1937). Dunken, *Ann. Physik* **41**, 567 (1942). Ward and Tordai, *J. Sci. Instruments* **21**, 143 (1944). Harman, *U. S. Bur. Mines Circ.*, No. 7351 (1946). (3) Capillary rise method: Brown, *Proc. Phys. Soc. London* **53**, 233 (1941). Jones and Frizzell, *J. Chem. Phys.* **8**, 986 (1940). Leaute, *Compt. rend.* **201**, 41 (1935). Natelson and Pearl, *J. Am. Chem. Soc.* **57**, 1520 (1935). Achmatov, *Kolloid-Z.* **66**, 266 (1934). (4) Bubble pressure method: Hazlehurst, *J. Chem. Education* **19**, 61 (1942). (5) Pendant-drop method: Andreas, Hauser, and Tucker, *J. Phys. Chem.* **42**, 1001 (1938). (6) Ripple method: Tyler, *Phil. Mag.* **31**, 209 (1941). (7) Sessile-bubble method: Wheeler, Tartar, and Lingefelter. *J. Am. Chem. Soc.* **67**, 2115 (1945). (8) General discussion of methods and theory: Ferguson, *Endeavour* **2**, 34 (1943). Antonoff, *Ann. Physik* **35**, 84 (1939); *J. Phys. Chem.* **47**, 463 (1943). Compare Brillouin, *J. Phys. Radium* **9**, 462 (1938). Dervichian, *Compt. rend.* **201**, 333 (1935). Picon and Mangeot, *J. pharm. chim.* **2**, 145 (1942). An interesting titration method using the interfacial tension as an indicator is described by Dubrisay, *Compt. rend.* **198**, 1605 (1934).

The methods for determining surface tension fall into two general classes, the static and the dynamic. The static methods measure the tension of a surface which has remained unchanged for an appreciable period of time and which is in equilibrium with the interior. The dynamic methods measure the tension of a surface which is *in process of being expanded or contracted.* Dynamic surfaces may or may not be in equilibrium with the interior depending on how rapidly equilibrium is attained and how rapidly the surface is being changed. In the case of pure liquids there is usually no significant difference between static and dynamic surface tension. In the case of solutions there may be large differences. In general the concentration of solute at a surface or interface differs from the concentration in bulk. After formation of a fresh interface a finite time (sometimes a surprisingly long time of the order of days) is required for equilibrium to be established.

Most of the static methods depend on one of two principles. The first is a measure, either direct or indirect, of the pressure differential across a curved interface due to the surface free energy and the curvature. The second involves the actual slow extension of the surface or interface and a

---

[31] Harkins, in Weissberger, *Physical Methods of Organic Chemistry*, Interscience, New York, 1945, Vol. I, chap. 6.

simultaneous measurement of the force necessary to bring about this extension. The dynamic methods depend on observing the form of waves or oscillations in a rapidly moving body of liquid. The restoring force acting on a deformed liquid surface is due to the free surface energy. The shapes of waves, jets, etc. are determined by the interplay of this restoring force and the forces of deformation. The latter may be controlled or measured, whereby the shape of the moving liquid surface becomes a measure of the surface or interfacial tension.

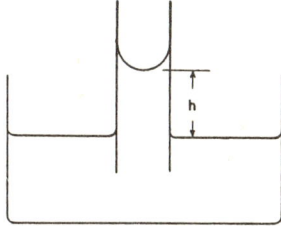

Fig. 10.   Capillary rise in cylindrical tube.

(b) *Static Methods*

CAPILLARY RISE

In considering the pressure differential across a curved surface due to the surface tension we have seen that the height to which a liquid will rise in a capillary tube is determined by the radius of the tube, the contact angle at the line between the tube wall, the air, and the liquid being measured, the surface tension of the liquid, and its density. This phenomenon affords one of the most accurate and widely used means of determining surface tension. Adam's treatment of the theory is excellent and unambiguous. The force which causes the liquid to rise in a capillary tube is not a *pull* on the column of liquid at the line of contact. It is equivalent to such a pull, but in actuality it is a *push* from the convex side of the surface, caused by the excess pressure on the convex side. We have seen that the pressure difference across a curved surface is:

$$p_1 - p_2 = \gamma \left(\frac{1}{R_1} + \frac{1}{R_2} \right)$$

where R_1 and R_2 are the principal radii of curvature. Let us consider now the situation illustrated in Figure 10, with a cylindrical tube dipping into a large vessel of liquid. If the radius of the tube is sufficiently small the surface can be assumed to be spherical in shape, *i.e.*, its two radii of curvature are equal to each other.[32]   If the contact angle is zero or nil the radius of curvature of the surface is equal to the radius of the capillary tube, r. The pressure difference across the surface is then:

$$p_1 - p_2 = \gamma \left(\frac{1}{r} + \frac{1}{r} \right) = \frac{2\gamma}{r}$$

This pressure difference is balanced at equilibrium by the pressure difference between the top and bottom of the column of liquid standing in the tube, which is equal to $gh(D - D_A)$ where D is the density of the liquid, D_A is the density of air, and g is the gravitational constant.   In actual calculations D_A may be neglected.   Accordingly:

$$ghD = \frac{2\gamma}{r}$$

or: $$\gamma = \tfrac{1}{2}grhD$$

Refinements of the theory include considerations of the surface shape deviating from the spherical and the weight of the meniscus itself above the lowest point from which h is measured.   It is important in all practical work that the contact angle between the capillary wall and the liquid at the air surface shall be zero or nil, and this point should be checked.   The capillary tube itself should be uniform in bore and the radius must be known accurately.   A number of modifications of the capillary rise method have been used.   These include measurement of the pressure necessary to prevent any rise in the tube, the use of parallel plates instead of a tube, and the use of vertical plates set at a slight angle whereby the profile of the liquid assumes the form of a hyperbola.   The capillary rise method and its modifications are not well adapted to measuring interfacial tensions even though in theory they should be quite useful.   Accordingly they are hardly ever used for this purpose.

DROP WEIGHT OR DROP VOLUME METHOD

The weight and size of a drop of liquid which first forms and then falls from the end of a tube is a function of the diameter of the tube and the surface tension of the liquid.   Instruments which provide for forming drops

---

[32] This is connected with the fact that the ratio of peripheral length to surface area increases as the radius of the tube decreases.   Therefore, the surface forces acting from the periphery become relatively more important than the gravitational forces which act uniformly on the whole column.   The gravitational forces tend to form the surface into a flat plane.   The surface forces, responsible for the contact angle, tend to form it into a sphere.

slowly under precisely controlled conditions, and either weighing them or measuring the volume of liquid per drop, are called stalagmometers. The theory of this method is at least in part on an empirical basis. Harkins and co-workers are responsible for its most complete development. The weight of a drop falling from a carefully ground tip is given by the expression:

$$W = 2\pi r \gamma f(r/V^{1/3})$$

in which $f(r/V^{1/3})$ is an undetermined function of the radius of the tube divided by the cube root of the drop's volume. Since $W = mg$ or the mass times the gravity constant:

$$\gamma = \frac{mg}{2\pi r} \bigg/ f\left(\frac{r}{V^{1/3}}\right) = \frac{mg}{r} F$$

The quantity $1/2\pi f(r/V^{1/3})$ has been called F by Harkins and Brown, who have published[33] a table relating F to V/r^3. This table has been universally accepted as standard and is included in the *International Critical Tables*. Using an accurately made and calibrated stalagmometer the drop volume, drop weight, and tube radius are measured and γ can be calculated from the above formula by use of the table. The stalagmometric method is also excellent for measuring interfacial tensions. In this case the heavier liquid is placed in the tube and caused to form a drop into the other liquid rather than into air. Instead of the drop "weighing" mg under these circumstances the gravitational pull is given by $V(D_1 - D_2)g$. The formula is accordingly:

$$\gamma = \frac{V(D_1 - D_2)g}{r} F$$

where D_1 and D_2 are the densities of the two liquids and V is the volume of the drop.

MAXIMUM BUBBLE PRESSURE METHOD

A tube is dipped into a vessel of liquid to a definite depth, and air is forced down the tube to blow a bubble. The pressure of air required increases at first as the bubble grows. Then it reaches a maximum and thereafter diminishes as the bubble increases beyond a certain critical size. If the orifice is sufficiently small this maximum pressure occurs when the bubble is hemispherical. In actual practice deviations from this simple

[33] Harkins and Brown, in *International Critical Tables*, Vol. IV, McGraw-Hill, New York, 1928, p. 435

case begin to occur with very small bubbles and the necessary corrections become quite complicated.   Bashforth and Adams in 1883 published sets of tables relating the pressure differential across a curved surface due to the surface tension to the geometrical form of the surface.   The geometrical forms they considered were surfaces of revolution about a vertical axis.[34] These tables have been modified by Sugden to apply to the maximum bubble pressure method for measuring surface tension and are accepted as the standards in calculating the results of this method.   The experimental setup is comparatively simple and accurate determinations can be made rapidly.   The reader is referred to the original publications for details.

SESSILE DROPS OR BUBBLES AND PENDANT DROPS

The word "sessile" is defined by Webster as meaning "attached by a broad base."   In the sessile-drop method for determining surface tension a drop of liquid is placed on top of a horizontal plate and its shape, particularly its height, is carefully measured.   In the sessile-bubble method a bubble is formed from a tip under the surface of the liquid being studied, or on the underside of a plate.   In the pendant-drop method a drop is allowed to hang from the lower side of a horizontal plate.   In each case the measurement of form is made either with a cathetometer or from a photographic enlargement.   The drops or bubbles depart from spherical form by virtue of the gravitational force acting on them.   The degree of departure depends on the relative effectiveness of the opposing forces of gravitation and surface tension.   Modifications of the Bashforth and Adam tables are used in the calculations.   These three methods are very useful in measuring the surface or interfacial tension in solutions where this property changes with time.   Change of surface tension on ageing is not uncommon with solutions of surface active agents and these methods have been used to study the ageing effect.

RING METHOD

The ring method of measuring surface tension is without doubt the most widely used method in industrial laboratories, the du Noüy form of instrument being favored.   This method consists in measuring the force necessary to pull a ring of known diameter away from the surface of the liquid. The ring is usually made of platinum.   It is essential that the solid–liquid contact angle at the ring–liquid–air line be zero or nil.   The exact theory of the ring method and other detachment methods is quite complicated. Consider the schematic diagram in Figure 11, where R is the ring (viewed

[34] Bashforth and Adams, *An Attempt to Test the Theories of Capillary Action*, Cambridge Univ. Press, 1883.

in elevation), L is the liquid, and B is a balance for measuring the force necessary to detach R from the liquid. Since the solid–liquid contact angle is zero or nil, as R is raised the curvature at C is indicative of a pressure on the convex side of C. This in effect supports a mass of liquid approximately a hollow cylinder in shape and of height H. The ring detaches when this pressure is no longer sufficient to overcome the gravitational force. It is obvious that observation of the curvatures in the ring's vicinity, the height at which detachment occurs, etc. would be extremely difficult. The elementary theory considers that the liquid exerts a pull (P) on the ring equal to:

$$2\pi R_1\gamma + 2\pi R_2\gamma$$

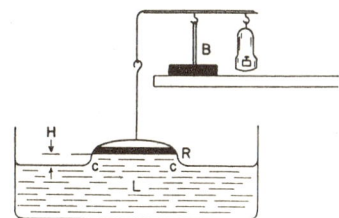

Fig. 11. Ring method of measuring surface tension.

where R_1 and R_2 are the inner and outer radii of the ring. Where the ring is large in relation to the gage of the wire, this can be reduced to:

$$\gamma = P/4\pi R$$

where P is the observed pull necessary to detach and R is the mean radius of the ring. The ring does not actually detach suddenly at the point of maximum force. As the ring is lifted from the surface the force increases with the distance and passes through a maximum somewhat before the point of detachment. When a torsion balance is used to measure the force, it is the maximum force which is automatically recorded. Harkins and co-workers have pointed out that use of the above formula to calculate surface tension gives erroneous results and have worked out tables of correction factors. Using the formula:

$$\gamma = PF/4\pi R$$

where F is a correction factor obtained from the tables, gives accurate results. The numerical value of F varies from about 0.75 to 1.45.

The du Noüy tensiometer, which is widely used to determine surface tension by the ring method, uses a torsion wire to measure the force of detachment. An indicator attached to the torsion wire gives readings directly in dynes per linear centimeter. The instrument is calibrated by using a liquid of known surface tension (usually pure water) and adjusting the indicator to show the correct reading at the point of detachment.

The ring method may be modified by using a long vertical plate or a straight piece of wire instead of a ring.   It may be used to measure inter-facial tension, but care must be taken that the contact angle between the ring and the liquid from which the ring is being pulled is zero.

MISCELLANEOUS STATIC METHODS

Adam has described an unusual method for estimating the surface ten-sion of natural waters which involves observing the spreading behavior of a carefully balanced oil mixture.   This mixture of water-immiscible liquids includes both spreading and non-spreading components.[35]

Waves or ripples on the surface of a liquid constitute a distortion of the surface and the restoring force involves the surface tension.   By initiating a wave of known frequency (for example, by means of a vibrating tuning fork), and observing the wave length the surface tension may be calculated from a formula developed by Lord Kelvin:

$$\gamma = \frac{\lambda^3 D}{2\pi T^2} - \frac{g\lambda^2 D}{4\pi^2}$$

where λ is the wave length, D the density, and T the period.   Brown has described the theory and experimental details of this method.[36]   Adam has pointed out that this method is rightfully classed as static since the surface is not being renewed as the measurements are being made.

Pfund and Greenfield have described a method particularly suited to viscous liquids which consists in allowing a jet of air to impinge vertically against the surface.   The depth of the depression formed in the liquid sur-face by the air current is a function of the surface tension.[37]

Meyerstein and Morgan have described an interesting method for meas-uring both surface and interfacial tensions in capillary tubes.   The surface tension effect is opposed by a centrifugal thrust, and the tube so arranged that it is either filled or emptied when the surface tension is exceeded by the centrifugal force.   The theory and full experimental details are pre-sented in the original paper.[38]

Bradley has described the theory and technique for measuring tension at the liquid–liquid–air contact line by measuring the diameter and thick-ness of lenses of one liquid floating on another.[39]

[35] Adam, *Proc. Roy. Soc. London* **B122,** 134 (1937).
[36] Brown, *Proc. Phys. Soc. London* **48,** 312 (1936).
[37] Pfund and Greenfield, *Ind. Eng. Chem., Anal. Ed.* **8,** 81 (1936).
[38] Meyerstein and Morgan, *Phil. Mag.* **35,** 335 (1944).
[39] Bradley, *Trans. Faraday Soc.* **36,** 999 (1940).

(c) Dynamic Methods

Observations on moving, changing liquid surfaces are quite difficult, and the dynamic methods of measuring surface tension have accordingly been studied to a far lesser extent than the static methods.   When a drop of liquid detaches from a suitable tip and starts to fall, it usually oscillates, changing its shape periodically from oblate along one axis to oblate along another.   The oscillation rate and changes in geometric form can be followed by the techniques of high-speed cinematography, and the surface tension can be calculated.[40]

A jet of liquid issuing from an elliptical orifice will tend to assume a circular rather than an elliptical cross section.   This tendency, which is a simple function of the surface tension, results in waves along the stream. From the wave length and form, the velocity of the stream, etc., the surface tension can be calculated.[41]

The dynamic surface tension can also be calculated by observation of the fan of liquid formed when two jets of liquid impinge on each other. In a variation of this method a liquid jet is caused to impinge on a solid surface and the bell which is formed is measured.[42]

The calculations in the dynamic methods are quite involved.   The results apparently check the static methods in the case of pure liquids.   In the case of solutions, particularly solutions of surface active agents, there is often a wide discrepancy.   This is due to the time required for the solute molecules to migrate to the surface and thereby lower the surface tension. If this time is greater than the time required to change the surface layer by mechanical agitation (under conditions of the experiment) a difference between the static and dynamic surface tensions will be observed.

[40] Hauser, Edgerton, Holt, and Cox, *J. Phys. Chem.* **40**, 973–88 (1936).

[41] Satterly and Strachan, *Trans. Roy. Soc. Canada* **29**, III 105 (1935).

[42] Buchwald and Koenig, *Ann. Physik* **26**, 659 (1936).   Bond, *Proc. Phys. Soc. London* **47**, 549 (1935).

CHAPTER 13

Surface Properties of Solutions

I. General Considerations

A given mass of a pure liquid can decrease its free surface energy only by decreasing its total surface area, *i.e.*, by assuming the form which exposes the least surface per unit of mass. This is because all the molecules in the liquid are of the same species and their force fields are identical. No matter which individual molecules are in the surface at any given instant, the pull on these molecules from the interior is always the same statistically. In the case of a solution conditions are quite different. Consider, for example, a simple binary solution of a solute A in a solvent B. Within this solution, both species of molecules are free to migrate and exert van der Waals' forces of attraction on their immediate neighbors. In general, the fields of attractive force exerted by the molecules A will be different from those exerted by B. These force fields, as we have seen, are responsible for the free surface energy, *i.e.*, they act to pull surface molecules into the interior. If the A molecules have stronger force fields than the B molecules, they will be pulled away from the surface at a greater statistical rate. The net effect will be a diminution in concentration of A molecules in the surface. In other words, the concentration of A in the surface will be less than in the bulk of the solution. Conversely if the A molecules have weaker force fields than B, they will tend to concentrate in the surface. The statistical accumulation in the surface of those molecules with weaker force fields results in a lowering of the free surface energy. Thus, a given mass of solution may lower its free surface energy not only by diminishing its total surface, but also by concentrating in the surface that component whose molecules have the weaker force fields. The tendency for these molecules to come to the surface is opposed by the thermal agitation and osmotic or diffusion forces. The latter forces act to prevent a complete usurpation of the surface layer by the weakly attracting molecules.

The concentration of one component of a solution at a phase boundary is called adsorption. Qualitatively, if a solution has a lower surface tension than the pure solvent, the solute is positively adsorbed in the surface. Conversely, if the solution has a higher surface tension the solute is negatively adsorbed, *i.e.*, it is more concentrated in the interior than in the surface. Most inorganic salts in aqueous solution are negatively adsorbed. The surface tension of these solutions is higher than that of pure water. Some substances, such as sodium hydroxide for example, raise the surface tension of water markedly. Most water-soluble organic compounds lower the surface tension of water, and the surface active agents show this effect to an extreme degree.

The quantitative relationship between the degree of adsorption and the lowering of surface tension was deduced by Gibbs, using the methods of thermodynamics.[1] The complete form of the Gibbs adsorption equation for the change in surface tension at constant temperature due to adsorption in a system of i components is:

$$(1) \qquad d\gamma = -\Gamma_1 d\mu_1 - \Gamma_2 d\mu_2 \cdots \Gamma_i d\mu_i$$

In this equation Γ_i is the "surface excess" of the ith component and μ_i is the chemical potential of the ith component. The surface excess is the concentration of the ith component in the surface phase minus the concentration in the interior.[2] The chemical potential, μ_i, of the ith component is:

$$\mu_i = RT \ln f_i N_i + \mu_i^0$$

where f_i is the activity coefficient, N_i is the mole fraction, and u_i^0 is the chemical potential of the pure ith component.

For dilute binary solutions, where the activity coefficient of the solute is unity, equation 1 reduces to:

$$(2) \qquad \Gamma_2 = -C_2 d\gamma / RT dC_2$$

where the subscripts refer to the solute (component number 2). This states that the surface excess of solute is proportional to the concentration multiplied by the rate of change of surface tension with respect to concentration. Equation 2 is the so-called approximate form of the Gibbs adsorption equation. It is much more widely used than the exact form (equation 1).

The verification of the Gibbs equation depends on accurate measurements of the surface excess or degree of adsorption of the solute. The experimental techniques involved in such measurements have proved to be

[1] For a condensed discussion and exposition of the Gibbs derivation, see Adam, *Physics and Chemistry of Surfaces*, Oxford Univ. Press, London, 1941; pp. 107–113. This contains ample references to the original and subsequent works.

[2] The region of surface phase and the exact boundaries used in calculating surface excess are rigorously defined in the original works.

quite difficult. One of the most direct methods used is that of McBain.[3] It consists in actually cutting off a thin layer of solution, including the surface, by means of a travelling knife and is known as the microtome method. The concentration of solute in the removed surface layer is then compared with that in the bulk of the solution. Agreement between experimental and theoretical results has been very good in some cases and poor in others. The widest discrepancies have occurred with solutions of surface active agents; these will be discussed in more detail later. The theoretical validity of the Gibbs equation is unquestionable, being based on thermodynamic reasoning. Discrepancies may be due to a difference between the actual molecular condition of the system being measured and our concept of that system. Lack of agreement between theory and experiment may also arise from a neglect to consider other variables (such as electrical variables, etc.) which may affect the system. In any event, the qualitative relationships among surface tension, adsorption, and concentration which are implicit in the Gibbs equation have never been seriously questioned. In effect there is a layer in the surface region which is far richer than the interior in solute molecules. In the case of surface active agents, this adsorbed layer is equivalent to a film of solute molecules in the surface. The study of surface films on liquids is an extensively developed subject and only a few aspects of it bear directly on a consideration of surface active agents. It will, nevertheless, be helpful at this point to touch upon the more pertinent phases of this subject very briefly.

For more complete reviews of the physical chemistry of surface films, see: Harkins, in Alexander, *Colloid Chemistry*, Vols. V and VI, Reinhold, New York, 1944 (these chapters constitute an extremely clear and complete presentation by one of the foremost investigators in the field of surface phenomena; the bibliography is large and includes a complete list of articles by Harkins and co-workers); *Pub. Am. Assoc. Advancement Sci.*, No. 7 (1939). For applications in biology and medicine, see Adam, *The Physics and Chemistry of Surfaces*, Oxford Univ. Press, London, 1941, chap. 2.

II. Surface Films of Insoluble Substances

We have seen that if a drop of a liquid A is placed on the surface of another liquid B, and the work of adhesion W_{AB} is greater than the work of cohesion of $A = 2\gamma_A$ then A will spread over the surface of B. W_{AB} is likely to be greater than $2\gamma_A$ if γ_B is high and γ_A is low. The surface tension of water is relatively high and that of many organic liquids is sufficiently low so that they spread over water quite readily. Practically all ordinary liquids will spread over clean mercury, which has a surface tension

---

[3] McBain and Humphries, *J. Phys. Chem.* **36**, 300 (1932). McBain and Swain, *Proc. Roy. Soc. London* **A154**, 608 (1936).

roughly seven times that of water. When the surface of the underlying liquid, water for example, is sufficiently large the spreading liquid will ultimately form a film one molecule thick on the surface. Special techniques have been devised for spreading monomolecular films (called monofilms) of substances which do not spread readily or spontaneously. For example, a solution may be made in a readily spreading volatile solvent. After spreading has occurred the solvent quickly evaporates leaving a monofilm of the desired solute on the surface. Monofilms are detected and studied by their effects on the surface characteristics of the underlying liquid. These effects may be classed as optical, mechanical, and electrical. During the course of development of this field of science, it has been found that the properties of monofilms can often be closely correlated with the structure of the molecules composing them. Accordingly, one of the newer and more fruitful techniques for studying the structure of complicated molecules such as proteins, sterols, and various polymeric substances is their transformation by

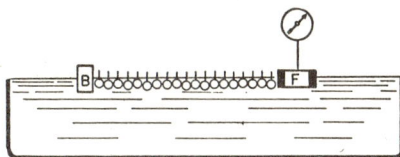

Fig. 12. Film balance.

appropriate methods into monolayers and the study of their properties in this state.

The principal technique used in studying monofilms involves spreading them on a clean water surface in a trough filled to the brim. The trough is fitted with movable barriers which confine the surface film. The pressure which the film exerts outward against the barriers is measured by means of a movable light float coupled with a device for measuring the force exerted against the float. This type of apparatus is called a film balance. Film balances have been highly refined and surface pressures as low as 0.001 dyne per centimeter have been measured. Surface pressure is due to the resistance which the floating molecules of the monofilm exert against being compressed. In Figure 12, B represents the barrier and F the movable float, which is connected with a torsion balance. The molecules in the film will, if compressed by moving B to the right, exert a pressure on the float F. The magnitude of this pressure is related to the surface tension in a very simple manner as follows:

Suppose that the float F is displaced to the right a distance dx under the force of the surface pressure π. Then the work done on F is:

$$W = \pi l dx$$

where l is the length of F. But if γ_c is the surface tension of the clean liquid surface and γ_m is the surface tension of the surface which is covered with monofilm, it is apparent that in moving F, an area ldx having the energy $\gamma_c ldx$ has been replaced by an equal area having energy $\gamma_m ldx$. The work done is:

$$W = (\gamma_c - \gamma_m)ldx$$

Equating the two expressions for the work:

$$\pi = \gamma_c - \gamma_m$$

or the surface pressure is equal to the difference in surface tension between the clean liquid and the liquid covered with monofilm. It is apparent that surface pressures of monofilms can therefore be measured by measuring the surface tension of the film-covered liquid in any conventional manner. Since the investigator is most often interested in the variation of surface pressure with the total area of the film, it is still necessary to have trough and barriers so arranged that the barriers may be moved and film area varied.

An alternate method of measuring the surface tension is the plate or Wilhelmy method. This consists in measuring the downward pull on a flat plate which is dipped vertically through the surface. This type of apparatus is called a vertical film balance.

Monofilms on liquid substrata may be regarded as constituting a two-dimensional state of matter and many illuminating analogies may be drawn between their behavior and the more familiar behavior of matter in three dimensions. The surface pressure of a monofilm is strictly analogous to the osmotic pressure of a solute. The barrier and float system corresponds to a semipermeable membrane. It is impenetrable by the floating molecules of the monolayer. It is permeated readily by the molecules of the underlying liquid since they are free to migrate into the interior and pass underneath the barriers. There is an overwhelming amount of evidence that a system of floating insoluble molecules such as the one we are considering always exists as a monomolecular layer when it has attained equilibrium. When the layer is compressed or confined into a space where its mass precludes arrangement in a monolayer it will form localized islands or floating lenses connected by regions of monofilm. If still less area is made available the lenses will, of course, occupy more and more of the available area until they consolidate to a layer of macroscopic thickness. These lenses may not be of visible thickness, but they are so thick that the force fields of their upper and lower boundaries are too far apart to interact.

If the mass of the monofilm is known, from having weighed the material before spreading it, the area occupied by each molecule in the film may readily be calculated from the total area and the Avogadro number. In plotting the results of surface pressure vs. area measurements, this calculated value of area per molecule is often used.

The viscosity of the surface layer bearing the monofilm is, in general, considerably different from the viscosity of the underlying liquid. Several techniques have been developed for measuring this surface viscosity. Probably the most direct method involves measuring the rate at which the film can be forced through a narrow opening or canal in the barrier. Other methods depend on the damping of oscillating rings, discs, and vanes in the surface layer. The rheological behavior of monofilms has been correlated to a certain extent with the type of film.

Monofilms can be detected and studied not only by the surface pressure effect, but also by their effect on the surface electrical potential. When two conducting phases are in contact, there is generally a difference of electrical potential established between them. This phase boundary potential is due to an excess of negative charges accumulating in one region and a corresponding excess of positive charges in an adjacent region near the boundary. Such a distribution of charge is called an electrical double layer. In many cases the phase boundary separates the two sides (positive and negative) of the electrical double layer. Many other instances are known, however, in which both sides of the double layer are within a single phase close to the boundary.

There are at least three general mechanisms by which phase boundary potentials can be established. The first is caused by a difference in escaping tendency between positively and negatively charged components of one phase. An example is the thermal emission of electrons from a heated metal in a closed space. This results in a potential difference between the metal and the surrounding atmosphere. The electrode potential obtained when a metal dips into an aqueous solution is another such example. Still another is the potential set up across a membrane dividing two ionic solutions when one ion is more easily passed through the membrane than another. In all these cases the phase boundary separates the two sides of the electrical double layer. The electrostatic forces existing between the two sides act over a long range by comparison with the van der Waals' forces of cohesion, adhesion, and solubility. The electrical double layer should therefore be regarded as quite diffuse. It is probably effective in many cases over distances an order of magnitude greater than an average molecular diameter.

A second way in which an electrical double layer can arise is the preferential adsorption at a surface or interface. Consider for example a solution containing both cations and anions. If the cations are preferentially adsorbed at the surface, this gives rise to an electrical double layer, wholly within the liquid phase, having the positive side nearer the surface and the negative side toward the interior. A potential established in this manner is called an adsorption potential. Although there is no tendency for the ions actually to migrate through the phase boundary this situation is *equivalent* to a higher escaping tendency on the part of the adsorbed ion. Electrical double layers due to adsorption are diffuse and the forces involved extend over ranges which are large in comparison with molecular diameters.

The third way in which an electrical double layer may be established is the orientation of a layer of dipolar molecules at a surface in such a way that similarly charged ends are pointed toward the surface and oppositely charged ends away from the surface. The primary double layer formed in this manner has, of course, a definite thickness equal to the vertical component of the distance between the poles in a single molecule. This layer can induce diffuse double layers on both sides of the boundary. The potential arising in this manner is difficult to distinguish operationally from an adsorption potential.

In order for phase boundary potentials to be detected and measured (and, if one adopts the operational point of view, in order for them to exist) both phases must be conductors. This, in the broadest sense of the word, means that they must allow a measurable migration of charged particles through them. The charged particles may be positive or negative ions or free electrons. Since air is normally non-conducting, it must be ionized in order to make a measurement of the surface potential. This is usually done by coating the air electrode with a radioactive substance. A cell is set up in which an ordinary calomel half-cell or a silver–silver chloride half-cell connected with a potentiometer dips into the liquid substratum in the trough. The air electrode is connected with an electrometer. The air electrode can be moved over the surface and the e.m.f. of the cell measured from point to point. In general, the e.m.f. of the cell will have a different value when a monofilm is present than when the liquid surface is clean. The difference between these two values of e.m.f. is defined as the surface potential of the film.

The optical effects of monofilms are much less frequently used in their study than the mechanical and electrical effects. The best-known optical effect of monofilms is their action in causing the elliptical polarization of light reflected at the surface. The degree of polarization is related to the type of film, and the uniformity of polarization can be used to check the film uniformity. A more convenient optical method of checking film uniformity and continuity consists in using an ultramicroscope technique. A dark-field illuminator is focused sharply at the surface using a strong light source.   When a true monofilm is present, no bright areas or spots are observed.   Lenses or regions of incompletely spread material are easily detected as illuminated regions.

A typical investigation of a monomolecular film on a liquid substratum involves first of all spreading the film in a trough fitted with the necessary apparatus.   The film may then be checked for uniformity by optical methods.   The film area is then varied by moving the barriers and the surface pressure is measured.   The surface potential is measured from point to point by scanning the surface with the air electrode.   It has been found by this type of study that monofilms can exist in a number of different states, strictly analogous to the various states of ordinary three dimensional matter (*viz.*, solid, liquid, gaseous, liquid crystalline, etc.).   These different states are evidenced primarily by the shape of the surface pressure *vs.* area curve.   As in the case of three dimensional matter each state corresponds to a different degree of attractive force between the molecules, *i.e.*, of lateral adhesion.   To be more exact it corresponds to a different balance between the three primary types of force acting on each molecule in the surface film.   These three forces are: (*1*) kinetic forces of ordinary thermal motion, (*2*) the attractive force between the surface film molecule and the molecules of the substratum liquid, and (*3*) the lateral adhesion forces.   It will be recalled that the relationship among these forces determines whether or not spreading will occur.   It also determines the type of film which will be formed when spreading takes place.

Adam distinguishes four main types of films: (*1*) *Gaseous films*, in which the molecules are separate and move about independently.   The lateral adhesion is very small and the surface pressure is approximately proportional to the area.   (*2*) *Condensed films*, in which the molecules are closely packed due to strong lateral adhesion.   The molecules are usually steeply oriented to the surface.   (*3*) *Liquid-expanded films*, which are still coherent but less so than the condensed films.   (*4*) *Vapor-expanded films*, which are more coherent than the liquid-expanded, but less coherent than the gaseous films.

Harkins lists six different types of surface films.   He includes the gaseous and Adam's "vapor-expanded" films in one class, and differentiates three types of condensed films, a liquid, a "superliquid" of very low compressibility, and a solid type.

Gaseous films can be considered from the theoretical point of view in much the same manner as three-dimensional gases.   In an ideal gaseous film the relationship between surface pressure and area is:

$$FA = kt$$

where k is the gas constant per molecule. At room temperature, expressing F in dynes and A in square angstrom units, the value of kt is about 400. Actually this value is never realized, although some films have been found which approach it at great dilutions.

In condensed films the surface pressure has a negligible value until a critical area is reached. It then rises suddenly to a fairly steady value. At areas greater than the critical area the surface potential varies widely and irregularly from point to point over the surface. At this limiting area the surface potential becomes uniform from point to point over the surface. These facts indicate that the condensed films are highly coherent and exist on the surface as large macroscopic "islands" of laterally adherent molecules. These islands exert very little pressure against the barriers, just as particles of colloidal dimensions exert very little osmotic pressure in the three-dimensional case. When these islands are brought together by reducing the area, the surface pressure which suddenly develops is a measure of the resistance of the coherent film to further compression.

The vapor-expanded and liquid-expanded films represent intermediate degrees of lateral cohesion among the surface molecules. The same spreading substance can sometimes form different types of film at different temperatures. Furthermore, one type of film may be transformed into another type in the same manner that ordinary phase changes take place in three dimensions. Changes in the type of film may often be brought about by adding a second insoluble film-forming component to the material which is already spread.

Condensed films are often characterized by having their molecules oriented with respect to the surface plane. In gaseous films, however, most of the evidence indicates that the molecules lie flat in the surface and have random orientation. Studies on the homologous series of straight-chain fatty acids illustrate the transition from gaseous to condensed films as the molecular weight increases. It may be noted at this point that the adsorbed surface layer of solute molecules in solutions of positively adsorbed solutes, including surface active agents, usually exhibits the structure of a gaseous monofilm, i.e., the molecules lie flat in the surface and have random orientation.

Relatively little work has been published on monofilms at liquid–liquid interfaces. They appear to have the same general properties as at the air–liquid interface, but to be more expanded.[4,5]

[4] Alexander and Teorell, *Trans. Faraday Soc.* **35,** 727 (1939). Alexander, *Trans. Faraday Soc.* **37,** 117 (1941).

[5] Devaux, *Compt. rend.* **202,** 1957 (1936). Pokhil, *J. Phys. Chem. U. S. S. R.* **13,** 301 (1939).

The main theoretical relationship between the study of monofilms and surface active agents is established by the high degree of surface adsorption of the latter. The relatively dense adsorbed layer is in effect a monofilm. In most cases of dilute solutions at the air–solution interface, this film is of the gaseous type and it is inadvisable to assume any high degree of orientation or any strong lateral adhesion. The film can and does affect the free surface energy, viscosity, optical properties, and surface potential.

Although the adsorbed surface films of most soluble surface active agents are of the gaseous type, condensed films have been observed. Stenhagen studied monofilms of sodium docosyl sulfate on 0.04 M phosphate buffer at pH 7.2 and on 0.01 N HCl. On both substrates the monofilms solidified at relatively low surface pressures and at an area of about 26 A^2 per molecule.[6]

The adsorbed layer or surface film of a surface active solute can be studied under certain conditions by means of the film balance. McBain, Ford, and Wilson have described[7] a modified trough type of film balance suitable for studying the soluble films formed by surface active agents. Using this instrument McBain and his associates[8] found that the formation of a "pellicle" or surface film depended on the concentration and the degree of compression. Adam and Adam and Miller[9] using a similar technique found that long-chain quaternary ammonium halides and the long-chain fatty acids on aqueous alkaline solutions both showed the formation of a gaseous type film.

At the liquid–liquid interface between two immiscible liquids, adsorbed films may also be formed. In general, simple monofilms at the liquid–liquid interface appear to be more expanded than similar films at the air–liquid interface.[10] It is apparently easier, however, to form condensed adsorbed films at the liquid–liquid interface than it is at the liquid–air surface. This is readily understandable on the basis of the stronger orienting forces at the interface. If either of the liquid phases contains a solute, and in particular a surface active solute, interfacial films may be formed by adsorption in the same manner as surface films. If both liquid phases con-

[6] Stenhagen, *Trans. Faraday Soc.* **36,** 496 (1940).

[7] McBain, Ford, and Wilson, *Kolloid-Z.* **78,** 1 (1937). See also McBain and Wilson, *J. Am. Chem. Soc.* **58,** 379 (1936).

[8] McBain and Spencer, *J. Am. Chem. Soc.* **62,** 239 (1940). McBain, Vinograd, and Wilson, *J. Am. Chem. Soc.* **62,** 244 (1940). See also McBain and Sherp, *J. Am. Chem. Soc.* **63,** 1422 (1941).

[9] Adam, *Proc. Roy. Soc. London* **A142,** 401 (1933). Adam, *Trans. Faraday Soc.* **32,** 653 (1936).

[10] Askew and Danielli, *Proc. Roy. Soc. London* **A155,** 695 (1936). Alexander and Teorell, *Trans. Faraday Soc.* **35,** 727 (1939).

tain readily adsorbed solutes, mixed interfacial films may be formed by adsorption. Adsorbed interfacial films of one or more components may range from the unoriented gaseous types to highly condensed types depending on the nature and concentration of the system's components and on the conditions of film pressure, temperature, etc. Interfacial films are of great importance in emulsification phenomena and there is ample evidence that a highly condensed interfacial film is necessary for a stable emulsion. This point will be discussed more thoroughly in the section on emulsions.[11]

III. Surface Tension Measurements of Solutions

(a) Effect of Concentration

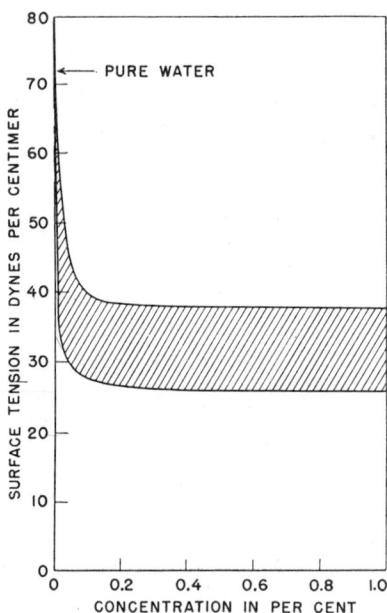

Fig. 13. Surface tension values for majority of surface active agents fall within shaded area on graph.[13]

When the surface tension of an aqueous solution is plotted as ordinate against concentration as abscissa any one of three different types of curve may be obtained, depending on the nature of the solute.[12] With many simple inorganic solutes such as sodium hydroxide, potassium chloride, potassium sulfate, and others, the surface tension rises with increasing concentration. This has been called a type II curve by McBain. In indicates a negative surface adsorption of the solute. The type I curve is one in which the surface tension falls with increasing concentration. The decrease in surface tension may be sharp or gradual. It is generally not linear, but is convex toward the origin. In the case of surface active agents the decrease is very sharp at low concentrations, then the curve levels off or "breaks" rather rapidly at a still relatively low concentration and continues to fall slowly as the concentration increases. Figure 13 taken from Fischer

[11] Fischer and Harkins, *J. Phys. Chem.* **36**, 98 (1932). Alexander and Schulman, *Trans. Faraday Soc.* **36**, 960 (1940). Alexander, *ibid.* **37**, 117 (1941). Schulman, *ibid.* **37**, 134 (1941). Schulman and Cockbain, *ibid.* **36**, 651, 661 (1940).

[12] McBain, Ford, and Wilson, *Kolloid-Z.* **78**, 1 (1937).

and Gans[13] illustrates the range within which the surface tension–concentration curves of most surface active agents fall.

A third type of curve, which is peculiar to certain surface active agents and which has been widely studied because of its anomalous character, is sometimes obtained. It is characterized by a sharp initial decrease of surface tension to a minimum, followed by a sharp short rise and a further gradual flattening. This type III curve is illustrated in Figure 14 (after Fischer and Gans). The type III curve apparently indicates a glaring

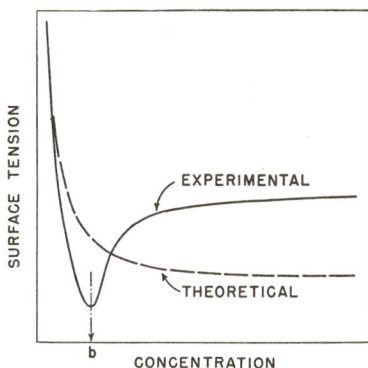

Fig. 14. Generalized curves showing experimental relation, frequently observed, and the theoretical relation between surface tension and concentration of surface active substances in aqueous solutions.[13]

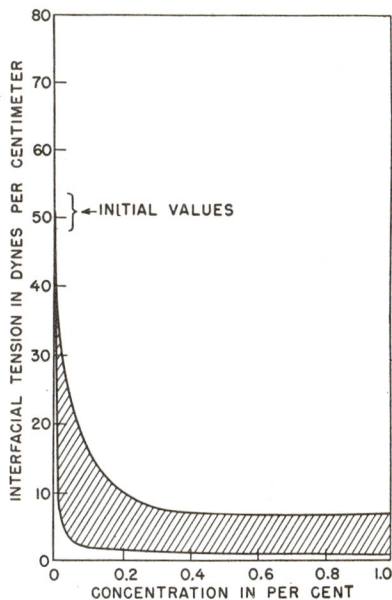

Fig. 15. Interfacial tension values between mineral oil and water for the majority of surface active agents fall within shaded areas.[13]

contradiction of the Gibbs adsorption law. On the rising portion of the curve the Gibbs law states that there is negative surface adsorption and yet the numerical surface tension values in this region are still extremely low. Type III curves are very commonly encountered in the case of surface active agents.

The interfacial tension–concentration curves of surface active agents fall within the range shown in Figure 15. It is apparent from this diagram that the actual values of interfacial tension, expressed in dynes per cm.,

[13] Fischer and Gans, *Ann. N. Y. Acad. Sci.* **46**, Art. 6, 371 (1946).

can become extremely small.   When this is the case, it is not unusual for
the two liquid phases to emulsify spontaneously and form an emulsion system of unusual stability.   Although only curves of type I are illustrated
in the diagram, type III curves are often encountered in liquid–liquid systems just as they are in liquid–air systems.[14]

The existence of the type III curve has led to serious questioning of the
validity of Gibbs' law, particularly by McBain and his co-workers,[15] and to
the suggestions of alternative hypotheses of surface adsorption.   It has
been pointed out by Alexander and other investigators[16] that the minimum
in type III curves corresponds approximately to that concentration at
which micelles begin to form in the bulk solution, i.e., the so-called critical
concentration.   This micelle formation lowers the bulk concentration of
the individual surface active molecules.   The micelles themselves may be
regarded as a second solute component of the system.   To be sure, they
are apparently in equilibrium with the individual molecules, but the rate
at which this equilibrium is attained is not definitely fixed.   There is ample
evidence that the true surface active species is the individual molecule, not
the micelle.   This interpretation offers a basis for at least partially reconcil-
ing Gibbs' law with the apparent anomalies exhibited by type III curves.

The evidence that micelle formation can affect the adsorption equilibrium
in a predictable manner has been strengthened by observations on simpler
solutes which are known to form dimeric molecules.[17]

Several complicating factors other than micelle formation should be
borne in mind in the consideration of type III curves.   In the case of soaps
and other ionizable surface active agents, there is the possibility that the
ionic equilibrium in the surface is different from that in the bulk of the
solution.   Electric charge effects are not taken into account in the classical
derivation of the Gibbs equation.   In the case of soap itself, the picture is
further complicated by the possibility of hydrolysis with the formation of
free fatty acid.   This free fatty acid is much more strongly adsorbed than
the neutral soap or the fatty acid anion.[18]   The effect of carbon dioxide in
the atmosphere above soap solutions on the surface tension has been noted
and studied by several investigators and has been correlated with changes

[14] For a more complete discussion of type III curves see "Anomalies in Surface
Tensions of Solutions" by Hauser in *Advances in Colloid Science*, Vol. I, Interscience
New York, 1942.   See also: Powney and Addison, *Trans. Faraday Soc.* **33**, 1243 (1937);
Lottermeister and Stoll, *Kolloid-Z.* **63**, 49 (1933).

[15] See McBain and Mills, *Reports on Progress in Physics* **5**, 30 (1939).

[16] A. E. Alexander, *Trans. Faraday Soc.* **38**, 54 (1942); *Nature*, **148**, 752 (1941).

[17] Cassel, *Naturwissenschaften* **22**, 60 (1934).   Dunken, Fredenhagen, and Wolf,
Kolloid-Z. **95**, 186 (1941).

[18] Adam, *Trans. Faraday Soc.* **32**, 653-6 (1936).   Neville and Harris, *Am. Dyestuff
Reptr.* **24**, 312 (1935).

in the pH of the solutions.[19] Preferential adsorption of free fatty acid on
the walls of the container may also upset the hydrolysis equilibrium of soap
and affect measurements of surface tension.[20] Nickerson has shown that
various hydrocarbons in contact with aqueous sodium oleate solutions can
sorb free oleic acid. The Gibbs equation accordingly fails because of
changes in the solute and in interfacial orientation.[21] The same effect has
been noted with aqueous sodium laurate solutions and heptane at pH values
below 11.[22]

A very important contribution to the problem of type III curves has been
made recently by Miles and Shedlovsky.[23] These investigators have shown
that the surface tension – concentration curves and interfacial tension –
concentration curves for very carefully purified sodium fatty alkyl sulfates
show no minima, i.e., they are type I curves. The addition of small amounts
of other surface active agents results in the appearance of a minimum in
the curve. For example, pure sodium lauryl sulfate gives a type I curve,
but a mixture of sodium lauryl sulfate with as little as one mole per cent
of sodium heptadecane-2-sulfate gives a type III curve. Sodium lauryl
sulfate to which a small amount of free dodecanol has been added also gives
a type III curve. These findings suggest that careful re-investigation
might show at least some of the recorded type III curves to be the result of
unsuspected impurities in the materials used. The relationship between
surface or interfacial tension and the concentration of a surface active solute
in non-aqueous systems has received relatively little attention. McBain
and Perry found that solutions of laurylsulfonic acid in Nujol give type III
curves. They also studied several other typical surface active agents in
non-aqueous solvents and obtained either type I or type III curves.[24] The
surface activity of soaps in mineral oils has been studied by Gallay and
Puddington,[25] who find it is closely related to the degree of dispersion of
the soap. Pamfilov and Starobinets have reported that the aliphatic alcohols from methyl to octyl are surface active in bromobenzene and that the
surface tension – concentration curves are analogous to those of surface
active agents in aqueous solution.[26]

[19] Lottermoser and Giese, Kolloid-Z. 73, 155, 276 (1935). Powney, Trans. Faraday
Soc. 31, 1510 (1935).

[20] Lottermoser, Fette u. Seifen 45, 595 (1938).

[21] Nickerson, J. Phys. Chem. 40, 277 (1936).

[22] Davis and Bartell, J. Phys. Chem. 47, 40 (1943).

[23] Miles, J. Phys. Chem. 49, 71 (1945). Miles and Shedlovsky, ibid. 48, 57 (1944).
Shedlovsky, Ann. N.Y. Acad. Sci. 46, 427 (1946).

[24] McBain and Perry, J. Am. Chem. Soc. 62, 989 (1940).

[25] Gallay and Puddington, Can. J. Research B21, 225 (1943).

[26] Pamfilov and Starobinets, J. Gen. Chem. U. S. S. R. 11, 501 (1941); Chem. Abstracts 35, 7268.

(b) Effect of Salts

The surface activity of the ionizable surface active agents is affected markedly by the presence of salts or other electrolytes in the solution. The surface and interfacial tensions are lowered by the addition of salts, but the form of the surface tension – concentration curve remains roughly the same. This effect is ascribed to the salt ion of charge opposite to the surface active species. These "gegenions" (ions of opposite charge) lower the critical concentration of micelle formation, which is an indication that they lower the repulsion between the surface active ions. Therefore, more of the surface active ions can be packed into the surface film and the surface tension is lowered correspondingly. The ions fall into a lyotropic series and vary in degree of effectiveness. In the case of a secondary alkyl sulfate Aickin[27] has found K^+ to be most effective $> NH_4^+ > Na^+ > Li^+$. The least hydrated ions are most effective in reducing surface tension. In the case of cationic surface active agents, the added anions are effective in increasing surface activity.[28] The presence of salts also lowers the surface tension of such substances as saponin and lecithin.[29,30]

(c) Variation of Surface Tension with Time

It has been known for many years that the surface tension of a freshly made soap solution falls steadily with time, and may require anywhere from a few seconds to several days before reaching a final equilibrium value. This effect was first demonstrated by Rayleigh,[31] who showed that the dynamic surface tension of a sodium oleate solution, measured by a vibrating jet technique, is only slightly lower than that of water. The static surface tension, measured by the capillary rise method, is about 25 dynes per centimeter. This time lag in attaining an equilibrium value of surface tension is a general property of the ionic surface active agents and has been studied by many investigators. It has also been observed in the case of saponin and certain proteins.[32] So far as the writers are aware, this effect has not yet been reported in the case of the simpler non-ionic detergents, but it is logical to predict its existence. Sodium laurate and sodium cetyl

[27] Aickin, J. Soc. Dyers Colourists **60**, 36 (1944).

[28] Robinson, Nature **139**, 626 (1937). Hill and Hunter, ibid. **158**, 585 (1946). Hauser and Niles, J. Phys. Chem. **45**, 954 (1941).

[29] Berthier and Boutaric, Compt. rend. **206**, 1760 (1938); Bull. soc. chim. **12**, 663 (1945).

[30] For theoretical discussions of the effect of salts on surface active agents, see: Powney and Addison, Trans. Faraday Soc. **33**, 1243 (1937); Cassie and Palmer, ibid. **37**, 156 (1941); Aickin and Palmer, ibid. **40**, 116 (1944).

[31] Rayleigh, Nature **41**, 566 (1890).

[32] Boutaric and Berthier, Bull. soc. chim. **6**, 804 (1939); J. chim. phys. **36**, 1 (1939); ibid. **42**, 117 (1945); Compt. rend. **220**, 730 (1945).

sulfate solutions attain surface equilibrium very slowly (several days) in concentrations below the critical concentration for micelle formation. At higher concentrations, where the bulk solution is composed largely of micelles, equilibrium is attained in a matter of minutes or seconds.[33] Even in the case of very dilute solutions, however, the rate of fall after the first few hours is very small and for all practical purposes the value obtained after such a time interval may be regarded as final.[34]

The time lag in attaining equilibrium at an oil–water interface when surface active agents are present has also been observed, although it was apparently overlooked until fairly recent years. It is quite pronounced and varies considerably with the type of oil used. The Nujol – water – Aerosol OT system requires several days to come to equilibrium, but when nitrobenzene is substituted for Nujol the final value of interfacial tension is reached in a few hours.[35] In the case of both surface tension and interfacial tension, the rate of attaining equilibrium is considerably increased by the presence of salts in the solution.

Since the lowering of surface tension is ascribed to the adsorption of solute molecules or ions in the surface layer, it is logical to suppose that a slow rate of attaining surface tension equilibrium is due to a correspondingly slow rate of adsorption. This has been checked experimentally in the case of hydrocinnamic acid using the microtome method mentioned previously.[36] If the active solute molecules migrated into the surface layer by normal unhindered diffusion processes, it can be calculated that equilibrium would be attained in a fraction of a second. Several explanations have been advanced for the anomalous slow adsorption, and empirical equations describing the effect have been developed. Doss postulates that, in the case of cetylpyridinium bromide, the first ions reaching the surface form an electrical double layer which acts as a potential barrier to the entry of new ions.[37] Other theories involve the penetration and reorientation of molecules into an already formed surface layer.[38]

[33] Lottermoser, Baumgurtel, et.al., Kolloid-Beihefte 41, 89 (1935). Kolloid-Z. 73, 155 (1935). Adam and Shute, Trans. Faraday Soc. 34, 758 (1938). Nutting, Long, and Harkins J. Am. Chem. Soc. 62, 1496 (1940); ibid. 63, 84 (1941).

[34] Dreger et.al., Ind. Eng. Chem. 36, 610 (1944). Miles and Shedlovsky, J. Phys. Chem. 48, 57 (1944).

[35] Alexander and Rideal, Nature 155, 18 (1945); Compare Alexander, Trans. Faraday Soc. 37, 15 (1941). Ward and Tordai, Nature 154, 146 (1944); compare Aubry, Compt. rend. 208, 2062 (1939).

[36] McBain, Ford, and Mills, J. Am. Chem. Soc. 62, 1319 (1940).

[37] Doss, Kolloid-Z. 86, 205 (1939); 87, 272 (1939). See also Doss, Curent Sci. (India) 4, 405 (1935).

[38] Ross, J. Am. Chem. Soc. 67, 990 (1945); J. Phys. Chem. 49, 377 (1945). Dognon and Gougerot, J. chim. phys. 40, 127 (1943). Addison, J. Chem. Soc. 1943, 535; 1944, 252, 477; 1945, 98; Nature 156, 600 (1945).

In conclusion, it should be emphasized that the theory of the surface behavior of surface active solutes is still in an active stage of development and there is still considerable disagreement among leading investigators as to the interpretation of the vast mass of data which has been accumulated. The foregoing brief discussion has not attempted to present all the contradictory points of view or their historical development.   McBain, one of the greatest originators and contributors in this field, has repeatedly pointed out the unsatisfactory state of current theories, particularly with regard to surface tension – concentration and surface tension – time data.   It is strongly recommended that the various review articles on this subject be consulted.   These articles for the most part contain ample references to the original literature.[39]

[39] Compare McBain, "Solutions of Soaps and Detergents as Colloidal Electrolytes," in Alexander, *Colloidal Chemistry*, Vol. V, Reinhold, New York, 1944.

CHAPTER 14

Bulk Properties of Surface Active Solutions

I. General Considerations

So far we have been considering only the surface properties of surface active solutions. It is logical to suppose that the unusual surface behavior of these substances would correspond to unusual properties of the interior or bulk of the solution. This is indeed the case. The surface and bulk properties of surface active solutions are quite closely related and many of the valuable technical properties of surface active agents derive from their bulk properties rather than from purely surface effects. The essential features of surface active solutions in bulk is the existence of colloidal-size particles or micelles formed by the spontaneous association of the ultimate molecules and/or ions of the solute. The micelles exist in thermodynamically stable equilibrium with these simpler ions and molecules. Most authorities agree at present that the pure surface effects are due to the unassociated ions or molecules, i.e., that micelles do not enter into the formation of surface or interfacial adsorbed films.[1] The equilibrium between micelles and molecular-size particles, however, requires that any factor affecting micelle formation should also affect the availability of molecules and ions for surface film formation. Hence, it affects the surface properties indirectly.

It has been recognized that concentrated soap solutions are colloidal in nature for almost as many years as the colloidal state of matter has been recognized. Exact studies of the nature of the soap colloid, however, date from the first publications of McBain and co-workers in the field beginning about 1911. They showed that soaps (i.e., the soluble salts of higher fatty acids) behave substantially like the simple inorganic electrolytes with regard to both conductivity and osmotic effects, provided the measurements are made in very dilute solutions. At somewhat higher concentrations the osmotic effects (namely, osmotic pressure, freezing point lowering, and vapor pressure lowering) begin to deviate markedly from ideal behavior, their values falling far below those calculated from Raoult's law. Such deviations from Raoult's law indicate that colloidal particles are being formed by the association of individual soap molecules or ions. The

---

[1] See, for example, Aickin, *J. Soc. Dyers and Colourists* **60**, 36, 41, 60, 286 (1944).

change in electrical conductivity at these higher concentrations, however, does not parallel the change in osmotic effects. The observed values of conductivity remain quite high and are much higher than would be expected from the osmotic behavior. That the high degree of conductivity is not due to hydrolysis of the soap was definitely proved shortly after the original observations were made. The discrepancy between osmotic and conductivity behavior led McBain to postulate the formation of a colloidal particle or micelle, formed by association of the long-chain fatty acid anions and bearing a high electrical charge. This picture has been amplified and elaborated, but remains unchanged in its essentials. Practically all the anionic and cationic surface active agents show the same general behavior and are generally referred to as colloidal electrolytes. Many dyestuffs and similar high-molecular-weight salts of organic acids or bases also fall into this class. It should also be noted that these substances often show the typical surface properties of the more usual surface active agents. Beside the charged or ionic micelles, there may exist, in solutions of colloidal electrolytes, micelles with relatively low net charge. The non-ionic surface active agents form only neutral micelles in solution. Their colloidal behavior is simpler of description because the electrical effects are absent. We shall not attempt to develop the micelle theory of surface active agents chronologically or from a historical point of view, but merely to present the more important evidence on which current views are based. Various phases of this subject have been ably and thoroughly discussed in numerous review articles within recent years.[2]

A wide variety of techniques has been used to study the various properties of surface active solutions, particularly those of the ionic type   The properties which have been investigated most thoroughly may be grouped into the following categories: (1) Electrical properties including conductivity, transport number, and electromotive force; these are, of course, not applicable to non-ionic substances. (2) Osmotic properties, particularly freezing point lowering. (3) Diffusion properties, including behavior on dialysis or ultrafiltration, and sedimentation. (4) Solubility and solubility–temperature relationships. (5) Viscosity. (6) Optical properties, including spectroscopic effects and particularly X-ray diffraction patterns. (7) Solubilization effects.

A satisfactory comprehensive theory should afford an explanation of all the data acquired by these various types of measurements and should

[2] McBain, "Soaps as Colloidal Electrolytes," in Alexander, *Colloid Chemistry*, Vol. V, Reinhold, New York, 1944. Ralston, *Ann. N. Y. Acad. Sci.* **46**, 351 (1946). Dervichian and Lachampt, *Bull. soc. chim.* **12**, 189 (1945). Freundlich, *Trans. Faraday Soc.* **31**, 4 (1935).

account for their interrelationships. Since the interrelationships between one type of data and another are particularly important, many of the more recent publications in this field report on two or more different effects. For example, conductivity and osmotic measurements are often presented and compared in a single article. It is convenient, however, to consider the various effects and their interpretation individually before proceeding to compare them.

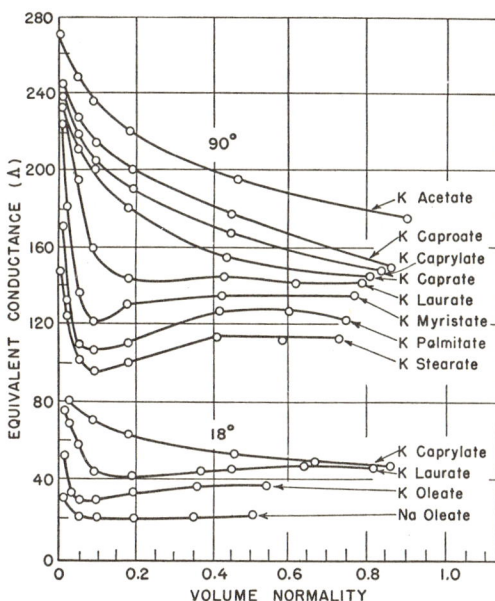

Fig. 16. Equivalent conductances of soap solutions at 18° and at 90°.[3]

II. Conductivity Effects

When the equivalent conductivity (usually designated by a capital lambda, Λ) of a colloidal electrolyte is plotted against the concentration, the shape of the curve differs radically from that of an ordinary salt. Furthermore, the curves for the various types of colloidal electrolytes are all qualitatively similar. Typical curves for the soaps[3] are shown in Figure 16.

For C_{12} and higher soaps, the curves slope sharply downward in dilute solution, pass through a minimum, and then rise again. The minimum becomes more pronounced as the molecular weight increases. The soaps in dilute solution apparently behave as half-strong electrolytes and the

[3] McBain, Laing, and Titley. *J. Chem. Soc.* **115**, 1279 (1919).

slope of the conductivity–concentration curve is rather steep even at great dilution.[4] There is, however, evidence that in the range below 0.006 N some hydrolysis occurs with the consequent formation of acid soap, resulting in anomalies.[5] A break in the slope of the curve at this concentration range apparently coincides with the point at which ionic micelles begin to form rapidly, i.e., the critical concentration.

Less ambiguous results have been obtained by studying those colloidal electrolytes which are not susceptible to hydrolysis. These include the

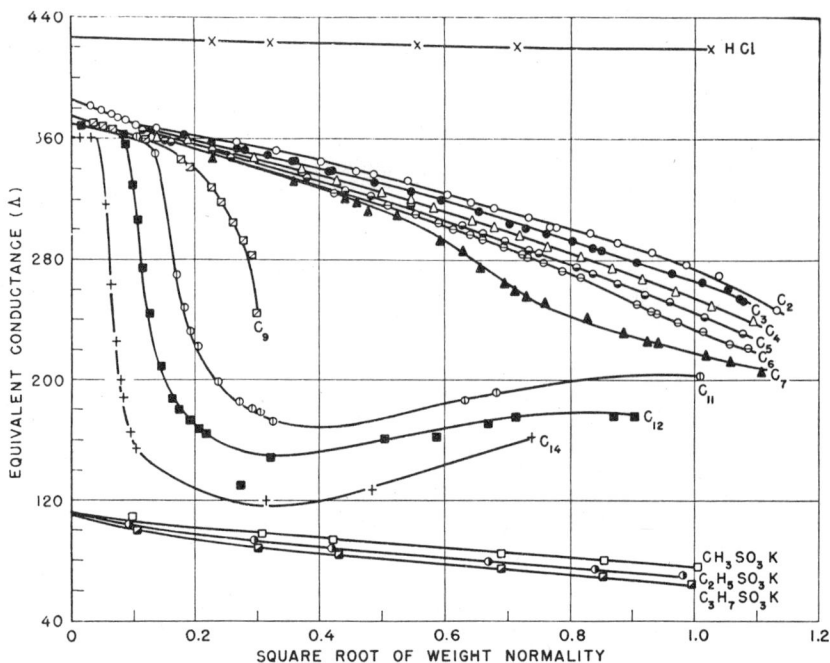

Fig. 17. The equivalent conductance of hydrochloric acid, various sulfonic acids, and some of their salts at 25°C.

long-chain alkyl sulfuric esters, alkane sulfonic acids, and the cation active long-chain ammonium salts.

For conductivity studies of alkyl sulfuric esters, see: Lottermoser and Püschel, Kolloid-Z. **63**, 175 (1933); Ward, J. Chem. Soc. **1939**, 522; Howell and Robinson, Proc. Roy. Soc. London **A155**, 386 (1936). For data on alkane sulfonic acids, see: McBain

---

[4] McBain, J. Phys. Chem. **43**, 671 (1939).

[5] Ekwall et al., Z. physik. Chem. **A161**, 195 (1932); Kolloid-Z. **94**, 42 (1941); **101**, 135 (1942).

and Betz, *J. Am. Chem. Soc.* **57**, 1905 (1935); M.E.L. McBain, Dye, and Johnston, *ibid.* **61**, 3210 (1939). Wright, Abbott, Sivertz, and Tartar, *ibid.* **61**, 550 (1929). For data on fatty ammonium salts, see: Hoerr and Ralston, *J. Am. Chem. Soc.* **65**, 976 (1943); **64**, 772 (1942); Ralston, Hoerr, and Hoffman, *ibid.* **64**, 97 (1942); Scott and Tartar, *ibid.* **65**, 692 (1943). These papers represent recent work using improved techniques. They cite many references to earlier work.

The straight-chain alkane sulfonates have been studied extensively and carefully and afford an excellent illustration on which to base theoretical discussion. Figure 17, from McBain, Dye, and Johnston,[6] shows the

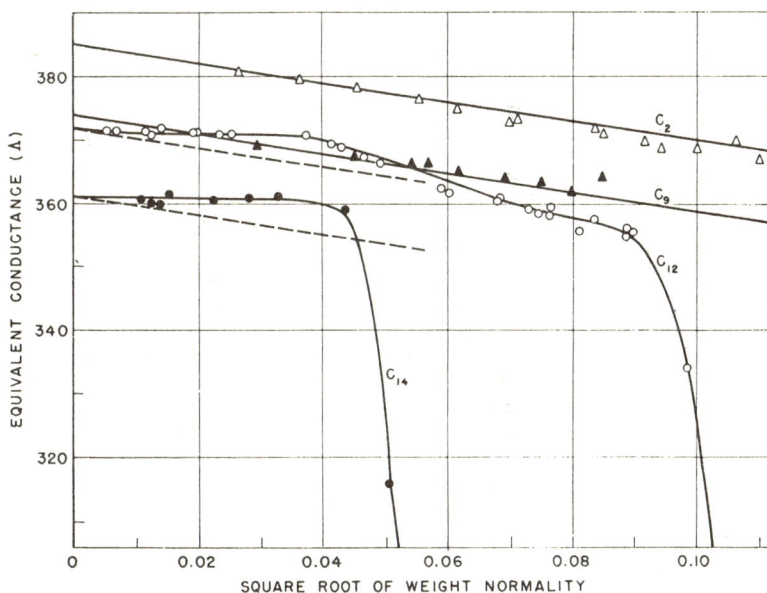

Fig. 18. Equivalent conductance values of lower alkane sulfonic acids at low concentrations.

equivalent conductance of the series of alkane sulfonic acids and some of their salts. Beginning with the C_{11} acid, the curve starts to follow the normal course, but breaks sharply downward at a relatively low concentration. The C_9 acid shows a much more gradual break and the lower acids show no break at all. This indicates that at 25°C. the lower acids do not form micelles even in concentrations as high as 1.0 N. Figure 18 shows the equivalent conductivity values (reported by McBain, Dye, and Johnston[6]) in the lower ranges only and demonstrates the sharpness of the break at

[6] M. E. L. McBain, Dye, and Johnston, *J. Am. Chem. Soc.*, **61**, 3210 (1939).

the critical concentration. The dotted lines represent the theoretical slope of a completely dissociated solute forming two ions as calculated by the Onsager equation.

Figure 19 shows similar plots at various temperatures for the sodium C_{14}, C_{12}, and C_{10} sulfonates according to Tartar et al.[6a] Figure 20 (also from Tartar et al.[6a]) shows specific conductance (in contrast to equivalent

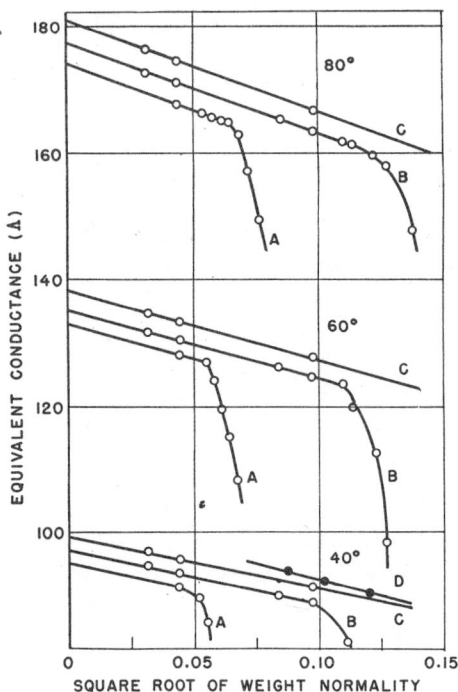

Fig. 19. Equivalent conductance against square root of weight normality for dilute solutions below critical concentration. Curves A, B, and C are for sodium tetradecyl, dodecyl, and decyl sulfonates, respectively. Curve D is for sodium chloride at 40° (curve is for $\Lambda - 64$ to bring to comparable range).

conductance) plotted against concentration for the same salts. This method of plotting shows the breaks in the curves more clearly.

The cationic colloidal electrolytes show the break at the critical point even more sharply. Figure 21 shows the equivalent conductivities of a series of fatty amine hydrochlorides, according to Ralston and Hoerr.[7]

[6a] Wright, Abbott, Sivertz, and Tartar, J. Am. Chem. Soc., **61**, 550 (1939).

[7] Ralston and Hoerr, J. Am. Chem. Soc. **64**, 773 (1942).

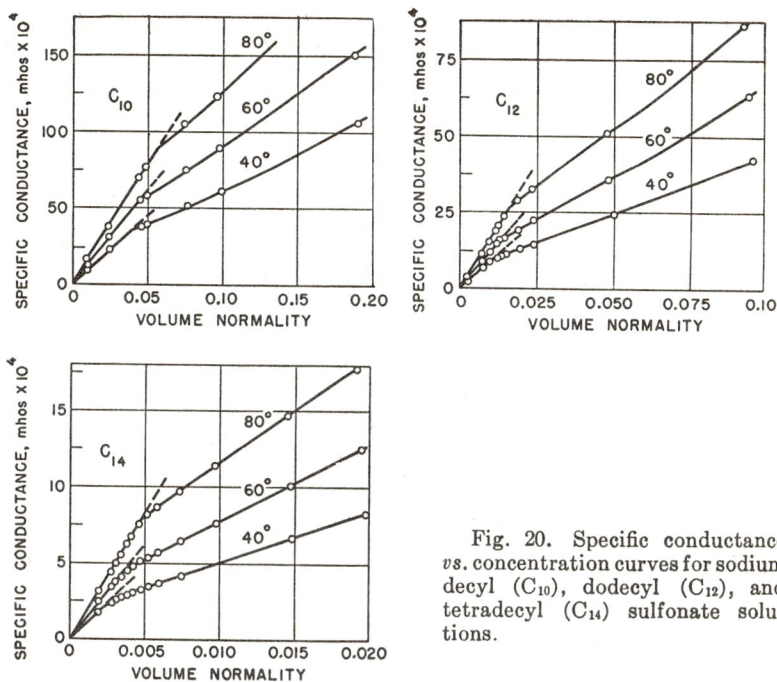

Fig. 20. Specific conductance *vs.* concentration curves for sodium decyl (C_{10}), dodecyl (C_{12}), and tetradecyl (C_{14}) sulfonate solutions.

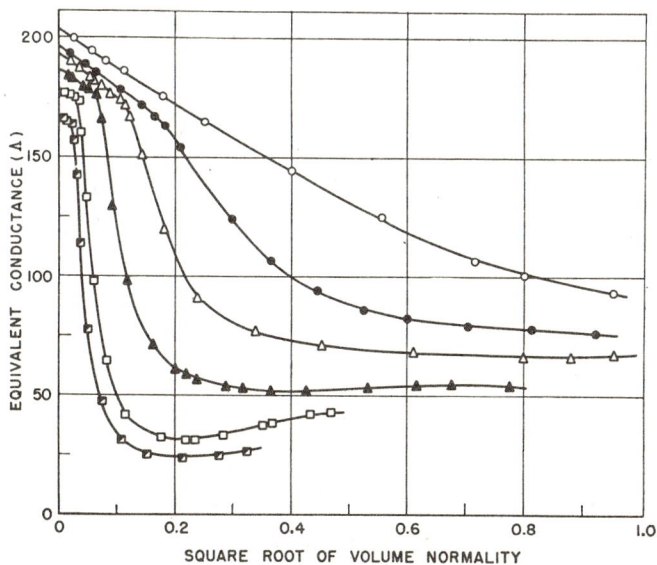

Fig. 21. Equivalent conductances of amine salt solutions at 60°: (○) $C_8H_{17}NH_2HCl$; (●) $C_{10}H_{21}NH_2HCl$; (△) $C_{12}H_{25}NH_2HCl$; (▲) $C_{14}H_{29}NH_2HCl$; (□) $C_{16}H_{33}NH_2HCl$; (▨) $C_{18}H_{37}NH_2HCl$.[7]

Figure 22 shows similar data for the C_{10}, C_{12}, and C_{16} trimethyl ammonium bromides.[8]

The critical ranges occur at lower concentrations and are sharper as the chain length increases. It is also evident that with the higher members of the series the first sharp dip in the curve is followed by a gradual leveling off through a minimum followed by a slow rise in equivalent conductivity. This minimum corresponds to a second major change in the bulk properties of the solution. One point of view, held by McBain and

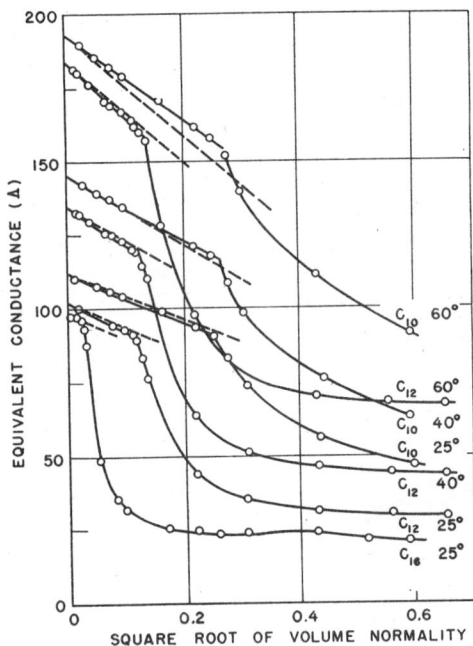

Fig. 22. Equivalent conductances of decyl-, dodecyl-, and hexadecyltrimethylammonium bromide solutions. Subtract 10 from ordinate for decyltrimethylammonium bromide.[8]

his school, is that large amounts of a new type of micelle begin to form in this region. This micelle is larger than the ionic micelle formed at the critical concentration and it contributes very little to the conductivity because it has only a small net charge. This is called the neutral colloid. Hartley's theory, based on the concept of a single micelle type, accounts for the minimum in the curve by postulating a change in the number of gegenions associated with the micelle. For example, in the case of sodium

[8] Scott and Tartar, *J. Am. Chem. Soc.* **65**, 696 (1943).

dodecane sulfonate, the ionic micelle is composed of the $C_{12}H_{25}SO_3^-$ ions. Associated with this micelle is a number of Na^+ ions (the gegenions). The number of Na^+ ions associated with a single micelle varies with concentration, temperature, etc.

The dependence of critical concentration on temperature is pointed out in Figure 23, taken from Ralston, Hoerr, and Hoffman.[9] In this series it is seen that as the temperature is increased, the slope in the pre-critical range becomes steeper and the critical point itself occurs at almost the same

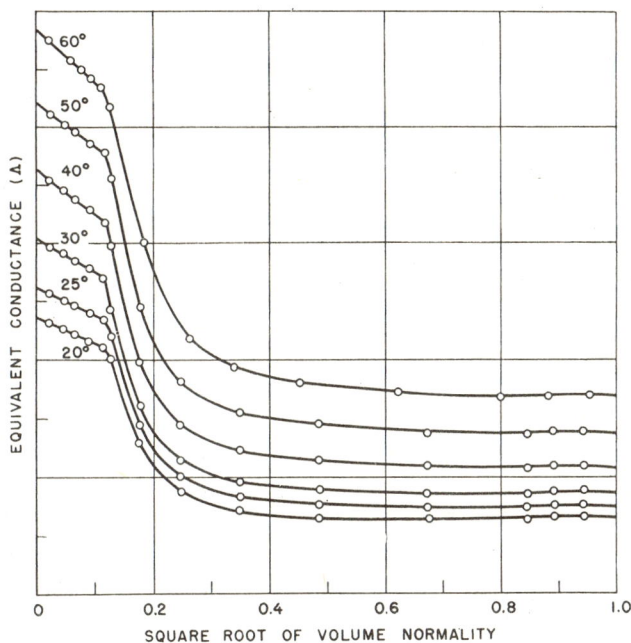

Fig. 23. Equivalent conductances of dodecylamine hydrochloride at various temperatures.[9]

concentration. It should be noted, however, that the data of Tartar *et al.* on alkane sodium sulfonates (*cf.* Fig. 20) show that in those particular compounds the critical concentration increases with increasing temperature. It is evident that temperature is not as important a factor as chain length in determining critical concentration. Tartar and co-workers[10] have also determined the conductivities of a series of straight-chain alkyl benzene sulfonates. By comparing these with the corresponding alkane

[9] Ralston, Hoerr, and Hoffman, *J. Am. Chem. Soc.* **64**, 97 (1942).
[10] Paquette, Lingafelter, and Tartar, *J. Am. Chem. Soc.* **65**, 686 (1943).

sulfonates, they find that the benzene ring is equivalent to about three and one-half straight-chain carbon atoms in its effect on the critical concentration.

We have seen that for higher members of all series of colloidal electrolytes the equivalent conductivity falls sharply past the critical point and reaches a minimum. Gonick[11] has recently shown that the shape of the curve in this minimum region can be represented by the empirical equation:

$$\Lambda = (A/\sqrt{c}) + B\sqrt{c}$$

where c is the concentration and A and B are constants. The interpretation of this relationship has not as yet been clarified.

III. Transport Numbers

When the transport number of the surface active ion of a colloidal electrolyte is plotted against the concentration, anomalous curves are obtained. These curves show a break at low concentrations indicative of a critical concentration for micelle formation. Below this concentration the transport numbers behave normally. After the break, they rise sharply to an abnormally high value, then level off. With higher members of the series, a maximum may be reached. The curves thus show the three ranges characteristic of the conductivity concentration curves.[12]

The cationic transport numbers for a series of amine hydrochlorides plotted against concentration are shown in Figure 24, taken from Hoerr and Ralston.[13] The critical concentrations and the concentrations at which neutral colloid formation begins to take place correspond to those in the conductivity curves. The actual heights to which the curves rise show an irregular order, which corresponds to the order of the salt densities. These data are interpreted in the same way as the conductivity data.[14]

IV. E.M.F. Measurements

The activity of the anion in n-dodecanesulfonic acid has been measured by Walton[15] both with and without salt additions. The curves obtained by plotting the activity coefficient against concentration are typical of

[11] Gonick, J. Am. Chem. Soc. 67, 1191 (1945).

[12] For transference data on soaps, see: McBain et al., J. Chem. Soc. 123, 2417 (1923); J. Phys. Chem. 40, 997 (1936); Laing, ibid. 28, 680 (1924). For data on lauryl sulfonic acid, see M.E.L. McBain, J. Phys. Chem. 47, 196 (1943).

[13] Hoerr and Ralston, J. Am. Chem. Soc. 65, 976 (1943).

[14] See Van Rysselberghe, J. Phys. Chem. 48, 62 (1944). McBain et al., ibid. 43, 1049 (1939); 47, 196 (1943).

[15] Walton, J. Am. Chem. Soc. 68, 1180 (1946).

colloidal electrolytes, showing a break at the critical concentration. Decrease in temperature and increase in ionic strength promote micelle formation as measured by this method.

V. Osmotic Properties

The effect of colloidal electrolytes in lowering the freezing point of their solutions has been studied as extensively as their conductivity behavior, and the results show the same general anomalies. The results of such

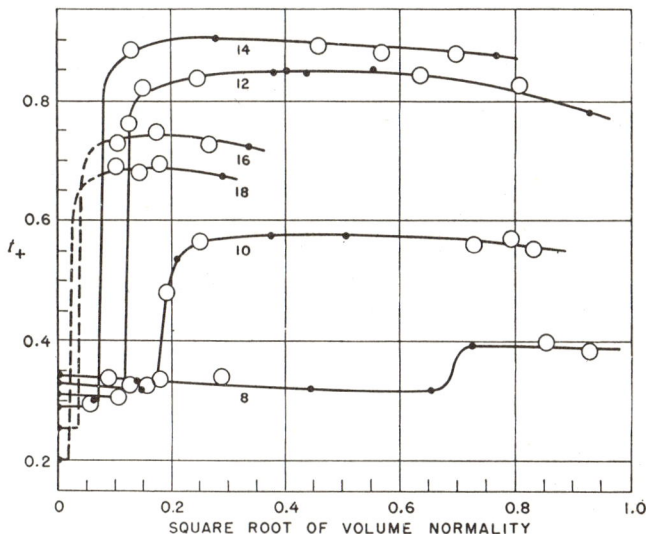

Fig. 24. Cationic transference numbers of amine hydrochlorides.[13]

studies are usually recorded graphically by plotting the osmotic coefficient g against concentration. The osmotic coefficient is equal to:

$$g = \theta/(2 \times 1.858\,M)$$

where θ is the freezing point depression and M is the molality. It is equal to unity for an ideal fully dissociated uni-univalent electrolyte. The technique for measuring freezing point depressions has recently been refined to such an extent that great accuracy is possible even in extreme dilutions.[16]

McBain and co-workers have studied a large number of colloidal electrolytes, typical of which are the Aerosols and Tergitols,[17] shown in Figures

[16] Scatchard, Jones, and Prentiss, *J. Am. Chem. Soc.* **54**, 2676 (1932). McBain and Johnston, *Proc. Roy. Soc. London* **A181**, 119 (1942).
[17] McBain and Bolduan, *J. Phys. Chem.* **47**, 94 (1943).

25 and 26.   In general, the curves of g vs. concentration start with a fairly straight portion similar to the conductivity curves.   They break sharply at the critical point and fall rather steeply, finally leveling off.   The higher members of the series sometimes pass through a shallow gradual minimum. This is not nearly so pronounced as the minima which occur in the conductivity curves.

Among other compounds which have been studied are the soaps and sulfonic acids,[18] higher fatty amine salts,[19] quaternary ammonium halides,[20] and alkylol amine salts of fatty acids.[21]

One of the most interesting developments in this field is the discovery by Brady[22] that all existing data on the freezing points of colloidal electro-

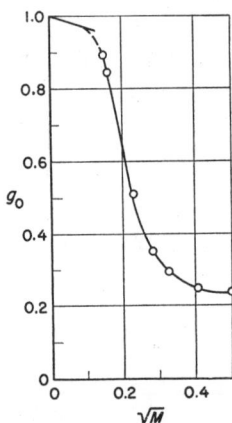

Fig. 25.   The osmotic coefficient of the wetting agent and detergent Tergitol 4 (sodium tetradecylsulfate).[17]

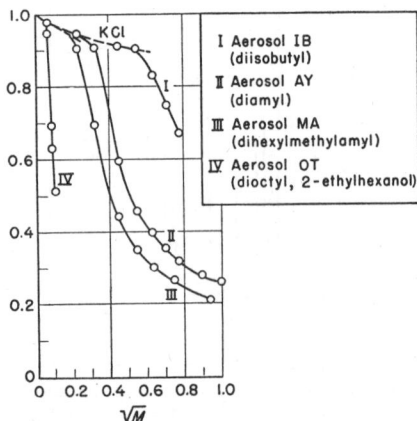

Fig. 26.   Osmotic coefficients of a series of the wetting agents and detergents, all esters of sodium sulfosuccinate.[17]

lyte solutions may be placed on one of three curves on the same diagram, provided that the concentrations are multiplied by a suitable factor to bring them to comparable values.   This is done by dividing the molality M by the molality $M_{0.5}$ at which the value of g is 0.5.   Thereafter g is plotted against log $(M/M_{0.5})$.   Depending on whether the substance in question is branched chain (including the oleic radical), straight chain, or cyclic, the curves will coincide with those in Figure 27.   The dotted line

[18] McBain and Barker, *Trans. Faraday Soc.* **31**, 149 (1935).

[19] Ralston, *et al.*, *J. Am. Chem. Soc.* **64**, 2824 (1942); **63**, 2576 (1941).

[20] Walton, Hiebert, and Scholtes, *J. Colloid Sci.* **1**, 385 (1946).

[21] Gonick and McBain, *J. Colloid Sci.* **1**, 127 (1946); Gonick, *J. Am. Chem. Soc.* **68**, 177 (1946).

[22] McBain and Brady, *J. Am. Chem. Soc.* **65**, 2072 (1943).

marked limit represents the curve for $d\theta/dM = 0$, where true solution can no longer exist and the system splits into two phases. The very low or high concentration ranges are not included because of interfering Debye-Hückel effects and hydration effects.

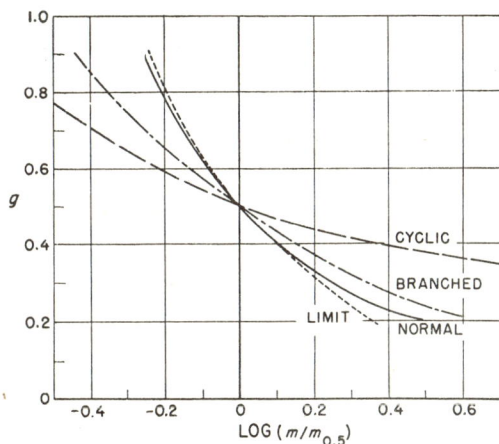

Fig. 27. Osmotic coefficients of all colloidal electrolytes brought to coincidence on one or another of three curves depending on whether they are straight chain, branched chain, or polycyclic, by dividing the concentrations for each by $m_{0.5}$, the molality at which $g = 0.5$. The dotted line represents the limiting curve at which the solution would fall into separate layers.[22]

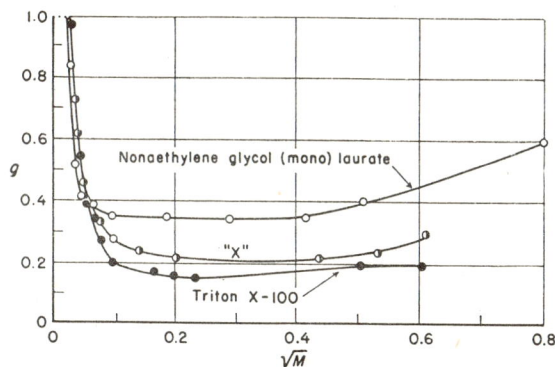

Fig. 28. Osmotic coefficients of non-ionic detergents.[23]

Osmotic effects of the non-ionic surface active agents have also been measured.[23] The curves closely resemble those of the colloidal electrolytes. Figure 28 shows the curves for three typical non-ionic detergents. It is interesting, and in accord with expectations, that the addition of salts has no effect on the freezing point lowering due to a non-ionic detergent.

Osmotic data for solutes of the surface active type in non-aqueous sol-

[23] Gonick and McBain, *J. Am. Chem. Soc.* **69**, 334 (1947).

vents have recently been studied.[24]   They show that association to micelles can take place in these systems as well as in aqueous systems.

VI. Diffusion and Dialysis

Measurements of the variation of diffusion coefficient with concentration have been aimed at determining the actual sizes and shapes of the micelles. This has also been investigated by ultrafiltration through graded porous membranes.   The reader is referred to the original articles for details.[25]

VII. Solubility–Temperature Coefficients

Murray and Hartley[26] have shown that the solubility of paraffin chain salts such as sodium cetane sulfonate varies with temperature in an unusual manner.   In low temperature ranges, the solubility is quite small but it increases slowly and regularly as the temperature increases.   Then within a rather narrow critical temperature range, the solubility begins to increase enormously.   Adam and Pankhurst[27] have recently extended these studies to include a number of cationic and anionic colloidal electrolytes.   The data are shown in graphic form in Figures 29, 30, and 31.   The rapid increase in solubility takes place in the temperature and concentration range where micelle formation begins.   In other words, the sharp upturn in the solubility–temperature curve takes place when the temperature–concentration relationship is at the critical range for micelle formation.   This indicates that the single ion is relatively insoluble whereas the micelle is highly "soluble."   It should be noted that since the micelle is a colloidal-size particle, it is not rigorous to compare its "solubility" with that of the ion or single molecule.   From the operational point of view, however, disappearance of the gross solid phase indicates the formation of a solution, regardless of the state of aggregation of the solute.

This critical solubility temperature increases as the chain length increases in each homologous series.   It varies in an irregular manner with the nature of the non-surface active ion.   Ralston and Hoerr[28] have studied the solubility–temperature relationships of dodecylammonium chloride in a mixed ethanol water solvent.   They note the sharp break in the solubility–temperature curve when the solvent consists preponderantly of water.   As the alcohol content of the solvent is increased, the break gradually dis-

[24] Gonick, J. Colloid Sci. 1, 393 (1946).

[25] Hartley and Runnicles, Proc. Roy. Soc. London A168, 420 (1938).   Lamm, Kolloid-Z. 91, 275 (1940); 91, 10 (1940); Chem. Abstracts 39, 3716.   E. L. McBain, Proc. Roy. Soc. London A170, 415 (1939); J. Phys. Chem. 48, 237 (1944).   McBain and Jenkins, J. Chem. Soc. 121, 2325 (1922).

[26] Murray and Hartley, Trans. Faraday Soc. 31, 183 (1935).

[27] Adam and Pankhurst, Trans. Faraday Soc. 42, 523 (1946).

[28] Ralston and Hoerr, J. Am. Chem. Soc. 68, 851 (1946).

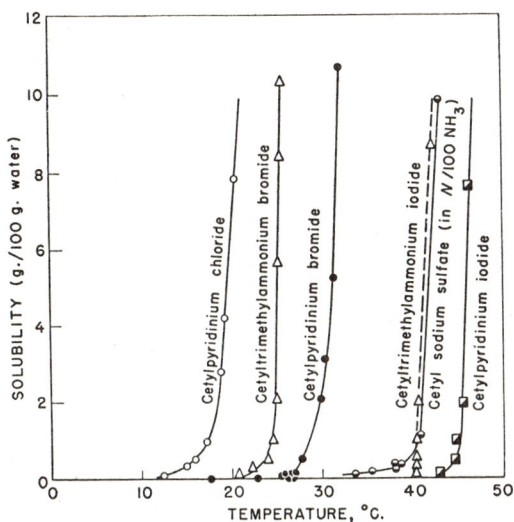

Fig. 29. Solubilities of detergents containing a cetyl group.[27]

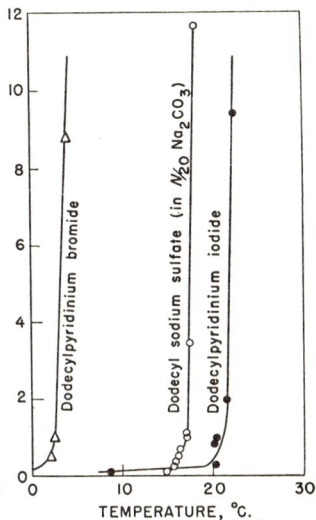

Fig. 30. Solubilities of detergents containing a dodecyl group.[27]

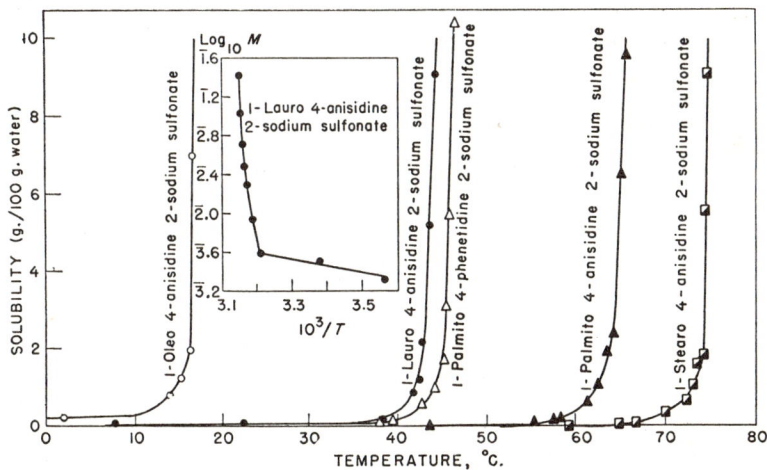

Fig. 31. Solubilities of miscellaneous detergents.[27]

appears.   This indicates that micelle formation takes place over a much wider range of both temperature and concentration in alcohol–water mixtures than in water.   The degree of association at any given concentration or temperature is also less in alcohol than in water.

Recently Tartar and Wright have published extensive solubility data on the series of straight-chain alkane sodium sulfonates from C_{10} to C_{18}.   Their results are shown in Figures 32 and 33.[29]

The curves show that for the higher members of the series the critical concentration range becomes narrower and the break is sharper.   Table II shows the critical concentrations and temperatures corresponding to the data in the curves.   The numerical values check well with values for critical concentration based on conductivity measurements.

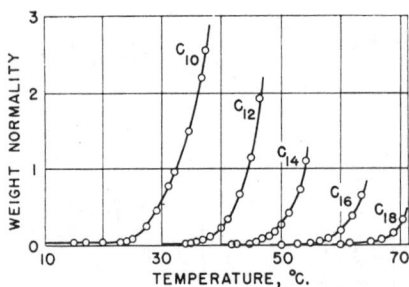

Fig. 32. Solubilities of the sodium salts of the higher alkyl sulfonates: C_{10} refers to sodium decylsulfonate, C_{12} to sodium dodecylsulfonate, etc.[29]

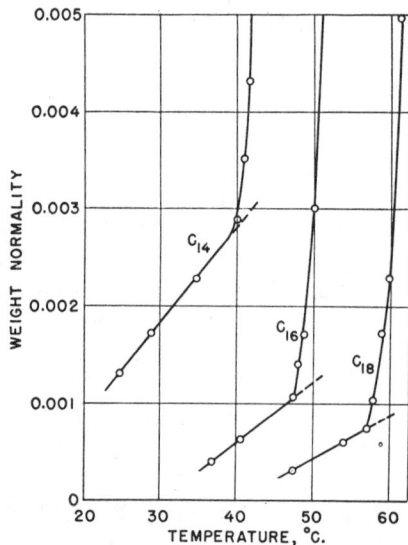

Fig. 33. Solubilities of the sodium salts of the higher alkyl sulfonates showing the breaks in the curves for the higher members.[29]

Viscosity–concentration curves for sodium dodecane sulfonate show a slight but definite break,[30] and this behavior is typical of other surface active solutes.[31]

According to Hess[32] the viscosity–concentration curves show a break at about the same concentration that X-ray diffraction patterns begin to appear.   This is considerably higher than the critical concentration.

[29] Tartar and Wright, J. Am. Chem. Soc. 61, 539 (1939).
[30] Wright and Tartar, J. Am. Chem. Soc. 61, 544 (1939).
[31] Philippoff, Kolloid-Z. 96, 255 (1941).
[32] Hess, Fette u. Seifen 49, 81 (1942).

VIII. X-Ray Effects

Mention has been made of the controversy as to whether more than one type of micelle exists. The X-ray diffraction patterns of soap solutions and other surface active solutions have been studied and constitute a strong argument for the existence of two types. At low concentrations no sharp X-ray patterns are detected but within a fairly sharp range, which may be called the "critical range for X-ray patterns" or $C_{k \text{ x-ray}}$, the patterns appear.[33] These patterns[34] show the micelle to be composed of lamellae made up of double layers of molecules arranged tail-to-tail as shown in Figure 34. Water molecules occupy the spaces between the layers of hydrophilic "heads" or polar ends. Figure 34 also shows how the lamellar micelles effect solution of solutes such as ethylbenzene which are normally insoluble in water. Table III shows the wide range between C_k according

TABLE II
Critical Concentrations for Solubilities[a]

Sodium salt of sulfonate	Critical concentration, N_w	Temperature, °C.
C_{10}	0.040	22.5
C_{12}	0.0098	31.5
C_{14}	0.0027	39.5
C_{16}	0.00105	47.5
C_{18}	0.00075	57.0

[a] Tartar and Wright, *J. Am. Chem. Soc.* **61**, 542 (1939).

to osmotic and electrolytic data and $C_{k \text{ x-ray}}$. It should be emphasized that these lamellar micelles differ from the smaller ionic micelles which form at lower concentrations. The former correspond to McBain's "neutral colloid." They are relatively poor conductors and are responsible for the solubilization phenomenon which will be discussed in the next section.[35]

IX. Miscellaneous Optical Effects

Sheppard and Geddes[36] found that the dye pinacyanol chloride, which is normally associated in aqueous solution to a dimeric form, changed its adsorption spectrum to that characteristic of the monomeric form when

[33] Hess, Philippoff, *et al.*, *Ber.* **B70**, 1800 (1937); *Naturwissenschaften* **26**, 184 (1938); **27**, 593 (1939); *Kolloid-Z.* **88**, 40 (1939); *Fette u. Seifen* **49**, 81 (1942). Stauff, *Kolloid-Z.* **89**, 224 (1939); **96**, 244 (1941).

[34] Harkins, Mattoon, and Corrin, *J. Colloid Sci.* **1**, 106 (1946); *J. Am. Chem. Soc.* **68**, 220 (1946).

[35] Ross and McBain, *J. Am. Chem. Soc.* **68**, 296 (1946).

[36] Sheppard and Geddes, *J. Chem. Phys.* **13**, 63 (1945).

cetyl pyridinium bromide was added to the solution.   This work has been extended by Harkins and co-workers,[37] who have studied this effect with several different surface active agents.   They find that the absorption spectrum changes in the critical range, and it serves as a good measure of

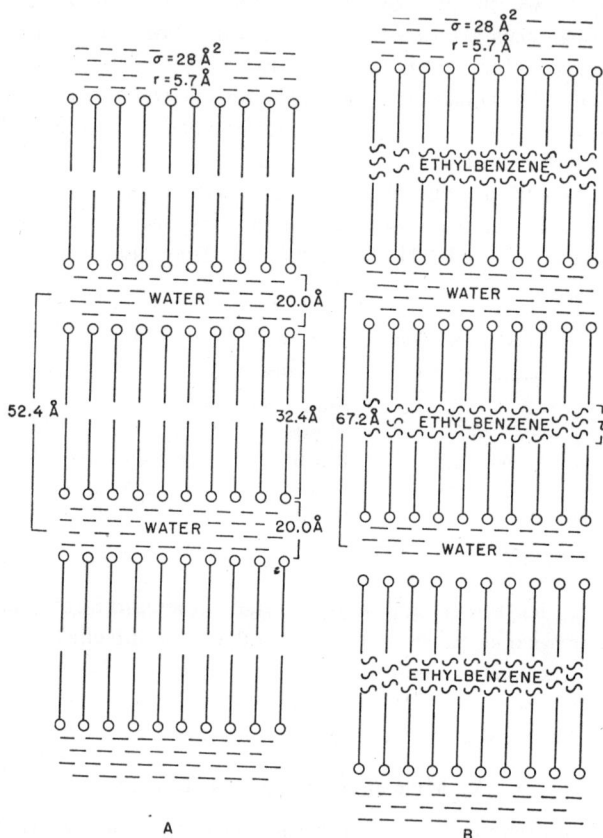

Fig. 34.   Idealized diagrammatic cross section of soap micelles in 15% aqueous potassium laurate.   (A) Without additive oil; (B) saturated with ethylbenzene. The actual structures are undoubtedly far less ordered.   The only spacings actually measured are the long and short spacings.[34]

the sharpness of the critical range.   From the data shown in Table IV, it is apparent that the spectrum changes occur in the "classical" critical range (*i.e.*, that corresponding to osmotic and conductivity data) rather than the "X-ray" critical range.

Another recently developed method for studying critical concentrations

[37] Corrin, Klevens, and Harkins, *J. Chem. Phys.* **14**, 216, 480 (1946).

is due to Klevens.[38] He has found that the refractive indices of surface active solutions change abruptly at the critical point. The measurements are made with an interferometer and the actual values of refractive index are read to seven significant figures. The critical values found by this

TABLE III

Differences in Critical Concentrations Based on Osmotic and Electrolytic Data and on X-Ray Investigations[a]

Soap	Temperature, °C.	C_k (% by wt.)	C_k, X-ray (% by wt.)
Na valerate	20	—	20–27
Na caproate	20	10	20–22
Na laurate	20	0.8	—
Na laurate	75	0.65	4.5–5.6
Na dodecyl sulfate	20	0.25	$\geq$4–5
Na tetradecyl sulfate	75	0.1	3.3
Na palmitate	70	0.084	7
Na oleate	20	0.003	$\geq$4–5

[a] Hess, *Fette u. Seifen* **49**, 81 (1942); *Fat Abstracts*, D: 122, p. 2, 1944.

TABLE IV

Critical Concentrations for the Formation of Soap Micelles as Determined by Pinacyanol in a Spectrophotometer[a]

Soap	Dye concentration, M	Temp., °C.	Critical concentration, M
Potassium dilinoleate	1×10^{-5}	25.8	2.5×10^{-4}
Potassium dehydroabietate	1×10^{-5}	25.8	2.5–3.2×10^{-2}
Potassium oleate	2×10^{-5}	25.8	7–12×10^{-4}
Potassium laurate	1×10^{-5}	25.8	2.15–2.20×10^{-2}
Sodium cetyl sulfate	1×10^{-5}	35.8	3.6–5.2×10^{-4}
Potassium myristate	1×10^{-4}	25.8	6.0–6.7×10^{-3}
Potassium laurate	1×10^{-4}	25.8	2.3–2.4×10^{-2}

[a] Corrin, Klevens, and Harkins, *J. Chem. Phys.* **14**, 480 (1946); *Fat Abstracts*, D: 122.2, p. 4, 1946.

method are again the low values corresponding approximately to those obtained by the conductivity method.

X. Solubilization Effects

The fact that concentrated soap solutions can actually dissolve (rather than emulsify) substantial amounts of water-insoluble organic liquids,

[38] Klevens, *J. Chem. Phys.* **14**, 567 (1946).

forming optically clear solutions, has been known for a long while.[39]   A tremendous impetus to the study of this phenomena has recently been furnished by the intensive development of emulsion polymerization.   It has been found that solubilization effects play a fundamental role in the production of synthetic rubber latexes by this process.[40]

The solubilizing effect of surface active agents is often referred to in technical literature and patents as "hydrotropy" or "hydrotropic action." It is, however, much more pronounced than the well-known hydrotropic effect of simple non-colloidal salts.   Very strong solutions of sodium xylene sulfonate (one of the best hydrotropic salts) will not solubilize as large a molecular proportion of oil as weak solutions of soap.   The salts solubilize by changing the intermolecular forces of the solvent, i.e., a 30 or 40% solution of sodium xylene sulfonate in water is essentially a different solvent than water itself, even though the sulfonate salt is molecularly dispersed.[41, 42]   Surface active agents, on the other hand, apparently solubilize by enclosing the oil molecules within their micelles and the effect becomes apparent at concentrations of the order of 0.1%.

The fact that solubilization is a true equilibrium effect, and that the solubilities of oils and oil-soluble dyes can be approached from either side of the equilibrium value was first pointed out as recently as 1936.[43]

The extent of solubilization for any given oil[44] varies with the concentration of the surface active agent in the manner shown in Figure 35, for Yellow AB and lauryl sulfonic acid.   The solubilization is expressed in terms of moles of Yellow AB per mole surface active agent.   It is seen that measurable solubility occurs at concentrations even below the classical critical concentration.   The curve rises fairly steeply and levels off.   At moderate concentrations, therefore, the weight of dye in solution is almost directly proportional to the weight of surface active agent in solution.

[39] McBain, in *Advances in Colloid Science*, Vol. I, Interscience, New York, 1942, cites a paper published in 1892 by Engler and Dieckhoff dealing with this phenomenon.

[40] Harkins and Stearns, *J. Chem. Phys.* **14**, 215 (1946).   For a thorough discussion and review of solubilization effects, see McBain, "Solubilization in Detergent Action," in *Advances in Colloid Science*, Interscience, New York, 1942.

[41] McBain, Merrill, and Vinograd, *J. Am. Chem. Soc.* **63**, 670 (1941).

[42] For a discussion of hydrotropy, see McKee, *Ind. Eng. Chem.* **38**, 382 (1946).

[43] McBain and McBain, *J. Am. Chem. Soc.*, **58**, 2610 (1936).   McBain and Woo, *J. Am. Chem. Soc.* **60**, 223 (1938).   McBain and O'Conner, *J. Am. Chem. Soc.* **62**, 2855 (1940).

[44] We shall use the general term "oil" for a water-insoluble, solvent-soluble, organic substance.   Actually much of the work has been done with oil-soluble dyes and other solids as well as with liquid organic compounds.

This leveling off of the curve is apparently a general phenomenon for all surface active agents. In many instances, however, it still continues to slope upward although not as sharply as in the low concentration ranges.[45] The curves vary somewhat in shape as well as in numerical values depending on the detergent used. Figure 36 shows the solubilization of dimethylaminoazobenzene in sodium rosinate and sodium laurate.[46] It should be noted that in this diagram actual weights of dye in solution are plotted as ordinates against concentrations of detergent as abscissae. The sodium laurate curve shows a sharp critical point but the rosinate curve does not.

The large variation in solubilizing power among different detergents is shown[47] in Table V. In this series the solubility is measured at only one

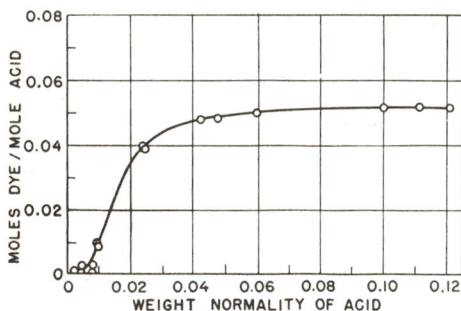

Fig. 35. Solubility at 25°C. of recrystallized Yellow AB in laurylsulfonic acid solutions.[41]

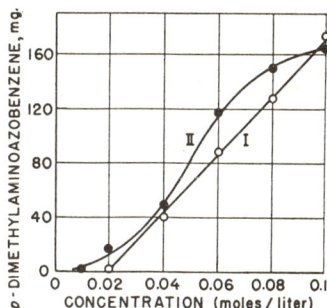

Fig. 36. Solubilization at 50°C. of p-dimethylaminoazobenzene in solutions of sodium laurate (I) and sodium rosinate (II).[46]

concentration of detergent. The results are expressed as milligrams of dye in a solution of one gram of commercial detergent per one hundred milliliters of water. These commercial detergents vary in their content of active material, which makes direct comparisons difficult. Nevertheless, it is apparent, on selecting those examples where the active content is known, that large differences in solubilizing power exist. The variation in solubilizing power among the homologous series of potassium soaps as well as the effect of added salts on solubilizing power are pointed out by the

[45] For several illustrations of this effect, see Green and McBain, *J. Phys. Chem.* **51**, 286 (1947).

[46] Kolthoff and Johnson, *J. Phys. Chem.* **50**, 440 (1946).

[47] McBain and Johnson, *J. Am. Chem. Soc* **66**, 9 (1944); compare McBain and Merrill, *Ind. Eng. Chem.* **34**, 915 (1942).

same authors in the same publication.   Figure 37 shows the solubilization of Orange OT:

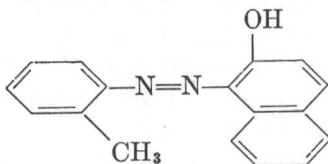

by the C_8–C_{14} potassium soaps.   The double curves for C_{12} and C_{14} indicate the results obtained by reaching equilibrium from both sides.   Figure 38

TABLE V

Solubilization in Milligrams of Orange OT per Hundred Milliliters Solution[a]

Detergent solution[b]	Dye, mg.	Type of detergent	Manufacturer
Aerosol 18	2.60	Disodium monosulfostearylsuccin-amide, 35%, + 65% water and slight turbid impurities	American Cyanamid and Chemical Co.
Aerosol OS	3.65	Sodium isopropyl naphthalene sulfonate, 85% active alkyl naphthalene sulfonate—fairly pure	"
Ammonium salt of Aerosol OT	1.44	Ammonium salt of dioctyl (2-ethylhexanol) ester of sodium sulfosuccinic acid	"
Zinc salt of Aerosol OT	6.56	Zinc salt of above	"
Alronol (100%)	3.00	Non-electrolytic detergent with trivalent nitrogen	Alrose Chemical Co.
Alronol (90%)	3.4	" " " "	
Nacconol NRSF	3.23	Sodium alkyl aryl sulfonate, not a sulfated fatty alcohol " " " "	National Aniline and Chemical Co.
Napco 1067	1.17	An alkyl aromatic sulfonate	National Oil Products Co.
Napco 1086-C	0.44	" " " "	
Orvus W.A. flakes	3.70	Sodium lauryl sulfate	Procter and Gamble Co.
Orvus W.A. paste	1.85	" " "	
Santomerse 1	1.25	Belongs to an homologous series of substituted aromatic sodium sulfonates (alkylated aryl sulfonate)	Monsanto Chemical Co.
Santomerse S	0.95	" " " "	
Sodium tetradecyl sulfate	0.13	Specially purified	E. I. du Pont de Nemours & Co.
Triton W-30	0.57	A sulfonated aromatic ether alcohol	Rohm and Haas

[a] McBain and Johnson, J. Am. Chem. Soc. **66**, 9 (1944); Fat Abstracts, E: 122, p. 1, 1944.
[b] 1 g. commercial detergent in 100 ml. water.

shows the unusual effect of added salt on the relationship between dye solubility and soap concentration.   Salt increases the dye solubility throughout the whole soap concentration range.   With salt present, the greatest relative solubility[48] is in the lowest concentration ranges and the curve falls before leveling off. This is just opposite to the effect without salt. Salt also lowers the soap concentration at which the relative solubility approaches constancy, *i.e.*, the range where the curve levels off.

The extent of solubilization also varies considerably for any given detergent with the type of oil used.[49] McBain and Richards have made a detailed study of the solubility of 36 different organic liquids in soap and three different synthetic detergents. They conclude that oils of low molecular weight are more readily solubilized, the solubility decreasing rapidly with increasing molecular weight or molar volume. Polar oils are more

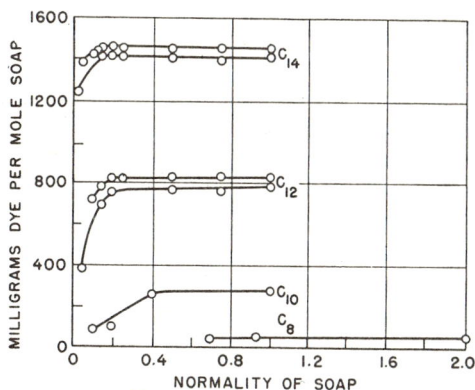

Fig. 37. The solubility of Orange OT at 25° in various molalities of aqueous solutions of a homologous series of pure potassium soaps.[47]

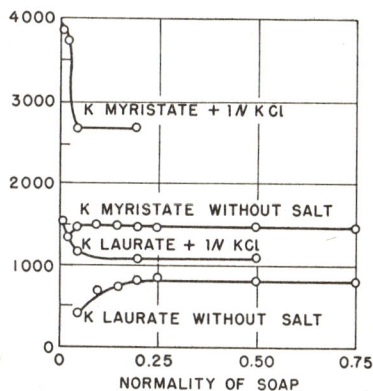

Fig. 38. The solubility of Orange OT at 25° in solutions of potassium laurate and potassium myristate with and without 1N KCl.[47]

readily solubilized than non-polar oils. The various detergents have parallel solubilizing power, differing from one another in degree. There are, however, numerous specificities. The C_{12} straight-chain detergents, for example, differ from one another, each favoring a different type of oil.

In a similar study made several years previously, Smith[50] investigated the solubilities of twenty-six organic liquids in 0.4 M sodium oleate solution. He found a rough correlation between the water solubility of the oils and

---

[48] We shall use the term "relative solubility" to indicate the ratio weight of dye to weight of soap in solution. Weight, of course, may be expressed as either grams or moles, provided the units are specified.

[49] McBain and Richards, *Ind. Eng. Chem.* **38,** 642 (1946).

[50] Smith, *J. Phys. Chem.* **36,** 1401 (1932).

their solubility in aqueous soap solution. Those oils having greater than 2% solubility in water had a soap solubility of greater than 10%. Solubilization also occurs in non-aqueous systems, and many instances have been studied at least qualitatively.[51] Palit and McBain have shown that soap greatly increases the mutual solubility of benzene and ethylene glycol. This is an instance of the use of "blending agents," a practice which has been well-known industrially for many years.[52]

There have been several theories advanced as to the mechanism of solubilization. Some have sought to explain the effect as an adsorption of the oil on the detergent micelles.[53] One of the main objections to this theory is that a hydrophobic substance adsorbed on the exterior of the micelle would tend to make the micelle thermodynamically unstable. Another theory postulates that the oil is held in solution in the hydrocarbon tails of the associated detergent molecules.[54] The fact that solubilized oils have far lower partial pressures than emulsified or dispersed oils points to an intimate association of oil and detergent, similar at least qualitatively to a solution effect.[55]

The theory which appears to have most evidence in its favor states that solubilization is accomplished by the lamellar micelles. We have seen that these micelles have alternate head-to-head and tail-to-tail spaces in their structure. The space between the head layers is filled with water. In a solution of pure detergent the intramicellar spacing between the head layers becomes greater as the dilution is increased, indicating that more water molecules have moved into this region. When an oil is introduced into the system, the oil molecules enter the *intra*micellar spaces between the tail layers and become essentially a part of the micelle. This is indicated by X-ray diagrams, which show an increased spacing of these layers when solubilized oil is present. A large amount of X-ray evidence has recently been published, all of which strongly supports this theory. X-ray studies have also been made which show that some solubilization may take place between the parallel long chains as well as between the layers of tail ends. This latter location of the dissolved molecules would be tantamount to a true "solubility" in the hydrophobic tails, as postulated by Lawrence.[54] It appears to be a secondary rather than a primary effect, however, and

[51] McBain, Merrill, and Vinograd, *J. Am. Chem. Soc.* **62**, 2880 (1940).

[52] Palit and McBain, *Ind. Eng. Chem.* **38**, 741 (1946).

[53] Lascaray, *Seifensieder-Ztg.* **61**, 156 (1934); Stauff, *Z. physik. Chem.* **A191**, 69 (1942); *cf.* Hoar and Schulman, *Nature* **152**, 102 (1943).

[54] Lawrence, *Trans. Faraday Soc.* **33**, 815 (1937).

[55] McBain and Johnston, *Proc. Roy. Soc. London* **A181**, 119 (1942).

most of the oil is sandwiched between the tail layers of the lamellar detergent micelles.[56]

XI. Theories of Micelle Structure

The fact that solutions of surface active agents form a unique type of colloidal system is now universally accepted, although it was regarded as highly radical when first proposed by McBain. The system is colloidal by many criteria, the most obvious of which is its osmotic behavior. It is unique in that the colloidal aggregates form spontaneously by the association of molecules or ions, and that these aggregates are in reversible thermodynamic equilibrium with the surrounding environment (solvent and molecularly dispersed solute). It is also generally accepted that the aggregates or micelles begin to form in large amounts only when a definite concentration range is reached, and that this range is relatively narrow. Bury and co-workers[57] have demonstrated, on the basis of the mass action law, that the concentration range within which micelles begin to form in large numbers should be fairly narrow. This is particularly true where the aggregate is fairly large (twenty or more ions) and is formed by the association of single ions. These investigators[57] suggested the term "critical concentration" for this range and it has been universally accepted.

We have seen, however, that when gross properties such as conductivity or osmotic coefficient are plotted against concentration sigmoid type curves are generally obtained. These can be considered to consist of three different ranges, a pre-critical range, a range in which the magnitude of the effect being measured changes rapidly with the concentration, and a third range where the curve levels off or even changes the algebraic sign of its slope. This indicates that the nature of the micelle itself changes with concentration, or else that there is more than one type of micelle present. In view of the large amount of data which has been accumulated using many different techniques it is not surprising that a completely satisfactory theory as to the structure of micelles is so difficult to develop. A major difficulty lies in the fact that the critical concentration values vary in several instances depending on which particular effect is being measured.

It is not within our limited scope to attempt a full discussion of micelle theory. There are at the present time, however, two leading schools of thought in this field, one represented by McBain and the other by Hartley.

---

[56] Harkins *et al.*, *J. Chem. Phys.* **13**, 534 (1945). Hughes, Sawyer, and Vinograd, *J. Chem. Phys.* **13**, 130 (1945).

[57] Grindley and Bury, *J. Chem. Soc.* **1929**, 679. Davies and Bury, *ibid.* **1930**, 2263.

McBain's views are summed up in his chapter in Alexander, *Colloid Chemistry*, Vol. V, Reinhold, New York, 1945, as well as in his more recent publications. Hartley's views are presented in *Aqueous Solutions of Paraffin Chain Salts*, Hermann, Paris, 1936. See also, *Kolloid-Z.* **88,** 22 (1939) and other publications of Hartley and co-workers. Tartar and co-workers have supported Hartley's views in many of their publications.

Hartley postulates essentially only one type of micelle, which is formed in substantial amounts at the critical range by association of ions. Below the critical range only free ions exist in the solution. The rapid fall in conductivity and osmotic coefficient is due to association of gegenions (ions of opposite charge) with the micelle. The later rise in the values of Λ and g is due either to dissociation of the gegenions or to Debye-Hückel effects (changes in the ionic atmospheres around the micelles). The micelle itself is roughly spherical in shape. This theory does not account for the large lamellar micelle deduced from X-ray diffraction data.

McBain holds that all the data can only be explained on the basis of different sizes and types of micelles. The two general types of micelle are the small, spherical, highly conducting micelle and the large, lamellar, poorly conducting micelle. This latter structure he identifies with the so-called "neutral colloid" which he recognized long before X-ray data confirmed its existence. McBain has repeatedly pointed out that this term is a misnomer, since these micelles actually conduct electrical current, although much more poorly than the ionic micelles.[58] The ionic micelles form from simple ion pairs and progressively larger aggregates even in very dilute solution. At the critical point, lamellar micelles begin to develop. They increase in size and amount until enough are present to produce an X-ray pattern.

The idea that several kinds and sizes of colloidal particles exist in detergent solutions, and that their relative concentrations depend on total detergent concentration, temperature, etc. is very logical and persuasive,[59] and it probably has more supporters than the single micelle type theory. Van Rysselberghe[60] has developed a mathematical treatment which interprets osmotic conductivity and migration data of dilute solutions on the basis of an "average inclusive micelle" the size and charge of which vary. This idea also seems to imply a wide range of micelle sizes and net charges.

At present, in view of the overwhelming X-ray evidence, there is practically no doubt as to the existence of lamellar micelles. In view of all

[58] M.E.L. McBain, *Trans. Faraday Soc.* **31,** 153 (1935); Laing, *J. Phys. Chem.* **28,** 673 (1924).

[59] Compare Stauff, *Kolloid-Z.* **96,** 244 (1941); *Naturwissenschaften* **27,** 213 (1939); *Z. physik. Chem.* **A185,** 45 (1939); **A187,** 107 (1940).

[60] Van Rysselberghe, *J. Phys. Chem.* **43,** 1049 (1939); **48,** 62 (1944).

previous evidence, the existence of a smaller and more highly conducting micelle must also be conceded.

XII. Relationship of Surface to Bulk Properties

Some qualitative relationships between surface activity and bulk properties may be readily deduced from the existing data. Those detergents, for example, which form micelles at lower concentrations, also reduce the surface tension at lower concentrations. The addition of salt to detergent solutions promotes micelle formation and also promotes surface activity. This is not surprising when we consider the forces involved from a very general point of view. A purely hydrophobic substance such as paraffin oil will not dissolve in water. No matter how intimately it is mixed with water, it will separate rapidly as a second phase. In a surface active agent of typical structure, the hydrophobic groups retain this tendency to form a second phase, whereas the hydrophilic groups tend to remain molecularly dispersed in the medium. Strong adsorption of detergent molecules in the surface is primarily responsible for surface activity. It occurs as a manifestation of the tendency to "escape" or form a second phase. Micelle formation, responsible for the anomalous bulk properties, is also a manifestation of the tendency to form a second phase, and we would therefore expect the two effects to correlate.

Quantitative correlations between surface effects and bulk effects have been pointed out in very few instances. McBain has shown that the surface tension minimum in type III curves occurs at a lower concentration than the critical concentration for conductivity or freezing point lowering. In other words, the surface anomalies occur in a lower concentration range than the bulk anomalies. The difference in concentration values is small, but definite. It has been used as a rebuttal to the explanation of type III curves in terms of micelle formation (see Chapter 13, Section III, a).

This lack of simple quantitative correlation between bulk and surface properties is likewise not surprising. In the surface layer of a solution there exist both water molecules and detergent molecules or ions. These are being subjected simultaneously to a highly unbalanced force field. In the interior of the solution, the force field in any small region is much more symmetrical. We should therefore expect that the tendencies toward orientation and association should differ quantitatively as between surface and interior.

The technologist, seeking a detailed molecular picture to account for all the various effects observed with surface active agents, will doubtless find several gaps and inadequacies in the theory. This is usually the case in actively developing fields of scientific endeavor. The theory in its present state, however, can clarify the technically important gross effects of surface active agents to an astonishing degree.

CHAPTER 15

The Gross Effects and Technical Evaluation of Surface Active Agents

I. General Considerations

So far we have been considering the effects of surface active agents in terms of basic physical measurements. Two of the most important basic effects are: (1) The effect in sharply lowering surface and interfacial tension. This is related to the strong surface or interfacial adsorption of individual surface active ions or molecules, and is fundamentally a surface property. (2) The effect of solubilization or the carrying of normally insoluble substances into thermodynamically stable solution. This is dependent on the presence of colloidal micelles in the interior of the solution and is essentially a bulk property of the solution system.

Several other properties and effects have been mentioned, all of which can be expressed in terms of fundamental physical measurements. These include surface potential, surface viscosity, and surface optical effects as well as the bulk properties of anomalous osmotic pressure, conductivity, etc.

From the technologist's point of view, however, surface active agents are more noteworthy for a group of typical effects which cannot readily be measured or expressed in terms of the basic physical units. We shall call these the gross effects. The most important gross effects can be grouped into five classes as follows: (1) Wetting, including as related properties rewetting, penetration, and spreading. (2) Foaming. (3) Dispersing, deflocculation and "protective colloid" action. (4) Emulsification. (5) Detergency. In most cases it is difficult to establish any high degree of correlation between the basic properties and gross effects in spite of the large amount of study devoted to both.[1] This is partly a reflection of the present inadequacy of our theories, but it is due more to the complex nature of the gross effects. Some of them, such as wetting and foaming, are essentially surface effects and apparently have little direct connection with the micellar organization of the system. Others, such as detergency, are extremely complex and involve a large number of the fundamental properties of both bulk and surface. Another source of difficulty lies in finding suitable methods of measuring the gross effects. The measure-

[1] See Ackley, *Ann. N. Y. Acad. Sci.* **46**, 511 (1946).

ments which are usually applied have for the most part been developed on a purely empirical basis and in many instances are difficult to reproduce. Nevertheless, since the technical utilization of surface active agents is based on their gross effects, they are usually evaluated in terms of these properties. A detailed consideration of the general nature of wetting, detergency, emulsification, etc. is accordingly of fundamental importance for the efficient utilization of surface active agents.

The technologist who is using a surface active agent for a specific purpose is primarily interested in its efficiency for that purpose alone. A laundry chemist, for example, is concerned with the effectiveness of the detergent when used in his machines with the type of soiled fabrics he must clean. The chemist in a textile-finishing mill may be interested in dye-leveling power for various types of dyestuffs. He may also want to know how well a given surface active product promotes wetting of gray cotton at various temperatures and/or in caustic soda solutions. A field trial (or the full laboratory equivalent of a field trial) is therefore the ultimate method of evaluating a surface active agent from the user's point of view. The manufacturer of detergents, on the other hand, obviously cannot anticipate the thousands of specific uses in which his product may be applied. Unless he is designing a new surface active compound or formulation to fulfill certain specific needs, he usually studies the gross properties listed above. Measurements of these properties by a series of more or less arbitrary tests can often be correlated quite well with field performance in actual industrial uses. The evaluation of surface active agents is in itself rapidly becoming an extensive and important branch of technology. The tendency among the more progressive laboratories carrying on this work is toward more diversified and more numerous tests under a wider variety of conditions. The tendency is also toward simulating as closely as possible the actual conditions of usage in the field. Until recent years, surface active products were often loosely classified as wetting agents or detergents or emulsifying agents, etc., with the strong implication that these classifications were mutually exclusive. These terms originally arose on the basis of certain large-scale uses in the textile industry. A surface active agent which was well suited to wetting out gray cotton, for example, but was not very efficient in scouring operations was classed as a "wetting agent." One which showed excellent performance in preparing yarn lubricants or finishing emulsions would be classed as an "emulsifying agent." Today the use of surface active agents has extended far beyond the confines of the textile industry, and these simple classifications are apt to be more misleading than helpful. A substance such as sulfonated castor oil, for example, which is a poor detergent for textile scouring, may be excellent as a cleaning agent for human hair.   Surface active products are therefore best described by means of a long series of performance specifications.   The simplest gross property, and the one we shall consider first, is wetting power.

II. Wetting

The basic mechanism of wetting and spreading on plane solid surfaces has already been discussed.   Wetting in this case depends on the contact angle at the solid–liquid–air boundary line and direct correlations can be easily made between wetting times and contact angles.   In practice, however, it is often difficult to measure contact angles directly and consequently some direct measurement of wetting must be devised.   This is particularly true in the case of fabrics, yarns, felts, and other fibrous masses where wetting characteristics are of the greatest importance.

The two most widely used wetting tests on fabrics are the Draves test and the Seyferth-Morgan or canvas disc test.   In the former a standard 5-g. skein of gray cotton yarn is held submerged in the test solution by means of a weight.   As the solution wets the yarn air is progressively displaced and the buoyancy decreases until, at the end point of the test, the skein sinks.[2]   In the canvas disc test the disc is held under the surface and the liquid penetrates the fabric until it sinks of its own weight.[3]   Both these tests are quite duplicable within limited sinking time ranges.   Below about five seconds the timing is necessarily inaccurate.   Above about three minutes results are difficult to duplicate.

Fischer and Gans have pointed out the close correlation between Draves wetting time, surface tension of the wetting agent, and contact angle of the wetting agent against paraffin wax.   This is illustrated in Figure 39.[4]   The roughly parallel curves indicate that under these test conditions the wetting action is uncomplicated and resembles the wetting of a plane hydrophobic surface by a liquid.   As the bath is used repeatedly to wet out successive skeins the wetting agent is removed from solution by adsorption on the cotton.   Both the wetting time and surface tension increase correspondingly as this exhaustion takes place, as shown in Figures 40 and 41.

Despite the apparently simple picture of fabric wetting indicated above, certain complicating factors are often present.   The end point of the Draves test, or the canvas disc test, is taken when the skein sinks.   At this point the skein is by no means completely wet, i.e., each individual fiber is not wetted along its whole length nor has all the occluded air been dis-

[2] Draves and Clarkson, *Am. Dyestuff Reptr.* **20,** 201 (1931).   Draves, *ibid.* **28,** 421 (1939).

[3] Seyferth and Morgan, *Am. Dyestuff Reptr.* **27,** 525 (1938).

[4] Fischer and Gans, *Ann. N. Y. Acad. Sci.* **46,** 389 (1946).

placed.   Only enough air has been displaced so that the net specific gravity of skein, air, and attached weight (if any) is just greater than unity.   The textile processor, however, is interested in having the fabric completely wet, *i.e.*, in having complete penetration of the bath into the spaces between individual fibers.

The time required to accomplish complete wetting, and its relationship to Draves sinking time, has been studied by Gruntfest, Hager, and Walker.[5]

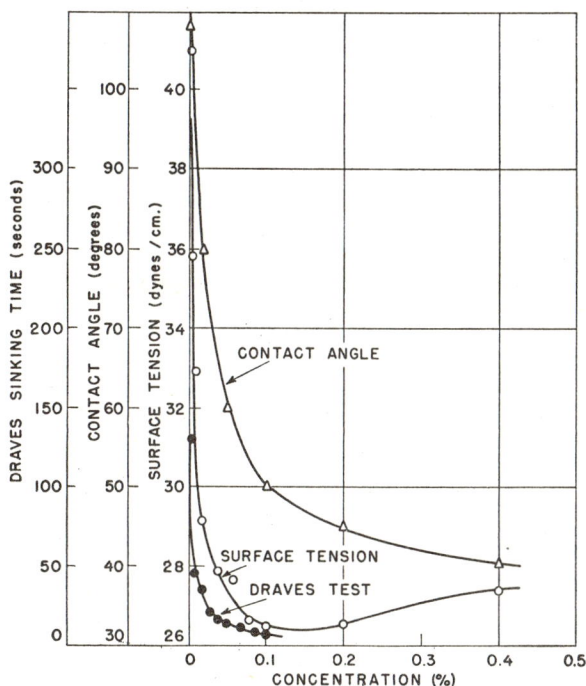

Fig. 39. Curves showing Draves test sinking times, surface tension values, contact angle measurements in relation to concentration of reagent (sodium salt of octyl ester of sulfosuccinic acid).[4]

In this important investigation the fabric sample to be wet was attached to the bulb of a specially designed hydrometer which was then immersed in the wetting solution.   The buoyancy of the fabric was calculated from the hydrometer reading, and was then plotted against time of immersion. The startling results are shown in Figures 42 and 43.   It is apparent that the more concentrated baths which show the shorter Draves sinking times do not wet the sample as thoroughly as the more dilute solutions, and that

[5] Gruntfest, Hager, and Walker, *Am. Dyestuff Reptr.* **36**, 225 (1947).

complete wetting is only achieved (under these particular test conditions) after a very long interval.

This effect is doubtless connected with the close proximity and tight packing of fibers in the yarn interior. The yarn interior constitutes in

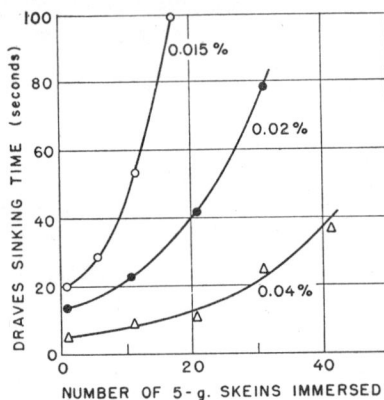

Fig. 40. Increase in Draves wetting times as a function of number of cotton skeins immersed per liter of solution. (Sodium salt of octyl ester of sulfosuccinic acid; temperature 30° ± 1°C.; 6-g. hook used for Draves test.)[4]

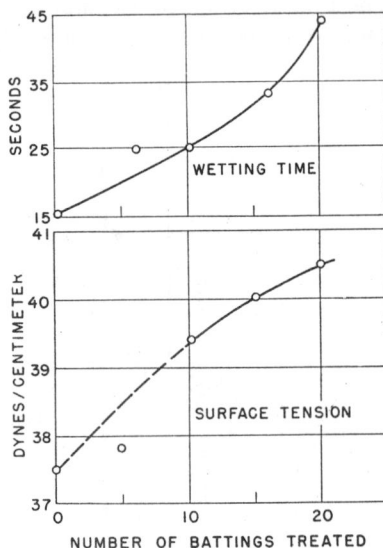

Fig. 41. Change in wetting times and surface tension values on bath depletion.[4]

effect a system of capillary spaces. Washburn has derived an equation for the rate at which a liquid penetrates a capillary, as follows:

$$dl/dt = (r\gamma \cos \theta)/(4l\eta)$$

where r is the radius of the capillary, γ is the surface tension, θ the contact angle with the capillary wall – air interface, η is the viscosity, and l the distance of penetration into the capillary.[6] From this equation it is seen that no penetration occurs if the contact angle is 90° or greater ($\cos \theta$ is zero or negative). When θ is zero the rate of penetration is proportional to the surface tension – viscosity ratio. When θ is between zero and 90°, high surface tension actually promotes penetration. This implies that surface

[6] Washburn, *Phys. Rev.* **17**, 273 (1921).

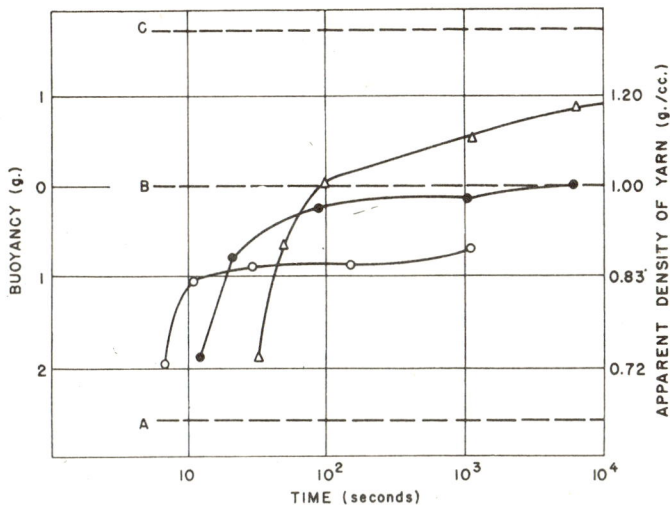

Fig. 42. Representative time–buoyancy curves for 5-g. skeins of gray cotton yarn immersed in wetting agent solutions. Dotted lines are drawn at buoyancies corresponding to: A, sinking of Draves skein with 3-g. copper weight; B, sinking of unloaded material; C, completely wet, clean cotton.[5] O = 0.05% wetting agent solution; ● = 0.03% wetting agent solution; △ = 0.02% wetting agent solution.

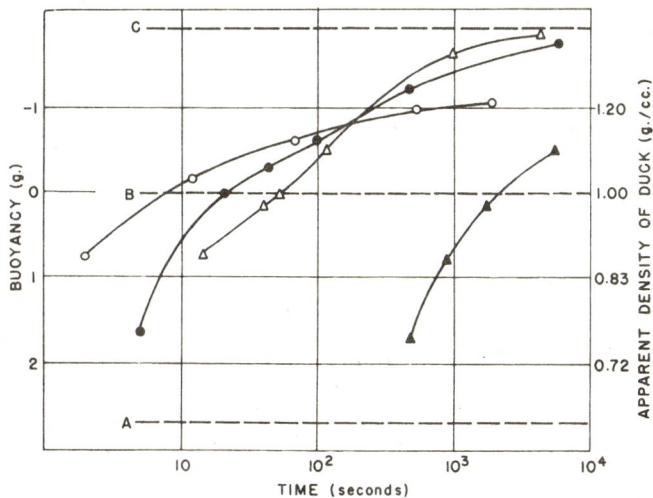

Fig. 43. Representative time–buoyancy curves for 5-g. samples of No. 10 Mt. Vernon duck immersed in wetting agent solutions. Dotted lines have same significance as in Figure 42.[5] O = 0.10% wetting agent solution; ● = 0.06% wetting agent solutions; △ = 0.02% wetting agent solution; ▲ = distilled water.

active agents will aid the penetration of a *capillary* mass only where the contact angle at the capillary walls must be reduced to below 90°. This has actually been borne out by experiments[7] on the penetration of wetting agent solutions into dry wood. A textile fabric represents a system which is partially capillary in nature. The spaces between the yarns are wide enough so that capillarity is no factor in wetting. In the yarn interior, however, the fibers are often packed so tightly that capillary effects become important.

It should be noted in this connection that surface active agents have no primary effect in promoting penetration into the fibers themselves. The so-called "capillaries" of a cellulose fiber, for example, are of molecular dimensions and are actually intermicellar spaces in the cellulose structure. They allow the passage of small individual molecules only, which migrate into the cellulose by diffusion rather than convection. A surface active solution will not penetrate a sheet of cellophane any more rapidly than pure water. In fact the penetration will usually be slowed down just as if salt or any other solute were present.

The resistance of fabrics to wetting is of great importance in the designing of water-repellent fabrics, that is, fabrics of porous structure which will shed rainfull rather than allowing it to penetrate. This problem has been studied and analyzed in truly masterful fashion by Baxter and Cassie.[8] They show that the hydrostatic pressure necessary to force water through a dry fabric depends on the structure of the fabric and on the advancing contact angle at the fiber–water–air boundary. The receding contact angle and the fabric structure determine the resistance to wetting. The optimum structure is one in which the yarns are fairly loose, evenly spaced, and rigid enough to resist drawing together. The fabric also should have a low bulk density. Structure is the determining factor in water repellency provided both advancing and receding contact angles are more than 90°. This is evidenced by studies on duck feathers as compared with proofed fabrics of various constructions.

In the wetting of continuous solid surfaces the degree of smoothness or roughness of the surface is important even though the roughness is not great enough to introduce capillary effects.[9] In these cases a "roughness factor" may be used to correct the adhesion tensions calculated from the contact angle and surface tension of the wetting liquid.

The comparative wetting power of surface active solutions and other liquids is often tested by observing their wetting action on hydrophobic

[7] Stamm and Petering, *Ind. Eng. Chem.* **32**, 809 (1940).

[8] Baxter and Cassie, *J. Textile Inst.* **36**, T67 (1945).

[9] Wenzel, *Ind. Eng. Chem.* **28**, 988 (1936).

solids such as sulfur or carbon black. This is at best a semi-quantitative test, but differential wetting tests between water and various organic liquids on several different solid substrata yield highly practical information as to wetting power.[10]

A tremendous amount of actual data on the wetting power of various surface active agents is available in manufactuers' brochures. These are usually given in terms of a Draves or canvas disc test. The wetting and spreading properties of the pure fatty acid soaps have been extensively studied by Cupples under a variety of conditions.[11]

It is well known that the wetting power of such substances as the lower alkyl naphthalene sulfonates can be measurably improved by the addition of moderate amounts of certain water-insoluble solvents. The Draves sinking time of dibutylnaphthalene sodium sulfonate, for example, can be reduced to less than half its original value by adding diamylphenol or *tert*-octylphenol. The aliphatic ketones and ethers of medium molecular weight, such as methyl hexyl ketone, are also very effective. The addition agent in these cases is probably solubilized to a certain extent in the bath. It is at least so finely dispersed as to be invisible. The practical advantage of these preparations appears to be limited by the relatively rapid exhaustion of the insoluble additive from the bath by adsorption on the fabric. To the authors' knowledge, no exact studies of this effect have been published, although several commercial wetting preparations are made by adding "boosting" solvents to the simple wetting agents. In this connection it should be noted that grey cotton is wetted instantly by most organic liquids. A Draves skein or a grey canvas disc placed in a beaker of benzene, alcohol, Cellosolve, or most other common solvents will sink before it can be timed. It is possible that the suspended booster acts by rapidly wetting the fabric and forming a surface which is in itself more readily wet by the solution.

III. Rewetting

The property of rewetting or "wetting-back" is important in certain phases of textile processing. Most persons have had the experience of using a freshly purchased bath towel and finding that it is non-absorbent. After one or two thorough launderings the towel becomes perfectly satisfactory. The initial non-absorbency is due to residual sizing and finishing materials on the fabric. A rewetting finish may be used to improve the initial absorbency of the towel. Rewetting is also important in shrinkproofing cotton fabric by Sanforizing. This process quite literally compresses fabric in its own plane by submitting it, in a thoroughly moistened

[10] Davis and Curtis, *Ind. Eng. Chem.* **24**, 1137 (1932).
[11] Cupples, *Ind. Eng. Chem.* **28**, 60, 434 (1936); **29**, 924 (1937); **31**, 1307 (1939).

condition, to the rubbing action of a series of friction elements or "shoes." Before entering the zone of mechanical action the previously dried fabric is sprayed to effect the necessary wetting out. If prior to previous drying the fabric is treated with a rewetting agent, the wetting-out step preliminary to Sanforizing may be greatly facilitated. A rewetting agent may be defined, accordingly, as a substance which may be applied to a fabric and dried on, and will thereby make the fabric readily wettable by water. It is in effect a water-absorbent finish.

Not all wetting agents are good rewetting agents. When a fabric is treated with a wetting agent solution and dried, it usually has good absorbency for several hours after drying, but this absorbency decreases with age. After a few days it may have disappeared completely. With a good rewetting agent the absorbency persists for several weeks or even months. Not all of the basic factors involved in rewetting are known, but a few of the principal ones have been generally recognized. In the first place the rewetting finish must contact the underlying fiber intimately, i.e., it must in itself be a good wetting agent. This is necessary since it must not only be easily wet iself, but must also act as a bridge and conduct the water to the clean absorbent-fiber surface. Secondly, it must, after drying, present an easily wettable surface. Many dry powdered wetting agents, notably some of the alkyl aryl sulfonates, are in themselves rather difficult to wet with water. The powder tends to float on pure water in the same manner as sulfur or carbon black until enough has been dissolved to act as a wetting agent for the remainder. This may be due either to the deposition of oily hydrophobic impurities on the powder surfaces, or to the crystal structure of the powder itself. If the crystals are so arranged that the hydrophobic groups are outward, a water-repellent surface may be expected. It is an empirical fact that those wetting agents which are liquids or non-crystalline solids when dry usually are good rewetting agents.

IV. Dispersion, Deflocculation, and Suspending Action

When a very finely divided solid is immersed in a liquid medium it is generally a difficult matter to obtain a uniform dispersion, i.e., one in which each individual particle is separate and completely surrounded by liquid. There is usually a strong tendency for the particles to remain in the form of clumps or aggregates. This formation of aggregates is called flocculation or agglomeration, and the dispersion of such aggregates, deflocculation. If the solid particles are heavier than the liquid they tend to settle by gravity. The laws governing the rate of settling and the various factors which affect it have been well worked out, particularly in connection with the development and use of the ultracentrifuge. Particles of smaller diameter settle more slowly, and aggregates or flocs settle as if they were large individual particles.   Therefore, a deflocculated suspension will generally settle more slowly than one in which the particles are agglomerated, and suspending action is very closely related to dispersing action.

A well-dispersed suspension of finely divided particles settles with surprising slowness.   A dilute suspension of bentonite in water, for example, may take several days to settle.   A well-made paint or lacquer will sediment badly only after prolonged storage even though the pigment may be many times heavier than the vehicle.   This unexpected stability is due, among other causes, to small convection currents set up by accidental vibrations or by localized temperature differentials.

Dispersion phenomena are of great technical importance in non-aqueous systems as well as in aqueous systems.   In either case, the fundamental phenomena involved and the roles played by surface action agents are probably similar.   Among important examples of non-aqueous dispersion systems are paints, varnishes, and printing inks.   Among the aqueous systems are water paints, drilling muds, writing inks, lime soap suspensions in textile treating baths, vat dye preparations, etc.

Degree and effectiveness of dispersion may be observed directly by microscopic examination.   It may also be measured by sedimentation rates and volumes, by filtration through graded filters, and sundry other means.   In many instances, correlations have been worked out between the degree of dispersion and the rheological properties of a suspension system so that rheological data can be used as a measure of dispersion.

When a solid is wetted by a liquid, the degree of attraction between solid and liquid is inversely proportional to the free energy of the solid–liquid interface.   It may be measured within limits by the contact angle in the presence of air or in the presence of a second immiscible liquid.   The magnitude of the solid–liquid attractive force may also be indicated by the heat of wetting and other thermal measurements.   In order for a liquid medium effectively to disperse an aggregate of solid particles it must completely wet each particle.   In other words, the wetting forces must be sufficient to overcome the solid-to-solid adhesion forces.   The solid–liquid interfacial free energy must be lower than the over-all free energy of the *effective* solid–solid interface.   There is as yet relatively little knowledge of the solid-to-solid adhesion forces in actual practical systems.   Wetting agents are added to lower the solid–liquid interfacial free energy.   In doing this they are adsorbed at the interface and statistically oriented so as to give a system of minimum free energy.   It has been well established that solid particles tend to agglomerate when they are in a medium which does not wet them easily.   It is difficult, for example, to form a well-dispersed paint from a drying oil and a hydrophilic pigment.   To achieve this purpose such pigments are often coated with polar substances, such as the metallic soaps, which, after being adsorbed, present an oleophilic surface to the oil.   A large amount of work has been done in this field, but it is beyond our present scope to consider it in detail.[12]

In a recent investigation which typifies modern methods, Damarell and co-workers have studied the effect of various surface active agents on xylene dispersions of carbon black and calcium carbonate.[13]   Sedimentation rates and volumes were measured together with the degree of adsorption of surface active agents on the suspended particles.   Charge effects were measured by cataphoresis experiments, and the number of particles per unit volume of sol by ultramicroscopic observation.   In every case they found that the most strongly adsorbed surface active agents showed the greatest suspending power.   Added water caused flocculation, presumably by displacing the adsorbed dispersing agent.   The suspensions appeared to be charged even when water was absent but the origin of the charge was not accounted for.

Agglomerates of finely divided solid particles may be formed even when the solid is actually being precipitated in a liquid medium.   A well-known example occurs on the addition of a soluble calcium salt to an aqueous soap solution.   The lime soap which precipitates consists of curds and flocs which are aggregates of very small individual particles.   These may be dispersed by addition of a suitable lime-resistant surface active agent. If sufficient surface active agent is present in the soap solution before the lime is added the lime soap will appear in the form of a well-dispersed deflocculated suspension.[14]   The ability to peptize lime soaps is used as an index of dispersing power of those surface active agents which are lime resistant and compatible with soap.   The general method consists of making up a series of mixtures containing varying proportions of sodium oleate and the detergent being tested.   These mixtures contain approximately 10% total soap-plus-detergent in distilled water.   5 ml. of each mixture is then added to 45 ml. of hard water (usually 200 p.p.m. hardness as CaO).   This is called the first dilution and it usually results in a turbid but well dispersed sol.   5 ml. of the first dilution is then added to 45 ml. of

[12] For a complete review and discussion, see Fischer and Gans, "Dispersions of Finely Divided Solids in Liquid Media," in Alexander, *Colloid Chemistry*, Vol. VI, Reinhold, New York, 1946, Chap. 14.   See also Mattiello, *Protective and Decorative Coatings*, Vol. IV, Wiley, New York, 1944.

[13] Damarell and Urbanic, *J. Phys. Chem.* **48**, 125 (1944).

[14] Boedeker, *Melliand Textilber.* **13**, 436 (1932).

hard water forming the second dilution.   This is a severe test since there is now more than enough lime present to precipitate all the soap.   Further-more, the total soap-plus-detergent concentration is of the order of 0.1%. The results are expressed as the percentage detergent in the soap–detergent mixture which is just sufficient to prevent flocculation on the second dilu-tion.   This varies considerably with the nature of the detergent.   The Igepals and Igepon T are very effective.   Igepon T (100% active) shows a value of about 8%.   Sulfated coconut fatty alcohol has a value of about 40%.   The petroleum-derived long-chain alkyl aromatic sulfonates are much less effective, most of them having values in the range of 70–90%.[15]

A recently described method for measuring the hardness of water is based on determining the turbidity of a standard soap – fatty alcohol sul-fate mixture when diluted with the water in question.   Enough fatty alcohol sulfate is present to insure a well-peptized lime soap, so that the nephelometric readings are duplicable.[16]

In the prepration of vat dye pastes it is customary to dissolve the coarse dye particles in concentrated sulfuric acid and reprecipitate the dye by pouring into a large volume of water.   The particle size and degree of dis-persion obtained can be modified and controlled by adding surface active agents to the precipitating bath.   In these and similar systems the surface active agent is adsorbed on the small particles relatively soon after they form from their nuclei and before they have a chance to grow or agglomer-ate.   This may be considered a form of protective-colloid action although the term "protective colloid" has been used very loosely to describe widely dissimilar effects.   Another well-known example of this effect is the use of surface active agents in electroplating baths.   Here they serve to modify the crystal size of the depositing metal, altering the roughness or brightness of the finish.

McBain has drawn a sharp distinction between protective colloid action and suspending action.[17]   In protective action the individual particles are coated with an adsorbed film of colloid.   Protective action may accord-ingly be exerted by non-electrolytes as well as electrolytes.   Suspending action may be brought about merely by imparting an electrical charge to the particles.   The presence of an adsorbed colloid is not necessary.   In this case the rate at which charged particles sediment is slowed down by the repulsive action of the charge of the gegenions associated with each

[15] For details of this method, see Baird, *PB 32565*, Office of Technical Services, Dept. of Commerce, Washington, D. C.

[16] Saifer and Clark, *Ind. Eng. Chem., Anal. Ed.* **17**, 757 (1945).

[17] McBain, in *Advances in Colloid Science*, Vol. I.   Interscience, New York, 1942, pp. 99–142.

particle. Soap and other ionic surface active agents combine the two effects. When they are adsorbed they impart a charge to the particle and also coat it with a film of protective colloid. The over-all suspending power of surface active agents is a function of the concentration and generally shows at least one optimum. It also varies with the presence of excess alkali or other ions. Figures 44 and 45 illustrate these effects.[18]

According to McBain it is the protective-colloid action which allows carbon black – soap dispersions to pass through a filter paper. The soap is adsorbed on the particles of the dispersion and also on the filter paper

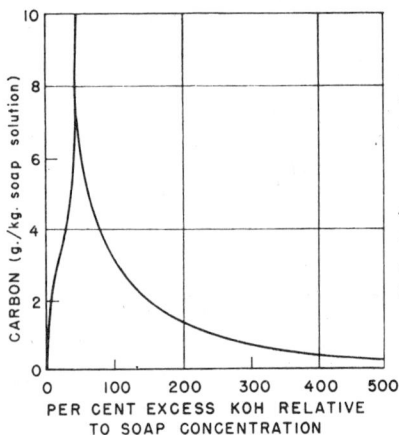

Fig. 44. Influence of added alkali on suspending power of 0.0125 N_w (or 3.22%) solutions of potassium myristate at 15°C.[17]

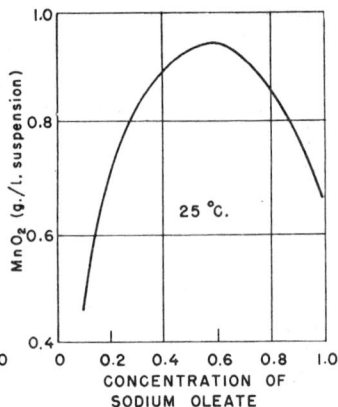

Fig. 45. Suspension of manganese dioxide particles by aqueous sodium oleate solutions.[17]

itself. A simple charging of the carbon particles by adsorption of a non-colloidal ion will not make them capable of passage through filter paper, although it will have a definite suspending action.

Dispersion and suspending action are of primary importance in detergency and will be considered further in the section dealing with that subject.

V. Foaming

One of the most conspicuous properties of many surface active agents is their ability to produce foam. The technological importance of foaming power has probably been overemphasized, because it has become falsely identified in some quarters with detergent efficiency. For some purposes,

[18] McBain, *loc. cit.*; Fall, *J. Phys. Chem.* **31**, 801 (1927).

however, such as ore flotation, fire-fighting compositions, certain cosmetic preparations, foamed concrete, etc., it is of primary importance. In many applications, on the other hand, foam is a secondary phenomenon and purely incidental to the desired working of the detergent. Sometimes, as in distillations and evaporations, it is actually a serious disadvantage.

Since foaming power is highly prized a large amount of industrial research is devoted to the study and improvement of this property. It has also been extensively investigated from the scientific as well as the technical point of view.[19]

Foam or lather may be defined as a measurably stable honeycomb structure of air cells whose walls consist of thin liquid films. The simplest foam structure is a single bubble on the surface of a liquid. This consists of a single liquid film of roughly hemispherical shape. In most ordinary multicellular foams the individual films are more nearly planar. Because of their large surface per unit mass these films are inherently unstable geometric forms.

It is generally conceded that pure liquids do not foam. Nevertheless, many different types of solutions beside those containing surface active agents have marked foaming power. Surface tension might be expected to be the primary force tending to destroy any individual film, and it is natural to predict that solutions of low surface tension would be the best foamers. It is indeed true that the best foamers are found among the surface active agents. But it is also true that certain inorganic salt solutions, having surface tensions actually higher than that of pure water, can form persistent foams. There is much other experimental evidence that surface tension is by no means the only important factor in determining ease of formation and stability of foams.

Foams are formed by introducing air or other gas under the surface of a mass of liquid, or by agitating the surface so that it breaks and encloses air. They also may be formed by liberation of a dissolved gas, as in the foaming of beer and other carbonated beverages.

The mechanism by which foams disappear or are destroyed spontaneously is relatively simple. The liquid drains out from between the two parallel surfaces of the individual film, thereby causing the film to become progressively thinner. At a certain critical thickness, the film collapses. Evaporation will, of course, hasten the thinning of foam films, and most foams show greater stability when stored in closed containers. Foams may also be broken mechanically, by deformation of the individual bubbles, or by the addition of "foam-killing" reagents. Compact stable masses of

[19] For an extensive review, see Berkman and Egloff, *Chem. Revs.* **15**, 377 (1934).

foam have very interesting gross mechanical properties.   Under some conditions they resemble a liquid and under others a gas.[20]   Foams vary widely in bubble size, and in the ratio of liquid to gas per unit volume of foam. These quantities may vary with the manner in which the foam is prepared, but they are also functions of the composition of the liquor itself.   Certain surface active agents are noted for producing small-bubbled or "creamy" foams, while others produce large-bubbled or "beady" foams.   Similarly, foams are often described as "dry" or "wet" depending on the liquid–gas ratio.   The dry foams have thinner individual films.   A stable wet foam becomes dryer as the liquid drains out of it.

In studying the over-all foaming characteristics of a given solution, several individual properties may be noted and measured.   Bubble size (under comparable conditions of formation) and the liquid–gas ratio, more often called the "foam density," have already been mentioned.   The foam volume, foam stability, foam drainage, and the ease with which foam is produced (foam initiation) are among other important properties.   Such properties are not necessarily directly related to each other.

In studying foams it is of primary importance to devise apparatus which can give reproducible results.   Fortunately this is not difficult to do.   In many cases simple shaking by hand of a partly filled closed vessel will give foams sufficiently reproducible for quantitative study.   Mechanical shaking is probably the most widely used method for evaluating the foaming properties of surface active agents in technical laboratories.   Another widely used method for producing foams consists of beating air into the solution with a mechanical egg beater such as the Sunbeam Mixmaster. Foam volume is measured in devices of this type by observing the height to which the foam rises, and stability by the rate at which it falls.   Drainage and density measurements, if desired, must be made separately. Foaming properties vary considerably with temperature, and it is important to have adequate temperature control in all accurate foam measurements.

A more elaborate type of foam meter, well designed for both technical and scientific work, has been described by John Ross and Miles.[21]   It consists essentially of a jacketed measuring cylinder in which a portion of the solution to be tested is placed.   The foam is formed by allowing a second portion of solution to stream in from a fixed height through a standard orifice.   Foam height measurements obtained with this device are astonishingly reproducible.

---

[20] Siehr, *Kolloid Z.* **77**, 27 (1936); **78**, 156 (1937); **85**, 70 (1938).
[21] Ross and Miles, *Oil and Soap* **18**, 99 (1941).

Another widely used general method of producing foams for purposes of study consists in blowing a stream of air or gas into the solution. The orifice for admitting air may be a single tube or a header containing multiple outlets, such as a sintered-glass plate. The volume of air introduced per unit time is controlled and measured, as is the temperature. A number of variations of this type of instrument have been described. Schütz[22] has used the blowing method to produce a single layer of bubbles completely covering the liquid surface. He measures the stability in terms of "foam

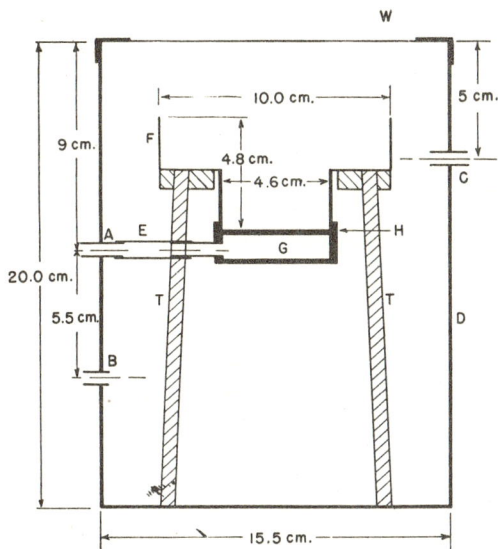

Fig. 46. Foam time apparatus. Air enters through inlet E and proceeds through three small holes in plate H generating foam from solution in cup F.[22]

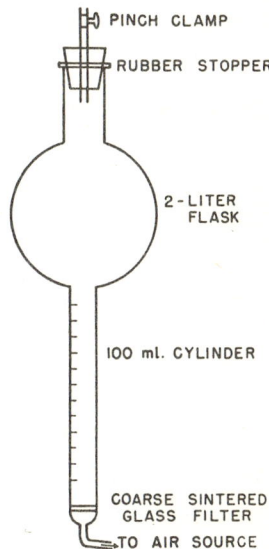

Fig. 47. Foam drainage apparatus.[22]

time" which is defined as the time required for half the surface area to become free of bubbles. The apparatus is illustrated in Figure 46. Merrill and Moffat[23] have used the blower method to produce foam in an apparatus designed to measure foam drainage accurately. Foam drainage is measured by the amount of liquid which will drain from a given volume of foam in a given time. The apparatus is illustrated in Figure 47. Air is blown through the liquid until it is all converted to foam. The air stream

[22] Schütz, *Trans. Faraday Soc.* **38,** 85 (1942); Compare Merrill and Moffett, *Oil and Soap* **21,** 170 (1944).
[23] Merrill and Moffett, *loc. cit.*

is then stopped and the liquid begins to drain back into the lower portion of the tube where its volume may be measured.   The authors found good correlation between drainage time and "foam time" for a number of different soaps and synthetic detergents.

Foam density may be measured directly, by weighing a given volume of foam, provided the foam is sufficiently stable to handle.

The stability of single bubbles at the surface of a liquid has been studied and correlated to a limited extent with foaming properties.   This method offers the advantage that it can be applied to liquids covered with insoluble monomolecular layers.[24]

For most technical purposes foaming data, particularly foam volume (or height) and foam stability, are expressed in comparative terms, *i.e.*, an arbitrary standard material is used in a particular apparatus and the materials to be tested are compared with this standard.   Foam volume, density, and drainage data are readily expressed in absolute terms, but the absolute expression of foam stability is more difficult.   Bikerman[25] has used a dynamic foam meter to this end.   The instrument is essentially a blower type, with a tall graduated cylinder as the foam container.   Air or other gas is blown through the solution at a constant rate until the foam has risen to a maximum height.   At this point the foam is breaking on top at the same rate as new bubbles are being formed at the bottom by incoming air.   The foam height is read directly and the foam volume is calculated. It should be noted that Bikerman defines the foam volume as the difference between (*1*) volume occupied by liquid and foam, and (*2*) volume of the liquid at rest. ꞏ He finds the foam volume, v, to be proportional to V/t, the volume of air passed through the septum per second, and writes the equation:

$$\Sigma = vt/V = v/r$$

where r is the rate of air flow, and Σ is a constant having the dimensions of time. Σ is accordingly characteristic of the solution in question and is a simple measure of its foam stability.   It is in fact equal to the average lifetime of a bubble in the foam.   It has been demonstrated, however, that Bikerman's equation does not hold true with wide variations in air velocity.   The value of Σ evidently depends to a certain extent on the size distribution of individual bubbles in the foam.[26]

[24] Hardy, *Proc. Roy. Soc. London* **A86,** 627 (19.12).   Talmud and Suchowolskaia, *Z. physik. Chem.* **154,** 277 (1931).   Ruyssen and Verstraete, *Meded. Kon. Vlaamsche Acad. Wetensch., Letteren Schoone Kunsten België, Klasse Wetensch.* **4,** No. 9, 39 pp. (1942); *Chem. Abstracts* **38,** 3889 (1944).

[25] Bikerman, *Trans. Faraday Soc.* **34,** 634 (1938).

[26] Hazelhurst and Neville, *Ind. Eng. Chem.* **33,** 1084 (1941).

The dynamic meter is best applied to solutions of relatively low foaming power. The highly foaming surface active agents show far too large a value of Σ to be measured conveniently.

Sydney Ross[27] has shown that foam stability may be expressed by two exact mathematical expressions, one for the average lifetime of gas in the foam, L_g, and the other for the average lifetime of liquid in the foam, L_l. These are:

$$L_g = \frac{l}{G_0} \int_0^{G_0} t\, dG \qquad \text{and} \qquad L_l = \frac{l}{V_0} \int_0^{V_0} t\, dV$$

where G and V are the volumes of gas and liquid, respectively, in the foam. These expressions are, of course, merely definitive. They cannot be used for practical calculations unless the functional relationships of t to G and t to V are known.

The mechanism of foam drainage has recently been investigated by Miles, Shedlovsky, and John Ross.[28] They find that drainage may be expressed by the equation:

$$kt = V + a \log V + b$$

where V is the volume of liquid remaining in the foam at the time t and k, and a and b are constants. The rate of flow of liquid through foam decreases with decrease in bubble size and with increase in surface and bulk viscosities. Drainage of foams is analogous to the creaming of emulsions, and is not a measure of foam stability any more than creaming is a measure of emulsion stability.

A detailed mathematical analysis of the factors involved in foam formation and stability has been presented by Dyakonov, and checked by air-streaming experiments on soap solutions.[29]

Many accurate measurements of the foaming properties of various solutions, surface active and non-surface active, have been reported and most of the investigators have been able to draw definite generalizations from their data. Brady and Ross,[27] working with non-aqueous systems, have shown that foam stability is affected by bulk viscosity, foam height, initial film thickness and bubble size, and the ratio of final to initial quantity of liquid in the foam (drainage).

Data on the action of anti-foaming agents has also been reported, but

[27] S. Ross, *J. Phys. Chem.* **50**, 391 (1946); **47**, 266 (1943). Ross, *Ind. Eng. Chem., Anal. Ed.* **15**, 329 (1943). Brady and Ross, *J. Am. Chem. Soc.* **66**, 1348 (1944). Clark and Ross, *Ind. Eng. Chem.* **32**, 1594 (1940).

[28] Miles, Shedlovsky, and J. Ross, *J. Phys. Chem.* **49**, 93 (1945).

[29] Dyakonov, *J. Tech. Phys. U. S. S. R.* **12**, 302 (1942); *Chem. Abstracts* **37**, 5299.

no simple relationships have been noted.[30] Possibly the most extensive work on non-aqueous systems has been reported by King[31] who studied 47 different commercial detergents in a variety of organic solvents. In many of these cases foams are formed comparable in volume and stability to those of aqueous systems. Foaming was usually but not necessarily accompanied by lowered surface tension. Solvents of relatively high surface tension appeared to give better foams with the surface active solutes. Vapor pressure and bulk viscosity were found to affect foam stability but not foam initiation. The formation of non-aqueous foams resistant to high temperatures has been studied using solutions of various oils and soaps in paraffin wax.[32]

Aqueous systems have, of course, been studied much more extensively. Foulk and co-workers have investigated the foaming of boiler waters and salt solutions. They have evolved a theory of foam formation which includes solutions having higher surface tension than water (negatively adsorbed solutes).[33] Among other substances of slight surface activity whose foaming properties have been studied are glues[34] and various organic compounds of low molecular weight, such as acetic acid, ethyl ether, etc.[35] Constable and Erkut have recently reported on the foaming of amyl, butyl, heptyl, and octyl alcohols in aqueous solutions. They find that saturated solutions of these substances do not foam, but unsaturated solutions have marked foaming powers. A maximum is reached at about 50% saturation.[36]

The fact that foam properties vary with the nature of the gas phase is well recognized. Carbon dioxide foams behave differently from air foams in fire-fighting compositions,[37] for example, and nitrous oxide improves the foam of soap and whipping cream.[38]

The foaming properties of soaps have been studied very extensively, but in relatively few cases have the raw materials and experimental condi-

[30] Ross and McBain, *Ind. Eng. Chem.* **36**, 570 (1944).

[31] King, *J. Phys. Chem.* **48**, 141 (1944).

[32] Hartner-Seberich, *Ges. Abhandlungen Kenntnis Kohle* **11**, 628 (1934); *Chem. Abstracts* **29**, 6745.

[33] Foulk and Miller, *Ind. Eng. Chem.* **23**, 1283 (1931).

[34] Sauer and Aldinger, *Kolloid. Z.* **88**, 329 (1939).

[35] Sasaki, *Chem. Products* **2**, 143 (1939); *Chem. Abstracts* **34**, 1223; *Bull. Chem. Soc. Japan* **13**, 669 (1938); *ibid.* **14**, 3, 63, 107, 250 (1939).

[36] Constable and Erkut, *Rev. faculté sci. univ.   Istanbul* **9A**, No. 1, 69 (1944); *Chem. Abstracts* **40**, 1376.

[37] Amsel, *Oel u. Kohle* **38**, 293 (1942).

[38] *Perfumery Essent. Oil Record* **37**, 147 (1946).

tions been completely controlled. Miles and John Ross[39] have recently studied the soaps from a series of highly purified fatty acids. For each soap the foam height and foam stability varies with pH, and there is an optimum pH value at which maximum stability is obtained. The optimum pH varies with chain length of the soap, from pH 6 for C_{10} to pH 11 for C_{18}. Foam stability increases with the amount of undissociated fatty acid present, and foam height increases with increasing temperature in the range 27–82°C. In general, the calcium and magnesium soaps do not act as foam breakers for the sodium soaps, important exceptions being the oleates and ricinoleates. Merrill and Moffett[40] have shown that the foam stability of natural mixed fatty acid soaps is greater than that of pure products and that stability may be increased by a wide variety of electrolytes, gums, and organic liquids. Foaming power also varies with concentration of the soap and in most instances reaches a maximum at fairly low concentrations.[41]

It is well known that mixtures of soaps with the fatty alkyl sulfates or alkyl aromatic sulfonates do not foam as well as the individual constituents alone. Such mixtures attained large-scale use during the war as salt water and hard-water soaps. Winsor[42] has studied mixtures of several isomeric sodium tetradecyl sulfates with sodium stearate, and finds that the foam height vs. soap concentration curve shows a minimum. There is also a distinct time effect, the foaming power increasing as the solution is aged. These phenomena are attributed to a displacement of the initially formed detergent surface film by stearate ions. It has also been shown that the addition of substances such as ethylcellulose increases the foaming power of soap–detergent mixtures, particularly in hard water.[43]

The foaming of organic dyestuff solutions has been extensively studied by Breitner,[44] using an air-streaming method. He has drawn empirical correlations between solute concentration in bulk and in the foam, bubble size, bubble stability, and bubble wall thickness.

When two or more solutes are present in the same solution, their ratio in the foam will generally be different than in the bulk. This is to be expected because of the different surface adsorptions of the solutes. The

[39] Miles and J. Ross, *J. Phys. Chem.* **48**, 280 (1944).

[40] Merrill and Moffett, *Oil and Soap* **21**, 170 (1944).

[41] Anon, *Seifensieder-Ztg.* **69**, 20 (1942).

[42] Winsor, *Nature* **157**, 660 (1946).

[43] McDonald, *Soap Sanit. Chemicals* **21**, No. 12, 41 (1945); compare Ruckman, Hughes, and Clark, *ibid.*, **19**, No. 1, 21 (1943).

[44] Breitner, *Kolloid-Z.* **100**, 335 (1942); **101**, 31 (1942); *Chem. Abstracts* **38**, 3530.

concentration differential is often so marked that the solutes can be separated analytically by forming and removing foam from the solution. Mixed fatty acids in the form of soaps have been separated in this way, and the method is particularly effective in separating proteins from dilute solutions.[45] The effectiveness of separation varies not only with the nature of the constituents but also with pH, rate of foam formation, bubble size, temperature, and other factors. In any individual case the optimum values for these controlling factors must be worked out.

There has been considerable theorizing as to the fundamental molecular features of foam structure, but no simple picture has yet been brought forth which accounts for all the facts. There are vast differences in the stability of foams which sometimes correspond to minor differences in composition of the foaming solution. Individual films of certain soap solutions have been preserved for over a year, and many investigators have studied the so-called "black" films of soap. These are the extremely thin films formed on prolonged drainage. They are only about 50 Å in thickness, which means that as great a proportion of their mass is in the surface layer as in the interior.[46] This represents an extreme of stability. On the other hand, the same soap solutions with a small amount of excess alkali or salt may give foams of very poor stability.

Adam has pointed out that film stability depends on the rapidity with which the surface tension can change to accommodate local strains in the film and stress differences which may arise among different localities in the film. For a film to stretch without rupture the surface tension should increase as stretching occurs. The most stable foams are accordingly obtained at those concentration ranges at which surface tension is varying rapidly with concentration. The speed with which solute molecules can migrate to and from the surface should also have a great influence on foam stability. As a soap film is stretched the spaces between the molecules become larger. Since soap molecules diffuse slowly these spaces are filled with water molecules rapidly coming up from the interior. This results in the rise of surface tension without which the film would break. The reverse effect occurs when the film is compressed or contracted.[47] Matalon[48] has recently reported an interesting series of experiments using

[45] Dubrisay, *Compt. rend.* **194,** 1076 (1932). Ostwald and Siehr, *Kolloid-Z.* **79,** 11 (1937); *Chem.-Ztg.* **61,** 649 (1937). Schütz, *Nature* **139,** 629 (1937); *Trans. Faraday Soc.* **42,** 437 (1946).

[46] Perrin, *Ann. phys.* **10,** 160 (1918). Lawrence, *Soap Films,* Bell, London, 1929.

[47] Adam, *The Physics and Chemistry of Surfaces,* Oxford Univ. Press, London, 1941.

[48] Matalon, *Compt. rend.* **222,** 1213 (1946).

a modified form of the du Noüy surface tension apparatus on a series of soap solutions. The ring is initially submerged in the liquid just under the surface. It is then pulled through the surface slowly until it has broken free; meanwhile force is plotted against distance of travel. The curve so obtained rises linearly at first, then shows a slight dip, and finally levels off to a plateau portion which is maintained till the break. Maximum length of the plateau portion corresponds to maximum foam stability of the solution and is related to the ease of diffusion from surface to interior.

Foulk's balanced-layer theory of liquid film formation has already been mentioned. In essence this qualitative theory states that liquid films are formed by the approach of two already formed liquid surfaces. In the case

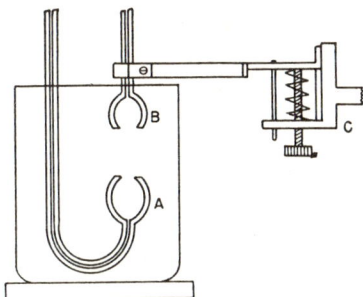

Fig. 48. Apparatus for bringing two bubbles together under surface of liquid.[33]

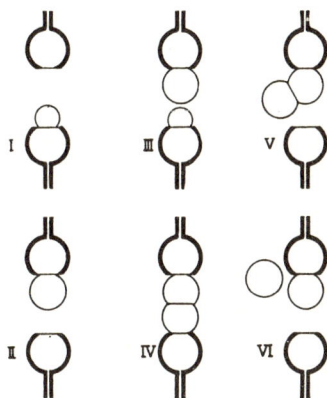

Fig. 49. Sketch of successive steps of two-bubble experiment.[33]

of solutions the mechanical force causing the approach of the liquid surfaces meets with resistance when the separating layer of liquid becomes very thin. A state of equilibrium is finally reached, and if the thin layer of liquid is stable it forms a foam unit. The source of resistance to close approach of one liquid surface to another lies in the difference of solute concentration between surface and interior. In pure liquids there is no resistance as two liquid surfaces approach; hence there is no foam. In solutions the *concentration differential* represents a state of thermodynamic equilibrium, *i.e.*, a state of minimum free energy for the system. When two surfaces approach very close to each other the effect is to establish between them a region of *uniform* concentration equal to the surface concentration. To do this requires work. The net effect is that the resistance to approach increases as the proximity increases. This effect can be demonstrated and studied in an apparatus designed to bring two bubbles together under the surface of a liquid, as shown in Figure 48.   The bubble is formed at air outlet, A, and on rising is caught in the bell, B.   A second bubble can then be blown at A, and its approach to the first bubble studied, as in Figure 49.   Here the effect obtained with strongly adsorbed solutes is shown.   With pure liquids the two bubbles coalesce showing that there is no resistance to the approach of the two liquid-air surfaces.[49]

These ideas, based essentially on the classical theory of surface tension and adsorption, have been extended by Michaud[50] to account for a variety of phenomena observed with foams and individual liquid films.

The essential problem of foam stability, however, remains to be worked out.   To date no completely satisfactory molecular picture of liquid film structures, accounting for their persistence, has been advanced.[51]

VI. Emulsification

(a) General Considerations

The ability of surface active agents to promote the formation and stabilization of emulsions is one of their outstanding characteristics.   The technical importance of emulsions and the tremendously wide scope of their application has made emulsification the subject of far more study than any other gross surface active effect.[52]

An emulsion can be defined as a suspension of liquid particles within a second immiscible liquid phase.   The suspended or dispersed phase is in the form of particles which may vary quite widely in size.   In some emulsions the particles may be so large as to be visible with the naked eye.   In others they may be in the range of less than 0.5 μ in diameter.   Emulsions in which the dispersed phase is very finely divided and present in small percentages differ markedly in both properties and structure from the more usual type of large particle emulsions.   An example is the very fine suspension of oil particles found in the water condensed from steam engines.   Another example well-known to organic chemists is the fine suspension of water-immiscible liquid in water which often is obtained as the condensate in steam distillations.   This type of emulsion seldom involves surface

[49] Foulk et al., Ind. Eng. Chem. **33**, 1086 (1941); **35**, 1013 (1943).

[50] Michaud, J. chim. phys. **41**, 147 (1944).

[51] Compare Sir Wm. Bragg, Proc. Roy. Inst. G. Brit. Advance Copy, Mar. 3, 1939; Chem. Abstracts **33**, 5721.   Dervichian, Corps Gras, Savons **2**, No. 6, 164 (1944); Chem. Abstracts **40**, 5271.

[52] Of the numerous publications dealing with emulsions, the reader is referred to Clayton Theory of Emulsions, Blakiston, Philadelphia, 1943, for a comprehensive, authoritative, and excellently presented treatment of the field as a whole.

active agents. It is often called an "oil hydrosol" to distinguish it from the normal type of emulsion with which we are concerned. Oil hydrosols are analogous to true sols of hydrophobic solid particles. In true emulsions the particle size is usually in the range of 0.2–50 μ, *i.e.*, easily visible with a microscope. They are therefore analogous to suspensions of solid particles in the same size range rather than to sols.

In order to be called an emulsion a dispersed liquid-in-liquid system must also have a reasonable degree of stability. If a mixture of pure benzene and pure water is shaken vigorously a dispersed system will be formed but the two liquids will very rapidly separate out into layers. It is questionable whether the highly dispersed system initially formed should be called an emulsion. If a small amount of soap is added before shaking, the dispersed system will be a true emulsion and will separate quite slowly. The degree of stability of an emulsion is one of its most important and fundamental properties, and a certain minimum degree of stability must obtain for the system to be classed as an emulsion. Stability is measured by the rate at which droplets of the dispersed phase coalesce to form larger and larger masses of liquid which separate by gravity and form a separate bulk layer. We have, therefore, set two arbitrary limitations on emulsion systems, one regarding particle size and the other regarding stability. A third limitation regarding composition may also be set, again with hazy rather than sharp boundaries. We have seen that pure single-component liquids do not form stable foams. Similarly, there are few if any cases known in which two pure immiscible liquids will form a stable emulsion, although they may form stable oil hydrosols. We may distinguish between emulsions and oil hydrosols by pointing out that emulsion systems not only consist of two liquid phases but also contain at least three components. The third component usually is soluble in one or both of the liquids and functions as the emulsifying agent. One or more emulsifying agents may be present. In some cases the emulsifying agent need not be a soluble substance but may be a very finely divided solid insoluble in either liquid phase. The surface active compounds in general function as emulsifying agents and are of primary interest in this discussion.

In almost all practical emulsions, one of the liquid phases is aqueous and is called the water phase. The other, of necessity, has only a limited solubility in water and is usually referred to as the "oil" phase regardless of its chemical constitution. Emulsions are classed into two types; oil-in-water (abbreviated O/W) and water-in-oil (abbreviated W/O), depending on which is the dispersed phase. The type of emulsion formed depends on the nature of the system's components, on the manner in which the emulsion is formed, and on the relative quantity of each component present.

It is possible in many cases to convert an O/W to a W/O emulsion or vice versa with relatively minor changes in the system as a whole. This phenomenon is called "inversion." One of the important characteristics of any given emulsion is the ratio of the volume of one liquid phase to the volume of the other liquid phase. This is called the "phase volume ratio." Although it is sometimes possible to bring about inversion merely by changing the phase volume ratio, this is less important in determining emulsion type, than is the nature of the emulsifying agent.

It is usually an easy matter to determine the emulsion type by any of several different techniques. A drop of the emulsion may be stirred gently in a larger volume of water. If it disperses easily it is an O/W emulsion, if not it is a W/O. A small amount of oil-soluble dye may be added and stirred into the emulsion. If the bulk of the emulsion becomes colored, the outside phase is oil. Another test of type is the rough measurement of the electrical conductivity. O/W emulsions will in general conduct as well as the isolated aqueous phase. In contrast, W/O emulsions will show high electrical resistance.

Emulsions can sometimes be quite complicated systems involving, besides the two liquid phases, one or more solid phases and dispersed air or foam. Of special interest are the so-called multiple emulsions. In a typical multiple emulsion the outside phase will be aqueous and the first internal phase will be in the form of oil droplets. Each of these oil droplets, however, will have dispersed in it smaller droplets of water.[53] Since our discussion is primarily concerned with the role of surface active agents in emulsification we need only consider the simpler types which serve adequately for purposes of illustration.

The average particle size of the dispersed phase is an extremely important characteristic of emulsions. It is related to the rate of creaming and the rate of breaking (stability) of the emulsion. Other things being equal the emulsion having the smaller particle size will exhibit the slower creaming rate and have greater stability. Fully as important as particle size itself is the size distribution, which is a quantitative measure of the uniformity of particle size. Since large particles, having a smaller phase boundary area per unit mass, are thermodynamically more stable than small particles, they tend to grow at the expense of small particles by coalescing with the latter. Therefore, an emulsion in which the particle size is uniform tends to be more stable than one with a wide size distribution. Particle size and size frequency are usually measured by direct microscopic observation.

Another important basic property of an emulsion is its electrical proper-

---

[53] Parke, *J. Chem. Soc.* **1934**, 1112.

ties. In practically all O/W emulsions the particles bear an electric charge. The intensity of charge and the zeta potential, as determined by measurements of electrophoresis, are important factors in the stability of emulsions. The aqueous particles in W/O emulsions are uncharged. The oil medium is non-conducting, and if an electrical double layer is formed in the neighborhood of the phase boundary it lies entirely within the aqueous phase.

In most emulsions there is a difference in density between the phases, and the dispersed phase tends to migrate by gravity either to the top or bottom of the containing vessel. The best-known example of this is the case of whole milk. On standing the dispersed fat globules float to the top and form a region where the emulsion contains a much higher volume ratio of fat globules. This fat-rich emulsion is called cream. Cream is still emulsified, for the fat globules have not coalesced, and cream exhibits the properties of a typical O/W emulsion. In emulsion technology the term *creaming* means a change in local phase volume ratio due to gravitional settling of the dispersed phase. It is clearly distinguished from "breaking," which means a coalescence of the dispersed particles. In creaming little if any such coalescence occurs. The rate of creaming is governed within limits by Stokes' law, and therefore depends on the density differential, the viscosity of the external phase, and the particle size. The importance of particle size is shown by the well-known example of homogenized milk. After milk has been homogenized by mechanical treatment, *e.g.*, by passage through a colloid mill, separation of a distinct cream layer will not occur during any reasonable time of standing. Homogenization of milk effects a reduction of particle size of fat globules and a narrowing of their size distribution range.

Possibly the most important property of emulsions from the technical point of view is their stability. Regarded as a simple system of two liquid phases, an emulsion is thermodynamically highly unstable. Consider a single globule of oil, 1 cm. in diameter, suspended in water. If this globule is split up into globules 1 μ (10^{-4} cm.) in diameter the total oil–water interface will be increased by a factor of 10,000. The resulting emulsion system therefore has 10^4 times as much free interfacial energy as the original non-emulsified system and is correspondingly less stable. We have assumed in the above example that the interfacial tension is the same in both systems. Actually emulsions cannot be formed unless the interfacial tension is first reduced to a relatively low value, but for any positive value of interfacial tension a highly disperse system is unstable with respect to a non-dispersed one. When two globules of oil suspended in water collide they will normally coalesce. In an emulsion, where an emulsifying agent is present, there will be relatively few instances of coalescence as compared with the number of collisions.   This ratio of coalescence to collision is an inverse measure of the emulsion stability.   The stability of any emulsion probably never reaches 100%, but can approach this perfect figure very closely under ideal conditions.   .The ratio of coalescence to collision is itself influenced by the nature of the protective film around each globule, which will be discussed in detail below.   It is also influenced by the violence of the collisions.   The latter factor can depend on temperature, on agitation or mechanical treatment, on the viscosity of the medium (high viscosity will slow down the motion of the globules and decrease the violence of the collision), and on the charge of the particles.   Stability also will depend on the frequency of collision, which in turn depends on phase volume ratio, temperature, electric charge, agitation, and viscosity.

In most technical emulsions a high degree of stability under ordinary storage conditions is desired.   Under conditions of use, however, it is most often desired that the emulsion break.   Emulsions are often employed, for example, as a convenient means for introducing an oil phase into a system, i.e., as a means for diluting oils.   This is the case in emulsion insecticides, emulsified floor waxes, spinning oils for textiles, etc.   In these cases the emulsion must be broken and the water phase removed, mechanically or by evaporation, before the desired result is achieved.   The stability and the conditions under which the emulsion may be broken are accordingly two of the most important gross characteristics of emulsions.   Stability is generally understood to mean the time an emulsion can stand under ordinary conditions of storage without breaking.   Accelerated tests of stability are usually made by increasing the gravitational effect, as by centrifuging,[54] and measuring the rate of separation of the inner phase.   Stability may also be correlated in certain instances with size–frequency analysis, and may be judged by changes in the size–frequency distribution.[55] Stability to elevated temperatures and chemical reagents, and resistance to mechanical action are also important.[56]

(b) Formation of Emulsions

The formation of emulsions is closely related to their stability.   In order to form an emulsion the oil phase and the water phase must be agitated together in the presence of one or more emulsifying agents.   The efficiency of emulsification depends among other factors on the type and

[54] Merrill, *Ind. Eng. Chem., Anal. Ed.* **15**, 743 (1943).

[55] Berkman, *J. Phys. Chem.* **39**, 527 (1935).

[56] Ohl, *Spinner u. Weber* **52**, No. 49, 9 (1934); *Chem. Abstracts* **29**, 8340.   Klingmann, *Bitumen* **5**, 206 (1935); *Chem. Abstracts* **30**, 2740.   Bochko, *Masloboino Zhirovoe Delo* **11**, 489 (1935); *Chem. Abstracts* **30**, 928.

degree of agitation, and the manner in which the emulsifying agent is introduced. The primary function of agitation is to break up of both phases so that the one which will become the inner phase is able to form small globules. The lower the interfacial tension the lower is the amount of mechanical energy needed to break up the phases. The various types of colloid mills and homogenizers are carefully engineered devices to obtain maximum shearing action on the fluids and facilitate the formation of fine, uniform globules. In some systems in which the interfacial tension is extremely low, spontaneous emulsification will occur with no external shaking. The phases are mixed by convection currents caused by diffusion and small temperature differentials. A solution of palmitic acid in Nujol (a highly refined petroleum white oil) will spontaneously form an emulsion if placed on an aqueous caustic soda solution. The alkali reacts with the palmitic acid at the phase interface, forming the soap (which acts as an emulsifying agent) *in situ*. The heat of reaction and the diffusion cause mixing of the phases and consequent emulsification.[57] It is noteworthy that Nujol will not spontaneously emulsify if placed on top of an aqueous sodium palmitate solution. Many other similar examples of spontaneous emulsification have been noted in systems of very low interfacial tension.[58] In most cases, however, agitation is needed for emulsion formation. It may be simple stirring with a rod or paddle, or it may involve mechanical devices. One of the most effective methods of forming emulsions is forcing the two liquids together through a narrow orifice or nozzle, the turbulent flow causing intimate mixing and dispersion.[59] Too severe agitation can cause breaking of the emulsion as fast as it is formed. An example of how different types of agitation can have opposite effects is afforded again by milk. Milk may be homogenized by being subjected to vigorous shearing for a short period of time. Cream, which is a concentrated milk emulsion, can be broken into butter by slow churning for a longer period of time. Ultrasonic waves have been used to supply the agitation necessary for forming emulsions, and under proper conditions they also serve as effective emulsion breakers.[60]

O/W emulsifying agents are usually more easily prepared if the emulsify-

[57] McBain and Woo, *Proc. Roy. Soc. London* **A163**, 182 (1937).

[58] Schulman and Cockbain, *Trans. Faraday Soc.* **36**, 651 (1940). Kling and Schwerdtner, *Melliand Textilber.* **22**, 21 (1941). Pospelova and Rehbinder, *Acta Physicochim. U. S. S. R.* **16**, 71 (1942); *Chem. Abstracts* **37**, 2245.

[59] Auerbach, *Chem. Tech.* **15**, 107 (1942). Clay, *Proc. Acad. Sci. Amsterdam* **43**, 852, 979 (1940).

[60] Söllner *et al.*, *Trans. Faraday Soc.* **31**, 835, 843 (1935); **32**, 556, 616, 1537 (1936). Krüger, *Pharm. Ind.* **1938**, 331. Brandt, Freund, and Hiedemann, *Kolloid-Z.* **77**, 103 (1936).

ing agent is formed *in situ*, as illustrated above.   If this is not possible the emulsifying agent should be dissolved or dispersed in the oil and the mixture added with stirring to the water.   Since most good O/W emulsifying agents are more soluble in water than in oil, they will diffuse rapidly into the water phase and tend to promote dispersion of the oil.   A less efficient method involves dissolving the emulsifying agent in the water and then adding the oil with stirring.

Emulsifying agents may be classed broadly into those which promote O/W emulsification and those which promote W/O emulsification.   The O/W emulsifiers are generally more hydrophilic than oleophilic, *i.e.*, they are more soluble in water than in oil.   They include the long-chain polar – non-polar compounds with which we are mainly concerned, and also the more complex hydrophilic colloids such as gums, starches, proteins, etc., which are known to be readily adsorbed at phase interfaces.[61]   The polar – non-polar compounds which are more oleophilic tend to promote W/O emulsification.   Examples are the fatty acids, fatty alcohols, sterols, etc.   High polymers which are primarily oil soluble but which contain polar groupings, such as ethylcellulose and the oil-modified alkyd resins, also promote W/O emulsification.   Of the solid-powder emulsifying agents, those which are hydrophilic in nature, more easily wet by water than by oil, promote O/W emulsification.   Among these are clays, hydrous oxides, etc.   Solids which are hydrophobic in nature, such as carbon black, powdered lignite, etc., promote W/O emulsification.   This action is easily explained by considering the contact angle at the solid–oil–water boundary.   With a hydrophilic solid the contact angle in the water is less than 90° and the oil–water interface between two solid particles is concave toward the oil.   The over-all effect of large numbers of such particles is one of making the total interface concave toward the oil, *i.e.*, formation of an O/W emulsion.[62]   The symmetrically opposite effect is caused by hydrophobic solids, and W/O emulsions are formed.

The behavior and utility of a large number of solid emulsifying agents has been surveyed by Bennister, King, and Thomas.[63]

The efficiency of an emulsifying agent may be estimated by adopting a standard water–oil mixture and standard technique, and observing the properties of the resulting emulsion.   Stability is usually measured, and a particle size and size frequency determination is made.   The latter may be

[61] Woodman, *J. Soc. Chem. Ind.* **53**, 57T, 115–16T (1934); **54**, 70T (1935).   Tice, *J. Am. Pharm. Assoc.* **24**, 1062 (1935).

[62] Carrière, *Chem. Weekblad* **27**, 638 (1930).   Bennister, King, and Thomas, *J. Soc. Chem. Ind.* **59**, 226 (1940).

[63] Bennister, King, and Thomas, *J. Soc. Chem. Ind.* **59**, 226 (1940).

determined indirectly by measuring the turbidity and light-scattering properties of the emulsion. In a similar manner, by fixing all variables except the one in question, the efficiency of emulsifying machinery, the emulsifiability of various oils or aqueous solutions, and the effect of physical factors such as temperature, etc., can be estimated.[64]

Another important property of emulsions is their viscosity. Emulsions may have a viscosity as low as that of the external phase alone. This is usually the case with very dilute emulsions. The viscosity generally increases as the concentration of the inner phase increases, and may become so high that the system is virtually solid. In regions near the inversion point (among those systems which give W/O in one concentration range and O/W in another) the curve of viscosity vs. phase volume ratio is apt to show a maximum. In moderately concentrated emulsions the viscosity is not proportional to the rate of shear as it is in normal liquids. This phenomenon, which is called anomalous viscosity, is quite common among disperse systems and solutions of large polymeric molecules. High viscosity in general increases the stability of emulsions since it lowers the frequency of collisions between dispersed globules and also the energy of each individual collision.

The viscosity of emulsions can be affected to an astonishing extent by relatively small changes in the nature and concentration of the emulsifying agent. As an example, an emulsion of 10 ml. Nujol in 40 ml. water with 75 mg. of sodium cetyl sulfate and 140 mg. of cholesterol present as the emulsifying agents possesses excellent stability and is quite fluid. The same emulsion with the cholesterol replaced by cetyl alcohol is also quite stable but has the consistancy of cold cream.[65]

(c) Theory of Emulsification

A satisfactory theory of emulsion structure should be able to explain all the important behavior characteristics of emulsions, namely: (1) Formation. (2) Stability. (3) Breaking and inversion. (4) The role of emulsifying agents and other chemical factors such as pH and non-surface active ions. (5) The influence of physical factors.

A great number of individual basic effects enter into emulsion phenomena; and it is only within recent years that a satisfactory integrated view of emulsion behavior has been developed. No attempt will be made in this limited discussion to follow chronologically the historical development of emulsion theory.

[64] Dobrowsky, Kolloid-Z. **95**, 286 (1941). Cohan and Hackerman, Ind. Eng. Chem., Anal. Ed. **12**, 210 (1940). Martin and Hermann, Trans. Faraday Soc. **37**, 25 (1941). Cataline, Reinish, and Jeffries, J. Am. Pharm. Assoc. **34**, 33 (1945).

[65] Neville, Am. Dyestuff Reptr. **34**, 534 (1945).

The two most important factors necessary for a stable emulsion are the existence of an interfacial film which may be regarded as an envelope around each dispersed globule, and low interfacial tension between the oil and water phases. The importance of the interfacial film is paramount, for example, in the case of emulsions of mercury or Wood's metal in aqueous media.[66] Of secondary importance in most instances are the electric charges on the particles, caused by adsorbed ions, and the phase volume ratio. We have seen that the effect of surface active agents in lowering interfacial tension is inextricably connected with their positive adsorption at the interface. In other words formation of an interfacial film is a prerequisite to marked lowering of interfacial tension. It is primarily the physical and chemical nature of the interfacial film, formed by adsorption of emulsifying agents from either or both phases, which controls the formation and stability of emulsions.

The behavior of certain adsorbed films can be deduced from microscopic observations by a technique first described by Mudd and modified by Nugent.[67] It consists in carefully placing a drop of a pure oil and a drop of an O/W emulsion on a slide and bringing them together by lowering a cover slip on them. The behavior of the already emulsified droplets as the larger pure oil–water interface contacts them can vary by stages from an instant coalescence to a repulsion and complete retardation of the advancing interface. This corresponds to differences in the resistance of the film to disruption and in the degree of hydration of the film.

There is ample evidence that in many instances the interfacial film in emulsions is of such thickness as to be visible under the microscope, rather than being monomolecular in nature. In such cases the film is sometimes highly solvated.[68] The primary importance of the interfacial film in emulsification was early recognized by Donnan and particularly by Bancroft.[69] The latter investigator regarded the interfacial film as an actual diaphragm separating the oil and water phases. This diaphragm would have a definite interfacial tension against the oil on one side and against the water on the other side. The side with the higher interfacial tension would be concave and the system would be W/O or O/W depending on which of the tensions was higher. Inversion would occur when, for any reason, the side of lower

[66] Kremnev, *Kolloid-Z.* **67**, 171 (1934). Tartar *et al.*, *J. Am. Chem. Soc.* **58**, 782 (1936). Bondy and Söllner, *Trans. Faraday Soc.* **32**, 556 (1936). *J. Phys. Chem.* **42**, 1071 (1938).

[67] Mudd and Mudd, *J. Gen. Physiol.* **14**, 733 (1930–31). Nugent, *J. Phys. Chem.* **36**, 449 (1932).

[68] Wilson and Fries, *Colloid Symposium Monograph* **1**, 145 (1923). Holmes and Cameron, *J. Am. Chem. Soc.* **44**, 66 (1922). Serrallach and Jones, *Ind. Eng. Chem.* **23**, 1016 (1931).

[69] Donnan, *Z. physik. Chem.* **31**, 42 (1899). Bancroft, *J. Phys. Chem.* **17**, 514 (1913).

tension acquired a higher tension than the other side.   This theory is highly persuasive.

The idea of a relatively thick interfacial film has- been further developed in more recent years by Roberts.[70]   According to the theory of this investigator the interfacial tensions and also the electrokinetic potentials differ on each side of the boundary.   The difference in interfacial tensions may be demonstrated experimentally as the difference in readings on a du Noüy tensiometer depending on the direction in which the ring is drawn through the interface.   In general, the existence of both W/O and O/W emulsion types and particularly the phenomenon of inversion indicates that curvature in one direction results in lower free energy than curvature in the other. This is readily accounted for if one assumes a film of macroscopic thickness with different interfacial tensions on the two sides.   It is more difficult to visualize if an interfacial monofilm is postulated.[71,72]

The most recent view on the role of interfacial films in emulsification has been developed largely by Schulman and co-workers, although many other investigators have contributed.   It has been found that the most stable emulsions of the O/W type are formed when there are two emulsifying agents present, one of which is primarily an oil-soluble W/O emulsifier and the other a water-soluble O/W emulsifier.   Such a case exists, for example, when sodium oleate or other soap is used at a pH low enough to permit hydrolysis. The actual emulsifying agent is then a mixture of soap and free fatty acid. Additional similar examples are furnished by cholesterol and sodium cetyl sulfate, and by higher fatty amines and their hydrochlorides.   Not all pairs of water-soluble and oil-soluble agents work well.   In order to function as an optimum emulsifying couple the two agents must form a stable complex monofilm at the interface.   The two molecular species must exist side by side in the monofilm.   This can result from the penetration of one molecular species into a homogeneous monofilm of the other species.   In the case of a good emulsifying couple the two components interact to form a condensed film of remarkable stability.   This is accompanied by an extreme lowering of the interfacial tension.   The two effects being concomitant, neither can be regarded as causing the other.   For O/W emulsions of maximum stability three other conditions must be met: (1) There must be an excess of the water-soluble agent present in the bulk of the aqueous phase.   (2) The interfacial film should be of the condensed liquid type, so that the concentration of the complex is high and the film is easily reformed on distortion.

[70] Roberts, J. Phys. Chem. 36, 3102 (1932).

[71] Hildebrand, J. Phys. Chem. 45, 1303 (1941).

[72] For descriptions of typical type reversal phenomena, compare Lachampt, Compt. rend. 220, 317 (1945), and Wilson, J. Chem. Soc. 1934, 1360.

(*3*) The oil droplets should be electrically charged, the sign of the charge being immaterial.

W/O systems are characterized by a rigid or solid-condensed film at the interface. The particles bear no effective charge. Since the oil is a non-ionizing medium, no ionic atmosphere or gegenions can be built up within it. Consequently, even if the water droplets possessed a charged interfacial film, it would have little effect in preventing coalescence. In practically all W/O emulsions the water droplets are of irregular shape rather than spherical. This is evidence of the rigidity of the interfacial film. According to this picture inversion of an O/W emulsion takes place in two stages. In the first place the charge on the oil particles is neutralized, allowing the particles to come together and coalesce. This may be done by the addition of appropriately charged adsorbable ions. If, at the time of coalescence, the elements are present for forming a solid film around the water "sacks" which are trapped by the coalescing oil droplets a W/O emulsion will result.

Schulman and Rideal, *Proc. Roy. Soc. London* **B122**, 29, 46 (1937); *Nature*, **144**, 100 (1939). Schulman and Cockbain, *Trans. Faraday Soc.* **35**, 716 (1939); **36**, 651, (1940). Schulman and Stenhagen, *Proc. Roy. Soc. London* **B126**, 356 (1938). Alexander and Schulman, *Trans. Faraday Soc.* **36**, 960 (1940). For further discussions relating to this theory and its application to specific systems, see also: Davis and Bartell, *J. Phys. Chem.* **47**, 40 (1943); Winsor, *Nature* **157**, 660 (1946); Aickin, *J. Soc. Dyers Colourists* **60**, 41 (1944); King, *Trans. Faraday Soc.* **37**, 168 (1941); Addison, *Nature* **144**, 249 (1939).

It should be noted that this theory is not adequate to explain all emulsion phenomena, although it is very satisfactory within its scope. The effects of electric charge, zeta potential, and phase volume, which are apparently of minor importance if the interfacial film is sufficiently strong, are of primary importance in many other systems.[73] It is logical to assume that the formation of interfacial films would affect both interfacial tension and electrophoretic mobility in corresponding degree. Powney and Wood, however, have found that the mobility of Nujol droplets is reduced to zero by concentrations of lauryl pyridinium iodide far too low to affect interfacial tension.[74] The charge on O/W emulsion particles can often be reversed, sometimes merely by increasing the *p*H, without affecting stability.[75] It has been concluded that when no chemical reaction takes place between the electrolyte and the emulsifying agent (*i.e.*, the film-forming compound) electrical charge does not determine type or stability of the emulsion.[76]

[73] Verwey, *Trans. Faraday Soc.* **36**, 192 (1940); *Chem. Weekblad* **36**, 800 (1939).

[74] Powney and Wood, *Trans. Faraday Soc.* **36**, 57, 420 (1940); **37**, 152 (1941).

[75] Dickinson, *Trans. Faraday Soc.* **37**, 140 (1941).

[76] King, *Trans. Faraday Soc.* **37**, 174 (1941). King and Wrzeszinski, *ibid.* **35**, 741 (1939); *J. Chem. Soc.* **1940**, 1513.

For the student of surface active agents primarily interested in the specific role of these products in emulsion phenomena, our main findings may be summarized as follows: (*1*) The science and technology of emulsions is extraordinarily complex, and in many of its phases is not at all concerned with surface active agents. The basic physical factors in emulsification and emulsion stability include mechanical working, viscosity, phase volume ratio, electric charge, ionic atmosphere, particle size distribution, and other factors, as well as the formation of interfacial films. (*2*) The surface active agents are of importance primarily because they form interfacial films. The formation and behavior of these films is governed more generally by van der Waals' forces than by coulombic forces, although in the complex emulsion systems the latter forces may greatly influence the nature of the film. The possible effect of primary valence forces, *i.e.*, chemical reaction, on the film is obvious. (*3*) In some cases the formation of films by surface active agents may be of secondary importance. In most technical emulsions, however, it is the major factor controlling stability, type, and ease of formation. (*4*) The most effective interfacial films are generally not formed from a single component, but are complex or multi-component films in which one component differs markedly from the other in its polar – non-polar balance. (*5*) The effectiveness of the surface active emulsifying agents can be predicted at least roughly by considering their polar – non-polar balance, the nature of their surface and interfacial films (if known), and their sensitivity to chemical influences.

VII. Detergency and Detergent Baths

(a) General Considerations

Detergency in its broadest sense merely means cleaning, and a detergent therefore is any agent which cleans. This definition, however, is far too inclusive for practical purposes. The term detergency as ordinarily used implies cleaning the surfaces of a solid object by means of a liquid bath in such a way that the cleaning process involves a physicochemical action other than simple solution. A spoon or a glass which has a deposit of dried sugar syrup on it may be cleaned by rinsing with water. The water simply dissolves the sugar residue leaving the surface of the utensil clean. Solution effects of this type are usually not considered pertinent in discussions of detergency, although there are several borderline cases. In dry cleaning, for example, oil stains are removed from fabrics by the solvent action of naphtha or carbon tetrachloride. It is questionable whether this action should be called detergency, although the removal of solid soil or hydrophilic soil in a dry-cleaning bath is certainly an example of true detergency. Detergency is generally considered to be an unusually enhanced cleaning effect of a liquid bath caused by the presence in the bath of a special agent, the detergent.   The detergent acts by altering the interfacial effects at the various phase boundaries within the system.

Physicochemical effects of interfacial energy, electrical charge, and adhesion are of primary importance in the consideration of detergency.   There are accordingly some borderline cases which might be considered detergency even though no liquid bath is present.   The erasing of a pencil mark with a rubber eraser is a typical example.   Here the primary action is one of abrasion or erosion.   Cleaning by simple abrasion is definitely outside the scope of detergent effects, but in erasing a pencil mark a further action takes place.   The graphite particles in the mark are abraded from the paper surface, together with some of the paper fibers in their neighborhood.   Some rubber particles are also abraded from the eraser, and these rubber particles adhere tightly to the graphite which is thus separated from the underlying paper.   The rubber–graphite adhesion complex is non-adherent to the paper and can readily be brushed or blown away leaving a clean paper surface.   The intimate contact afforded by the abrasive action facilitates formation of the rubber–graphite adhesion complex, but the complex formation itself involves surface forces.   For purposes of this discussion, however, we shall consider the typical detersive system to consist of the following elements: (1) a solid object to be cleaned, called the substrate; (2) "soil" or "dirt" attached to the substrate and to be removed in the washing process; and (3) a liquid "bath" which is applied to the soiled substrate.

The substrate may vary almost without limit in the configuration of its surface.   It may be smooth and flat as in the case of a glass window, or a painted wall.   It may, on the other hand, present a highly complex surface as in the case of a textile fabric.   Complex surface configurations are difficult to clean because the soil may be held mechanically even though it has been physicochemically detached from the surface.   This is particularly true in fabrics where soil may be mechanically held not only between yarns but also between individual fibers within a single yarn strand.   Substrates may also vary in the wettability of their surfaces.   Some may be hydrophilic, such as glass or wood.   Others, such as paint, some metals, oilcloth, etc., may be hydrophobic.   Both the physicochemical nature of the substrate surface and its geometry are important factors in the over-all detergent effect.

The soil, in most practical detergent systems, is even more variable than the substrates and usually less is known of its chemical and physical nature.   The soil may consist of a single solid or liquid phase but usually two or more phases are present, intimately and randomly mixed with each other and haphazardly disposed over the substrate surfaces.   Solid soils occur rather commonly, examples being dry clay or soot on fabric.   A more typical soil, however, will contain a water-insoluble oily or greasy material together with solid dirt. Some of the soil components or phases may be either soluble in the solvent component of the bath or susceptible to solubilization by the surface active agents in the bath. The latter effect is a particularly important factor in detergency. As mentioned above, it is generally understood that if all the soil can be dissolved, *i.e.*, molecularly dispersed, in the solvent component of the bath acting alone, the system does not afford a true example of detergency.

The bath in a detersive system is almost invariably a solution, the solute consisting of the detergent. Other solute components may also be present. If these contribute to the detergent effect they are called "builders." In some special cases the bath may be a suspension rather than a solution. A suspension of bentonite in water can exert a considerable cleaning effect in certain cases, and here the bentonite can be considered a true detergent. The action of suspended bentonite as a builder for a soluble detergent component is well known.

The over-all detersive effect in any given system depends first of all on the nature of the system itself, *i.e.*, on the composition and physicochemical and geometrical properties of substrate, soil, and bath. Detergency is however, a dynamic effect. It consists essentially of the removal of soil, from the substrate and segregation of the soil, usually by suspension or emulsification, in the bath. The detersive effect also depends, therefore, on all those factors which affect physicochemical interaction. The most important of these are temperature, degree of mechanical agitation, and duration of treatment. In most practical systems we do not have complete and detailed knowledge of all these factors. In laundering, for example, the nature of the soil is practically unknown. The same is true in rug cleaning, automobile washing, etc. In a few instances the nature of the soil is a well-known and constant factor. Such is the case in the boiling out of cotton and rayon piece goods in a textile mill, or the scouring of woolen yarns. But in these cases there are also unknown factors. The geometrical disposal of soil in the fabric, for example, is almost impossible to ascertain.

(b) Evaluation of Detergency

It is always a difficult and complicated matter to study an effect such as detergency which involves numerous factors. This is particularly true where full control over every factor cannot be attained. From the purely scientific point of view, it is desirable to know the absolute and relative importance of each factor affecting the process. This can sometimes be achieved in the case of simple systems. More often, however, the study is limited to a few more easily controlled factors. Among these the *bath composition*, particularly the nature of the solute detergent, is often the most interesting. It is obvious that a surface active product cannot be referred to as a "good detergent" or "poor detergent" unless the system is specified. But in such cases it is understood that the system is at least loosely specified by context or implication. In any event a *sine qua non* of detergency studies is a method of measuring the extent to which soil has been removed in the process. This involves measurements of both the initial and final amounts of soil present. The other prime requisite in detergency studies is control over the conditions of washing. We are assuming here that the compositions of substrate and bath are known.

There are two general categories of tests for determining detersive efficiency of a given bath. The most accurate is a field trial under conditions of actual use, *e.g.*, if laundry technologists are interested in trying a new detergent for white cotton goods they must ultimately test its efficiency by using it in the laundry wheels on a commercial load of soiled goods. Many wool-scouring plants test new detergents by scouring wool with them under their usual operating conditions. In tests of this nature the soiled substrates are not absolutely standard, but they fall within a definite range and are "standard" from a practical point of view. Over a long series of trials they usually give reproducible averages. Methods for estimating the soil removal in practical cleaning operations vary considerably in their accuracy. In laundry work soil removal is usually estimated visually. The same is true in commercial dishwashing and glassware washing in bars and restaurants. In wool-scouring plants, on the other hand, the degree of cleaning is measured very accurately by extracting and weighing the residual wool grease on the fibers.

One shortcoming of these practical tests is the large scale on which they are run. To avoid this, tests are often run in pilot plant equipment, or in laboratory scale duplicates of commercial equipment. Another disadvantage of the large-scale practical test is the fact that the operating conditions are usually fixed, either by custom or by limitations of the equipment. A new detergent bath to be evaluated may require somewhat different conditions to deliver its best performance. Accordingly it may be rejected, even though under other equally practical conditions it might outperform the standard bath. To overcome these defects of purely practical tests a number of artificial detergency tests have been devised. The better-designed testing systems of this nature have not only been used for measuring the detergent power of a bath. They have also become indispensible instruments for studying detergency as a physicochemical phenomenon, that is for discovering and weighting all the individual factors which contribute to the detersive effect. The great preponderance of published work in this field deals with the cleaning of textiles and our discussion for the most part will be directed to this class of substrates. The artificial detergency tests fall into two classes, the first of which may be called semi-practical. This type of test is carried out on a small scale using artificial washing conditions, but the sample to be cleaned is taken from a full-scale plant lot of soiled goods.

A large number of laboratory scale semi-practical tests have been described, only a few of which need be considered in any detail. They are particularly important in the textile industry. Detergents for scouring raw wool may be tested on wool which has been thoroughly carded by hand to remove burrs, straw, and other loose vegetable matter. This wool has a remarkably white and clean appearance, but it is extremely greasy to the touch and actually contains up to 30% or more of natural wool wax. The detergent bath is made up to appropriate strength, usually about 0.05% detergent and 0.3% soda ash. Weighed samples of the wool are washed in two or three successive baths, being squeezed out between baths. Gentle agitation is used in the baths and the temperature is kept at about 120°F. After the final bath the wool is rinsed, squeezed out, and air-dried. After being dried, a carefully weighed sample of the clean wool, usually 5 or 10 grams, is extracted with petroleum ether in a Soxhlet apparatus. The extracted grease is weighed. For satisfactory scouring it should amount to less than 1% of the weight of wool.

Detergents to be used for scouring greige woolen piece goods are evaluated in a very similar manner using swatches of actual greige woolen or worsted. The "soil" in this case consists of the oily lubricant which had been added to the clean fiber to facilitate spinning and weaving operations. The amount and composition of such lubricant present on the wool is accurately known and controlled.

In the preparation of cotton and rayon piece goods for dyeing it is very important to clean the fabric thoroughly so that it will be hydrophilic and absorbent upon entering the dye bath. This is accomplished by a thorough scouring process which not only removes the sizing material but also, in the case of cotton, the natural wax which is present on the fiber. Detergents for this purpose are sometimes tested on actual greige fabric. The fabric is washed with the detergent under conditions similar to those encountered in actual plant operation. Temperature, time of treatment, bath concentration, degree of agitation, and liquor length are carefully controlled. The "liquor length" is the ratio of the bath weight to the weight of goods treated. In most scouring operations it is of the general order of twenty to one. After being washed, rinsed, and dried, the fabric test strips are partly immersed in water. The height to which the water rises by capillarity or wicking action and the rate of rise are then observed.   These quantities indicate the extent to which hydrophobic impurities have been removed, *i.e.*, the efficiency of cleaning.

Another simple and serviceable method for evaluating the detergent power of a surface active agent has recently been described by Kornreich.[77] It consists essentially jn washing a sample of raw silk in the boiling detergent solution for a fixed tim e.   The material is rinsed hot and is then stained with Shirlastain A, a mixture of dystuffs developed by the British Cotton Industry Research Association and marketed by *Imperial Chemical Industries Ltd.* This stain gives widely different shades on silk fibroin (the clean fiber) and serecin (the gum).   The degree of gum removal therefore corresponds to a series of intermediate shades assumed by the washed and stained sample. This method is particularly adaptable for controlling the soap content of textile mill washing liquors.   It has the same limitations as other specific detergency tests in that the results are not directly translatable to other detersive systems.

The evaluation procedures outlined above are called semi-practical because the materials to be cleaned, *i.e.*, the soil–substrate complexes, are taken from actual practical systems.   The completely artificial tests afford control over the soil and the methods of application, as well as over the washing conditions.   They are much more widely used than the semi-practical tests, particularly by detergent manufacturers and marketers.   This is easily understood in view of the difficulty of obtaining representative soiled samples for semi-practical tests, even though the latter give a much better indication of the expected field performance.   The artificial tests vary considerably among themselves, particularly with regard to soiling procedures and washing conditions.   The methods for estimating soil removal are less varied.

The method which is most widely used in this country for determining detergency involves the use of a mechanical washing device called the Launderometer.   This consists of a thermostatted bath fitted with a rotating clamp for holding a series of Mason jars.   The samples of standard soiled cloth are placed in the Mason jars together with a measured amount of the detergent bath to be tested and a number of small stainless-steel balls.   The balls provide mechanical agitation and a light beating action as the jars are rotated.   The bath temperature is fixed at the desired level.   The rate of rotation, and the number and size of the balls control the degree of agitation.   The washing may be carried out for any length of time, ten to thirty minutes being the usual span.   After washing the samples are removed from the jars and may be rinsed in the machine or by hand as desired.   They are then dried and the degree of soil removal is estimated.   This may be done

---

[77] Kornreich, *Textile Mfr.* **72**, 271 (1946).

on a comparative or an absolute basis. If one of the detergent baths is adopted as standard, samples from the other baths may be compared visually with the sample from the standard bath. A more elegant and absolute method of grading is the measurement of the degree of whiteness (or blackness) of the washed samples and also of the unwashed control sample with a reflectance photometer. This has the great advantage that the actual weight percentage of soil removed in the washing procedure can be estimated since it is a function of the increase in whiteness on a sample which has been properly soiled. There are several photometers suitable for this purpose available commercially. They employ a photoelectric cell as the sensitive element, and usually a block of magnesia as the standard 100% white element for comparison. A simple type of reflectometer has recently been described by Hurwitz.[78]

The Launderometer is a trade-marked instrument manufactured by *Atlas Electric Devices Co.* of Chicago.[79] It has been widely adopted because it is well designed to control conveniently the mechanical and physical factors which affect the washing process, notably time, temperature, and degree of agitation. Actually any other device by which these three variables may be controlled can be used for carrying out artificial detergency tests. Washing may be carried out in beakers, stirring by hand, in the same manner that dyeing tests are often carried out. This seemingly crude method was used at one of the most important detergency evaluation laboratories in the world, that of the *I. G. Farbenindustrie*. Actually a skilled technician can attain fully as good reproducibility with this method as with a Launderometer. Washing tests are often carried out in single tumble jars using pebbles, wooden or metal balls, or ordinary rubber stoppers as the frictional elements. Such devices simulate the action of the Launderometer.

The Research Committee of the American Association of Textile Chemists and Colorists has developed a laboratory washing machine called the Detergency Comparator. It is essentially a miniature washing dolly similar to those used for scouring woolen piece goods in textile mills. It is particularly adapted to measuring detergency on woolen piece goods and gives results which closely duplicate plant performance.[80]

Another widely used device is a miniature commercial laundry wheel, or a small-size home washing machine.[81]

[78] Hurwitz, *Am. Dyestuff Reptr.* **36,** 83 (1946).

[79] For a complete description of the Launderometer, see *Year Book*, Am. Assoc. Textile Chemists and Colorists, 1944, p. 149.

[80] See "A Study in Detergency," a paper presented by the Northern New England Section of the A.A.T.C.C., *Am. Dyestuff Reptr.* **36,** 91 (1947).

[81] See, for example, Rhodes and Brainerd, *Ind. Eng. Chem.* **21,** 60 (1929); Gehm, *Seifensieder-Ztg.* **68,** 159 (1941).

There is considerably more latitude in the preparation of standard soiled goods than in the mechanical devices for washing.   A large number of soiling formulas has been published.   The variation among them corresponds to the various "natural" soils they are intended to simulate.   The American Society for Testing Materials has recommended the following soiling formulas for goods to be tested in the Launderometer.[82]

For cotton: 6 g. Oildag (a commercial graphite–mineral oil suspension), 1.7 g. Wesson Oil (an edible vegetable oil), and 3000 ml. carbon tetrachloride.
For worsted: 0.5 g. lampblack, 7.5 g. edible tallow, 25 g. Nujol (medicinal mineral oil), and 4000 ml. carbon tetrachloride.

In each case the carbon tetrachloride acts merely as a solvent or carrier for the soil.   The goods are impregnated with the soiling suspension, squeezed out, and dried.   They may be passed through the soiling mixture several times until the evenness and degree of soiling is satisfactory.   Soiling may be carried out mechanically, using a small pad roll and treating long, narrow ribbons of the fabric which are later cut into individual swatches.[83] With these soils hard-water detergency tests may be made using 50 to 500 or more p.p.m synthetic hard water containing both calcium and magnesium in the ratio of 60 to 40.   The detergents to be tested are usually used in 0.1 to 0.4% concentration.

Most of the soils recommended for use with the Launderometer or similar washing devices consist of lampblack, graphite, carbon black, or similar dark, solid, insoluble material together with a mineral oil, a glyceride oil, and a volatile solvent such as carbon tetrachloride, which serves merely as a carrier to aid in depositing the soil on the fabric.   A soil intended to test detergents for scouring gray cotton has been described which includes lampblack, an edible shortening, Nujol, and wheat starch.[84]   India ink and mixtures of India ink with vegetable and mineral oils have also been used as soils.[85]   Natural products such as lanolin, egg yolk, egg albumen, cocoa, soot, and milk have also been described.

The standard soils and washing procedures used by the *I. G. Farbenindustrie* laboratories have been described in detail by Baird, Brown, and Perdue.[86]   For detergents designed to scour wool in textile mills the soil

[82] Crowe, *Am. Dyestuff Reptr.* **32**, 237 (1943).

[83] Van Zile, *Oil and Soap* **20**, 55 (1943).

[84] Bacon, *Am. Dyestuff Reptr.* **34**, 556 (1945).

[85] Jaag, *Tech.-Ind. schweiz. Chem.-Ztg.* **25**, 331 (1942); *Chem. Zentr.* **1943**, I, 2254; *Chem. Abstracts* **38**, 4339.

[86] Baird, Brown, and Perdue, *PB 32565*, Office of Technical Services, Dept. of Commerce, Washington, D. C.

consists of a mineral oil – glyceride oil mixture containing a standard amount of Sudan Red 7B, an oil-soluble dye with no affinity for wool. After washing, the amount of residual oil is estimated by extracting the wool with carbon tetrachloride, diluting to a standard volume, and measuring colorimetrically. Washing is carried out by hand in beakers or in a miniature washing dolly, similar to a full-scale wool-scouring machine. In a more severe washing test a mixture of mineral oil, lanolin, and black lead is used as the soil. This is printed onto the fabric in a design of large circular spots. The results are compared visually or photometrically with those obtained from a standard detergent.

For cotton washing and household wool washing the soil consists of tallow and mineral oil mixed with "natural road dust" or an "artificial road dust." The former consists of actual street sweepings ground and sifted. The latter is made of 96.5 parts clay, 2 parts black iron oxide, 1 part yellow iron oxide, and 0.4 part lampblack. This soil is applied in carbon tetrachloride suspension by means of a padding mangle. Results obtained by washing samples soiled in this manner are judged by photometric measurements. Washing tests are run at various temperatures and liquor lengths, with soft and hard waters, and with various builders as well as the detergent to be tested present in the bath.

A special soil is used to test spotting detergents and special spot removing compositions. This soil consists of lubricating grease, lanolin, graphite, and spindle oil. 1 ml. of this mixture is applied to the fabric in such a manner as to produce a spot about 2–3 cm. in diameter. The detergent is applied by hand; the time required for complete eradication of the spot measures the efficiency of the detergent.

It should be noted that the length of time elapsed between soiling and washing as well as the conditions under which the soiled goods are stored has a marked effect on the ease with which soil is removed. This is particularly true with soils containing glyceride oils. In general, the older the soiled goods, the more difficult it is to clean them. For optimum reproducibility it is recommended that the goods be washed not less than 24 hours nor more than one week after soiling.

Among other soils which have been recommended is the solid dirt recovered from the naphtha used in commercial dry cleaning.[87]

The results obtained with solid-containing soils, measured by the reflectometer method have a low order of reproducibility. Even when a large number of reflectance measurements are made on each sample, and the results carefully averaged there is a variation of about ± 10–15%

[87] Ringeissen, *Teintex* **8**, 31 (1943); *Chem. Abstracts* **38**, 6120.

in the calculated detersive efficiency. The detersive efficiency is defined as:

$$100 \times \frac{\text{units of soil removed by test detergent}}{\text{units of soil removed by standard detergent}}$$

The "units" of soil removal are taken as proportional to the reflectance measurements. It is accordingly a comparative measurement between two detergents. The reproducibility in terms of unit soil removal for any single detergent and soiled substrate is of course much better.

It has been recommended that the soil remaining after washing be estimated not only by reflectance but also by extraction and weighing. Measurements of this type have demonstrated that in many cases the solid dirt and oily dirt are removed selectively. A sample may, for example, lose almost all the oily dirt so that an extraction measurement shows that a high degree of cleaning has been attained. It may at the same time have retained most of the solid dirt so that a reflectance measurement indicates relatively little cleaning.[88]

Possibly the most accurate method so far devised for evaluating detergents to be used on fabrics is that of Woodhead, Vitale, and Frantz.[89] Although quite lengthy and laborious, this method gives a standard deviation in the calculated detersive efficiency of only ±3.4%. Twenty lots of standard soiled cotton fabric are used. The fabric is soiled as described above for Launderometer tests. Twenty sets of swatches are actually washed in a single experiment, each set consisting of twenty individual swatches, including one selected at random from each lot of soiled fabric. A home type washing machine is used and is filled with 10 gallons of tap water (of known hardness; any other type of water may also be used) at 110°F. Soap is added to make a 0.3% solution. The machine is started and fifteen clean towels are added one by one. Then the sets of soiled swatches are added one by one in rapid succession. Washing is continued for twenty minutes after adding the last set. The swatches are separated from the towels and are rinsed together in lukewarm tap water. They are then squeezed out and air-dried.

The "spread," or units of soil removed, is calculated from blackness measurements made with a photoelectric reflectometer. The measurements are expressed in terms of percent black as compared with an arbitrary 100% black standard. A swatch from each set including the control is selected at random and measured in a group with similarly selected swatches from other sets. This is repeated until all the swatches have been examined.

[88] Holland and Petrea, *Am. Dyestuff Reptr.* **32**, 534 (1943).
[89] Woodhead, Vitale, and Frantz, *Oil and Soap* **21**, 333 (1944).

Reading a swatch consists of taking ten measurements, five on each side. Thus over 4000 individual reflectance measurements must be made for each complete experiment. The standard deviation for tests with a single soap is of the order of ±0.6%. For

$$\text{detersive efficiency} = 100 \times \frac{\text{spread of unknown soap}}{\text{spread of known soap}}$$

the standard deviation is about ±3.4%. This method will therefore detect differences between two soaps which actually differ by less than 5% in their true detergent power. It may also be used to indicate differences in the efficiency of various washing machines by running tests in the different machines with a single soap.

A variety of other methods for calculating the amount of soil removed and expressing it in mathematical terms has been described, some of which are more complicated than the inherent accuracy of the method justifies.[90]

Another important modification of this general type of detergency test consists in submitting each sample to a series of washings rather than a single washing. The reflectance after each washing may be measured as well as the ultimate whiteness after the final washing. This method of multiple washes is particularly useful in testing detergents intended for commercial laundry use.[91]

Relatively less attention has been devoted to the evaluation of detergency on non-textile materials, and only a few examples need be considered here. The cleaning of metals prior to plating and finishing is of great commercial importance, and it is often difficult to judge the relative merits of detergents designed for this purpose. Morgan and Lankler[92] have devised an ingenious method for determining the residual oil on a metal surface by photographing the fluorescence under ultraviolet light. The metal prior to the test cleaning is soiled with a fluorescent oil or an oil containing a fluorescent dye. This method not only indicates the quantity of residual oil but also shows how it is distributed.

The cleaning of ceramic and glass surfaces in connection with commercial dishwashing has been studied extensively by Wilson and Mendenhall.[93] In this case the "soil" consists not only of adherent oily, proteinaceous, and starchy food residues, but also of films built up by repeated application of hard-water solutions of the usual alkaline cleaners. The visual effect of

[90] See, for example, Heron, *Textile Mfr.* **71,** 253 (1945).

[91] Vaughn, Vittone, and Bacon, *Ind. Eng. Chem.* **33,** 1011 (1941). Kind, *Fette u. Seifen* **45,** 383 (1938).

[92] Morgan and Lankler, *Ind. Eng. Chem., Anal. Ed.* **14,** 725 (1942).

[93] Wilson and Mendenhall, *Ind. Eng. Chem., Anal. Ed.* **16,** 251 (1944).

these films is measured by means of a specially designed photometer. Washing is carried out on glass plates in a commercial home type dishwashing machine.

Other methods for studying and evaluating dishwashing detergents have been discussed in detail by Gilcreas and O'Brien.[94] The closely related and extremely important problem of washing milk bottles has also been studied extensively. A comprehensive review has been presented by Liddiard.[95] The general problems involved in evaluating dishwashing detergents have been discussed recently by Tiedman.[96]

It cannot be emphasized too strongly that only an actual detergency test will predict the detersive performance of a surface active agent. As the theory of surface activity has evolved there have been numerous attempts to correlate detergency with simpler and more easily measured effects. It has become increasingly apparent that simple effects such as surface tension, interfacial tension, wetting power, etc. all are factors in detergency. Considered individually, however, any close correlation with detergency occurs only under limited conditions. There is still a tendency on the part of many workers in allied fields to judge or predict the detergent power of a product on the basis of its interfacial tension, dispersing power, or other easily measured property. Possibly the most widely held among these fallacious views is that detergency is directly related to foaming power. Another is that detergency can be predicted by surface tension measurements.[97] The present state of development of fundamental theory, however, is such that laboratory tests for detergency should be empirical, and should simulate actual operating conditions as closely as possible.[98]

For reviews dealing with evaluation of detergents, see: Götte, *Chem. Ztg.* **68**, 86 (1944). Kraus, *Wollen- u. Leinen-Ind.* **57**, 265 (1937). Oesterling, Univ. Microfilms (Ann Arbor, Mich.) Pub. No. 368 (1941). Sisley, *Corps Gras, Savons* **1**, 66 (1943). Chwala, *Fette u. Seifen* **48**, 435 (1941).

(c) Fundamental Mechanisms and Factors Affecting Detergency

The numerous published studies of detergency have been developed from several different points of view. Some have been aimed at developing reliable testing and evaluation procedures, and these have already been considered. Others have proceeded to adopt one or more testing procedures and used these to compare the effectiveness of various detergent baths.

[94] Gilcreas and O'Brien, *Am. J. Pub. Health* **31**, 143 (1941).
[95] Liddiard, *Dairy Inds.* **5**, 7, 36, 67, 104, 131, 138 (1940).
[96] Tiedman, *Soap Sanit. Chemicals* **23**, No. 4, 48 (1947).
[97] Hetzer, *Fette u. Seifen* **44**, 425 (1937).
[98] Compare Whitehead, *Ciba Rev.* **1945**, No. 49, 1789.

Bath composition is probably the most interesting variable involved in detergency, particularly from the practical point of view. It is still only one of the factors affecting the detersive process, however, and the importance which should be assigned to it can only be established if all the other factors are known and recognized.

We shall therefore first consider some of the attempts to arrive at a comprehensive theory of detergency. For the most part, these involve a resolution of the over-all detersive effect into a series of simpler basic or gross effects such as solubilization, suspending action, lowering of interfacial tension, etc. Furthermore, they involve a thorough mechanistic analysis of the dynamic detersive systems under consideration. The mechanism of detergency varies considerably from system to system and only in the simpler ones is it understood with any degree of completeness. Various investigators have emphasized different points of view and some of these can be considered individually before attempting an integration.

Quite early in the modern study of detergency (from the colloid chemical point of view) it was recognized that the emulsifying power of the bath and the protective-colloid action of the bath on the suspended dirt particles were of great importance. Deflocculating or dispersing action was also explicitly recognized as a major factor in detergency.[99, 100] The importance of surface and interfacial adsorbed films was also recognized.

McBain has summarized the earlier theories of detergency and given abundant references in *Advances in Colloid Science*, Vol. I, Interscience, New York, 1942, p. 99 *et seq.*

Spring[101, 102] first showed that detergency could be regarded as the effect of forming adsorption compounds between soap and dirt, and soap and fabric, at the expense of the previously existing adsorption complex consisting of dirt and fabric. The essential action was considered to be one of making the dirt surface more hydrophilic by adsorbing a layer of detergent. Rhodes and Brainerd extended this idea in what may be regarded as the first typical modern detergency study.[103] They showed that detergency involves adsorption equilibria by washing soiled and unsoiled goods together and showing that the dirt redistributed itself between the bath and the soiled and unsoiled goods. Furthermore, it was shown that there is a definite maximum degree of soil removal attainable in any single bath, and the time necessary to attain this maximum may be quite lengthy. They further pointed out

[99] Goldschmidt and Weissman, *Kolloid-Z.* **12**, 18 (1913).
[100] Zsigmondy, *Ann.* **301**, 29 (1898).
[101] Spring, *Kolloid-Z.* **4**, 161 (1909).
[102] Spring, *Kolloid-Z.* **6**, 11, 109, 164 (1910).
[103] Rhodes and Brainerd, *Ind. Eng. Chem.* **21**, 60 (1929).

that the soil is redeposited from the bath onto the fabric, *i.e.*, that the removal and redeposition of soil come to a dynamic equilibrium. It has recently been proved that in this type of system the rate of soil removal is directly proportional to the residual removable soil content of the fabric.[104] Results of these studies may be expressed in graphic form by plotting per cent soil removal against duration of washings. Heron[105] has extended the washing times out to several hours using a variety of detergents. He finds that detergents may be divided into two categories, those which give maximum soil removal within about 10 minutes, and another class which requires about four hours to give maximum soil removal under the same conditions.

The removal of the natural grease from wool fibers in a detergent bath has been studied in a very ingenious manner by Adam.[106] He actually observed the removal of grease from single fibers under the microscope, and found that the original continuous layer of grease rolled up into comparatively large globules which detached themselves from the fiber. Mild agitation aided this process. The main function of the detergent in this system appears to be the alteration of the contact angle at the fiber–grease–bath interface. To check this theory Adam devised an apparatus to measure the advancing and receding contact angles of a single wool fiber at the air–fiber–bath interface. This was done with both clean and greasy fibers. When the bath was pure water the contact angle at the fiber–grease–bath interface was greater than 90°. Addition of detergent to the bath lowered the contact angle to 0°. It was concluded that the advancing angle was the more important factor in detergency. Among other factors considered of importance by Adam are dispersion and emulsification of the grease once it is removed, wetting of the fibers, and penetration of the bath into the yarns of a fabric. He points out also that interfacial tension between the soil or grease and the bath may sometimes be only of relatively minor importance in detergency. Cetanesulfonic acid and cetylsulfuric ester show almost the same surface and interfacial tensions, but the latter is a far better detergent than the former. Furthermore, the interfacial tension between Igepon T solutions and oil becomes extremely low in the presence of calcium or lanthanum ions, but the detergency is not improved.

A similar study of detergency on oiled single fibers has been made by Kling, Langer, and Haussner.[107] Their excellent photomicrographs are shown in Figure 50. They report that the detersive mechanisms are similar on wool, viscose rayon, and cuprammonium rayon. The rate of oil removal

[104] Vaughn, Vittone, and Bacon, *Ind. Eng. Chem.* **33**, 1011 (1941).
[105] Heron, *Textile Mfr.* **71**, 253 (1945).
[106] Adam, *J. Soc. Dyers Colourists* **53**, 121 (1937).
[107] Kling, Langer, and Haussner, *Melliand Textilber.* **25**, 198 (1945).

depends primarily on the nature of the oil. Those oils which contain free fatty acid are removed most rapidly, followed in order by neutral glyceride oils and mineral oils. Mathieu[108] has emphasized the importance of the interfacial monomolecular layer of detergent in the washing process. He points out particularly its applicability in considering Adam's experiments.

Tomlinson has emphasized the importance of Adam's views, but also points out that suspending power of the bath is equally important. In the case of a soil containing both oil and solid particles, the oil, as it separates in globules from the fiber, may retain the solid.[109] The effect of the adhesion of solid dirt particles to fabric has been studied by Powney and Noad[110] in a very revealing series of experiments. They immersed cotton fabric in various baths containing suspended particles of ilmenite and then measured the amount of ilmenite deposited on the fabric. The presence of simple alkalis such as sodium carbonate or sodium hydroxide in the bath increased deposition. This effect was attributed to a sodium ion effect rather than a pH effect. Silicates and phosphates, on the other hand, particularly sodium hexametaphosphate and tetrasodium pyrophosphate prevented deposition very effectively. This is attributed to adsorption of the anion. Soaps also prevented deposition, being much more effective in this respect than the fatty alkyl sulfates.

The same problem has been attacked in a slightly different manner by Palmer and Rideal.[111] Particles of marble were coated with a hydrophobic coating by immersing in a solution of stearic acid in gasoline. The coated particles were then suspended in dilute solutions of several different synthetic detergents. The suspension cell was coated on the bottom with a layer of wax. After an interval of settling the number, n, of particles adhering to the bottom was counted. The cell was then inverted and the number, n', of particles which remained stuck to the wax layer was counted. An "adhesion number," $N = 100(n'/n)$, was then assigned to the system and proved to be characteristic of the bath.

Aickin studied the detergency toward oil-soiled wool of a series of secondary fatty alkyl sulfates. Removal of oil from the fibers and also the dispersion of removed oil in the bath is affected markedly by the presence of electrolytes therein. Adsorption of the detergent by the fibers also plays a part in the detersive action.[112] The power of the hydrated colloidal surface active agents to form gel-like protective films on both soil particles and

[108] Mathieu, *Corps Gras, Savons* **1**, 168 (1943).
[109] Tomlinson, *Mfg. Chemist* **15**, 159 (1944).
[110] Powney and Noad, *J. Textile Inst.* **30**, T157 (1939).
[111] Palmer and Rideal, *J. Chem. Soc.* **1939**, 573.
[112] Aickin, *J. Soc. Dyers Colourists* **60**, 60, 170 (1944).

fibers has been emphasized as a major factor in detergency by both Seck and Lindner,[113] and also by Hall.[114]

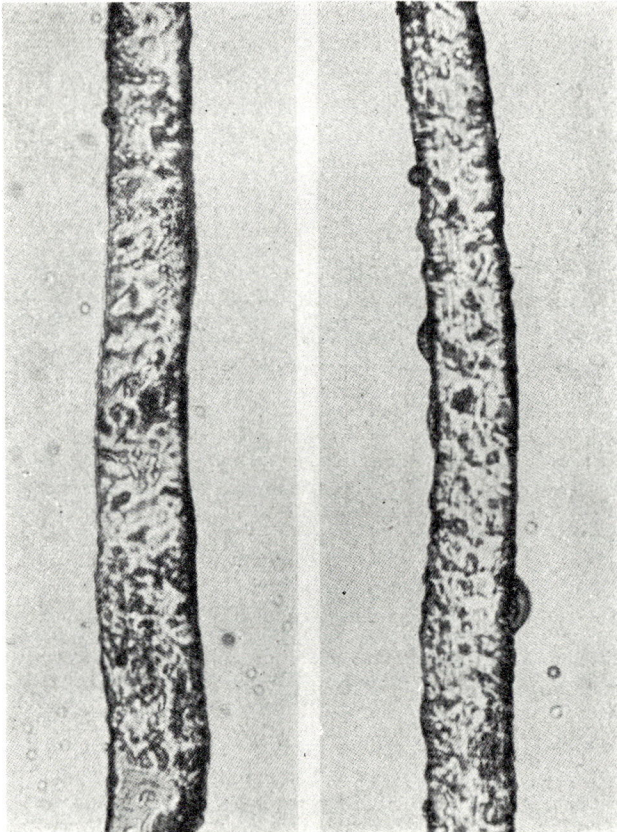

Fig. 50.   Part A.

Fig. 50 (in 4 parts).   Detergency experiments with oiled wool fibers.[107] Parts A and B, castor oil; Parts B and C, neutral glyceride oil (unspecified origin).

Neunhoeffer[115] has advanced some interesting speculations as to the nature of the adsorption of detergents by soils.   It is postulated that in a pair or group of associated soap ions the paraffin chains are juxtaposed and held

[113] Seck, *Fettchem. Umschau* **42,** 120 (1935).   Lindner, *Fette u. Seifen* **43,** 214, 253 (1936); **44,** 47 (1937).
[114] Hall, *Am. Dyestuff Reptr.* **27,** P612 (1938).
[115] Neunhoeffer, *Kolloid-Z* **107,** 104 (1944).

together by the "pressure" (kinetic bombardment) of water molecules. The free ionic ends of the soap molecules tend to repel one another and a strain results.   On contact with a hydrophobic surface the paraffin chain

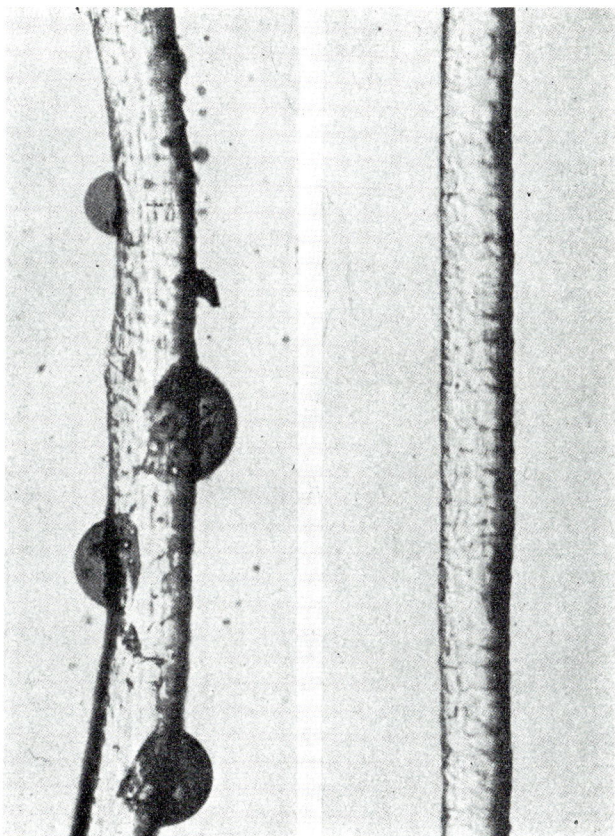

Fig. 50.   Part B.

ends can separate and the strain is thus relieved.   This accounts for the wetting powers of paraffin chain salts.

Solubilization as a factor in detergency has been emphasized by McBain, but its relative importance in practical detersive systems has not been fully ascertained.   McBain also emphasizes protective-colloid action, suspending power, and base exchange as major factors in detergency.[116]

[116] McBain *et al.*, *Chem. Products* **4**, 19 (1941).   McBain, in *Advances in Colloid Science*, Vol. I, Interscience, 1942, p. 99 *et seq.*

Base exchange or the ionic effect is without doubt extremely important, and is closely related to such phenomena as the effect of builders and of pH. Cotton and rayon both contain free carboxyl groups which are capable of exchanging cations.   Wool is amphoteric and within a fairly wide pH range near its isoelectric point is capable of firmly sorbing either anions or cations.

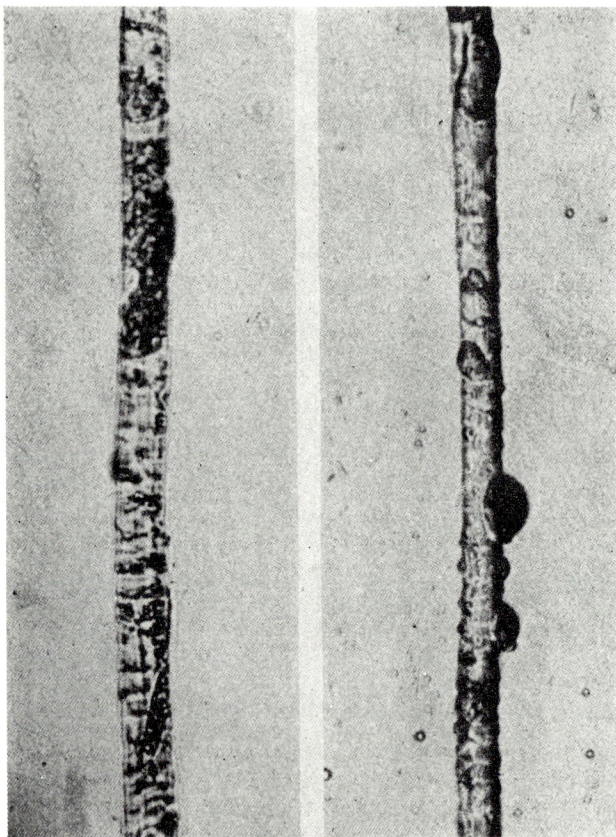

Fig. 50.   Part C.

The base exchange properties of various solid soils are less well known but fully as important.   It is readily understood how ions which are strongly adsorbed by either soil or substrate, and which enter into a base exchange reaction during the washing process, can be a fundamental factor in soil removal.[117,118]

---

[117] Chwala, *Österr. Chem.-Ztg.* **40**, 39 (1937); Chwala and Martina, *ibid.* **40**, 270 (1937); *Chem. Abstracts* **31**, 2441, 6080.

[118] Henk, *Seifensieder-Ztg.* **66**, 2 (1939).

Sisley[119] is one of the few investigators in this field who considers foaming to be an important factor in detergency. He points out that the detergent is more concentrated in the foam films and that the building and breaking

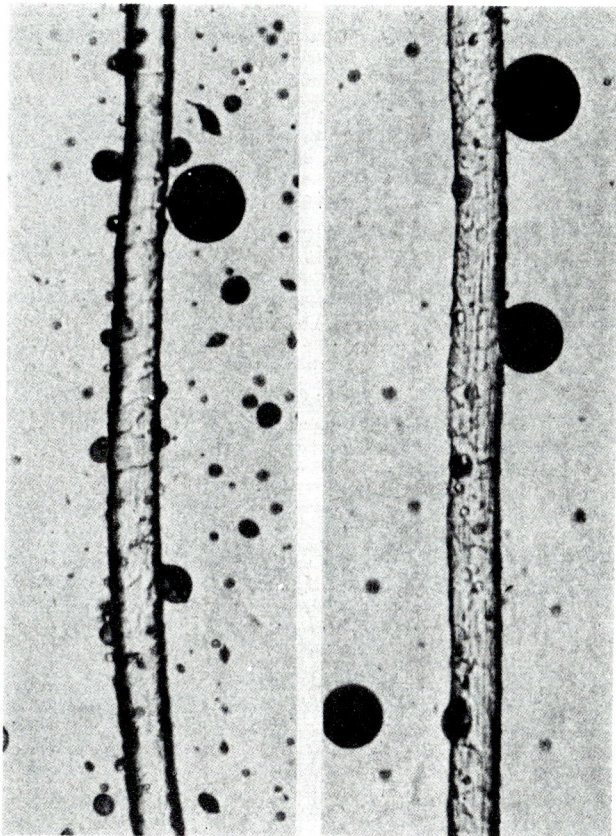

Fig. 50. Part D.

of foam in the presence of soiled goods constitutes an effective mechanical action.

The great importance of mechanical work in removing soil under ordinary washing conditions was understood qualitatively long before the study of detergency became a branch of science. The most primitive washing methods involve beating and pounding of the soiled fabrics. The effect of mechanical action has recently been studied quantitatively by Bacon and

[119] Sisley, *Corps Gras, Savons* **1,** 66 (1943).

Smith.   They suggest that detergents may be compared by determining
the difference in mechanical work (energy input into the washing machine)
necessary to achieve the same degree of soil removal.[120]

The studies considered above constitute only a small proportion of those
which have been published, but they illustrate most of the important variables in detergency with the exception of bath composition.   Before proceeding to attempt a systematic presentation of the factors affecting detergency, two more lists of important variables, each by a well-known worker
in the field may be considered.   Chwala[121] gives the following list: (1)
Wetting. (2) Adsorption. (3) Surface tension. (4) Interfacial tension.
(5) Electrical charge (in the case of ionic detergent). (6) Foaming power.
(7) Emulsifying and dispersing power. (8) Protective-colloid action.   And
(9) Age of bath prior to washing.

Sisley[122] in an excellent analysis of the detersive process lists the following
factors in three categories:

A. (1) Wetting power. (2) Foaming power. (3) Emulsifying power.
(4) Dispersing and deflocculating power. (5) Solubilizing and solvent
power. (6) Protective action against redeposition. (7) Resistance to calcium ions.

B. (1) Nature of the detergent bath. (2) Nature of the substrate surface. (3) Nature of the soil. (4) Composition (hardness) of the water
used and its effect on the substance to be cleaned.

C. (1) Chemical nature of the detergent. (2) Concentration of detergent. (3) Temperature. (4) Degree of mechanical action.

The following analysis attempts to segregate the factors affecting detergency and the molecular mechanisms effective in accomplishing cleaning.
For the sake of simplicity we shall consider the best-known type of detersive
system, namely, a fabric bearing both liquid and solid soil, in an aqueous
bath containing a detergent and other ionogenic components.

The detersive process consists in taking a soiled substrate, treating it
with a detergent bath by immersion with or without simultaneous mechanical action, withdrawing it from the bath, and finally rinsing off the residual
bath liquor.   In ordinary detergent processes in which the liquor length is
of the order of 10 to 1 or higher, rinsing gives only second-order effects.
To demonstrate this we can consider a soiled piece of fabric weighing 1
pound being washed in 10 pounds of detergent bath.   On withdrawal the

[120] Bacon and Smith, Abstracts of the American Chemical Society Meeting at
Atlantic City, April 1947, Division of Colloid Chemistry.
[121] Chwala, *Textilhilfsmittel*, Edwards Brothers, Inc., Ann Arbor, 1943.
[122] Sisley, *Corps Gras, Savons* 1, 66 (1943).

fabric will retain about 1 pound of the used bath, *i.e.*, about 10% of the removed soil, assuming that the soil was distributed evenly throughout the bath. The extremes of rinsing action can be represented as: (*1*) Complete removal of all residual bath liquor and soil, or (*2*) Complete redeposition of all soil in the residual bath liquor. In our example, the difference between these two extremes would correspond to only a 10% difference in the over-all soil removal. In actual practice the rinsing operation rarely involves any appreciable redeposition of soil on the fabric. For purposes of this discussion it can be neglected, although it does become important in some special cases (*cf.* section on shampoos in Chapter 19). The total detergent effect therefore depends on:

(I) Nature of the substrate.
(II) Nature of the soil.
(III) Composition of the bath, which determines both its bulk and surface properties.
(IV) Physical and mechanical conditions of washing, which include temperature, duration of treatment, and type and degree of mechanical action.
(V) The relative quantities of soil, substrate and bath present in the system.

Total detergent effect can be expressed very simply as the removal and segregation of soil from the substrate by the bath. This action is conditioned solely by the five primary variables listed above. It is the subvariables and the relationships among them which form an intricate and difficult pattern.

The stability of the substrate–soil complex, which may be termed the I–II complex since it depends on I and II above, is conditioned by:

(I–II-A) Pure interfacial adhesional forces between substrate and soil, *i.e.*, van der Waals' forces of attraction. These are most important where the soil is liquid in nature, as in the case of oil on wool. They can be expressed in this example in terms of the free surface energy of the wool–oil interface. The van der Waals' forces may also be important in the case of solid soils but, as we have seen above, very little is known about this type of solid-to-solid adhesional force.

(I–II-B) Electrostatic or coulombic forces between substrate and soil. These are particularly important in the case of solid soils, and probably not so important in the case of liquid or oily soils. They may, however, condition the adhesion of solid soil particles to the oily or waxy coating on a fiber where the soil is mixed.

The question as to where a division may be made between solid and liquid soils immediately arises since many practical soils are greasy, waxy, or plastic in nature. We shall consider a soil phase to be liquid when its rigidity is so low that it *may be rapidly deformed by its own phase boundary forces*, acting together with the mechanical forces of agitation in the system. By rapidly we mean well within the time limits of the washing procedure.

(I–II-C) Purely mechanical juxtaposition of soil and substrate. This is the major factor in the case of coarse solid soils and might be illustrated by the example of sand in a swimming suit. It may also be an extremely effective factor in the case of finally divided soil especially where the presence of the bath has minimized I–II-A and I–II-B above.

Factors of primary importance in the composition of the bath itself include:

(III-A) Chemical nature and concentration of the surface active solute or solutes.

(III-B) Chemical nature and concentration of other solutes which are not surface active but which may either help detergency, be inert, or hinder detergency. Substances of this nature which promote detergency are called builders.

(III-C) Nature and concentration of non-solutes or suspended matter (other than soil which has come from the article to be washed) in the bath. This includes additives such as clay or bentonite, or emulsified solvents such as pine oil, kerosene, etc.

(III-D) pH of the bath.

It should be noted that the bath composition may effectively change if it is allowed to age before being used. This may be due to chemical interaction of the components. A physical effect may also be involved due to the time required for equilibrium to be established in the adsorbed surface layer and among the micelles. We shall assume, however, that baths are allowed to come substantially to inner equilibrium before being used and shall accordingly not consider this variable.

(III-E) The bath composition invariably changes during the washing process. The most obvious changes are: (III-E-1) The addition of suspended solid soil. (III-E-2) The addition of suspended and/or emulsified liquid soil. (III-E-3) The addition of solubilized solid and/or liquid soil. The bath composition may also change by: (III-E-4) Adsorption of one or more components from the bath by the substrate. (III-E-5) Adsorption of one or more components from the bath by the soil particles (liquid or solid). (III-E-6) Solution or leaching of soluble material from substrate or soil. (III-E-7) Chemical interaction of one or more bath components with soil, substrate, or with soluble products leached from these.

The physical conditions of washing (IV above):

(IV-A) Temperature. (IV-B) Duration of treatment. (IV-C) Degree and type of mechanical action.

are important because they influence the interaction between the bath and the soil–substrate complex. The probable modes of interaction bath, substrate, and soil can be listed as I–III interactions, between substrate and bath, or II–III interactions between soil and bath. They are as follows:

(I-III-A) Wetting of substrate by bath. This is necessary to establish the contact without which interaction of any sort is impossible. It is in fact difficult to imagine how detersion can take place without some degree of wetting. Wetting implies penetration into the yarns of fabric so that each individual fiber is wet by the bath at those points where it is not covered by soil.

(II–III-A) Wetting of soil by bath.   The same considerations apply here as in the wetting of substrate.

(I–III-B) and (II–III-B) Adsorption of bath components on substrate surface and on soil surface.   These adsorbates may be monofilms of surface active agents, inorganic ions, or particles of colloidal dimensions.   Adsorption of ions on the fiber can modify electrostatic attractive forces for the soil.   Adsorption of ions on the soil can help prevent its redeposition and increase the extent to which it is *suspended* by the bath.   Adsorption of colloid on both soil and fiber enables the soil to migrate out of the fabric, much as adsorbed colloid enables a carbon black suspension to pass through a filter paper (see Section IV of this chapter).   Adsorption of surface active agents and of colloid on liquid soils promotes formation of a stable emulsion of the liquid soil in the bath.

(II–III-C) Solubilization of soil by the bath.   This may be quite important in the case of liquid soils and some solid soils.   Only those solid soils which can be solubilized by detergent solutions are affected in this manner.   Practically none of the usual mineral soils are in this category.

To summarize, partial or complete soil removal can be accomplished by:

(*1*) The interfacial energy relationships of bath, soil, and substrate.   In the case of liquid soils this can be called preferential wetting of the substrate by the bath.

(*2*) Adsorption of colloid on soil and substrate.

(*3*) Adsorption of ions on soil and/or substrate resulting in a diminution of electrostatic attractive forces.

(*4*) Solubilization of the soil in the bath.

Soil segregation (and the prevention of redeposition) can be accomplished by any or all of the four factors above since these result in emulsification of liquid soils, suspension of solid soils, and a lowered attraction of both liquid and solid soil particles for the fiber.

It should be noted that this lowered attraction between soil and fiber is sometimes a much more important consideration than a high degree of dispersion of the soil in the bath.   Thus, it is often observed that a bath with very little soil-suspending power will show excellent detergency.   The dirt, after having been detached from the fabric, may again contact the latter without adhering to it.   It is important to mention, however, that well-dispersed soil is much more easily rinsed than flocculated but nonadherent soil.   This is because well-dispersed soil cannot be held so easily by purely mechanical means.

It is readily seen from the above considerations that bath composition is the main key to detergency, and a great number of detergency studies have been made in which over-all detersive effect is measured against changes in bath compositions.   It is usually difficult to pick out the particular molecular mechanism which is being affected by a bath change, *i.e.*, whether ionic adsorption, colloid adsorption, or interfacial energy relationships are being altered.   The studies are therefore primarily of practical interest.   Nevertheless many of them have contributed greatly to our intimate knowledge of detersive mechanisms.

(d) Detergent Effects of Various Baths

It has been repeatedly emphasized that total detersive effect depends on all four of the primary variables previously mentioned, namely: composition of soil, substrate, bath, and conditions of washing.   In studying the effects of various baths we must control the other three variables, and for a direct comparison of baths they should be held constant.   These elementary precautions have not always been observed in some of the published studies on detergents.   Even where they have been observed it is difficult to achieve complete reproducibility in soils and substrates.   Unless otherwise specified in subsequent discussion "detergency" will refer to the cleaning of fabrics soiled with a mixed solid–liquid soil in an aqueous bath.

Some important commercial detergent baths contain no organic surface active agent whatsoever.   The best example is the mixture of alkaline silicates and phosphates used for machine dishwashing.   These are of secondary interest, however.   Practically all baths for washing fabrics contain one or more surface active agents.   They may also contain other soluble components, and in some instances insoluble suspended components, which are not in themselves surface active.   If these contribute to the detergent effect they are called builders, otherwise they are called fillers or diluents. Builders are of extreme importance in practical detersive operations and there are very few baths used in which builders are not present.   In general, the builder effect is not highly specific.   In other words, a substance which is a builder for one type of surface active agent will usually "build" other similar types.   There is, of course, a certain degree of specificity among builder–detergent combinations.   Before considering the action of builders, however, the action of the pure surface active agents should be considered.

One of the most important concepts in considering the gross effects of surface active agents is the concept of synergism.   This term is applied when a mixture of two or more surface active agents shows a greater efficiency than the sum of the effects of the two components used alone. Examples of synergism are quite commonly encountered in wetting and emulsification as well as in detergency.[123]   Synergism is distinguished from

---

[123] See, for example: Macaluso, *Ind. Eng. Chem.* **36**, 1071 (1944); Datta and Das Gupta, *Soap Perfumery, Cosmetics Trade Rev.* **8**, No. 4, 34 (1935).

builder action in that each synergistic component is in itself surface active. To understand this in detail let us consider two detergents X and Y. We measure their detersive powers separately under standard conditions on a standard soiled fabric and express the results in terms of per cent soil removal. Suppose 0.1% concentration of X has a rating of 20% soil removal, and 0.1% Y also has a rating of 20%. If 0.1% of a 50:50 XY mixture (i.e., 0.05% X and 0.05% Y) has a rating of higher than 20%, then X and Y are synergistic. Now consider a builder B with detergent X. 0.1% X rates 20%. B in any concentration rates 0%. If a mixture containing 0.1% X together with any reasonable concentration of B has a rating higher than 20%, then B is a true builder.

The relative detergent action of the various fatty acid soaps has been investigated many times. In general, the C_{16}, C_{18} saturated and C_{18} unsaturated soaps have better detergent power than the lower soaps, particularly at higher temperatures.[124] Soaps of elaidic acid have been reported to be better detergents than the isomeric oleic acid soaps.[125] In ordinary soaps the effect of the cation is considerable but it is not as great in most instances as the effect of the anion. The triethanolamine soaps, for example, fall in the same order of effectiveness, $C_{18} > C_{16} > C_{14}$, as the potassium or sodium soaps, although each triethanolamine soap is less effective than the corresponding sodium soap.[126] Rayner[127] and other investigators have shown that soap formed in situ, by coating the soiled fabric with oleic acid and then washing in a soda ash solution, has several times the detergent power of a soap bath of similar concentration. This phenomenon is closely analogous to the superiority which is attained in emulsion formation when the emulsifying agent is dissolved in the oil.

The detergency of rosin soaps and mixtures of fatty acid soaps with rosin soaps has been extensively studied. Rosin soap by itself is a poor detergent, but up to 50% can be incorporated into fatty acid soaps without seriously impairing the detergency of the latter.[128] The detergency of soap nut powder, a preparation containing saponins, has been studied by Sarin and Uppal, who found it to be far less effective than soap.[129]

---

[124] Davidsohn, Am. Perfumer, May, 1945, p. 52. Chernichkina et al., Allgem. Oel- u. Fett-Ztg. 30, 294 (1933). Treffler, Soap Sanit. Chemicals 20, No. 4, 29 (1944).

[125] Bertram and Kipperman, Öle Fette Wachse 1936, No. 6, 1; Chem. Weekblad 32, 624 (1935).

[126] Kulkarni and Jatkar, J. Indian Inst. Sci. 21A, Pt. 34, 395 (1938). Fiero, J. Am. Pharm. Assoc. 28, 284 (1939).

[127] Rayner, Chemistry & Industry 1934, 589.

[128] Pohle and Speh, Oil and Soap 17, 214 (1940). Van Zile and Borglin, Oil and Soap 21, 164 (1944).

[129] Sarin and Uppal, Ind. Eng. Chem. 33, 666 (1941).

The detergent powers of the various commercial synthetic surface active agents have been reported from time to time in journal articles.[130]   Many manufacturers give detergency data on their products in descriptive brochures.   The latter sources are usually more accurate and up to date because commercial products often undergo changes in composition and in percentage of active content.

The detergency of Santomerse, alone, with soap, and with builders, has been reported thoroughly by Harris and co-workers.   These investigators were particularly interested in producing a detergent mixture suitable for use in sea water and extremely hard waters.   Similar studies on mixtures of soap with other synthetic detergents were made during the war with the purpose of developing an optimum all-purpose detergent in bar form for the armed services.   These mixtures show little if any synergism in soft water, but they wash well in hard water or sea water, in which pure soap is completely ineffective.[131]   Miles and Ross[132] have shown that a decrease in detersive efficiency and foaming occurs when soap – fatty alkyl sulfate mixtures are used in hard water.   This is attributed to the formation of a mixed calcium salt of the two anions.

(e) Influence of pH on Detergency

It has been known for a long time that the addition of mild alkalis such as soda ash aids the detersive action of soap.   Soap itself is decomposed by acid and hydrolyzes to a certain extent even in pure water.   This hydrolysis lowers the concentration of the effective detergent by converting it into the form of free fatty acid, devoid of detergent power.   Tomlinson has recently studied the detersive efficiency of soap at various concentrations and has demonstrated a minimum efficiency under those conditions at which hydrolysis is at a maximum.   He has further shown experimentally that agents which suppress the hydrolysis eliminate this minimum in the efficency curve.[133]   Alkaline materials are excellent for suppressing hydrolysis and one of the functions of the usual alkaline soap builders consists of of maintaining a high pH value in the bath.   Rhodes and co-workers, Morgan, Snell, and others have determined the pH at which soap gives the

[130] Compare Franz, *Melliand Textilber.* **16,** 277 (1935); **18,** 918, 1003 (1937).

[131] Harris, *Soap* **19,** No. 8, 21; No. 9, 29 (1943); *Rayon Textile Monthly* **26,** 77, 142 (1945).   Harris and Brown, *Oil and Soap* **22,** 3 (1945).   McDonald, *Soap Sanit. Chemicals* **21,** No. 12, 41 (1945).   U. S. Army Specification OQMG No. 100A.   Ruckman, Hughes, and Clarke, *Soap. Sanit. Chemicals* **19,** No. 1, 21 (1943).

[132] Miles and J. Ross, *Ind. Eng. Chem.* **35,** 1221 (1943).

[133] Tomlinson, *J. Soc. Dyers Colourists* **63,** 107 (1947).

greatest detergency under simulated laundering conditions, and find it to be in the range of 10.5 to 11.[134]

The buffering value of the builder, *i.e.*, its ability to maintain the bath at the desired pH, is of great importance. The buffering action can be measured by titration with acid in most cases. Alkalis such as trisodium phosphate, sodium carbonate, sodium sesquicarbonate, sodium metasilicate, etc. do not give excessively high pH values even when present in fairly high concentrations. They maintain these pH values well as the concentration is lowered, thereby maintaining optimum washing conditions. Caustic soda, on the other hand, gives too high a pH even when a few tenths per cent is present. This can be reduced to an ineffective low value by small amounts of acid. Snell and co-workers have shown that too high a concentration of caustic soda lowers the detersive efficiency of soap markedly.[135] An optimum alkaline builder for soap should accordingly give the correct pH value at the washing temperature, and should have a high titratable alkali content,[136] in other words function as a buffer. For accurately measuring the pH of soap and builder solutions the antimony electrode is said to be particularly useful, although the more common glass electrode can be used.[137]

The effect of pH on the action of the various synthetic detergents in the absence of buffers or builders has not been studied so thoroughly. There is no doubt that pH is a factor in the detersive efficiency of these products since it affects their interfacial properties against soils and substrates, and also their adsorption and electrostatic behavior.

(f) Builder Effects Other Than pH

Although control and maintenance of the pH value of the bath is one of the primary functions of a builder it is by no means the sole function. Some builders for use with soap are effective because they soften the water of the bath by one of two mechanisms. In the first place the calcium and magnesium ions can be precipitated as insoluble salts as is accomplished when soda ash or trisodium phosphate is used in the bath. The undesirable ions can also be effectively removed by the "sequestering" action of certain com-

---

[134] Rhodes and Bascom, *Ind. Eng. Chem.* **23**, 778 (1931). Rhodes and Wynn, *ibid.* **29**, 55 (1937). Snell, *ibid.* **25**, 1240 (1933). Morgan, *Can. J. Research* **8**, 429, 583 (1933).

[135] Snell *et al.*, *Soap Sanit. Chemicals* **19**, No. 11, 63 (1943).

[136] Vaughn, Vittone, and Bacon, *Ind. Eng. Chem.* **33**, 1011 (1941). Snell, *Food Inds.* **13**, No. 10, 48 (1941). Bergell, *Seifensieder-Ztg.* **66**, 406 (1939). Schuffelen and van der Hulst, *Chem. Weekblad* **38**, 134, 231, 491 (1941).

[137] Wulff, *Fette u. Seifen* **48**, 388 (1941). Dole, *Am. Dyestuff Reptr.* **30**, P231 (1941).

plex phosphates, of which the best-known is sodium hexametaphosphate, sold under the trade name of Calgon. It has the formula $(NaPO_3)_6$ or $Na_6P_6O_{18}$. This compound ionizes in aqueous solution to form complex anions containing sodium, for example:

$$Na_6P_6O_{18} \rightarrow 2\,Na^+ + Na_4P_6O_{18}^=$$

These anions can then react with calcium ions according to the equation:

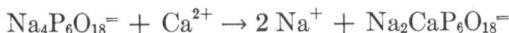

$$Na_4P_6O_{18}^= + Ca^{2+} \rightarrow 2\,Na^+ + Na_2CaP_6O_{18}^=$$

thus effectively removing calcium ion from the solution. Miles and John Ross[138] consider water softening to be the most important action of builders when soap is used as the detergent, and compare the various builders on this basis alone.

The complex phosphate and the silicate builders can act in at least two other ways, by modifying the adsorption of the detergent on substrate and/or soil, and by acting as suspending agents or peptizing agents. Colt and Snell have studied the effect of various builders in modifying the adsorption of soap on wool, rayon, cotton, and silk.[139] It has also been shown that sodium hexametaphosphate, sodium tetraphosphate, and sodium tripolyphosphate and sodium metasilicate have a definite suspending action on dispersed solid particles.[140]

There is yet another way in which builders of the inorganic salt type may act, namely, by modifying the interfacial energy relationships at the substrate, soil, and bath phase boundaries. We have already seen that the presence of salts, particularly sodium sulfate and sodium chloride, has a marked effect on the surface and interfacial tensions of most surface active agents. It also affects their solubilizing properties. Sodium sulfate and sodium chloride in small amounts definitely improve the detergency of the anionic sulfonate and sulfate type detergents. In very small amounts they improve the detergency of soaps. If larger amounts are used with soap the beneficial effect is reversed, probably due to the formation of excessively large micelles and to actual salting out of the soap.

Numerous patents have been issued on soap compositions containing builders, and many studies of the effects of various builders on soap detergency are available.[141]

[138] Miles and J. Ross, *J. Am. Oil Chemists' Soc.* **24**, 23 (1947).

[139] Colt and Snell, *Oil and Soap* **17**, 33 (1940).

[140] Hatch and Rice, *Ind. Eng. Chem.* **31**, 51 (1939). Liddiard, *Chemistry & Industry* **1941**, 480. Gillet, *Soap* **15**, No. 1, 25. (1939). Kling and Schmidt, *Seifensieder-Ztg.* **66**, 626 (1939). Bergell, *ibid.* **66**, 385, 445 (1939).

[141] Vaughn and Vittone, *Ind. Eng. Chem.* **35**, 1094 (1943). Venkataraman *et al.*, *J. Indian Chem. Soc., Ind. & News Ed.* **2**, 81–94 (1939). Widaly, *Seifensieder-Ztg.*

The effects of the alkaline builders on the lime-resistant anionic synthetic detergents are less pronounced than in the case of soap. They have been reported by Harris.[142]   Thomas and Brown have claimed that the detergency of the usual anionic synthetic detergents is improved by alkaline carbonates, phosphates, and silicates provided that chlorides and sulfates are substantially absent.[143]   Snell has used a new type of sodium borate containing 1.1 to 2 moles Na_2O per mole of B_2O_3 as a builder for these detergents, and Quimby has disclosed the use of acid triphosphates.[144] Numerous other patents covering mixtures of carbonate, phosphate, and silicate builders with both soaps and non-carboxylic anionic detergents have been issued.[145]

(g) Solvents as Additives to Detergent Baths

In some instances water-immiscible solvents are added in substantial amounts to aqueous detergent baths containing a surface active agent as the main detersive ingredient.   This is commonly done when the soil is extraordinarily heavy and is greasy in nature, for example, in removing paint or tar from raw wool.   The solvent is usually added to a concentrated stock solution of the soap or detergent, in which it forms a stable emulsion. Several commercial preparations of this general nature are made and marketed.[146]   The solvent remains emulsified in the diluted, ready-to-use bath.   It probably functions by dissolving the soil, thus rendering it fluid and easy to remove by emulsification in the aqueous phase of the bath. It should be noted that a truly solubilized solvent will not have this effect. Consequently, the solvent should be present in sufficient quantity so that it cannot all be solubilized.   Among commonly used solvents are pine oil, terpene solvents, aromatic solvents, and particularly hydroaromatic solvents such as hexalin and tetralin.[147]

70, 102 (1943).   Oesterling, Am. Dyestuff Reptr. 27, 617 (1938).   U. S. Pats. 2,342,786, Bornemann and Huber, vested in Alien Property Custodian; 2,310,475, Thomas and Oakley to Lever Bros. Co.

[142] Harris, Oil and Soap 23, 101 (1946).   Compare Aickin and Palmer, Trans. Faraday Soc. 40, 116 (1944).

[143] Brit. Pats. 551,616; 547,688; to Lever Bros. Co. and Unilever Ltd.

[144] U. S. Pats. 2,376,096, Snell to Foster D. Snell, Inc.; 2,383,502, Quimby to Procter and Gamble Co.

[145] Brit. Pats. 443,487; 435,317; French Pat. 781,142; to Henkel et Cie.   French Pats. 835,274, to Lever Bros. Co. and Unilever Ltd.; 760,236, to I. G. Farbenindustrie. Ger. Pat. 623,403, to Deutsche Hydrierwerke A.-G.

[146] See, for example, Hetzer, Monatschr. Textil-Ind. 57, 317 (1942).

[147] Vallance Soap, Perfumery & Cosmetics 11, 998 (1939).   Ohl, Allgem. Oel-u. Fett-Ztg. 35, 337 (1938).

(h) Colloidal Dispersions as Additives to Detergent Baths

A large number of bath additives which form true colloidal dispersions in the bath have been proposed and are used.   The oldest and best known of these are bentonite and various other colloidal clays.[148] They can be used with soap and other anionic detergents or with non-ionic detergents.   Their main function consists of adsorbing the soil as it is removed and preventing it from redepositing on the fabric.   With coarser clays, the localized abrasive action caused by mechanical agitation during washing also plays a part.   Bentonite is also well known for its emulsifying power and it helps segregate liquid soils by emulsifying them in the bath.   When bentonite is used with cationic detergents it is itself deposited firmly on the fibers being washed.   This occurs because the fibers first adsorb the hydrophobic cation, thus reversing their surface charge.   They then have an electrostatic attraction for the bentonite particles.   Other adsorbent hydrophilic solids such as magnesium silicate and silica aerogel have also been proposed as additives to detergent baths.[149]

A second group of additives which sometimes improve detergency comprises amino acids, peptides, and proteins.   The amino acids and lower peptides are said to be effective in reducing the surface tension of surface active agents.[150]   Proteins such as casein, glue, albumen, etc., are excellent protective colloids and thus aid in removing and segregating soil.[151]

Lecithin has also been added to soaps, and is said to improve the emulsifying and foaming power.[152]

More recently a group of higher fatty amides and nitriles have been proposed as additives to the anionic sulfonate and sulfate detergents.   They are said to improve the detergency of these compounds significantly when used in quantities of 10 to 30%.   The substances specified include the higher fatty amides, ethanolamides, dialkyl and alkylol amides, morpholides, and nitriles, as well as the lower acyl derivatives of higher fatty amines.[153]   It is probable that these easily emulsified substances act by adsorbing solid soils, thus enabling them to be segregated.

By far the most widely heralded colloidal additive for detergent baths in recent years has been sodium cellulose glycolate, a soluble, strongly

[148] Compare Lesser, Soap Sanit. Chemicals 21, No. 10, 37 (1945).

[149] Belgian Pat. 447,464, to Zellowolle- und Kunstseide-Ring G.m.b.H.   U. S. Pat. 2,257,545, Curtis to Monsanto Chemical Co.

[150] Muto, J. Biochem. (Japan) 33, 1 (1941).   Ger. Pat. 734,337, to I. G. Farbenindustrie; Chem. Abstracts 38, 1304.

[151] Wittka, Allgem. Oel- u. Fett-Ztg. 38, 309 (1941); Chem. Abstracts 37, 2203.

[152] Lederer, Seifensieder-Ztg. 60, 919 (1933).

[153] U. S. Pats. 2,383,525–6; 2,383,737–40; Tucker, Richardson et al. to Procter and Gamble Co.

hydrophilic colloid made by etherifying cellulose with sodium chloroacetate. It was developed and used on a large scale in Germany during the war under the name of Tylose HBR. The manufacture of this material has been described in detail by Brown and Traill,[154] and its applications and practical uses by Hoyt.[155]

According to German claims Tylose HBR can replace two to three times its weight of soap when used as a soap extender, i.e., a mixture of one part soap with one part Tylose HBR has detergent power equivalent to three to four parts of soap alone. It is also said to be remarkably effective in improving the detergency of the anionic sulfonates and sulfates, particularly for soiled cotton goods. The improvement in washing action is attributed to prevention of soil redeposition. Thereby the development of a grey color in white cotton goods on repeated laundering is counteracted. Tylose HBR is also said to prevent the build-up of ash content in fabrics on repeated washing. It is reported to have exhibited particular merit when used with Igepon T or Mersolate, greatly increasing their effectiveness in hard water and sea water. About 20 to 50% of Tylose HBR based on the weight of the detergent was used in these mixtures, together with the usual amounts of soda ash, phosphates, and metasilicate. The developers of this product consider that its action is primarily that of a protective colloid, and that it is by far the most effective protective colloid, in normal detergent baths, of any so far discovered. At the present writing these claims have been at least partially substantiated in American laboratories.

A great many other substances of similar colloidal nature have been proposed and patented as synergists for detergents. Among water-soluble cellulose and starch derivatives are: (1) sodium starch glycolate,[156] (2) oxidative degradation products of cellulose, starch, and hydrocellulose,[157] (3) methyl-, ethyl-, and hydroxyethylcelluloses,[158] (4) cellulose ethyl ether β-sulfonate, i.e., the cellulose ether of isethionic acid,[159] (5) complex polymers formed by condensing urea and formaldehyde with water-soluble cellulose ethers.[160]

Water-soluble salts of polyacrylic acid and co-polymers of acrylic acid

[154] Brown and Traill, *PB 8007*, Office of Technical Services, Dept of Commerce, Washington, D. C.

[155] Hoyt, *PB 3865*, Office of Technical Services, Dept. of Commerce, Washington, D. C.

[156] U. S. Pat. 2,335,194, Nüsslein and Pauser, vested in Alien Property Custodian.

[157] Austrian Pat. 140,042, to Oskar Janser.

[158] U. S. Pat. 2,347,336, Seyferth to *Allied Chemical and Dye Corp.* French Pat. 805,718, to *Kalle and Co.* Compare Smith, *Am. Perfumer* **38**, No. 3, 45 (1939).

[159] Ger. Pat. 741,823; U. S. Pat. 2,184,171; Sponsel and Daimler to *Kalle and Co.*

[160] U. S. Pat., 2,237,240, Sponsel to *Kalle and Co.*

containing carboxyl groups have also been proposed as washing aids. They presumably act in the same way as the cellulose derivatives.[161]

Another important detergent additive, which probably acts as a protective colloid or a suspending agent, is the type of sodium silicate called water glass. This should not be confused with the alkaline builder silicates which are either orthosilicate $(2\,Na_2O:SiO_2)$, metasilicate $(Na_2O:SiO_2)$, or sesquisilicate $(3\,Na_2O:2SiO_2)$. Water glass has a much higher ratio of silica to soda, usually about $1\,Na_2O:3\,SiO_2$. It is often considered to be a filler rather than a synergist, but there is evidence that in some cases it increases the detersive action of both soaps and the anionic synthetic agents.[162]

VIII. Miscellaneous Gross Effects

There are several minor gross effects characteristic of most surface active agents aside from the major ones just considered. Among those which are of considerable technical importance are specific catalytic effects, the dye-leveling or adsorption-modifying effect, and the bactericidal effect. Although each of these actions may be brought about by agents which are in no sense surface active, nevertheless a great number of surface active agents can produce or promote them. Furthermore, they are not related to the other more general gross effects.

The specific catalytic effect or surface active agents is best exemplified in the Twitchell process for splitting fats. In this process the fat is heated to a high temperature with water in the presence of an anionic acidic surface active agent: The reaction is purely one of hydrolysis, the fat reacting with water to form glycerol and free fatty acid:

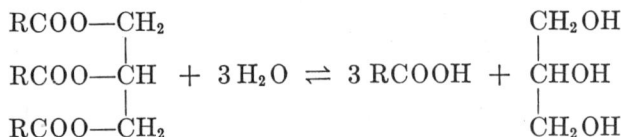

$$
\begin{array}{ll}
RCOO-CH_2 & CH_2OH \\
\quad\quad | & \quad\quad | \\
RCOO-CH + 3\,H_2O \rightleftharpoons 3\,RCOOH + CHOH \\
\quad\quad | & \quad\quad | \\
RCOO-CH_2 & CH_2OH
\end{array}
$$

This reaction follows the law of mass action, and even though high temperature favors formation of the ester the reaction proceeds toward the right because glycerol is removed from the reaction sphere. The system consists of two phases. The glycerol which is formed dissolves and is diluted in the water. In the technical procedure the water–glycerol mixture is removed and replaced by fresh water three or four times. In the first stage about 60% of the fat is hydrolyzed and at the end of the third stage over 90% has been hydrolyzed. Vigorous agitation is employed in the process

[161] U. S. Pat. 2,327,302, Dittmar to *E. I. du Pont de Nemours & Co.*; Ger. Pat. 690,951; Brit. Pat. 369,915; to *I. G. Farben-industrie.*
[162] Foulon, *Monatschr. Textil-Ind.* **51**, 104 (1936).

and the reaction mixture is in the form of a relatively unstable O/W emulsion. It is well known that intimate mixing will promote any two-phase reaction merely by increasing the area of contact between the two reacting phases. It is also known that hydrogen ions catalyze establishment of the esterification equilibrium. One would therefore expect that all acidic emulsifying agents would have roughly the same effect in the Twitchell process. Such is not the case. There are large differences in catalytic effect among the various anionic surface active acids. The better ones include those alkyl aromatic sulfonates made by condensing a straight-chain fatty acid with naphthalene and subsequently sulfonating. These were originally used in the early development of the process. The reagents now favored are certain petroleum sulfonates recovered in the sulfuric acid process for refining mineral oils.

The specific catalytic effect of several different emulsifying agents on the hydrolysis of amyl acetate by aqueous alkali has been studied by King and Mukerjee.[163] They found that in every case the emulsified system reacts much more rapidly than the agitated but unemulsified system. Furthermore, the rate of reaction is not proportional to the interfacial area when different agents are used but is characteristic of the agent itself.

Another example of the catalytic action of surface active agents is the promoting effect of saponin on the hydrogenation of ethylene with colloidal palladium. This has been studied by Zolin,[164] who reports an increase of tenfold in the reaction velocity at room temperature. He attributes the increase to the presence of copious foam, postulating that the reaction takes place in the walls of the foam bubbles. In this connection Nasini and Mattei have studied various reactions of the double bond in oleic acid while the latter is in the state of a monofilm on water. A film balance of the Adam type was used to follow reactions of halogen addition and oxidation by potassium permanganate.[165]

The presence of surface active agents has been found to decrease the current efficiency of certain electrochemical reductions of organic substances at lead, cadmium, and mercury cathodes. The cationic surface active agents are effective in impeding reduction even in very low concentrations. Larger concentrations of anionic agents are required to produce the inhibiting effect, whose magnitude passes through a maximum as the concentration of anionic agents is increased. The cationic surface active agents act by being adsorbed on the cathode and preventing access of the organic substance being reduced.[166]

[163] King and Mukerjee, *J. Soc. Chem. Ind.* **57**, 431 (1938).
[164] Zolin, *J. Phys. Chem. U. S. S. R.* **5**, 1299 (1934).
[165] Nasini and Mattei, *Gazz. chim. ital.* **70**, 635 (1940); *Kolloid-Z.* **101**, 113 (1942).
[166] Proudfit and France, *J. Phys. Chem.* **46**, 42 (1942).

It would be expected that the surface active agents, by virtue of their unique surface and bulk properties, would strongly affect the sorption of other adsorbable material by various substrates. The best example of a large-scale application of this effect is in the dyeing of both animal and cellulosic fibers.

The most widely used dyes for wool are the so-called acid dyes. They are for the most part aromatic sulfonic acids of the azo or anthraquinone series, soluble in water. They are applied to the wool by immersing the latter in a hot, slightly acidic aqueous solution of the dyestuff. The dyestuff molecules diffuse into the intermicellar spaces of the fibrous protein and are firmly fixed on the reactive groups of the wool, which chemically is an amphoteric high polymer containing both free amino and free carboxy groups. At pH values below the iso-electric point (about pH 4.8) the amino groups combine readily with anions such as the dyestuff sulfonic acid ions. The reaction is therefore a simple one of salt formation. Different anions have differing degrees of affinity for the basic groups of the wool, and hence will be adsorbed to differing extents. Of the different acids which may normally come in contact with the wool, the dyestuffs are the most strongly adsorbed; strong mineral acids of low molecular weight such as sulfuric acid are moderately well adsorbed; and the organic carboxylic acids are feebly adsorbed. The usual types of surface active sulfates and sulfonates are more strongly adsorbed than sulfuric acid but less strongly adsorbed than the commercially satisfactory dyestuffs. In general, large anions of strong acids will be more strongly adsorbed and will remain fixed to the wool even if the pH in the surrounding aqueous medium is raised far above the iso-electric point.

In the actual dyeing process there is often a strong tendency for the dye to go on the fabric unevenly, some areas of the fabric becoming more strongly colored than others. This is due to one or more of several different factors: (1) uneven wetting of the fabric; (2) too rapid adsorption of the dye; (3) poor circulation of the dye liquor, resulting in concentration gradients throughout the liquor and differing concentrations of dye in contact with the fabric from area to area; (4) a very rapid and uneven surface adsorption of the dye on the fabric, probably due to coulombic forces and occurring on the surface of the fibers. This is called the "strike" and is clearly distinguished from the slower true dyeing where the dye penetrates into the interior of the fiber. Areas where the strike is heavy will be dyed a deeper shade than areas where the strike is light.

Aside from their wetting action, the dye-leveling agents used in acid dye baths for wool tend to prevent the strike and thus aid level dyeing. Their main mode of action, however, is in slowing down the actual rate of true dyeing, *i.e.*, the net rate at which dyestuff anions are taken up by the basic groups of the wool.   This retardation occurs because the dyestuff anions and the detergent anions compete with each other for positions on the basic groups.   We have here a case of two reactions, each having a rate relatively close to that of the other.

Leveling agents act somewhat differently in the dyeing of cellulosic fibers. In the case of cellulose, whether it is dyed with direct colors or vat colors, the dyestuff is not held to the fiber by coulombic linkages.   The dyestuff diffuses into the interior of the fiber and is adsorbed on the cellulose micelles, presumably by secondary valence forces (sometimes called dispersion forces in the quantum mechanical theory of valence) or van der Waals' forces. The strength of these bonding forces determines the wash-fastness of the dye, *i.e.*, the ease with which it may be removed by leaching or washing. As in the case of wool dyeing any set of conditions which slows down the rate of adsorption of the dye will result in more level dyeing.   There are relatively few substances which are strongly and permanently adsorbed by cellulose.   The direct dyes and the soluble leuco forms of the vat dyes are among the few soluble substances in this class.   None of the leveling agents, so far as is known, act by competing with the dyestuff for available points of adsorption on the cellulose.   They slow down the rate of adsorption by altering the physical state of the dye solution so as to make the dye itself less available for adsorption.   Direct dyes and leuco vat dyes normally exist in aqueous solution in the form of individual molecules, although there is a certain amount of association to form larger aggregates or dye micelles. These are strictly analogous to the micelles formed by surface active agents themselves.   An equilibrium exists between individual dye molecules and dye micelles.   Only the former are available for adsorption by the fiber.   The action of a leveling agent is the promotion of the formation of dye micelles, thus lowering the concentration of individual dye molecules and consequently the rate of dyeing.

For a more detailed discussion of the dye-leveling effect, see Chwala, *Textilhilfsmittel*, Edwards, Ann Arbor, 1943, pp. 344–359.

The effect of surface active agents in destroying or inactivating microorganisms has received widespread attention in recent years, following the discovery of the powerful germicidal action of certain cationic detergents. Although there are vast differences in effectiveness among the various surface active agents, a certain degree of germicidal power is possessed by most of the known classes with the possible exception of the non-ionic agents. The actual mechanisms by which the microorganisms are killed or inactivated are not completely clear, although they appear to be better understood than might be expected.   The first step is doubtless a strong but reversible adsorption of the detergent ions by the cells.   Following this there may be either a denaturation or a precipitation of the cell proteins by the detergent ions.   Most ionic detergents have a strong affinity for proteins, and can combine with them in stoichiometric quantities as well as alter their spatial configuration.   The detergent ions may also inactivate the enzyme systems which operate in cell metabolism.   Finally certain surface active agents may actually cause cytolysis, the rupture and disintegration of the cell.

For a complete review of the physiological and bactericidal effects of surface active agents, see Valko, in *Ann. N. Y. Acad. Sci.* **46**, Art. 6, 451 (1946), and Hotchkiss, *ibid.* 479.

CHAPTER 16

Relationship of Surface Activity to Chemical Constitution

I. General Discussion

In the early development stages of most fields of organic chemical technology one of the prime problems is the correlation of molecular structure with other properties, particularly with physical properties. The patterns for such correlations have been fairly well established in numerous different fields, and certain very general rules have become apparent. In the first place it is never certain before the problem has been attacked that any useful correlations can be established. Even when the particular property is simple, unequivocal, and easy to measure it may not correlate in a simple manner with chemical constitution. The melting point and the crystalline structure, for example, correlate poorly with chemical constitution. On the other hand, such properties as the refractive index and the parachor have been so satisfactorily correlated that they can actually be used in elucidating unknown structures. The same is true of the absorption spectrum, the Raman spectrum, and sundry other properties.

A second general rule concerns the complexity of the property to be correlated with molecular structure. Increasing complexity of the effect to be correlated usually increases the difficulties encountered in establishing useful relationships. This is exemplified by the field of chemotherapy. This science is largely based on tracing relationships between chemical constitution and physiological activity of one type or another. Before useful correlations can be established in chemotherapy highly restricted individual modes of physiological activity must be selected, and quantitative methods for their measurement developed. The effect of chemical structure on such restricted properties as local anesthesia, soporific effect, anti-trypanosome effect, anti-malarial effect, etc. can then be studied. Even when this approach is followed, a third generalization concerning correlation soon becomes apparent. Simple correlations can usually be established only within limited series of structures. For example, in the study of antiseptic action using *Eberthella typhosa* as a test organism it is possible to show a regular variation in killing power within a series of alkylated phenols. Similar regularities may occur in a series of long-chain quaternary ammonium compounds, but the law governing the variation in killing power of the phenols will not apply to the quaternary compounds. Nor will it account for the killing power of such unrelated compounds as mercuric chloride, sulfanilamide, or penicillin.   The relationship between chemical constitution and physiological activity accordingly reduces to a group of empirically derived correlations.   Each of these relates the variation of a limited physiological effect to a variation within fairly narrow limits of a definite structural type.

The same considerations apply in full degree to surface active agents, and the problem is still further complicated by a number of other factors. In the first place most surface active compounds are difficult to purify since they do not crystallize readily.   The commercial products are invariably made from non-homogeneous raw materials.   Those derived from the natural fatty acids can be separated into their pure components by laborious procedures, but those derived from petroleum are quite inseparable from any practical point of view.   We have seen, moreover, that both the basic and gross properties of surface active agents can be affected to a remarkable degree by small amounts of foreign substances, both organic and inorganic.   In studying correlations these factors must be taken into account and the compounds must either be rigorously purified or they must be brought to a comparable and reproducible state of contamination.

Limiting and defining the particular surface active property to be correlated constitutes a second fundamental problem in establishing correlations with molecular structure.   With the basic effects such as surface tension or solubilizing power, this problem is not a particularly difficult one.   With the gross effects, on the other hand, it is a considerable task.   Arriving at a precise definition of the categorical nouns "detergency," "emulsifying power," etc., would in itself be no easy matter, and in correlating these inadequately defined properties with molecular structure one must be doubly careful.   It is still quite common to hear surface active agents classed, on the basis of the gross effect most commonly used in industry, as "detergents," "wetting agents," or "emulsifying agents."   One of the earliest and most widely popularized relationships of structure to surface activity had to do with the chain length of the hydrophobic portion of the surface active molecules.   It stated that below C_8 no appreciable surface activity is observed; from C_8 to C_{12} the products are wetting agents; and from C_{14} to C_{18} they are detergents.   When we come to examine the real meaning of this obvious oversimplification we find first of all that it applies in the case of normal primary aliphatic sulfate sodium salts.   It refers to aqueous solutions of about 0.1% concentration.   By "wetting agent" is meant a product which wets grey cotton at 25°C., and by "detergent" is mant a product which washes wool (soiled with carbon black and Nujol) at 50°C.   Nothing is said about the rate at which wetting action falls off with increased chain length, nor about the way in which the sinking time *vs.* temperature curves behave as the chain length is increased, nor about the variation in salt effect as chain length is increased.

In spite of the large amount of work which has been reported only the merest beginning has been made in establishing scientifically sound correlations between structure and surface activity.   This is particularly true when the various gross properties are considered.   Many more data are available when basic effects are considered, particularly surface tension, interfacial tension against medicinal-type paraffin oil, and solubilization. The surface tension – concentration curves for several homologous series of surface active agents have, for example, already been noted above. There is a strong possibility that some interesting quantitative relationships could be established in this field using only the data which are now available. Attempts to do this, however, appear to have been remarkably infrequent. The general types of relationships which have been established are quite limited and several of them have already been pointed out in the chapter on basic effects.

II. Relationships between Basic Effects and Chemical Structure

Probably the earliest successful attempt to correlate the surface tension of dilute aqueous solutions with structure of the solute was made by Traube.[1]   He showed that in any homologous series the concentration required for equal lowering of surface tension diminishes threefold for each additional CH_2 group.   Langmuir has demonstrated that this indicates an arithmetical increase in the work of adsorption for each CH_2 group.   For several homologous series the energy of adsorption can be expressed by the equation:

$$\lambda = \lambda_0 + 625n$$

where λ_0 is a constant characteristic of the end group and n is the number of carbon atoms in the chain. λ is measured in calories per mole.   This rule holds fairly well only in dilute aqueous solutions.   With the surface active agents it is only applicable in the range of extreme dilution before the sharp drop in the surface tension *vs.* concentration curve occurs.

The surface tension – concentration curves of a series of lower alkyl benzene sulfonates, specially prepared and purified, have been studied by Neville and Jeanson.[2]   They found that sodium benzene sulfonate itself lowers the surface tension proportionately to the concentration, but its

[1] Traube, *Ann.* **265**, 27 (1891); *Ber.* **17**, 2294 (1884); *J. prakt. Chem.* **31**, 177 (1885).
[2] Neville and Jeanson, *J. Phys. Chem.* **37**, 1000 (1933).

homologues show considerable deviations in the curve. Among isomers, the xylene sulfonate is more active than the ethylbenzene sulfonate, and cymene sulfonate is more active than butylbenzene sulfonate. o-Toluene sulfonate is more active than the para isomer. None of the compounds in the range examined shows strong surface activity, and measurements were made on solutions of as high as 10% concentration to demonstrate these effects.

Kimura and Taniguti[3] have studied the surface tensions of the C_{12} to C_{18} normal primary fatty alkyl sodium sulfates at 60°C. The concentrations at which the surface tensions are a minimum at this temperature are C_{12} 0.25%, C_{14} 0.1%, C_{16} 0.01%, C_{18} 0.005%. The salt effect is pronounced with these products, as has been pointed out by numerous investigators.[4]

Merckel[5] studied the surface tensions of aqueous solutions of the fatty acids, alcohols, esters, and urethans. He finds in all these series that a linear relationship can be demonstrated between the slope of the surface tension – concentration curves and the number of carbon atoms in the chain.

Some interesting relationships have been demonstrated among the unsaturated and substituted fatty acid soaps with regard to their surface and interfacial tensions. Surface activity is increased by the presence of a plurality of double bonds in the fatty acid radical. Surface activity is also increased by halogen substitution in the fatty acid radical as in dibromoricinoleate or α-bromolaurate soaps. Soaps of highly hydroxylated fatty acids and α,ω-dicarboxy long-chain acids, on the other hand, are less effective surface active agents, as is evident from the fact that such substances give higher values of interfacial tension. This is attributed to a change in orientation of the interfacial film.[6]

Numerous studies have been made in which the surface and/or interfacial tensions of purified commercial surface active agents have been compared with one another. Bertsch, one of the great pioneers in developing the commercial fatty alkyl sulfates, showed that esterification of the carboxylic group increased the surface activity of sulfated ricinoleate products. Among such esters the amyl is most active.[7] Purified samples of Igepon A, Igepon T, and Gardinol WA have been compared with regard to several

[3] Kimura and Taniguti, *J. Soc. Chem. Ind. Japan* **42**, No. 89 (1939); *Chem. Abstracts* **33**, 5941.

[4] Compare Tartar and Cadle, *J. Phys. Chem.* **43**, 1173 (1939).

[5] Merckel, *Kolloid-Beihefte* **45**, 413 (1937).

[6] Cavier, *Compt. rend.* **213**, 70 (1941). Szegö and Malatesta, *Atti V congr. nazl. chim. pura applicata, Rome* **1935**, 569 (1936); *Chem. Abstracts* **32**, 21.

[7] Bertsch, *Melliand Textilber.* **11**, 779 (1930).

different gross and basic effects.[8] Similar studies have been made with a series of commercial soaps.[9] The interfacial tensions in oil–water systems containing various fatty monoglycerides has recently been reported by Feuge.[10] An extremely valuable compilation of surface and interfacial tension data for over a hundred commercial surface active agents has been made by Cupples. These lists include chemical composition, trade name, etc. Although no attempt is made to relate the surface properties to the structure, these data afford an excellent and convenient source on which such studies could be based.[11]

One of the most significant effects of structure on surface activity has been brought out by Hartley as the result of his studies with the alkyl ethers of mono- and dihydric phenol sulfonates.[12] It is well known that the surface tension – concentration curve levels off rather sharply after the critical concentration for micelle formation is reached. This is due to the fact that micelles as such are not effective in reducing the surface tension. They act merely as reservoirs for the active single ions or molecules. Hartley postulated that if micelle formation could be inhibited the surface tension (and interfacial tension) should be depressed to a considerably lower value with increasing concentration. One way in which we might expect to reduce the extent of micelle formation is by decreasing the symmetry of the molecule as a whole. This can be done by branching the hydrophobic chain or by using two short chains rather than a single longer one. It was actually demonstrated, by comparing the sulfonate of hexadecyl phenyl ether with the sulfonate of resorcinol dioctyl ether, that at identical single-ion concentrations the single long chain is the more effective surface active agent. The single-ion concentrations of both test compounds are equal when the total concentrations are well below the critical point. At higher total concentrations, however, the hexadecyl compound begins to form micelles and the surface tension becomes relatively constant. The dioctyl compound continues to lower the surface tension to values well below those obtainable with the hexadecyl analogue. In the actual experiments interfacial rather than surface tensions were measured and the stalagmometer method was used. Ethers of hydroquinone, catechol, and the cresols were studied as well as those of phenol and resorcinol. These results correlate well with those of Neville and Jeanson mentioned above, and establish the

---

[8] Weltzien and Ottensmeyer, *Monatsch. Seide u. Kunstseide* **40**, 504 (1935).

[9] Holiday, Kelly, and Rising, *J. Am. Pharm. Assoc.* **29**, 367 (1940).

[10] Feuge, *J. Am. Oil Chemists' Soc.* **24**, 49 (1947).

[11] Cupples, U. S. Dept. of Agriculture, Bur. Entomology and Plant Quarantine, E-426 (April, 1938), and E-504 (June, 1940).

[12] Hartley, *Trans. Faraday Soc.* **37**, 130 (1941).

point that increasing the degree of asymmetry of the molecular structure enables lower values of surface tension to be attained, although higher total concentrations must be used to attain them.

Aside from surface and interfacial tension few other basic effects have been extensively related to structure. The effect of structural variation on solubilizing power has been reported by McBain and Richards.[13] The structure of the material being solubilized is a much more important factor in this effect than the structure of the surface active agent.

An interesting study of the diffusion of various soaps and synthetic surface active agents through cellophane membranes has been made by Kröper.[14] Among the fatty acid soaps the presence of double bonds or hydroxyl groups favors diffusion. Increasing the chain length from C_{12} to C_{18} lowers the diffusion when the higher alkyl groups are present in either the anion or the cation. The same effect is obtained with the homologous fatty alkyl sulfates and fatty acyl taurides. These results correlate well with the degree of micelle formation. It is to be expected that only single ions or molecules would pass through the membrane and that even fairly small micelles would be retained.

III. Relationships between Gross Effects and Chemical Structure

Despite the relatively unsatisfactory methods available for measuring the various gross effects many excellent studies have been made of their dependence on chemical structure.

The detergency of the C_{12} to C_{18} fatty alkyl sulfates has been reported by Götte.[15] A typical detergency test was used in which cotton soiled with carbon black and oil was washed in glass tumble jars at 60°C. in buffered 0.1% solutions of the detergents. For all the detergents soil removal was optimum at a pH value of about 10 and was poorest at pH 4–5. Under the test conditions, at optimum pH value, the C_{16} compound was the best detergent followed in order by $C_{14} > C_{18} > C_{12}$. The foaming power was also measured and was found to vary considerably with temperature. At 60°C. the order of foaming power was the same as the order of detersive power. At 40°C. the C_{14} compound was the best foamer and at 20°C. the C_{12} compound was best.

In a more recent discussion the same author points out that wetting power is optimum when the chain is branched and when the solubilizing group is near the center of the molecule.[16] This parallels the results of

[13] McBain and Richards, Ind. Eng. Chem. 38, 642 (1946).
[14] Kröper, Fette u. Seifen 44, 298 (1937).
[15] Götte, Kolloid-Z. 64, 222, 327 (1933).
[16] Götte, Die Chemie 57, 67 (1944).

Hartley with regard to interfacial tensions and is explained on the same basis.

The dialkyl esters of sulfosuccinic acid are among the most powerful wetting agents for raw cotton known. It is noteworthy that they have the double-chain structure with the sulfo group in the center The wetting power of a long series of these products, including the monoesters and mixed diesters as well as the symmetrical diesters has been studied by Caryl.[17] The Draves test was used to measure wetting power, and the results are expressed as the concentration of wetting agent necessary to give a Draves sinking time of 25 seconds at 30°C. The diesters are in general more effective wetters than the monoesters. No great difference is apparent between symmetrical and unsymmetrical diesters, provided both groups are of approximately the same size. The best wetting agents of this series are those in which the alkyl groups are branched octyl or nonyl radicals. The commercial product Aerosol OT, di(2-ethylhexyl)sulfosuccinate sodium salt, rates close to the top of those tested. The highest rating is shared by two compounds, the di(1-methyl-4-ethylhexyl) and the 2-ethylhexyl, 1-methylheptyl esters.

Venkataraman and co-workers have prepared and studied a long series of compounds of the generic formula:

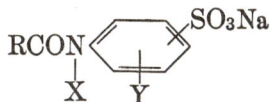

$$RCON \underset{X \quad Y}{\overset{}{\bigcirc}} SO_3Na$$

where R stands for a long-chain fatty radical, X is hydrogen or a short alkyl radical (methyl or ethyl), Y is hydrogen, methyl, chlorine, methoxy, or ethoxy in various positions, and the position of the sulfo group relative to the other nuclear substituents is varied. Some naphthalene derivatives were also prepared and tested.[18] The wetting power of these compounds was measured by the Draves test and by a modified Herbig test.[19] In some instances the lime soap dispersing power was also measured.

In general those compounds where the sulfonic group is in the ortho position to the hydrophobic group proved to be the best wetters, followed in order by the corresponding meta and para compounds. Shorter wetting times were observed with products containing mixed coconut fatty acid radicals rather than the oleic acid radical. The X substituent, methyl or

[17] Caryl, *Ind. Eng. Chem.* **33**, 731 (1941).

[18] Venkataraman *et al.*, *J. Soc. Dyers Colourists* **53**, 91 (1937); **55**, 125 (1939); **57**, 41 (1941); *J. Sci. Ind. Research (India)* **3**, 193 (1944).

[19] Forster, Uppal, and Venkataraman, *J. Soc. Dyers Colourists* **54**, 465 (1938).

ethyl, had little effect on the wetting power, but methoxy or ethoxy substituents in the nucleus exerted a favorable effect. The main factors, however, appeared to be the nature of the fatty acyl radical and the position of the sulfonic group relative to it.

This work illustrates the necessity for extensive and complete physical measurements in attempting to relate structure to gross performance. The selection, preparation, and purification of the compounds were exemplary, but it is apparent that the significant results could have been greatly extended if the physical tests had been more complete. In addition to the simple Draves tests, a series of wetting time *vs.* concentration curves and wetting time *vs.* temperature curves might have been plotted. This would, of course, have involved an even more extensive study but it would certainly have given a much more complete picture of performance. The importance of making wetting and detergency tests at various temperatures has often been mentioned and cannot be overemphasized.

A comparison between the normal fatty alkyl sulfates and sulfonates has been made by Evans.[20] The former products, under comparable conditions, are superior in both wetting and detersive power.

One of the best planned studies attempting to relate structure to surface activity is that of Dreger, Keim, Miles, Shedlovsky, and Ross.[21] These investigators prepared a series of sodium alkyl sulfates derived from the following twelve secondary fatty alcohols containing unbranched alkyl groups: 6-undecanol, 7-tridecanol, 2-pentadecanol, 4-pentadecanol, 6-pentadecanol, 8-pentadecanol, 9-heptadecanol, 10-nonadecanol, 2-undecanol, 2-tridecanol, 2-heptadecanol, and 2-nonadecanol. The sodium sulfates of the primary normal alcohols 1-decanol, 1-dodecanol, 1-tetradecanol and 1-hexadecanol were also prepared and purified for comparison. The solubility, surface tension, foaming power, wetting power, and detergency of all these products were measured. Full precautions were taken to minimize possible errors in the measurements, and the percentage error inherent in the method was ascertained in all cases. The canvas disc wetting test and the Launderometer detergency test were used. Foaming power was measured by the pour method of Ross and Miles.[22] The results (compare Fig. 51) may be summarized as follows[23]:

The surface tension becomes lower with increasing molecular weight, and also as the sulfate group is moved nearer the center of the chain. The

[20] Evans, *J. Soc. Dyers Colourists* **51**, 233 (1935).

[21] Dreger, Keim, Miles, Shedlovsky, and J. Ross, *Ind. Eng. Chem.* **36**, 610 (1944).

[22] J. Ross and Miles, *Oil and Soap* **18**, 99 (1941).

[23] For a complete discussion of this work and a comparison with supplementary work by the same and other investigators, see Shedlovsky, *Ann. N. Y. Acad. Sci.* **46**, Art. 6, 427 (1946).

behavior of the secondary salts follows Traube's rule fairly closely, but this is not true of the primary salts.

The largest volume of stable foam is obtained with the 17–9 compound, i.e., the C_{17} compound with the sulfate group on the number 9 carbon atom. In the C_{15} series the foaming power increases as the sulfate group is shifted toward the center. In the 2-sulfate series maximum foam is obtained

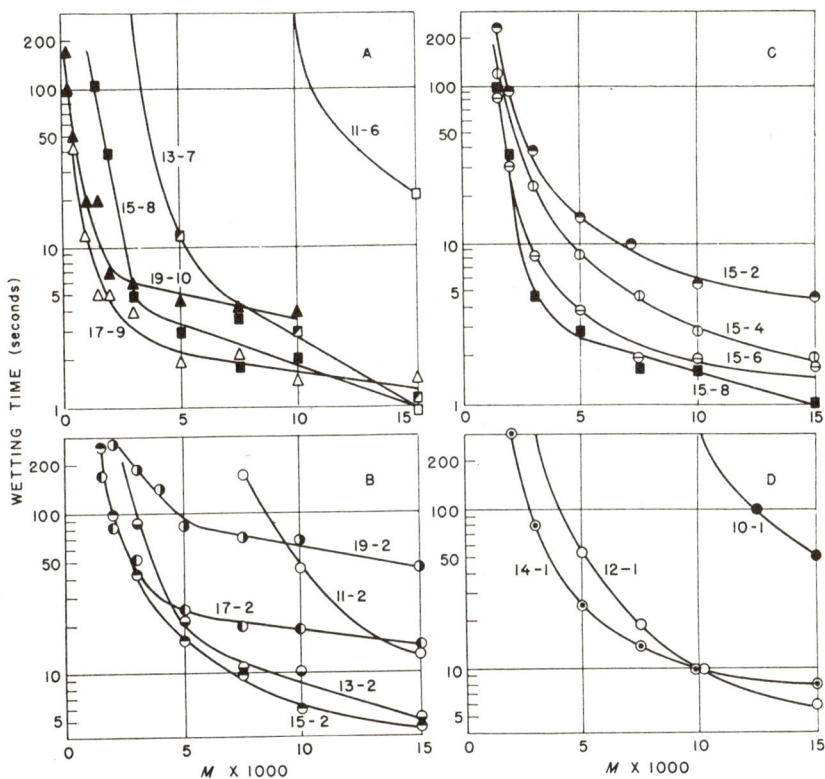

Fig. 51. Canvas disc wetting time *vs.* concentration for aqueous solutions of sodium alcohol sulfates at 43.3–46.7°C.[21] A, sodium *sym-sec*-alcohol sulfates; B, sodium *sec*-alcohol sulfates; C, isomeric sodium *sec*-pentanol sulfates; D, sodium *n*-alcohol sulfates.

with 15 carbon atoms. The order of foaming power in the 2-sulfate and 1-sulfate series is:

$$10 - 1 < 11 - 2 < 12 - 1 = 13 - 2 < 14 - 1$$

$$= 15 - 2 > 16 - 1 > 17 - 2 > 19 - 2$$

Wetting power is also at a maximum when the sulfate group is in the center of the molecule.   The results are shown in Figure 51.

Detergency is poorest when the sulfate group is in the center of the molecule and is maximum for the normal primary compounds.   The C_{17} and C_{19} compounds are effective detergents at much lower concentrations than the lower members of the series.

Some studies relating constitution to the less usual gross effects have appeared from time to time.   An interesting example is afforded by Cavier,[24] who has related the hemolytic activity of a number of surface active agents to the degree to which they are adsorbed by charcoal.

In conclusion it may be stated that Hartley's theory appears to hold very well for surface and interfacial tensions, and also for wetting power.   We have seen that wetting power for raw cotton (and probably for other waxy or hydrocarbon-like surfaces) can be closely correlated with surface tension, so this relationship is not surprising.   Foaming power, emulsification, and detergency are more complex and further removed from the simple basic effects.   Correlations in these fields are limited at present to simple series of homologues or analogues.   It seems plausible that further strides will depend on sounder, more fundamental methods for expressing and measuring the gross effects themselves.

[24] Cavier, *Compt. rend.* **216,** 255 (1943).

PART III

PRACTICAL APPLICATIONS
OF SURFACE ACTIVE AGENTS

CHAPTER 17

General Considerations and Classification of Uses

Surface active agents are useful primarily because of their peculiar gross effects which have been discussed in Chapter 15 of Part II. The fatty soaps as a class possess as well-balanced an array of these properties as any other single class of surface active agents available today, and they are still the most important economically. In considering the present use of soap, we must not overlook the historical fact that the virtually complete absence of competitive surface active agents during many centuries has ensured the soaps a very wide field of use. Through the centuries (including well over a hundred years of our modern industrial era), the householder and artisan have become familiar with the properties of the soaps. Rule-of-thumb methods have been gradually worked out for using soaps so as to utilize their valuable properties and avoid their weaknesses. As a result, the soaps came to be used, and still are used on a very large scale in many different industries and trades as well as for personal and household uses.

In terms of modern requirements, however, the soaps have certain deficiencies. Perhaps the most troublesome is their well-known sensitivity to various common chemical agents, particularly acids and all but a few metallic ions, including those which cause hardness in water. Furthermore, with respect to one of the major gross effects, namely wetting and penetrating power, the soaps are decidedly inferior to various synthetic surface active products available today. Introduction of the newer surface active agents for the purpose of overcoming these deficiencies of soap has stimulated and coincided with the development of a whole host of new fields of application. In these newer applications, where soap never was applicable, the new synthetic agents have scored their most dramatic successes. Moreover in certain of the fields once dominated by the soap, the newer products have been found to be superior, and the soaps are being displaced more or less rapidly. In other fields, soaps are being supplemented or improved, and their use is being expanded by the admixture of synthetic surface active agents.

The development of large-scale uses of synthetic surface active agents has not been achieved without considerable effort. In the first place, new methods often had to be worked out for using the new products. Merely working out such methods, however, was not enough. Prospective users, traditionally accustomed to thinking in terms of soaps, often regarded with suspicion the new products' strange properties and their different methods of use. Considerable effort had to be directed to demonstrating the advantages of the new products and to teaching the methods of using them. Consequently, most of the important producers of surface active agents, have instituted and maintained application research laboratories.

It may be safely stated that the present-day extensive use of surface active products is due more to the activities of such laboratories than to any other single factor. In the course of their work, the application research laboratories survey industry after industry, for the purpose of seeking out those unit operations where surface active agents might be applicable. They cooperate closely with technologists in the consuming industries, learning the specific problems involved and the difficulties to be overcome. They develop methods of testing and evaluating their product for the particular use in question, and finally they attempt to develop the optimum application techniques whereby their product may be utilized to its full potential. The application research laboratories of detergent manufacturing companies are sometimes limited, however, by the range of products with which they must work. When a company, for example, produces only long-chain fatty alkyl sulfates, the commercial function of the application laboratory is to find uses for these specific products. In doing so new fields may be discovered where fatty alkyl sulfates show only fair performance whereas another type, possibly a non-ionic or an alkyl aryl sulfonate, shows excellent performance. In such cases the discovered use may not be publicized at all, and its introduction may be delayed appreciably.

Certain industries, for example textile-finishing mills, that consume large amounts of surface active agents for well-established uses, often have excellent facilities for evaluating new products which may be offered to them. They are also in the forefront in developing new uses within their particular sphere of operations. There are relatively few industries, however, in which surface active agents constitute a large segment of the total purchases or in which they are indispensable, and even these industries do not always make the best choice of the available products nor utilize them to best advantage. The more efficient consumer companies often set up their own laboratories or secure the services of independent laboratories to advise them on problems involving surface active agents. At the present time, however, there is little doubt that many consumers of surface active agents are using neither the optimum product nor the most efficient application methods. This might be expected in a new, fast-growing field in which promotional activity and publicity often play a decisive role.

Outside the research laboratory, the development of a new use is often the result of the ingenuity of a well-informed craftsman or artisan. As an example, we may cite the use of wetting agents in photographic-developing baths to insure rapid uniform contact between the bath and the film. Possibly another example of this sort is the use of wetting agents in the water sprays for removing wallpaper, or for removing paper labels from used bottles, etc. In these cases, in which the user is seldom versed to any extent in detergent technology, the optimum product often is not used. Once a reasonably satisfactory product has been introduced, it usually continues in favor unless a deliberate sales effort is made to displace it.

Where the performance of several different products is in the same general range, and when the use is on a sufficiently large scale to merit careful consideration, selection is normally made on the basis of "money value." Money value may be defined as degree of effective action divided by cost. Cost is relatively easy to pin down, although it often involves more than the simple figure of cents per pound delivered. The degree of effective action, on the other hand, is often extremely difficult to analyze and rate. Individual instances of this will be considered when actual uses are discussed.

In many fields of application, however, money value is not the primary consideration. This is obviously true where the products are used in medicine or pharmacy. It is also true in the cosmetic field, where dark color or objectionable odor may rule out products which are inexpensive and functionally excellent. In household uses many products characterized by excellent performance have failed commercially because they did not resemble the highly advertised soap products in foam, cloudiness of solution, etc. Even in the cost-conscious textile industry, products of optimum money value may not be accepted for a variety of unexpected reasons. One instance may be cited of a synthetic detergent which showed a money value about 50% higher than that of soap for a particular scouring operation in a certain textile mill. Although it cost considerably more than soap only about one-fifth the poundage was needed in the scouring bath. The workmen, however, were accustomed to adding a definite measure of soap to the bath at stated intervals and could not be convinced that the much smaller measure of new detergent could possibly do the job. They insisted on using several times the necessary amount, and the total cost consequently was prohibitive.

These and similar circumstances complicate the technology and economics involved in the utilization of surface active agents. Problems posed by this situation have been approached in two ways by the successful workers in this field. One approach strives to develop a single ideal surface active agent suitable for all purposes. The other approach is directed to

finding an optimum product for each specific use.   The desired product
sometimes may be found among those presently available, or it may have
to be synthesized specially to fit specific requirements.   Of the two approaches, the latter is certainly more logical, but in actual practice a com-
promise between the two is almost always effected.   A single ideal surface
active agent is a little more than a figure of speech.   It is no more realizable
or even desirable than a single ideal type of motor vehicle or a single ideal
metal.   It is true, however, that some of the cheaper tonnage detergents
e.g., the alkyl aryl sulfonates or Igepon types, can be used effectively in a
very wide variety of applications.   Just as a station wagon body on an automobile is more versatile than a funeral coach or a tank truck, so certain
surface active products are more versatile than other, more highly specialized types.   In many instances, a study of the money values will favor the
low-cost versatile product even though the specialized "tailor-made" product may have better absolute performance.   In other cases, the special
requirements fulfilled by tailor-made products will justify considerable
additional cost.   The fact that there are several dozen established successful types of surface active agents on the market, varying in price from a few
cents to almost two dollars per pound (on a 100% active basis) indicates
that each one of these has at least a few applications where it has gained
preference over the others.

Sources of information concerning uses of surface active products are
often too vague and generalized to be of value.   The disclosures of product
patents on new surface active agents invariably state that the product can
be used as "an assistant in textile-treating baths," "a detergent," "a wetting
and foaming agent," etc.   A few of the more recent patents disclose actual
test and performance data on the claimed products in order to clearly
establish their utility and to teach the art as required by basic patent law.
Such data are more apt to be found in the numerous process patents dealing
with the utilization of surface active agents than in product patents.   Beside patents and trade journals, sales literature issued by manufacturers
and vendors of surface active agents constitutes another extremely useful
and complete source of information.   Practically all the larger producers
and most of the smaller ones publish pamphlets or bulletins describing in
detail the various uses of their products.   Such publications are frequently
revised to keep them up to date.   They represent a summary of much of
the work of the manufacturers' application research laboratories together
with that portion of their customers' field results which is not confidentially
withheld.   Numerous review articles have also appeared from time to time
on the various uses of surface active agents.[1]

[1] Compare Lenher, *Chemistry & Industry*, **1941**, 497.   Ackley, *Ann. N. Y. Acad.
Sci.* **46**, Art. 6, 511 (1946).

Since surface active agents are used mostly for the various gross effects which have been discussed it would appear logical to select the gross effects as a basis for classifying uses. Thus we would have uses wherein wetting was the desired action, cleaning uses, emulsifying uses, etc. There are several disadvantages to such a classification, however. In the first place, a vast number of practical applications involve more than one simple gross effect. In the second place, some very important applications of surface active agents depend directly on basic properties such as adsorption and charge phenomena not readily classified under any one gross effect. An example of this is the use of cationic detergents in fixing delustering pigments on fabrics. Other applications may be based on chemical properties, as exemplified by the use of cationic detergents to improve the fastness of direct dyestuffs. In still other cases the lubricant properties may be of primary interest, as in the case with textile-softening agents. Another disadvantage of classification on the basis of gross effects is that it juxtaposes many industrial arts which otherwise may be completely foreign to one another. Fire-fighting compositions, for example, have very little in common with bubble baths or shampoos and yet all three applications utilize surface active products characterized by high foaming power. Nevertheless, the gross effects are basic to any consideration of practical uses. The largest number of uses are based on the wetting, emulsifying, and detergent power. In many cases a combination of wetting and detergent power is desired. Emulsifying ability, on the other hand, need not be combined with other effects in order to afford wide practical applicability. For some applications protective-colloid action and suspending action may be desirable in themselves, apart from other effects. Foaming ability is sometimes useful by itself, as in fire-fighting compositions, but more often it is desirable in combination with high detersive action. Wherever possible in the following discussions, an effort will be made to point out the various gross effects (and/or basic effects, such as solubilization, etc.) involved in the practical usage under consideration.

Classification on the basis of the industries in which surface active agents are used also involves difficulties. The main one is the large number of industries, trades, or arts in which these substances are used only to a very limited extent. However, there are certain industries, e.g., textile and cosmetic manufacturing, in which surface active agents are of such importance that it is both logical and convenient to discuss all uses within such industries as a unit. We shall accordingly consider first of all those industries in which surface active agents are used in many different unit processes. Following this a number of other industries will be discussed in which surface active agents play a less important role. In conclusion a number of miscellaneous uses will be discussed.   These will be classified on the basis of the major gross effect involved.

As in Part I of this volume (Chapters 2 to 10), no attempt will be made to present an exhaustive discussion.   The list of specific uses to which surface active.agents are put is currently expanding at a very rapid rate, and any compilation would thus become incomplete within a short time. The various types of uses which may be served by surface active agents are, however, of more permanent importance.   Familiarity with the types or classes of utilization will enable the technologist to judge how and where he may profitably try surface active agents in his own field.   Since the particular gross effect involved is obviously very closely related to the type of utilization, such effects will be emphasized rather than the details of particular individual uses.

CHAPTER 18

Applications in the Textile Industry

I. General Considerations

The various unit operations of the textile industry offer numerous opportunities for advantageous use of surface active agents. As a consequence, a larger number of such products is used in textile processing than in any other industry. Beginning with the invention of the sulfated oils about 1870, and continuing to the present decade, almost all new surface active products have been developed with a view toward specific textile applications. The adoption of these new products has followed an interesting and erratic course, conditioned by economic considerations and by a well-justified conservatism on the part of the textile mills as well as by the technical merits of the products themselves.

Until about twenty-five years ago all wet textile processing was geared to the use of soaps. The sulfated oils were also used wherever applicable, particularly for their wetting power and their ability to impart a desirable finish to cottons. The sulfated oils, however, were rightly regarded as supplements rather than substitutes for soap. The most serious limitation of soap is its inability to function in acidic solutions and in the presence of heavy-metal ions. Over a period of generations the textile industry had developed techniques which by-passed these disadvantages and utilized the soaps to their fullest potential. Finishing mills were, for example, located in areas where ample supplies of soft water were available. When zeolite water-softening systems were introduced they were widely adopted by mills whose water supply was unfavorably hard. The normal gradual improvement of the equipment used in textile finishing also contributed to more efficient utilization of soap.

The synthetic detergents were first introduced in Germany, where their enthusiastic reception can be attributed to the following factors. In the first place, many of the industrial water supplies in central Europe are extremely hard and the cost of softening them is correspondingly high. Secondly, the price differential between soap and the synthetic products has generally been rather small in Germany due largely to the relatively high cost of fats. Still another favorable factor was the amazingly high degree of integration and organization within the German chemical industry.   This enabled a wide variety of newly developed surface active agents to be profitably manufactured in relatively small poundages.   The close cooperation between the textile mills and the chemical manufacturers also hastened the adoption of the new products.

The advent of synthetic surface active agents rapidly stimulated the development of new efficient textile-treating processes impossible with soap. As a rule each type of synthetic surface active agent has some outstanding property which may profitably be utilized.   It may be high wetting and penetrating power, a specific effect in dyeing, compatibility with salts or other chemicals, high affinity for fibers, low affinity for fibers, etc.   The more important of these specific properties and the ways in which they have been utilized will be discussed in detail below.

The adoption of synthetic surface active agents by American textile mills has largely followed the development of these new processing economies which the synthetic agents make possible.   In short, they have been adopted for processes in which soap never was used or else where soap did an admittedly poor job.   Within very recent years the synthetic agents have gained considerably more ground, and in some mills they have even displaced soap for alkaline scouring operations, an application in which soap is quite satisfactory.   The increased efficiency and lower cost of some of the new detergents has thus enabled them to surpass soap in money value even where soap is at its best.

The purposes for which water-soluble surface active agents are used in textile processing may be grouped into the following general classes: (1) *Wetting-out*.   The purpose here is merely to wet the goods thoroughly in the processing liquor.   (2) *Scouring*.   In scouring or washing processes foreign matter must be removed from the goods and the surface active agent serves primarily as a detergent.   (3) *Dyeing, dye leveling, dye fixing, and stripping*.   In these applications wetting and penetrating power are important but other specific effects are also necessary.   All good wetting agents are not necessarily good dye bath auxiliaries.   (4) *Finishing*.   Some surface active agents are in themselves excellent agents for conferring a soft full hand to fabrics.   More often, however, they are used in conjunction with oils, starches, gums, resins, etc. to make up the finishing treatments.

It should be noted that the word "finishing" is used in at least two different senses by textile technologists.   In its broadest meaning, "finishing" refers to any and all treatments applied to the fabric after it has left the loom (or knitting machine, as the case may be).   Material coming from the looms still contains the sizes and/or fabricating lubricants.   Such material is known as "gray goods" or "greige goods," the spelling depending on whether cotton or rayon is being discussed.   Further treatments which may be applied to gray goods include kier-boiling, bleaching, dyeing, printing, etc. Performing these operations is the function of a finishing mill, as opposed to a gray mill which spins and looms the fabric. In a narrower sense, however, the term "finish" refers to a chemical process, product, or dressing which is applied to the fabric in its final processing stages in order to confer special properties. Thus, there are softening finishes, crease-resistant finishes, shrink-resistant finishes, water-repellent finishes, etc. In still another sense, "finish" may refer to a style of cloth normally produced from an appropriate basic weave by mechanical and/or chemical processing as: a moire finish, a chintz finish, etc. The exact meaning of the term is, however, usually obvious from the context.

(5) *Emulsifying and dispersing.* There are several operations of both routine and special nature in which fabrics are treated with water-insoluble materials in aqueous dispersion. The surface active agent in these cases serves primarily as an emulsifying or dispersing agent. Other auxiliary purposes may also be served. In the application of spinning lubricants to cellulose acetate yarns, for example, surface active agents may exert an anti-static effect. In the application of dulling pigments to rayon, they may change the electrostatic charge, thus improving adherence of the pigment particles to the fibers. In the preparation of vat dye printing pastes, they not only disperse the pigment but also promote wetting and penetration, thus insuring good contact when the paste is applied to the fabric.

(6) *Miscellaneous special purposes.* These cannot readily be classed in any of the above groups.

It is in some ways advantageous to discuss the numerous individual applications under the above-mentioned headings, since certain methods of utilizing surface active agents remain the same no matter what fiber is being processed. Wax–aluminum water-repellent emulsions, for example, may be applied to cotton, rayon, or wool in about the same manner. The scouring processes for viscose and acetate yarns are quite similar to each other. The dye leveling effects are similar whether cotton is being treated with direct colors or wool is being treated with acid colors. It is also instructive, however, to classify textile processes on the basis of the fibers used. In this case, wool, cotton, and rayon processing would be discussed separately. This approach is particularly appropriate in considering certain important processes which involve surface active agents and which are carried out on only one type of fiber. As examples of these we have the fulling of wool, the mercerizing of cotton, the dyeing of acetate rayon with dispersed colors, etc. In the following discussion, neither scheme of classification will be rigidly followed, since surface active agents rather than textile-treating processes are at the focus of attention.

It would be impossible in the limited space available to discuss in full detail all textile applications of surface active agents. The textile processing industry is by no means highly standardized. Considerable variation exists in methods used by different finishing mills in carrying out the "standard" operations of scouring, bleaching, dyeing, etc. Many mills have developed unusual short cuts in processing. As a typical example, rayon fabrics have been desized, scoured, and dyed simultaneously by padding on a mixture of enzymes, dyestuffs, and wetting agents, storing briefly, and passing directly to the dye beck.

Another factor contributing to the complexity of wet textile processing is the large number of differently constructed grey fabrics which must be handled. When this is multiplied by the number of different designs and finishes which may be specified by the mill's customer the possibility for variation becomes almost limitless. Accordingly, each new lot of goods may involve a new series of problems for the textile mill technologist and may be handled in a manner considerably different from previous lots. We shall therefore consider only the more basic operations and those in which surface active agents are highly valuable.

Before proceeding to the discussion of specific applications, another point of general interest may be brought out. There are several hundred brand-named surface active agents representing the score or more major structural types. Most of these are valuable in more than one process although a few of them are "tailor-made" for one single use or group of uses. It is logical to suppose that, for any specific use, careful tests would reveal an optimum agent superior to all others being offered in the trade. In actual practice, however, such careful and complete evaluations are seldom carried out. A product which is brought to the mill chemist's attention, and which proves to be satisfactory, is often adopted without making an exhaustive search for better products. This is justified to a limited degree, at least, by the fact that there are many products and even many different structural types which deliver about the same order of performance and money value.

A good example of this situation is encountered in the scouring of raw wool. There are at least five different types of detergent which have been used profitably on a commercial scale for this important operation. Beside soap, they include the polyethylene oxide non-ionic agents, the alkyl benzene sulfonates, the fatty alkyl sulfates, the Igepon type, and the fatty diethanolamide or Ninol type. Individual mills make their choice on the basis of local conditions and can usually point out good reasons for their own particular selection. No single product, however, is so outstanding that it would be chosen unequivocally by every mill. A still more striking example is in the field of carbonizing wetting agents. Here the usual wide variety of anionics is used and some mills even use certain cationic types.

Although the above-described situation is encountered in general throughout the whole gamut of finishing operations, there is little doubt that considerable savings could be effected in many mills by devoting more care to selection of the most efficient agents for various processes. There is an understandable tendency for the more aggressively merchandised products, especially those which are backed up by ample data from the manufacturer's application research laboratory, to get the call over worthier but less well-known competitors. In view of this situation no attempt will be made in the following discussion to rate the various commercial products in their order of current preference. The various types which have been used in each application will be discussed, and wherever any type has proved to have outstanding merit this fact will of course be pointed out.

II. Wool Processing

(a) Scouring

Raw wool, as received by the mill, may contain 30 to 50% or more of wool grease, in addition to smaller amounts of dirt, straw, burrs and other vegetable matter, pitch, paint, etc. The first step in processing is to wash or "scour" the loose raw wool. This is done by passing it through a series of bowls containing detergent solutions in which the wool is gently agitated. The wool is then rinsed thoroughly. Scouring is generally carried out at a relatively low temperature (40 to 55°C.) and with as little agitation as possible, in order to avoid felting or matting. Soap and soda ash were formerly used almost universally for this operation. Soda ash alone is quite effective with some wools. This is the case where the wool grease has a high acid number. The soda ash converts the free fatty acids to soaps and these serve to scour off the unsaponified portion of the grease. It is well known that any aqueous alkalis are detrimental to the wool fiber, the amount of damage being roughly proportional to the strength of the alkali and the temperature of treatment. It is quite easy to dissolve wool completely in hot caustic soda solutions. In the normal soap – soda ash scour a wool fiber may loose as much as 15 to 20% of its ultimate tensile strength. Nevertheless, soda ash is so effective a cleansing adjunct in wool-scouring baths that it is still used almost universally.

Wool, an amphoteric protein with an isoelectric point at about pH 4.8, although easily damaged by alkalis suffers little damage from the action of even fairly strong acids. The washing of wool in acid solutions, particularly at the isoelectric point, at which damage is at a minimum, has received much attention in Germany. This procedure is only possible, of course, if detergents which are effective in the acid pH range are used. The polyethylene oxide non-ionic agents and the Igepons have been most prominently mentioned.[1,2]   Processes have even been proposed in which wool is washed in acidic solutions of carbonizing strength.   The wool can then be passed directly from the scouring bowls to the carbonizing ovens. These processes have apparently not been adopted to any significant extent in the United States.[3]

Many of the anionic and non-ionic synthetic detergents scour wool effectively at much lower concentrations than are necessary with soap. Such synthetic agents therefore approach and may even surpass soap in money value.   Furthermore, they require much less soda ash in the bath, thus affording the advantages of lower alkalinity.   Since they are lime resistant, the scoured wool is recovered free of lime soap and accordingly has a somewhat different hand and appearance.   This may be favored or disfavored, depending on the processor's preferences.   The presence of lime soap in small amounts is generally not detrimental since it is removed in the subsequent carbonizing.

The lime-resistant synthetic detergents are particularly advantageous in scouring pulled wool.   This is wool taken from the hides of slaughtered sheep with the aid of a depilating agent.   Since lime is a common ingredient of depilating compositions, many pulled wools contain a high percentage of calcium salts which seriously impedes the action of soap in the scouring baths.

One of the potential disadvantages of the anionic sulfates and sulfonates is the ease with which they are adsorbed by wool.   Since these detergents are anions of strong acids, they are readily taken up by the wool protein, even at pH values well above the isoelectric point.[4]   The scouring baths accordingly tend to become depleted more rapidly than when soap is used. The non-ionic agents are adsorbed to a much lesser extent and consequently are better able to exert their full scouring potential.   One disadvantage of most synthetic detergents, as contrasted with soap, is the difficulty of separating the valuable wool grease from the spent scouring liquors.   When soap is used as the detergent the emulsified wool grease may be separated by acidifying the liquors and thus precipitating the soap as fatty acid. When acid-stable synthetic agents are used, the liquors must be centrifuged or subjected to even more costly treatments.

In spite of these disadvantages the synthetic detergents are widely used because of their high efficiency at low temperatures and the possibility of using low concentrations of alkali.   The fact that synthetic detergents are

[1] Meyer, *Melliand Textilber.* **20**, 355 (1939); *Deut. Wollen-Gewerbe* **70**, 1280 (1938). Gara, *Kém. Lapja* **2**, No. 4, 10 (1941); *Chem. Abstracts* **35**, 8304.

[2] Reumuth, *Klepzig's Textil-Z.* **44**, 260 (1941).

[3] Ger. Pats 639,535; 641,190; French Pat. 772,085; to *Deutsche Hydrierwerke, A.-G.*

[4] Steinhardt, Fugitt, and Harris, *J. Research Natl. Bur. Standards* **26**, 293 (1941).

easier to rinse from the fabric than soap is also a major advantage.[5] It should be noted that at least two other processes for cleaning raw wool have been used aside from washing in an aqueous bath. The most important of these is "solvent scouring," which is essentially a dry-cleaning operation. The wool grease is extracted with an appropriate solvent such as naphtha or trichloroethylene. This process eliminates the dangers of felting and alkali degradation. It has been used from time to time by some of the larger mills but has not been widely adopted. A more spectacular process which has been described consists in chilling the raw wool to a very low temperature. The wool fiber itself remains quite flexible at this temperature, but the grease becomes brittle and friable. It is removed from the fiber by beating or threshing, and is recovered as a powder which sinters and melts to a grease on warming to room temperature.[6]

(b) Depainting

Sheep which are raised on large ranches are generally branded for identification by marking with paint or road tar. When these animals are shorn the painted areas of their fleece are generally separated and are sometimes sold as "paint clips." The removal of paint or pitch from wool is a difficult operation, and the depainted wool is not useful for yarns. Such wool is usually mixed with noils (the waste fiber from the carding operation) and made into felts. Depainting may be carried out entirely with non-aqueous solvents, with solvent emulsions, or with strong aqueous detergent solutions. The emulsion system is quite widely used and gives good results. Alkali metal or amine soaps are the usual emulsifying agents. Aromatic or terpene solvents are preferred because of their high solvent action on the types of paint usually encountered. Strong solutions of the polyethylene oxide non-ionic detergents are said to be effective pitch removers even when no solvent is present. These detergents are less effective on paint.[7]

(c) Carbonizing

Most raw wool contains a considerable quantity of foreign vegetable matter in the form of straws and burrs which cannot be removed by scouring. The purpose of the carbonizing operation is the elimination of these

---

[5] Canadian Pat. 359,059, Lenher to Canadian Industries, Ltd. Gutensohn, Klepzig's Textil-Z. 44, 478 (1941). Finck, Melliand Textilber. 16, 209 (1935). Brit. Pat. 510,158, to I. G. Farbenindustrie Ger. Pat., 616,797, to I. G. Farbenindustrie. Sitomer and Vilbrandt, Proceedings of the Virginia Academy of Science, 1940–41, in Virginia J. Sci. 2, 206 (1941).

[6] Pettinger, Soap 11, No. 7, 31 (1935). Hünlich, Appretur-Ztg. 31, 113 (1939); Chem. Abstracts 33, 7577.

[7] Gutenshon, Klepzig's Textil-Z. 44, 54 (1941). U. S. Pat. 2,340,977, Nüsslein and Gutensohn to General Aniline and Film Corp.

cellulosic impurities from the wool. The wool, after having been scoured, is treated with a dilute solution of relatively non-volatile strongly acidic material. Sulfuric acid is usually used, although aluminum chloride is favored by some mills. The acid-soaked wool is then passed through flues where the excess water is driven off. Thereafter, it is baked at a relatively high temperature to promote charring and embrittlement of the burrs and other cellulosic materials by the strong acid. The wool itself is practically unharmed by this treatment, although a slight degradation of the keratin does take place. After being removed from the baking flues the wool is beaten and shaken to pulverize the charred vegetable matter, thereby causing the latter to separate as a dust. Carbonizing may be carried out on raw-wool stock or on finished piece goods. It is not unusual practice, when only a small amount of vegetable matter is present, to spin and weave the wool into fabric before carbonizing.

It is very important in all carbonizing that the burrs and straw be completely saturated with acid. Therefore a wetting and penetrating agent is almost always added to the bath. The wetting agent must be stable to the acid and must not form objectionable decomposition products during the baking process. The lower alkyl naphthalene sulfonates are widely used for this purpose in spite of the fact that they are readily adsorbed by the wool. Since they are no more readily adsorbed than the carbonizing acid itself, no disadvantage is incurred. Some mills employ cationic wetting agents for carbonizing baths and many of the non-ionic detergents are also satisfactory. The wetting agent used in carbonizing may sometimes have an effect on the felting properties of the wool. In general, the anionic agents are said to cause more felting than the other types. The evenness with which the wool fiber itself is wet during carbonizing may also have an effect on its behavior during subsequent dyeing.[8]

(d) Oiling

As a necessary preliminary to the spinning and weaving processes, the wool fiber must be properly lubricated or oiled. Vegetable oils, particularly olive oil, were formerly used for this purpose but they have been largely supplanted, except for special purposes, by the less expensive mineral oils. Blends of mineral oil with minor proportions of vegetable oils are also quite widely used. The oil is generally applied in the form of an aqueous emulsion by spraying directly onto a lap of wool. About 5% oil, based on the weight of wool, is normally added. Some mills make up their own emulsions from an oil base and various emulsifying agents.

[8] Creely and Le Compte, *Am. Dyestuff Reptr.* **30**, 247 (1941). Rendell, *Textile J. Australia* **16**, 138 (1941). Fischer, *Melliand Textilber.* **21**, 469 (1940).

More often, however, the mills purchase a ready-made "wool oil," which is a mixture of oil plus emulsifying agent, *i.e.*, a "soluble oil" type of composition. The wool processor merely mixes the "wool oil" with about four or five times its weight of water to obtain an emulsion suitable for spraying onto the goods.

Since there are several rather exacting requirements which good wool oils should possess, their formulation is not so simple as might be expected. The base oils must have the correct viscosity and lubricating characteristics and the surface active agents with which the oils are blended must be capable of converting them into stable aqueous emulsions. One of the most important properties of a satisfactory wool oil is its easy removability by scouring after its purpose has been served. The ease with which a wool oil may be scoured depends on the nature of the oil base and the emulsifying agent as well as on their relative proportions. Mineral oils are considered more difficult to remove the glyceride oils; high ratios of emulsifying agent to oil base promote easy scouring. The so-called "self-scouring" wool oils, which may be largely removed by rinsing in warm water, usually contain high proportions of glyceride oils and an emulsifying agent which is also a good detergent for wool. Another requisite of a wool oil is non-injurious character with respect to the machine parts with which it comes in contact. Such parts may be made from leather, wood, or plastics, as well as from steel. Some emulsifying agents, for example, although satisfactory in every other respect, cause the leather aprons and conveyors of the carding machines to stiffen or crack. Stability of the wool oil emulsions is also an important factor in their application. If an emulsion is too unstable it will break in the feed tanks and the wool will be unevenly oiled. An excessive degree of emulsion stability, on the other hand, impedes the desired deposition of a film of oil on the fibers. All the various factors which influence emulsion stability and breakage are operable in wool oil emulsions. Oil-soluble stabilizers such as the fatty alcohols may be added to the oil phase if an increase in stability is desired. If anionic emulsifiers are used the deposition may be increased by operating in the acid pH range. With cationic emulsifiers, an alkaline medium promotes deposition of the oil.[9]

The most commonly used emulsifying agents for wool oils are the soaps. The potassium and amine soaps of oleic acid, which are readily incorporated into glyceride oils or hydrocarbon oils, are inexpensive and generally satisfactory. Beside using the soaps themselves as emulsifying agents it is

[9] Palmer and Blow, *J. Textile Inst.* **29,** 91 (1938). Brit. Pat., 436,956; Ger. Pat., 702,567; to *Kammgarnspinnerei Stohr und Co. A.-G.* U. S. Pat. 2,143,986, Kling and Götte to *H. Th. Böhme A.-G.*

common practice to add substantial quantities of free oleic acid to the oil. This greatly improves the scouring properties. Soda ash is almost always used in the scouring bath and serves to saponify the oleic acid, which then aids in removing the unsaponifiable portion of the oil.[10]

Petroleum sulfonates of the mahogany type, which are used in large tonnages for making wool oils, have the great advantages of being completely soluble in the oil as well as being inexpensive. These sulfonates are often mixed with auxiliary emulsifiers to improve the scourability of the oil.[11]

Sulfated glyceride oils are excellent emulsifying ingredients for wool lubricants. These emulsifiers are used at present to a lesser extent than formerly, although they still find favor as auxiliary emulsifiers.[12] The fatty acyl taurines and fatty acyl sarcosines have also been recommended as wool oil emulsifiers. They have the advantage of being good detergents, thereby improving the scouring properties of the oil. They are difficult, however, to incorporate into mineral oils and the emulsion must be prepared just before use.[13]

A number of cationic emulsifiers have been recommended for use in wool oils, among them being the oil-soluble long-chain glyoxalidines. The cationic emulsifiers, however, are apt to cause difficulty if the subsequent scouring is carried out in normal fashion with soap or other anionic detergents.[14]

The polyethylene oxide non-ionic agents have been particularly favored in Germany as wool oil emulsifiers. Under the name of Servitol OL a wool oil based on this type of non-ionic detergent was widely used during the war years for making fine yarns.[15]

A large number of oil-soluble auxiliaries have been included in the non-aqueous components of wool oils to improve their scourability and lubricating properties. Oleic acid and the fatty alcohols have already been mentioned in this connection. Other substances suitable for this purpose include the fatty alkylolamides, fatty esters, and fatty ethers.[16]

[10] Brit. Pat. 487,949, to *Lister and Co. Ltd.* Ger. Pat. 709,382, to *I. G. Farbenindustrie.* Schmidt, *Teintex* **5**, 88 (1940).

[11] Brit. Pats. 444,851; 523,520; to *Standard Oil Development Co.*

[12] Austrian Pat. 141,026, to Chwala and Waldmann. Ger. Pat. 632,481 to *Fettsäure und Glyzerin Fabrik G.m.b.H.* Welwart, *Seifensieder-Ztg.* **63**, 549 (1936).

[13] Brit. Pat. 454,559; French Pats. 741,262; 44,384; to *I. G. Farbenindustrie.*

[14] U. S. Pat. 2,214,152; Brit. Pat. 524,847; Wilkes to *Carbide and Carbon Chemicals Corp.*

[15] Reumuth, *Klepzig's Textil-Z.* **44**, 1226 (1941). Schulz, *Deut. Wollen-Gewerbe* **72,** 752 (1940). Brown, *PB 63822*, Office of Technical Services, Dept. of Commerce, Washington, D. C.

[16] Brit. Pat. 400,681, to Speakman and Chamberlain. U. S. Pat. 2,164,237, to Garner. U. S. Pats. 1,985,687, Nüsslein and Ulrich to *I. G. Farbenindustrie*; 2,049,-043, Birkby to *Standard-I.G. Co.*

(e) Fulling and Piece Goods Scouring

After leaving the loom most woolen and worsted greige goods are fulled. Fulling consists essentially in mechanically working the thoroughly wet goods by means of rollers. This causes shrinking and felting of the goods to a limited degree, thus producing a denser stronger fabric. The fulling operation has been carried out in the past almost exclusively by means of relatively strong soap – soda ash solutions. It was generally believed that the alkali was necessary to achieve good felting. The soap was supposed to serve as a lubricant for the fibers, thereby preventing mechanical damage. It is now known that mechanical action is by far the most important factor in fulling and that alkali is unnecessary. If sufficient lubricant is present on the goods, satisfactory fulling can be carried out with water alone, although this is rarely if ever done in actual practice.

Since greige wool piece goods must eventually be scoured in order to remove the spinning lubricant, it is expedient to combine the fulling and scouring operations. When soap and soda ash are used in the fulling medium, for example, it is necessary only to flood the goods with water and continue washing in order to produce a clean, fulled and scoured piece. The enormous practical advantage of including a good detergent in the fulling bath is thus apparent. Although soap – soda ash is still the most widely used fulling agent, fulling procedures which utilize the newer synthetic detergents have been described and practiced. The list of detergents which may be used includes the Igepon type, the non-ionic agents, the fatty acylamino carboxy acids, and many others.[17] Aqueous emulsions of neutral oils are sometimes used as fulling assistants. In this case the emulsifying agent may be a soap or one of the suitable synthetic detergents.[18]

Washing and scouring operations are required at numerous other stages in the processing of wool. In the manufacture of worsted yarns, for example, the scoured raw stock is first oiled and then carded and combed. After combing it is generally "back-washed" before being re-oiled and spun. Yarns which are to be dyed must be thoroughly scoured before entering the dye bath. Some worsted yarns are sized before weaving, and the greige pieces in this instance are generally given a desizing-scouring treatment immediately after leaving the loom. The same general considerations apply in these operations as in all other wool washing. The lowest temperatures, lowest alkalinity, and least mechanical action capable of achiev-

---

[17] Krüger, *Deut. Textilwirt.* **7**, No. 16, 36 (1940). Klar, *Wollen u. Leinen Ind.* **61**, 371 (1941). Krüger, *Klepzig's Textil-Z.* **41**, 558 (1938). Rehmann, *Monatschr. Textil-Ind.* **53**, 57 (1938). Brit. Pat. 417,654, to *I. G. Farbenindustrie.*

[18] Brit. Pats. 505,020, to E. Franz; 512,927; to *H. Th. Böhme Chemische Fabrik.* Ger. Pat. 643,539, to A. Prescher.

ing the desired results should be used. The various synthetic detergents are accordingly being adopted for woolen and worsted scouring operations by more and more mills.[19]

The dyeing, printing, and stripping of wool will be discussed in a subsequent section.

III. Cotton Processing (Wetting and Scouring)

(a) Desizing

Cotton warp yarns, before being woven, are almost always heavily sized. The purpose of sizing is the coating and impregnation of the yarns with a tough, flexible protective film without which they could not be manipulated satisfactorily in the loom. The film of sizing material in effect converts the yarn into a monofil and protects it against mechanical damage during weaving. Cotton warp sizes are for the most part based on starches or dextrins, and their formulation is in itself an important technical art, which will be discussed in a separate section. Grey cotton fabric, as it leaves the loom, accordingly contains a resistant outer coating of dried amylaceous sizing material, together with the waxes and pectins originally present as a coating on the raw fiber. A few of the heavier cotton fabrics such as ducks and Osnaburgs are sometimes used in the grey state. By far the greater number of fabrics, however, must be submitted to finishing operations. The first finishing operation involving chemical treatment is the removal of the sizing material.

The most widely used method of desizing cotton fabrics consists in treating them with amylolytic enzymes. The fabric is passed through a bath containing a concentration of enzyme appropriate to the construction and size content of the goods. It is then usually rolled up to prevent loss of moisture and is stored long enough for the enzyme to exert its full hydrolytic effect. Alternatively, the goods may be allowed to soak in the enzyme bath while hydrolysis of the starchy size takes place. The enzyme derived from malt works most effectively at a pH value of about 6.0, whereas pancreatic enzyme is most effective at about pH 6.8–7.0. With either type of enzyme the time of soaking may vary from two hours to a full day. It is obviously important that the enzyme bath should thoroughly wet and penetrate the yarns. Since grey cotton is quite difficult to wet, a good wetting agent greatly increases the efficiency of the desizing process. The wetting agents used must be carefully chosen so as not to inactivate the enzyme or otherwise interfere with the hydrolytic process. The wetting

[19] Brandenburger, *Wollen u. Leinen Ind.* **61**, 343 (1941).   Ger. Pat. 702,229, to *I. G. Farbenindustrie.*

agents should preferably have high detergent power so that they may aid in removing partly hydrolyzed, undissolved clots of size from the fibers. Among those surface active products which have been used successfully in desizing are the Igepons, the sulfated fatty acylmonoethanolamines, the polyethyleneoxide non-ionic agents, and a variety of cation active types.[20]

Desizing of cottons has been accomplished by methods other than enzyme hydrolysis, but these are of minor importance. One such process involves treating the goods with per salts (*e.g.*, sodium persulfate) or active chlorine preparations, thus oxidizing the starch size. Another process consists essentially in treating the goods with an acid, thereby effecting an acid hydrolysis of the starch. In either case, wetting agents are highly advantageous and serve the same purposes as in enzyme desizing. [21]

(b) Kier Boiling, Open-Boiling, and Bleaching

The purpose of the boiling-out process is the removal of the naturally occurring waxy outer coating of the cotton fiber. Until this coating is removed the cotton is relatively water repellent, non-absorbent, and difficult to dye. If the raw cotton is boiled out (sometimes referred to simply as "boiled") in the form of loose fibers before spinning or in the form of yarns after spinning, it is only necessary to remove waxes and pectins present in the natural coating of the fiber. If, on the other hand, the cotton is boiled after weaving and desizing, as is the more frequently occurring case, the residues of hydrolyzed sizing material must also be removed. Raw stock and light-weight woven goods are generally boiled in large closed tanks which operate under slight superatmospheric pressure. These tanks are called kiers and the operation is known as kier boiling. Yarns and heavy woven fabrics are boiled in open tanks at atmospheric pressure.

In the kier boiling of lightweight fabrics the goods are loaded into the kier in rope form. After closing the vessel the bath liquor is circulated through the goods by means of an external circulating pump. The liquor is maintained at a temperature of about 220°F. or higher, and the treatment is usually continued for several hours. The essential cleaning agent in the kier liquor is caustic soda, which is present to the extent of about 1–2%. Other alkaline reagents such as the silicates and phosphates are often added together with the caustic. Since cotton readily suffers oxidative degradation if exposed to air in the presence of hot strong alkalis, it is imperative that the goods be thoroughly and uniformly saturated with the liquor.

[20] Molnar, *Monatschr. Textil-Ind.* **51**, 11 (1936). Ger. Pats. 671,900; 676,636; to *Kalle und Co.* U. S. Pat. 2,149,709, Rein to *I. G. Farbenindustrie.* U. S. Pat. 1,859,-084; Canadian Pat. 384,728; Frey, Józsa, and Gore to *Standard Brands.*

[21] Howell, *Dyer* **78**, 23 (1937). Ger. Pat. 423,464, to *Chemische Fabrik Pyrgos Ges.*

It might be thought that this requirement could be fulfilled by any number of good wetting agents.   There are, however, relatively few wetting agents which function effectively under such severe conditions of temperature and in the presence of such concentrations of alkali.   It is also desirable for the kier liquor to have good detersive properties but detergent action under kier boiling conditions is likewise not easily attained.   Many cotton processors use no surface active agents whatsoever in kier boiling.   It is more common practice, however, to add small amounts of the more stable detergents and/or solvents such as pine oil, etc.   Among the more highly favored detergents and wetting agents are the Igepons, the sulfated oleic esters, sulfated oils, and sulfated fatty alcohols.   It has been claimed that wetting agents can successfully replace part of the caustic soda, particularly if peroxygenated salts are present in the liquor.   This view, however, has not been widely accepted.   It should be noted that the peroxygenated salts are often used in kier boiling because of their bleaching action. Through their use it is possible in some instances to eliminate the necessity for a separate bleaching operation.   Inorganic adsorbents such as bleaching clays and silica gel have also been recommended for use as adjuncts to the kier boiling liquor.   Raw cotton and linters are boiled out in the same general manner as light fabrics.   Lower temperatures and higher concentrations of detergent are often used in the treatment of these unspun masses of fibers.[22]

The open boiling of yarns and heavy fabrics resembles kier boiling in that relatively strong alkali constitutes the major cleansing ingredient. The only reason that these goods are not kier boiled is that they are inconvenient to handle in a kier.   The open boil-out processes practically always utilize surface active agents in the bath.   The same types which function well in the kier are also favored for open boiling.   The soaps appear to be quite satisfactory for this operation and continue to enjoy widespread preference.[23]

Fabrics which are to be dyed (or printed in a pattern which leaves relatively little of the white background) do not require bleaching.   A large proportion of undyed fabrics, however, are bleached in order to convert the slightly yellowish natural color to a desirable clean white.   Bleaching

[22] Venkataraman et al., J. Indian Chem. Soc., Ind. & News Ed. 2, 81 (1939).   Marklen, Tiba 12, 835 (1934); 13, 19 (1935).   Ger. Pat. 702,557, to A. Kammer.   Brit. Pat. 403,009; French Pat. 755,541; to I. G. Farbenindustrie.   U. S. Pat. 2,048,775, Bolton to E. I. du Pont de Nemours & Co.

[23] U. S. Pat. 2,098,527, Stickdorn to Unichem.   Béha, Ind. textile 54, 560 (1937); Chem. Abstracts 33, 9000.   Dhingra, Uppal, and Venkataraman, J. Soc. Dyers Colourists 53, 91 (1937).   Hughes, Indian Textile J. 47, 219 (1937).

is generally carried out with either sodium hypochlorite or hydrogen peroxide, although numerous processes utilizing other oxidizing agents have been described and practiced. Some bleaching processes involve the use of two successive bleaching baths, one containing hypochlorite and the other containing peroxide. Wetting agents are quite widely used in the bleaching of heavy cotton fabrics to insure complete penetration of the goods by the bleaching liquor. In some instances, the goods are subjected to a wetting-out pretreatment in a bath containing a wetting agent but no active bleaching components.[24] It is more common practice, however, to add the wetting agent directly to the bleaching bath. In this case, it is essential that the wetting agent should be chemically resistant to the oxidizing action of the bleaching agent and should not interfere with the action of the latter component. The sulfated fatty alcohols, the sulfated esters or amides of oleic and ricinoleic acids, and products of the Igepon type are among those which have been used successfully in bleaching baths.[25]

Within recent years several processes have been developed whereby bleaching and scouring can be accomplished in one' operation. The scouring in these operations is substantially equivalent in its results to a thorough boil-out and a considerable saving in processing time may accordingly result from the combination process, particularly if it is set up on a continuous basis. The continuous peroxide bleaching process involves two distinct stages. In the first stage the goods are saturated with a solution containing about 3% caustic soda and are heated in this condition for about one hour. They are then passed through a scouring bath, and proceed to the second stage in which the bleaching is accomplished by means of hydrogen peroxide and sodium silicate. Wetting agents and detergents are used both in the initial saturation with caustic soda and in the scouring bath. They are selected for their effectiveness under these unusual and drastic conditions.[26]

Sodium chlorite ($NaClO_2$), a powerful bleaching agent, has also been recommended for combined bleaching–scouring operations. It may be used with the aliphatic sulfates, and sulfonates, or with the alkyl aryl sulfonates.[27]

[24] French Pat. 755,637, to *H. Th. Böhme A.-G.* U. S. Pat. 2,148,842, Sherman to *Geo. E. Sherman Co.* Bonnet, *Teintex* **3,** 350 (1938). U. S. Pat. 2,048,991 Butz *et al.* to *H. Th. Böhme A.-G.* Ger. Pat. 686,091, to *Oranienburger Chemische Fabrik A.-G.*

[25] Ger. Pat. 720,775, to *I. G. Farbenindustrie.* U. S. Pat. 2,141,189, Lind to *Henkel et Cie.* Canadian Pat. 382,353, Rooseboom to *Shell Development Co.*

[26] Mills, *Am. Dyestuff Reptr.* **35,** P388 (1946).

[27] U. S. Pat. 2,253,368, Dubeau to *Mathieson Alkali Works.* U. S. Pat. 2,383,900, Vincent, Dubeau, and Synan to *Mathieson Alkali Works.*

(c) Mercerizing

It was discovered by John Mercer in the middle of the nineteenth century that cotton fibers undergo a tremendous shrinkage when immersed in strong caustic soda solutions: Corresponding to this longitudinal shrinkage, there occurs a commensurate lateral swelling of the fiber. If the fiber is firmly held to its original length while being immersed in the caustic bath, and is subsequently freed of caustic by appropriate means, it will be found to have acquired a glossier, smoother appearance and a considerably higher tensile strength. Mercerizing is the process of treating cotton yarns or fabrics with strong caustic soda solutions while maintaining their original dimensions substantially unchanged. The treatment is generally carried out at room temperature and the caustic may vary in concentration from about 25 to 35%. The actual duration of immersion is generally less than one minute and thereafter the fabric, held in position by tenter frames or rollers, is showered to remove the excess of strong caustic. It then proceeds through washing and scouring tubs until freed of residual alkali.[28]

In order to achieve satisfactory results in mercerizing it is necessary for the fabric to become thoroughly and completely saturated with the caustic liquor within a very short interval after being immersed. If the cotton has previously been boiled out no serious wetting problem is encountered, since well-boiled cotton will absorb the caustic liquor very rapidly. It is sometimes expedient, however, to mercerize grey goods and these are often quite difficult to wet. The ordinary long-chain wetting agents, which function well in dilute aqueous solutions, are for the most part valueless in a mercerizing bath. Almost without exception they are either completely salted out of the bath or decomposed by the excess inorganic alkali. In some few cases where they are both soluble and stable they are nevertheless ineffective as wetting agents. There are, however, two major classes of specialized wetting agents which function effectively in mercerizing liquors.

The most important type of mercerizing wetting agent consists of cresylic acid (or other mononuclear phenol of low molecular weight) together with a small amount of additive of medium molecular weight. The presence of this additive is quite important, but its chemical composition may be varied over a wide range. Oxygenated compounds such as ethers, alcohols, and ketones are generally effective, and products such as Cellosolve or Carbitol have been effectively utilized in this connection. The amides and alkylolamides of C_4 to C_8 fatty acids and of naphthenic acids, as well

---

[28] For a complete discussion of the theoretical and practical aspects of mercerizing, see J. T. Marsh, *Mercerizing*, Chapman and Hall, London, 1941.

as other nitrogenous compounds, have also been used as additives for cresylic acid mercerizing wetting agents. These additives are often effective in astonishly small concentrations. As an example, 0.5% of cresylic acid in a mercerizing lye may be ineffective for wetting cotton, but 0.5% of a cresylic acid mixture containing 1% additive may be a fully satisfactory wetting agent. Thus, one part of the additive in 20,000 parts of lye makes the difference between poor wetting and excellent wetting.[29,30]

The other type of mercerizing wetting agent includes individual compounds which resemble the more usual wetting agents in having a polar – non-polar structure. The non-polar or hydrophobic portion of the molecule, however, is generally much shorter and of lower molecular weight. By thus decreasing the size of the non-polar residue, while maintaining the polar residue unchanged, the solubility of the molecule in mercerizing lyes is greatly increased. Examples of this class include such products as the C_4 to C_{10} branched-chain fatty acids, C_4 to C_{10} sulfated alcohols, substituted aminoacetic acids, aliphatic and aromatic sulfonamides, etc.[31] None of these substances, however, appear to have attained the commercial success of the cresol-based mercerizing assistants.

The dyeing, printing, and final finishing of cottons as well as rayons will be discussed in a later section.

IV. Rayon, Synthetic Fibers, and Miscellaneous Natural Fibers (Wetting, Scouring, Special Effects)

The synthetic surface active agents are probably used more extensively in the processing of viscose rayon, acetate rayon, and various other syn-

[29] For general reviews of mercerizing wetting agents, see Michel, *Tiba* 15, 363, 431, 491 (1937). Klumph, *Textile Bull.* 46, No. 4, 8 (1934). Replat, *Tiba* 13, 81 (1935). Walter, *Melliand Textilber* 12, 40 (1931). Hall, *Am. Dyestuff Reptr.* 22, 623 (1933).

[30] For patents on cresol-based mercerizing wetting agents, see Ger. Pats. 593,048; 614,913; 669,426; 674,894; Brit. Pats. 491,048; U. S. Pats. 1,851,393–4; 1,851,914; Lier to *Chemische Fabrik Sandoz*. Ger. Pats. 606,025; 710,959; Brit. Pat. 463,644; French Pats. 784,359; 771,856; U. S. Pat. 2,188,287; Münz to *I. G. Farbenindustrie*. Swiss Pats. 177,545; 200,662; 204,124; to *Ciba Co*. Ger. Pat. 703,605, to *Oranienburger Chemische Fabrik A.-G*. French Pat. 796,498, to *Deutsche Hydrierwerke A.-G*. Brit. Pat. 405,492 to *Erba Fabrik*. U. S. Pat. 2,000,559, Dunbar and Todd to *Imperial Chemical Industries Ltd*.

[31] U. S. Pat. 2,257,183, Münz and Trösken to *I. G. Farbenindustrie*. U. S. Pats. 2,157,294; 2,196,562, Lier to *Chemische Fabrik Sandoz*. U. S. Pats., 2,043,329, Münz and Bayer to *General Aniline Works*; 2,081,528, Brodersen to *I. G. Farbenindustrie*; 2,236,617, Brandt to *Colgate-Palmolive-Peet Co*. Gandhi and Venkataraman, *Current Sci.* 8, 367 (1939). Brit. Pat. 414,485, to *E. I. du Pont de Nemours & Co*. French Pat. 777,800 to *J. R. Geigy, S.A*. Swiss Pat. 179,041; Ger. Pat. 699,655; French Pat. 831,-462; Brit. Pat. 431,662; to *Ciba Co*. Ger. Pats. 545,190; 546,942; 547,646; to *I. G. Farbenindustrie*. French Pat. 834,169; Brit. Pat. 501,094, to *Rohm and Haas Co*.

thetic fibers than they are in cotton and wool processing. In the case of viscose rayon, for example, surface active agents have even been used in the preparation of the fiber itself. The addition of 0.1 to 2.0% of an anionic or a cationic surface active agent to the cellulose pulp from which viscose is made is claimed to facilitate steeping, shredding, and xanthation.[32] Salts of the sulfated low-boiling components of coconut fatty alcohols have also been added to the viscose itself before spinning.[33] Quaternary ammonium compounds of the type represented by laurylpyridinium chloride as well as surface active phosphonium compounds have been added to viscose coagulating baths for the purpose of preventing clogging or incrustation of the spinnerettes.[34] Cationic surface active agents of similar constitution have been used to clarify fouled spinning baths. These cationic detergents cause a flotation of the insoluble sulfur which forms in the bath after long-continued use, thus enabling it to be removed readily by skimming.[35] Cationic detergents have also been used in the regenerating baths as well as in the coagulating baths. They are said to have a particularly favorable effect in the process for making high-tenacity yarns if included in those baths where the stretching operation is carried out.[36] Processes have been proposed whereby mixed cellulose–protein fibers can be made by spinning a mixture of solutions of viscose and an alkali-soluble protein. The component solutions are immiscible and form a two-phase liquid system. Sulfated oils or cationic detergents have been suggested in such cases for forming and maintaining a finely dispersed liquid mixture.[37]

A wide variety of surface active agents, including such anionic detergents as the Nekals, Igepons, and sulfâted fatty alcohols as well as several cationic types, have been used in desulfuring baths for viscose rayon. It is claimed that they bring about a more rapid and complete elimination of sulfur from the fiber than is possible with water alone.[38]

Although some rayon is delustered by applying dulling finishes to the fabricated material, delustering is usually accomplished simultaneously with spinning of the fiber. To this end, a pigment (such as titanium dioxide) or a water-insoluble well-dispersed oil is incorporated into the viscose.

[32] U. S. Pats. 2,331,935–6, Schlosser, Gray, and Hallonquıst to *Rayonier, Inc.*

[33] Ger. Pat. 719,023, to *Böhme Fettchemie G.m.b.H.*

[34] U. S. Pats. 2,125,031; 2,145,527, Polak and Weeldenburg to *American Enka Corp.* U. S. Pat. 2,310,207, Bley to *North American Rayon Corp.*

[35] U. S. Pat. 2,336,778, Coşta and Kahler to *Manville-Jenckes Corp.*

[36] Ger. Pat. 744,891, to *I. G. Farbenindustrie.*

[37] U. S. Pat. 2,318,544, Thurmond, Jacokes, and Brenner to *American Enka Corp.* Japanese Pat. 130,410, to K. Hukaisi; *Chem. Abstracts* 35, 2011.

[38] French Pat. 766,877, to A. Stein. French Pat. 783,563; Brit. Pat. 446,757; to *I. G. Farbenindustrie*.

The pigment or oil must be finely subdivided and evenly distributed throughout the viscose before spinning. The naphthalene-formaldehyde sulfonates and certain cationic detergents have been recommended as dispersing agents for these delustering additives. In a similar manner the amine soaps and substances such as diglycol stearate have been used for delustering acetate rayon.[39]

Regenerated cellulose fibers of unusual properties have been made by saponifying preformed cellulose acetate yarns in an aqueous alkaline bath. In this process surface active cationic compounds as well as anionic agents of the sulfated oil type have been used as auxiliaries in the saponifying bath.[40] Surface active agents are also used in some of the processes which have been proposed for crimping and crepeing cellulose acetate fibers. In these processes the detergents may function as antistatic agents or as auxiliaries in the crepeing compositions themselves.[41]

One of the processes for manufacturing textile fibers from casein involves the preparation of a spinning solution containing casein which has been treated with sulfuric acid. Included in the spinning solution is a relatively high percentage of a surface active agent such as sodium dodecyl sulfate. When this type of spinning solution is extruded and coagulated, the resultant fibers are particularly amenable to stretching. "Stretchability" (the property of being susceptible to appreciable elongation without breakage) is particularly important in the manufacture of synthetics fibers, since stretching is one of the important methods for increasing fiber tenacity.[42] In spinning solutions of the type described the detergent forms a chemical complex with the protein molecules, thus weakening the forces which oppose their unfolding and alignment. The binding forces between the protein and detergent molecules are apparently electrostatic in nature. After the fiber molecules are oriented by stretching the detergent may be removed.[43] The fact that mahogany sulfonates can act as plasticizers for sheeted gelatin has been known for several years. The fundamental mechanism of this plasticizing action may possibly be related to the complex formation effect described above.[44]

[39] Canadian Pat. 352,255, Dreyfus and Whitehead to Camille Dreyfus. U. S. Pat. 2,373,712, Schlosser and Gray to *Rayonier, Inc.* French Pat. 846,014, to *I. G. Farbenindustrie.*

[40] U. S. Pats. 2,144,202; 2,208,857; Schlack to *I. G. Farbenindustrie.* Brit. Pat. 400,996, to Henry Dreyfus.

[41] U. S. Pats. 2,186,135, Childs to *Eastman Kodak Co.*; 2,089,240, Whitehead to *Celanese Corp. of America.*

[42] U. S. Pat. 2,403,251, Watson to *E. I. du Pont de Nemours & Co.*

[43] Lundgren, *Textile Research J.* **15**, 335 (1945).

[44] U. S. Pat. 1,902,304, Heckel to *Twitchell Process Co.*

In the spinning of alginate fibers considerable difficulty is often experienced because the freshly coagulated fibers tend to stick to each other. The inclusion of an emulsified oil in the acidic coagulating bath is said to be effective in counteracting this tendency. Emulsifying agents such as the fatty alkyl sulfates and the quaternary ammonium salts, which are effective in acidic media, are recommended for this purpose.[45]

Wetting and penetrating agents have been recommended as additives in various wet processes for preparing bast fibers such as flax and ramie. These fibers are generally separated by means of fermentation processes from the woody tissue in which they occur. Other processes involving a preferential chemical attack on the matrix may also be used. Wetting agents such as the sulfated oils or alcohols are also recommended for the wet-spinning of flax.[46]

The operation in which raw silk is prepared for knitting is generally called "soaking." It consists essentially in swelling and softening the serecin by immersion in a bath containing a lubricant and a penetrating agent. Sulfated oils, waxes, mahogany sulfonates and fatty alkyl sulfates have been recommended as useful components of silk soaking baths. The fatty alkyl sulfates are stated to be particularly useful as emulsifying agents for the oily components of the bath because their freedom from alkalinity minimizes the removal of serecin during the soaking operation.[47]

The bleaching and scouring of greige goods fabricated from rayon or other synthetic fibers differs considerably from the corresponding treatment of cotton goods. Viscose rayon goods are generally desized with enzymes and then "boiled-off," to remove the starchy and oily residues of the sizing compositions. The "boiling-off" or scouring process is carried out under much milder conditions than are necessary in the case of cotton. Since there are no refractory waxes or pectins present on the rayon fibers, caustic soda is not used, nor is it desirable, in the scouring bath. The scouring operation is seldom if ever carried out in kiers, and the bath generally carries a relatively high ratio of detergent to builder. The synthetic anionic detergents as well as the non-ionic types are quite widely used in boiling-off viscose rayon goods, and it is probable that a larger poundage of synthetic detergents is consumed in this operation than in any other single operation within the textile industry. The synthetic detergents may be used alone or together with soap. Sometimes when soap is used in the

[45] U. S. Pat. 2,371,717, Speakman to *Cefoil Ltd.* French Pat. 767,874–5, to T. Gohda.

[46] Brit. Pat. 470,612, Leblanc to *Société Robert Descendre.* Brit. Pat. 402,544, to *Les Fibres Textiles S.A.* French Pat. 835,085 to *S.A. fibre tessili Italiane.* Biczysko, *Melliand Textilber* **21**, 508 (1940).

[47] Johnson, *Textile World* **86**, 85 (1936). Neville and Marshall, *Ind. Eng. Chem.* **23**, 58 (1931). Tyler, *Soap* **10**, No. 4, 21 (1934). U. S. Pat. 2,067,888, Chamberlin to *National Oil Products Co.*

scouring train a small amount of a synthetic may be added to the rinse water for the purpose of eliminating lime soap streaks. Rayon is sometimes sized with compositions based on linseed oil rather than on starches or dextrins. In this case the boil-off conditions must necessarily be more severe, and soap is generally used together with appreciable quantities of alkaline builders.

The same considerations which hold true for viscose rayon scouring apply even more generally in the case of acetate rayons. The sizing materials are for the most part easily removable and in many cases a separate desizing operation is not even necessary. The synthetic detergents, since they are effective in neutral baths, are particularly desirable for scouring acetate. Baths of high pH value tend to saponify the surface of the acetate fiber, particularly when used at high temperatures.

The synthetic protein fibers are in general so sensitive to alkali that they cannot be scoured with soap. Neutral baths containing appropriate synthetic detergents are used for scouring this class of goods.

The scouring of natural silk is usually referred to as "degumming" because it involves removal of serecin, the naturally occurring gummy coating on the silk fiber. Degumming baths are usually made up to a fairly high pH value, with soap as the preferred detersive component. Wetting agents and/or sulfated oils are often added. Degumming baths which are based on sulfated oils alone have been described.[48]

Large quantities of woven viscose rayon goods (and also some cotton goods) are treated with resins of the urea-formaldehyde type in order to render them crease resistant and shrink resistant. The treatment consists essentially in impregnating the goods with a pre-condensate of urea and formaldehyde (or melamine and formaldehyde) together with a latent acidic catalyst. The goods are then dried and subsequently heated to a relatively high temperature. This heating or "curing," as it is called, activates the catalyst and causes the resin to form within the structure of the fiber. After the curing process the goods must be thoroughly scoured in order to remove the excess of resin, resin components, and catalyst. Since the goods contain acidic material when they emerge from the curing flues the synthetic detergents are generally superior to soap for this particular scouring operation.

V. Oiling and Sizing of Textile Fibers

Most fibers, with the exception of cotton, require lubrication at one or more stages during their conversion to yarn. The wool-spinning lubricants

[48] Springer, *Soap* **12**, No. 1, 29 (1936). White, *Am. Dyestuff Reptr.* **23**, 389 (1934). Volz, *Monatschr. Textil-Ind.* **50**, Trade Issue III, 93 (1935). U. S. Pat. 2,060,529, Reddish to *Emery Industries.* Brit. Pat. 415,027 to *British Celanese Ltd.*

have already been mentioned as end products which include surface active agents in their composition. The lubricants used in fabricating yarns from other fibers also may include surface active agents as normal constituents.

Continuous filament rayon which is to be used for knitting is generally wound on cones by the manufacturer. As a necessary preliminary to the cone-winding operation, the yarn must be properly lubricated, and the lubricant which is applied also conditions the yarn for knitting. Lubricants of this type are called coning oils, and they must meet several exacting requirements, of which the more important are: (1) Emulsifiability, since the oils are applied in the form of aqueous emulsions. (2) Ease of removal by scouring. (3) Ability to prevent or reduce the accumulation of electrostatic charge on the yarn during processing. The latter property is particularly important in the case of cellulose acetate yarns.

These three characteristics may be imparted to the lubricant by inclusion of appropriate surface active agents. The sulfated oils and esters and the amine soaps are generally used in this connection, although a large number of specialized products have been described. For the most part the specialized types have been designed primarily for efficient anti-static action rather than for emulsifiability or scourability. When the rayon is intended for weaving, i.e., for warp or filling yarns, the yarn lubricant must be compatible with the sizing or throwing compositions which are applied before weaving.[49]

Lubricants used with cellulose acetate yarns must be specially designed to minimize their tendency to build up large troublesome charges of static electricity during reeling or spinning. Since acetate yarns are also characterized by poor affinity for the usual types of warp sizings, a secondary function of the lubricant is the provision of an adequate bond between the yarn and the sizing composition. The sulfated oils and the amine soaps are used in acetate lubricants as well as in lubricants for viscose yarns. Several other types of surface active agents, however, are claimed to have advantageous over these older and more standard spinning oil components. Surface active products particularly effective as anti-static agents include various amine salts of the fatty alkyl sulfates and sulfonates, amine salts of the fatty acyl taurines, phosphated alcohols, phosphated partial fatty esters of alkylol amines, and sulfated oleic acid. Various phosphated glyceride oils, phosphinic acids, and phosphonic acids have also been pro-

[49] Deutsch, *Rayon Textile Monthly* **21**, 239 (1940). Corty, *Klepzig's Textil-Z.* **42**, 279 (1939). U. S. Pats. 2,096,705, to Heckethorn; 2,101,532, Conquest to *Armour and Co.*; 2,232,565, Segessemann to *National Oil Products Co.*; 2,151,711, Moscowitz to *L. Sonneborn Sons*; 2,176,402, Koch to *American Enka Co.*

posed as anti-static additives for acetate yarn lubricants. Lubricants for Nylon and Vinyon yarns must also have anti-static properties and the ability to improve the adhesion of sizings.[50]

"Throwing" is the operation of putting extra twist into silk and rayon yarns which are intended for knitting or for woven crepe fabrics. Prior to throwing, the yarns must be lubricated and this particular lubrication process is generally known as "soaking." Soaking compositions usually are based on the sulfated oils. If the yarns are to be used for filling in crepe fabrics, the soaking oils may also contain gelatin or a similar sizing material. Fugitive tinting colors, which are used to identify the yarns during weaving or knitting, are usually applied in the soaking bath.

The sulfated oils, esters and alcohols are often applied to staple rayon fibers before baling, so that they are at least partially lubricated when received at the spinning mill.

For successful spinning, twisting, or knitting operations the yarns must be soft, pliable, and well lubricated. In almost all weaving operations, however, the warp yarns must be stiffened as well as lubricated by a protective coating of a suitable sizing composition. The art of formulating sizing compositions is quite complicated and a detailed consideration would be well beyond the scope of this discussion. In essence, however, a good warp sizing must form an adherent, smooth, protective film on the yarn. The film must have the correct degree of stiffness without being so brittle as to flake off during weaving. It must also have good surface lubricity so as to minimize the frictional effects of harnesses, shuttles, and reeds. Finally, the film of sizing material must be easily removable, so that the fabric may be suitably prepared for finishing. In most cotton sizings, the principle film-forming ingredients are starches or converted starches (British gum or dextrin), with which vegetable gums and pectins are sometimes used as auxiliary film-forming components. For rayons, particularly acetate rayons, gelatin rather than starch may be used as the primary film-forming ingredient. Softening and lubricating components, such as tallow and other glyceride oils and fats, are generally used in sizings. Blending these softening and lubricating components with the fundamentally incompatible carbohydrate or protein film-forming substances requires the use of a surface active agent. The sulfated oils are generally used for blending

[50] U. S. Pat. 2,384,053 Thomas to *Celanese Corp. of America.* U. S. Pats. 2,328,600, Baggett, and 2,150,568, Whitehead, to *Celanese Corp. of America.* U. S. Pats. 2,176,-510, Robinson, 2,298,432, Thompson, 2,191,033, Faw, 2,197,930, Jackson and Faw, 2,289,760, Dickey and McNally, 2,233,001, Dickey, 2,191,039, Dickey, 2,256,380-1, Dickey, and 2,331,664, Dickey, to *Eastman Kodak Co.* Brit. Pats. 431,964; 463,548; 477,639; 514,134; to *British Celanese Ltd.* Brit. Pats. 346,912, to *Aceta Gesellschaft;* 488,945, to Finlayson and Perry; 526,683, to *Courtaulds Ltd.*

tallow in sizings based on starch or gelatin. The soaps as well as the sulfated alcohols and petroleum sulfonates are also used to a certain extent for this purpose.

Linseed oil has been used rather extensively as a sizing material for rayon yarns. Sizings based on linseed or other drying oils may be applied in emulsion form; a wide variety of emulsifying agents are said to give satisfactory performance in this application. They include the soaps, sulfated alcohols, polyethylene oxide non-ionic agents, etc. Polyvinyl acetate and other synthetic polymeric materials have also been used as yarn sizings and have been applied in conjunction with surface active agents. Sizings of this type tend to adhere tenaciously to the yarns and are difficult to remove by the usual desizing and scouring procedures. The surface active agents which are included in the sizings serve an important function in making them more easily removable.[51]

VI. Dyeing, Printing, and Stripping

(a) General Considerations

The processes by which solid colors or color patterns may be applied to fabrics are exceedingly numerous and diversified, and there are relatively few of these coloring procedures in which surface active agents are not employed at one or more stages. The surface active agents which are used as "dyeing assistants" (a generic term which does not specify the exact function they may fulfill) may serve one or more of the following purposes:—

(1) They may act simply as wetting agents to insure that the entering goods are thoroughly and uniformly wet by the dye bath. In this capacity they are often referred to as penetrating agents, since they cause the dye solution to penetrate into the interior of the yarns. It should be noted that they have little if any effect on the rate at which the dye molecules migrate into the fiber substance.

(2) They may function as suspending and penetrating agents when the coloring material is being applied as a suspension. When so used, the

---

[51] Furry, *U. S. Dept. Agr. Tech. Bull.*, No. 674 (1939). Wood, *Cotton (Atlanta)* **106**, No. 4, 92 (1942). Canadian Pat. 387,173, Whitehead to Camile Dreyfus. U. S. Pat. 2,081,180, Leupold to *National Oil Products Co.* Brit. Pat. 509,445, to *Courtaulds Ltd.*, and James Procter. U. S. Pat. 2,112,728, Morgenstern and Eggert to *Deutsche Hydrierwerke A.-G.* Canadian Pat. 411,469, Hathorne to *Emery Industries.* Brit. Pat. 346,267, Dreyfus and Taylor to *British Celanese Ltd.* U. S. Pat. 2,211,266, Gibello to *Société Nobel française.* Brit. Pat. 499,995, to *Société Nobel française.* French Pat. 769,977, to *Bianchini Ferier.* U. S. Pat. 2,218,506, Davis and Segessemann to *National Oil Products Co.*

active agents hold the pigment particles in an evenly dispersed state in the bath liquor, and promote uniform penetration of the dye into the yarns.

(3) They may function as dye levelling agents, acting according to either of the two mechanisms which have already been discussed.[52]

(4) They may act as solubilizing agents in the application of water-insoluble dyestuffs to acetate rayon (and certain other synthetic fibers). When so used, surface active agents are generally referred to as suspending or dispersing agents.

(5) Certain surface active agents can function as mordants or mordant auxiliaries. A closely allied function is that of increasing the water-fastness of fabric dyed with water-soluble dyestuffs.

(6) In special cases, for example, when dyeing and scouring are effected simultaneously (or successively in the same bath), the dyeing assistant may also serve as a detergent.

(7) When incorporated into printing pastes, surface active agents may aid in homogenizing and suspending the ingredients of the paste. They also may promote the rapid absorption of the paste by the fabric.

(8) When fabrics are printed or dyed with fast colors of the vat dye or azoic series, it is common practice to give them an "after-soaping" or "after-washing." This operation serves to remove the loosely held surface color and, in the case of printed patterns, thickeners and other ingredients having no color value. The after-soaping also serves to bring out the full brightness and strength of the dyestuff. After-washing, as the name implies, is simply a scouring treatment, usually of short duration and carried out at the boil. As in all scouring operations the detersive power of the selected surface active agent is of primary importance.

Textile materials may be dyed at almost any stage of mechanical fabrication. Wool, for example, can be dyed in the form of uncarded fibers, slubbing, top, yarn, or finished pieces. When raw stock (uncarded fibers), partially fabricated yarns (tops or slubbing), or yarns are dyed, the problem of penetration and even distribution of the dye liquor throughout the goods is a serious one. With piece goods, on the other hand, level dyeings are usually more easily attained. Spotty dyeing of piece goods, however, makes reworking of the piece imperative since the unevenness cannot be covered up by any future mechanical operations. Printing is almost always carried out on the finished piece. Faults in printed patterns are accordingly difficult or even impossible to correct. In some cases yarns may be printed in order to obtain special effects.

The class of dyestuffs which is chosen for application to a given fabric depends on the raw material from which the fabric is made and on the

[52] See Chapter 15, Section VIII.

fastness properties which may be required.   The choice of surface active dyeing assistants depends on their compatibility with the remaining components of the coloring system, their stability under the conditions of operation and their ability to perform the desired function or functions. In some dyeing operations only a very limited range of effective surface active assistants is available.   There are at present, for example, two classes of surface active substances which are outstandingly effective levelling agents for vat dyestuffs on cotton, namely, the quaternary ammonium salts and the polyethylene oxide non-ionic agents.   For many processes, however, the dyer has available a fairly wide array of assistants, any one of which may give satisfactory results.   This is particularly true where wetting, dispersion, or detergency are the only desired functions.

(b) Wool Dyeing

Almost all woolen fabrics are dyed either with acid colors or chrome colors.   The levelling agents usually employed with the acid dyes belong to the sulfate or sulfonate class and exert their levelling effect by competing with the dyestuff molecules for loci of adsorption on the fiber molecules (see pp. 380–383).   The sulfated alcohols and fatty esters, the fatty acyl taurines, the alkyl aromatic sulfonates, and other similar types may all be successfully used in dyeing wool with acid dyes.   In dyeing wool with chrome dyestuffs, particularly when applying dyes whose molecules contain chromium, the polyethylene oxide non-ionic and certain of the cationic detergents are said to be optimum levelling agents.   These substances act by increasing the degree of association of the dyestuff molecules and thereby lowering the effective rate at which they diffuse into the fiber.   Cation active assistants are also said to increase the wash fastness of woolens dyed with acid dyes.[53]

Although thorough scouring of wool goods is usually considered a necessary prerequisite for good dyeing it is sometimes possible to scour and dye wool fabrics simultaneously in the same bath.   It is claimed that this has been done by using an assistant of good scouring properties selected from the sulfate, sulfonate, or polyethylene oxide non-ionic classes.[54]

The cationic surface active agents are said to be effective levelling agents for basic dyestuffs on wool.   Their action is similar to the action of the anionic agents with acid dyes.[55]

[53] Valko, *Oesterr. Chem.-Ztg.* **40**, 465 (1937).   French Pat. 771,270; Brit. Pats. 439,890; 442,901; to *Ciba Co.*   U. S. Pats. 2,228,369, Schoeller to *General Aniline and Film Corp.*

[54] Rendell, *Textile J. Australia* **16,** 174 (1941).   French Pat. 712,121, to *I. G. Farbenindustrie.*   Schoeller, *Melliand Textilber.* **19**, 57 (1938).

[55] French Pat. 850,494, to *I. G. Farbenindustrie.*

(c) *Dyeing Cotton, Regenerated Cellulose, and Mixed Fabrics*

In dyeing cellulosic fabrics with direct dyes a wetting and penetrating agent is generally added to the dye bath. Some of the more widely used sulfated and sulfonated anionic surface active substances also exert a true levelling (*i.e.*, retarding) effect on the direct dyes. The most powerful retarders, however, are the polyethylene oxide non-ionic detergents, and these are generally preferred when poorly levelling dyestuffs must be used. Ampholytic detergents of the sulfonated acyl benzimidazole type are also said to be useful levelling agents for direct dyestuffs.[56]

Rayons are sometimes dyed and scoured simultaneously by using a powerful detergent in the dye bath. This procedure is facilitated if a high proportion of soap or sulfated oil has been used in lubricants and sizes previously applied to the goods.[57]

The two most widely used levelling agents for vat dyes, namely, the cationic and the polyethylene oxide non-ionic agents, have been mentioned above. The sulfonated acyl benzimidazoles are claimed to be effective levelling agents for vat dyes as well as sulfur dyes. . The addition of a cation active detergent and a fatty alkyl sulfate successively to the vat dye bath is likewise said to produce level effects and the organic base salts of fatty acids or fatty alkyl sulfates may also be used for this purpose.[58] These assistants are all used in the conventional procedure of vat dye application, which involves immersion of the goods in a solution of the reduced dyestuff. The assistant is dissolved in the dye bath along with the alkali and the reducing agent (sodium hydrosulfite is the reducing agent most widely used). Vat dyes are often applied, however, by the so-called pigment-padding process. In this process the goods are padded with a suspension of the unreduced dyestuff and are then immersed in a solution containing the alkaline reducing agent. In preparing the padding liquor (*i.e.*, the suspension of finely divided, insoluble, unreduced vat dye) a suitable surface active agent is practically indispensible. The alkylnaphthalene sulfonates, the sulfated fatty esters and amides, and other surface active agents of high suspending power are used for this purpose. The cation active detergents have also been used in pigment padding, and are readily adsorbed by the pigment particles, imparting to them a positive electric charge. The pigment is then strongly attracted to, and firmly held by, the negatively charged cellulosic fabric.[59]

[56] French Pat. 778,476, to *Ciba Co.*

[57] U. S. Pat. 1,979,188, Bouhuys to *American Enka Corp.*

[58] Brit. Pat. 441,296; French Pat. 839,086, to *Ciba Co.* Brit. Pats. 435,431, to *H. Th. Böhme A.-G.*; 470,346, Piggott and Woolvin to *Imperial Chemical Industries Ltd.*

[59] Ger. Pat. 598,441, to *Ciba Co.* Ranshaw, *Silk J. Rayon World* **18**, No. 207, 30 (1941).

The azoic dyes are next in importance to the vat dyes as fast-coloring materials for cellulosic fabrics. Dyestuffs of the azoic series may be applied by a number of different methods, the most widely used of which involves first padding the fabric with a naphthol, such as Naphthol AS (β-oxynaphthoic anilide), and then passing it through a solution of a diazotized aromatic amine. A diazo coupling reaction takes place which results in the formation of an insoluble dyestuff on the fabric. In preparing the naphthol solution with which the fabric is padded or otherwise impregnated, surface active agents, usually of the sulfated oil or sulfated ester type, may be used. The surface active products counteract the tendency of the naphthols to precipitate out of solution. Wetting of the fabric by the naphthol solution is also promoted. The diazo salt solution, or the "developing bath" as it is sometimes called, may also contain surface active agents. Here they not only exercise the same functions as in the naphthol liquor but also serve to control the size of the precipitated dyestuff particles, an effect which tends to produce dyeings of superior resistance to crocking. The polyethylene oxide non-ionic and cationic detergents of the Sapamine type are favored for use in developing baths.[60]

The after-soaping of vat and azoic dyeing has already been mentioned. Almost any of the synthetic detergents which have high detersive power may be used for this purpose, and soap itself is generally quite satisfactory.[61]

Basic dyes are sometimes applied to cellulosic fabrics which have previously been mordanted with tannin and antimony or with a synthetic mordant of the Katanol type. The cationic detergents are claimed to be useful assistants in applying the basic colors. Faster and less sensitive color effects are also said to be obtained with the anion active sulfated alcohols or sulfated fatty alkylolamides.[62]

It is sometimes desirable to color cellulosic fibers with dyestuffs for which they normally have little affinity. This is the case, for example, when mixed wool–rayon or wool–cotton piece goods are to be dyed. The cationic detergents are strongly adsorbed by the cellulosic fibers and modify the fiber surfaces in such a manner as to increase their affinity for most classes of dyestuffs. In other words, the cationic surface active agents may be used to mordant the cellulosic components of a mixed fabric prior to or during the dyeing operation. The process of altering cellulosic fibers so that they will take up acid dyestuffs is sometimes called "animalizing," because the acid dyes are normally taken up only by wool (of animal origin)

[60] Anon, *Ciba Rev.* **2**, 761 (1939). U. S. Pat. 2,030,859, Drapal to *General Aniline Works*.

[61] U. S. Pat. 2,222,285, Ellner to *General Aniline and Film Corp.*

[62] Brit. Pat. 442,292, to *H. Th. Böhme A.-G.* Ger. Pat. 701,863, to *I. G. Farbenindustrie.* U. S. Pat. 2,052,716, Lenher to *E. I. du Pont de Nemours & Co.*

or other proteinaceous fibers. Treatment with a cationic detergent is one of the simplest but also one of the most superficial methods of animalizing cottons and rayons.[63] A wide variety of long-chain quaternary ammonium compounds, non-quaternary nitrogen bases, sulfonium salts, and phosphonium salts, have been described as suitable mordants for acid dyes.[64] Chrome dyes may also be applied successfully to cellulosic fabrics which have been pre-treated with these compounds.[65] A mordanting treatment with cationic detergents is claimed to increase the affinity of various direct colors and vat colors for cotton.   In general, dyestuffs of these classes are substantive to cotton but individual members of these series may have less affinity than is required for complete satisfaction.[66]

Dyeings or prints made with direct colors often show a tendency to bleed in water. This can be counteracted by after-treating the colored fabrics with a solution of a cationic detergent. The mechanism by which the dye is "fixed" is probably related to the mordanting action described above. The fixation effect is also due, at least in part, to the formation of an insoluble salt of the dyestuff molecule with the cationic detergent.[67] Although the long-chain monofunctional cationic detergents are quite effective in increasing the water resistance of direct dyeings, they do not effect improvement in resistance to soaping. Direct dyeings may be markedly improved, however, with respect to soaping and washing resistance, by an after-treatment with certain polyfunctional quaternary ammonium compounds. These polymeric cation active products may be made, for example, by treating polyethylene polyamine with an excess of a lower alkyl halide or sulfate. Since these products do not contain a long alkyl chain, they can hardly be classed with the usual types of surface active agents.[68]

(d) Dyeing Cellulose Acetate

Dyeing acetate rayon differs fundamentally from dyeing natural fibers and the regenerated cellulose rayons; both these latter classes of fibers

[63] Hall, *Textile Recorder* **54**, 45 (1936). Thomas, *Rayon Textile Monthly* **20**, 639 (1939). U. S. Pat. 2,254,965, Kling *et al.* to *Patchem A.-G.*

[64] French Pats. 809,215; 783,008; Ger. Pat. 738,015, to *I. G. Farbenindustrie*. Brit. Pats. 443,022, to *H. Th. Böhme A.-G.*; 501,020, to *Courtaulds Ltd.*; 404,969, to *Imperial Chemical Industries Ltd.*

[65] Ger. Pat. 675,953; Brit. Pat. 488,099; to *I. G. Farbenindustrie*. French Pat. 788,634, to *Imperial Chemical Industries Ltd.*

[66] Brit. Pat. 448,272; Ger. Pat. 639,199; to *I. G. Farbenindustrie*.

[67] Rière, *Rev. gén. mat. color.* **44**, 93 (1940). Ger. Pat. 629,732; Brit. Pat. 466,157; to *I. G. Farbenindustrie*. Ger. Pat. 671,782, to *Ciba Co.* U. S. Pat. 2,201,814, Evans, Piggott, and Woolvin to *Imperial Chemical Industries Ltd.*

[68] Brit. Pats. 435,388; 440,488; 441,767; 460,961; French Pat. 852,031; to *I. G. Farbenindustrie*. U. S. Pat. 2,362,915, MacGregor to *Courtaulds Ltd.*

take up the coloring matter from aqueous solution, and the dyes must accordingly be in solution at the time they are being applied. Acetate rayon is considerably more hydrophobic in nature than the natural fibers and it has much less affinity for the water-soluble classes of dyestuffs. With few exceptions the dyestuffs which are used for coloring acetate rayon are water insoluble. They are, however, soluble in oxygenated solvents and soluble in the cellulose acetate itself. The mechanism by which these dyestuffs impart color to the acetate fiber appears to be closely related to a true solution phenomenon, whereas the soluble dyes become attached to natural fibers either through adsorption or ionic bonding.

Since acetate rayon dyes are applied from an aqueous suspension, the dyestuff must be very finely divided and well dispersed in order to obtain satisfactory results. The presence of an effective surface active agent in the dye bath is essential. Acetate dyestuffs are normally supplied by the manufacturer in the form of finely powdered mixtures containing a large proportion of soap or other detergent of high suspending power. It is customary, however, for the dyer to add a further quantity of surface active agent to the dye bath before treating the goods. The sulfated oils, sulfated esters, and amides, as well as the sulfated fatty alcohols, are widely used for this purpose. Less expensive dispersing agents, such as the alkyl naphthalene sulfonates, the Syntans (q.v.), lignin sulfonates, and petroleum sulfonates, are also said to give satisfactory results.[69]

Azoic colors are sometimes applied to acetate rayon and, since the fibers are quite sensitive to alkali, the commonly used naphthol components cannot be applied in the normal manner from a caustic soda solution. The naphthols may be applied from an aqueous suspension and the impregnated fabric can subsequently be treated with a diazo salt. As an alternative procedure the naphthol and the base may be applied simultaneously from suspension and the impregnated fabric can be treated with nitrous acid to bring about diazotization and coupling.[70]

In special instances some of the acid and direct dyes can be applied to acetate rayon. In such processes the cationic detergents are claimed to be useful auxiliaries. Anionic detergents may be used as auxiliaries when special reagents intended to swell the acetate fibers are also included in the dyebath.[71] Mixed·fabrics containing acetate rayon and wool may be

[69] Brit. Pats. 530,013, to British Celanese Ltd.; 428,935; to British Celanese Ltd. and Geo. H. Ellis. U. S. Pat. 2,042,751–2, Whitehead to Celanese Corp. of America. Brit. Pats. 428,767, to Ciba Co.; 401,735, to I. G. Farbenindustrie; 510,515, to C. S. Bedford, J. G. Bedford, and Storey.

[70] Brit. Pat. 506,740, to Ellis, C. F. Topham, and H. C. Olpin. Röhling, Melliand-Textilber. 20, 63 (1939).

[71] French Pat. 834,420, to I. G. Farbenindustrie. U. S. Pat. 2,291,052, Miller to E. I. duPont de Nemours & Co.

dyed from a bath containing an acid wool dye and an insoluble, suspended acetate rayon dye. The sulfated fatty alcohols are said to be satisfactory auxiliaries for this process. They act as levelling agents for the wool dye and suspending agents for the acetate rayon dye.[72]

Nylon may be dyed satisfactorily with the same types of dyestuff used for acetate rayon. The dyeing procedure and the surface active auxiliaries are similar to those described above. Nylon may also be dyed with acid dyes and even with direct dyes. In the latter case the cationic detergents are said to increase the dyeing rate provided they are used in low concentrations. In high concentrations the dyeing rate is seriously retarded.[73]

(e) Printing

Surface active agents in printing pastes perform at least two necessary functions: they act as homogenizing, suspending, or emulsifying agents for the other ingredients of the paste (notably the dyestuff itself) and they also serve to impart the necessary high wetting power to the paste. Since foaming tendencies in a printing paste cannot be tolerated, the choice of surface active agents which may be used is severely limited. In preparing vat dye printing pastes, for example, it is customary to add high percentages of short-chain dispersing agents rather than low percentages of the more effective long-chain compounds. The short-chain dispersing agents not only minimize the foaming but also have the added advantage of being extremely soluble.[74] Long-chain dispersing agents such as the amine salts of sulfated fatty alcohols have been used in making powdered vat dye preparations suitable for pigment padding.[75] The long-chain cationic detergents have also been utilized in this connection, as well as for mordanting fabrics before printing with Indigosol type colors.[76]

The after-washing of fast color prints (vat and azoics) is carried out in the same manner as the after-washing of fast color dyeings. Prints made with dyestuffs which tend to bleed may be after-washed with solutions of cationic detergents. As an alternative procedure they may be washed in a bath containing calcium or magnesium salts, which reduces the tendency of the colors to bleed. In this case an anionic detergent of high lime re-

[72] Ger. Pat. 646,001, to *I. G. Farbenindustrie.*

[73] Canadian Pat. 410,825, to *Canadian Industries Ltd.* U. S. Pat. 2,220,129, Stott to *E. I. du Pont de Nemours & Co.* MacGregor and Pugh, *J. Soc. Dyers Colourists* **61,** 122 (1945).

[74] Ger. Pat. 671,996; French Pat. 795,683; to *I. G. Farbenindustrie.* U. S. Pats. 2,145,193, Kern to *National Aniline and Chemical Co.*; 2,079,788, Chambers to *E. I. du Pont de Nemours & Co.* Chambers, *Am. Dyestuff Reptr.* **25,** 289 (1936).

[75] U. S. Pat. 2,318,439, Waldron to *E. I. du Pont de Nemours & Co.*

[76] French Pats. 832,962; 777,588; to *I. G. Farbenindustrie.*

sistance, such as a fatty alkyl sulfate, or an alkyl aryl sulfonate, is used as the active washing agent.[77]

(f) Stripping

The removal of a dyestuff from a fabric is called stripping or discharging. It is usually accomplished by destroying the dyestuff molecule with a strong reducing agent and washing out the colorless non-substantive intermediates which result from the reduction reaction. This method is effective with the majority of dyestuffs, including all those which contain the azo group as a chromophore. Sodium hydrosulfite and sodium formaldehyde sulfoxalate are the preferred reducing agents in most stripping processes. Penetrating agents and detergents are often added to stripping baths, and for this purpose usual sulfated and sulfonated types are generally satisfactory.[78]

The stripping of vat dyes is exceptionally difficult. This is due to the fact that reducing agents do not effect cleavage of such dyes into simpler non-substantive fragments. Instead, reducing agents convert vat dyes to the substantive, soluble leuco form in which they were originally applied to the fabric. As a consequence, equilibrium effects play an important role in stripping of vat dyes. When a vat-dyed fabric is immersed in an alkaline hydrosulfite stripping bath a small proportion of the leuco dyestuff leaches out into the bath until a state of equilibrium is reached with respect to adsorption. At equilibrium by far the greater portion of the leuco dyestuff is adsorbed on the fiber, and as a consequence the net stripping effect is negligible. An effective stripping assistant for vat dyes must effectively remove the leuco compound as it is formed and leached out into the bath, thus preventing it from being re-adsorbed on the fabric. The polyethylene oxide non-ionic detergents and the quaternary ammonium type cationic detergents accomplish this effect in the same manner by which they bring about levelling during the application of vat dyes. They cause the molecules of leuco dyestuff to agglomerate into micelles. This lowers the effective concentration of adsorbable leuco dyestuff molecules in the bath, thus shifting the adsorption equilibrium and removing more dyestuff from the fabric.[79] The sulfonated long-chain benzimidazoles and similar ampholytic detergents are also said to exert a stripping effect on vat dyes, presumably by the same mechanism.

[77] Smith, *Ind. Eng. Chem.* **31,** 40 (1939). U. S. Pat. 2,189,807, Lenher and Smith to *E. I. du Pont de Nemours & Co.*

[78] Zill, *Deut. Färber-Ztg.* **75,** 489 (1939). Ger. Pat. 701,845 to *Ciba Co.*

[79] Strobel, *Deut. Färber-Ztg.* **77,** 457 (1941). Brit. Pats. 422,466; 422,556; 444,169; French Pat. 771,349; to *Imperial Chemical Industries Ltd.*

VII. Finishes and Special Treatments

The final finishing operations, whether chemical or mechanical in nature, are of extreme importance in textile manufacture, and the surface active agents often play an indispensable role in the various chemical finishing treatments which may be applied to fabrics.

One of the most widely used finishing operations consists in the application of a "softening agent" to improve the hand and the draping properties of a fabric. The softening agents are of particular importance in finishing cottons and rayons. These fabrics tend to have a stiff harsh hand, particularly when they have been well scoured. The softening agents are essentially lubricants of high spreading and penetrating power. They apparently exert their effect by forming a thin coating on the yarns and on the individual fibers within the yarns. The net effect is the diminution of the internal friction between adjacent fabric elements, thereby making the whole structure more pliable. Softening agents are effective in relatively small quantities, and it is seldom that more than 2 or 3% (based on the weight of the fabric) is applied. An excess of softening agent is generally to be avoided since it may impart an undesirable greasy or oily hand to the fabric. Practically all softening agents are either soluble or dispersible in water and are applied in an aqueous medium. Many softening compositions are mixtures rather than single chemical individuals and they may include insoluble components such as mineral or glyceride oils. In well-prepared compositions of this type the insoluble components are thoroughly emulsified so that they may be evenly and uniformly distributed throughout the fabric.

Compositions which are intended primarily for softening may be divided into two general classes, those which are non-substantive or readily removable by washing, and those which are substantive or resistant to washing. The former class includes many of the better-known anionic surface active agents, whereas the substantive softeners are, for the most part, based on cation active compounds.

The sulfated glyceride oils are by far the most widely used non-substantive softeners. The sulfated derivatives of castor, olive, peanut, teaseed, and similar oils, as well as sulfated tallow, are all used in large quantities for softening cottons and rayons. There are, however, several other types of anionic surface active agents which have been proposed or used as softening compositions. The sulfated esters or amides of oleic acid, the higher sulfated fatty alcohols (e.g., sodium octadecyl sulfate), and the higher fatty acyl sarcosides or taurides are all good softening agents for certain types of fabrics. They may be used by themselves or in admixture with insoluble oils and waxes.[80] The monoesters of phthalic, hexahydrophthalic, and naphthalic acids with cetyl, oleyl, or octadecyl alcohol are claimed to be excellent softeners. These products may also be used as the basis for softening compositions containing additional oils, waxes, or higher fatty alcohols.[81] Hydrocarbon sulfonates prepared by the Reed reaction, sulfated red oil, and the sulfated fatty alcohols have also been used as carriers for insoluble, oily softening ingredients.[82]

Although almost all of the anion active softening agents are non-substantive, a few substantive types have been described. These are, for the most part, polynuclear aromatic aminocarboxylic acids substituted on the amino group by long-chain aliphatic residues. It is doubtful if these products have been used in any sizeable commercial quantities.[83]

The cationic surface active agents have, within recent years, gained widespread acceptance as softening agents for cellulosic fabrics. They possess the important advantage of substantivity, and the additional advantage of showing their effect even when very small quantities are applied to the fabric. Many of the cationic softeners have a tendency to alter the shade or diminish the light-fastness of goods dyed with direct colors, and considerable care must therefore be taken in their selection and application.[84] The cationic softeners are usually applied to the fabric from a weak acetic acid solution and therefore the primary, secondary, or tertiary bases may be used as well as the quaternary ammonium salts. A very large number of individual cation active substances have been proposed as softening agents and it is beyond the scope of this chapter to discuss them in detail. The more effective products generally include a n-octadecyl residue as the essential hydrophobic portion of the molecule. In compounds intended for softening, as in water-repellent compositions, the C_{18} straight-chain radical usually affords more satisfactory results than shorter-chain or branched-chain substituents.[85]

[80] U. S. Pats. 2,047,066, Glietenberg to *General Aniline Works*; 2,183,721, Lubs and Arnold to *E. I. du Pont de Nemours & Co.* Ger. Pat. 575,022 to *I. G. Farbenindustrie*. U. S. Pats. 1,981,108, Kalischer, Nüsslein, and Müller to *General Aniline Works*; 2,100,845, to Franz and Hardtmann.

[81] Ger. Pat. 680,774, to *Zschimmer und Schwarz Chemische Fabrik Dölau*. Swiss Pat. 177,226; French Pat. 812,285; Brit. Pat. 448,294; to *Ciba Co.* U. S. Pat. 2,175,101, Albrecht to *Ciba Co.*

[82] Brit. Pat. 550,195 to *British Celanese Ltd.* U. S. Pats. 2,180,133, Arnold to *E. I. du Pont de Nemours & Co.*; 2,334,764, Henke and Lockwood to *E. I. du Pont de Nemours & Co.*

[83] Swiss Pats. 205,898; 207,448–52; to *J. R. Geigy A.-G.*

[84] Rhode Island Section AATCC, *Am. Dyestuff Reptr.* 29, P686 (1940). Mosher, *Am. Dyestuff Reptr.* 28, 484 (1939).

[85] A few typical disclosures on cationic softening agents are: French Pat. 836,280, to *Zschimmer und Schwarz Chemische Fabrik Dölau*. Brit. Pat. 531,691, to *Imperial*

Certain long-chain compounds of cation active structure are difficult to dissolve in dilute aqueous solutions of weak acids. Products of this nature have been used as softeners by dispersing them with the aid of water-soluble non-ionic detergents. After deposition on the fabric this type of softener is said to have a high degree of substantivity.[86]

Softening agents of both the cationic and anionic types are frequently applied in conjunction with finishing agents which would tend, by themselves, to stiffen the fabric. Softening agents may be incorporated, for example, in urea-formaldehyde finishing treatments, cellulose-ether type permanent finishes, etc.[87] Wetting agents of the non-softening type, such as the Nekals or dodecylamine hydrochloride have been utilized in this type of finishing operation to improve the penetration and distribution of the active finishing ingredients.[88]

The long-chain anion active sulfates and sulfonates have been described as useful weighting agents for wool. They may be utilized for this purpose either alone or in conjunction with heavy-metal salts. The heavy-metal carboxylic soaps have also been used for weighting purposes, particularly on cellulosic fabrics.[89]

One of the major uses for surface active compounds in finishing operations is as a carrier or assistant in the application of coatings or other insoluble finishing compositions. Wax water-proofing emulsions, waxy impregnants for twine and rope, wax emulsions for use in ironing or calendaring, and similar compositions are almost always formulated to include a surface active emulsifying agent of the polar – non-polar type. According to the specific purpose for which they are intended, these substances may be anionic, cationic, or non-ionic in structure.[90]

Chemical Industries Ltd. Swiss Pats. 206,717, to *J. R. Geigy A.-G.*; 187,319–20 and 209,972, to *Ciba Co.* Brit. Pat. 490,637; U. S. Pat. 2,243,980, Rheiner and Link to *Sandoz Ltd.* Brit. Pat. 465,166; U. S. Pats. 2,168,253, Balle and Schulz; 2,185,427, Brodersen *et al.*; 2,195,194 Ulrich *et al.*; to *I. G. Farbenindustrie.*

[86] Brit. Pat. 511,545; Ger. Pat. 699,028; U. S. Pat. 2,197,464; Brodersen *et al.* to *I. G. Farbenindustrie.*

[87] Kuhlmann, *Klepzig's Textil-Z.* **39**, 679 (1936). French Pat. 852,460, to *Joseph Bancroft and Sons Co.* Brit. Pat. 509,079, to *Rohm and Haas Co.*

[88] U. S. Pat. 2,143,352, Koch and Eckelmann to *I. G. Farbenindustrie.* French Pat. 787,950, to E. Pollak

[89] Brit. Pats. 402,094; 401,477, to *I. G. Farbenindustrie.* Ger. Pat. 640,508, to *Chemische Fabrik Stockhausen et Cie.* U. S. Pat. 2,080,755, Whitehead to *Celanese Corp. of America.*

[90] U. S. Pats. 2,206,090, Haggenmacher to *Warwick Chemical Co.*; 2,285,579, Gröner to *Rohm and Haas Co.*; 2,139,343, Williams and Bone to *Ironsides Co.*; 2,374,931, Griffin to *Atlas Powder Co.* Brit. Pat. 498,771, to *Institute of Paper Chemistry.* Ger. Pat. 611,441, to *H. Th. Böhme A.-G.*

Water-insoluble finishes based on synthetic resins, such as the poly-acrylates, the polyvinyl esters, and the solvent-soluble urea-formaldehyde resins, are usually applied to fabrics from an aqueous dispersion. The sulfated fatty esters, sulfated alcohols, and the polyethylene oxide non-ionic detergents have all been used as dispersing agents in the application of these resinous finishes. The cationic detergents are said to be of par-ticular value in this connection because they improve the deposition and adhesion of the resin particles.[91]

The cation active detergents are effective assistants for the impregnation of fabrics with rubber latex. Their function in this process is similar to that which they perform in the resin- or wax-impregnating procedures. The surface active cations alter the charge relationship between the fibers and the dispersed rubber particles so that they attract rather than repel each other.[92] Anion active dispersing agents have also been utilized as assistants for impregnating fabrics with rubber. In this type of applica-tion the dispersing agents are applied primarily with compounded rubber dispersions rather than with pure rubber latex.[93]

Rayon may be delustered by distributing a finely divided, highly refrac-tive pigment over the fiber surfaces. This process is still practiced to a considerable extent although most rayon is now delustered during manu-factured by the inclusion of pigments in the viscose (or acetate) spinning mixture. The pigment, usually titanium dioxide, is applied to the fabric as an aqueous suspension prepared with the aid of a dispersing agent. The cation active dispersing agents exert the same effect in this application as they do in the application of other insoluble materials, and they cause a remarkable degree of adhesion between the fiber and the pigment particles. Some of the cationic detergents of higher molecular weight are in themselves delustering agents by virtue of their forming an adherent, refractive waxy coating on the fabric. Rayon may also be delustered by treatment with certain anionic detergents of the sulfate or sulfonate type in the presence of heavy-metal salts.[94]

[91] Brit. Pat. 512,187; U. S. Pat. 2,137,465; Thackston to *Rohm and Haas Co.* U. S. Pats. 2,343,089, Smith to *E. I. du Pont de Nemours & Co.*; 2,224,293, Finlayson and Perry to *Celanese Corp. of America.*

[92] Brit. Pat. 557,611, to H. E. Brew. · Katz, *Am. Dyestuff Reptr.* **28**, P671 (1939).

[93] Harold, *Am. Dyestuff Reptr.* **28**, 255 (1939). Brit. Pat. 451,622, to *International Latex Processes Ltd.* U. S. Pats. 1,958,821, to Haarburger; 2,238,165, Ellis and Stanley to *Celanese Corp. of America.*

[94] U. S. Pat. 2,101,251, Götte and Kling to *Böhme Fettchemie-G.m.b.H.* Ger. Pat. 736,970, to *Zschimmer und Schwarz Chemische Fabrik Dölau.* French Pat. 772,788 to *H. Th. Böhme A.-G.* French Pats. 48,888; 827,717; 834,971; to *I. G. Farbenindustrie.* Brit. Pats. 400,244; 412,929; 391,847; to *Imperial Chemical Industries Ltd.* Ger. Pats. 597,194, to *Chemische Fabrik Stockhausen et Cie.*; 669,299, to *Oranienburger Chemische Fabrik A.-G.*

The application of anti-mildew, moth-proofing, and bactericidal compositions to fabrics is often carried out with the aid of wetting or dispersing
agents, depending on whether the active agent is soluble or insoluble.
Compounds such as salicylic anilide or the chlorinated phenolic bactericides, for example, may be applied in the presence of effective dispersing
agents. Mothproofing agents of the water-soluble fluoride or fluosilicate
type may be advantageously applied with the aid of a wetting and penetrating agent.[95]

In fabrics which are loosely woven from smooth-surfaced yarns there is a
tendency for the goods to distort or "slip." Most of the special finishes
designed to overcome this tendency are formulated with surface active
agents, particularly with soaps made from rosin or modified rosin. These
products exert their effect by increasing the coefficient of friction between
the yarns, thus making it more difficult for them to slide over one another.
Sulfonated rosins and sulfonated terpene-phenol condensates, as well as
rosin esters, are also said to impart a slip-proof finish to fabrics.[96]

The rewetting effect and the use of rewetting agents in textile finishing
has been discussed in Chapter 15, Section III. Rewetting agents are used
for the most part in sanforizing, in the finishing of towelling fabrics, and in
the preparation of absorbent pads or surgical dressings.

[95] Engel and Gump, *Am. Dyestuff Reptr.* **30,** 163 (1941). Brit. Pat. 350,642, to
Imperial Chemical Industries Ltd. U. S. Pats. 2,225,352, to Roberts; 2,176,894,
Engels and Weijlard to *Merck and Co.*; 2,091,075, Landers to *Phillipp Bros.*

[96] U. S. Pats. 2,155,961; 2,228,712, Trowell to *Hercules Powder Co.* French Pat.
814,346; Ger. Pat. 671,276; to *I. G. Farbenindustrie.* Ger. Pat. 705,045, to *Chemische
Fabrik Stockhausen et Cie.* Brit. Pat. 560,020, to *Ciba Co.*

CHAPTER 19

Cosmetics and Personal Use

I. General Considerations

Surface active agents (including soaps in one form or another), are employed in numerous individual cosmetic preparations of which the more important ones can be grouped in the following categories: (*1*) Creams, including cold creams, vanishing creams, absorption base creams, and creams for special purposes such as suntan creams, depilatories, deodorants. etc. (*2*) Lotions. Like the creams, these may be specially formulated for a variety of purposes. (*3*) Make-up and facial preparations. (*4*) Shampoos, hair-waving preparations, and other preparations for the hair. (*5*) Shaving preparations. (*6*) Dentifrices. (*7*) Bath preparations. (*8*) Toilet soaps, including special cleansers for persons with sensitive skin. (*9*) Hand-cleaning compositions, including industrial hand cleaners, hand soaps for use in dispensers at public lavatories, etc.

The last two categories verge closely on the fields of medical uses and janitors' supplies, respectively. This is scarcely surprising in view of the function and action of surface active agents in such products. Many preparations which might be considered primarily as cosmetics have actual beneficial values for the skin and could therefore also be classed as dermatological remedies. The reverse is, of course, also true. A dermatological remedy which will aid in restoring health to a disordered skin and which is harmless enough to be dispensed without a prescription may, in fact, be sold as a cosmetic at cosmetic counters.[1]

II. Creams

Cold cream is essentially an emulsion of oils and water in about equal proportions. Such oils usually consist of a relatively large amount of mineral oil and/or petrolatum, together with a saponifiable wax such as spermaceti or beeswax. In making the emulsion, the oil mixture is heated and stirred with an aqueous alkali such as borax, which is used in an amount less than required to saponify all the wax. The soap formed by saponifica-

---

[1] For an excellent and complete discussion of cosmetics in relation to skin health, see L. Schwartz and Peck, *Cosmetics and Dermatitis*, Hoeber, New York, 1946. For an excellent and modern treatise on cosmetics in general, the reader is referred to DeNavarre, *The Chemistry and Manufacture of Cosmetics*, Van Nostrand, New York, 1941.

tion of the wax acts as the emulsifying agent, and the surplus wax remains as part of the oil phase of the emulsion. In some cold creams the saponifiable wax may be partly or wholly replaced by a glyceride oil such as sweet almond oil, peanut oil, etc. Stearic or other free fatty acid may also be used to replace more or less completely the previously mentioned saponifiable components. When free fatty acid is used in formulating cosmetic creams, triethanolamine may replace the inorganic alkali as the saponifying agent. Other alkylolamines are sometimes used, but triethanolamine is the most popular.[2]

˙ Cold creams may be either O/W or W/O emulsions, and it is not always easy to detect which type a given preparation may be. The true W/O types are sometimes called absorption base creams. This name refers to the fact that cosmetic creams and pharmaceutical ointments are often prepared by emulsifying water into certain commercially available oily materials known in trade circles under the names, "absorption base" or "ointment base." These substances belong to the class of oil-soluble emulsifying agents discussed in Chapter 8, Section II. These products, which are in fact W/O emulsifying agents are of two types, one being based on lanolin or lanolin alcohols and the other on such materials as sorbitol oleate, glycerol monostearate, etc. Special cold creams containing vitamin D, estrogenic hormones, cholesterol, or other special-purpose ingredients are often of the absorption base type. Cleansing creams, on the other hand, are usually of the O/W type.[3]

The partial fatty acid esters of polyhydric alcohols, particularly glycerol monostearate, are very widely used in making modern cosmetic creams. Such partial esters are commercially available in pure form and are sometimes used to replace part of the oil in the cream. More frequently, however, they are used in the form of self-emulsifying cream ingredients. For this purpose the partial esters are mixed with a synthetic surface active agent or a soap and sold as self-emulsifying products which can be mixed directly with water to give excellent creams. The synthetic surface active agents most widely used for this purpose are the fatty alkyl sulfates, particularly sodium cetyl sulfate, and cationic agents of the Sapamine type. The higher fatty alcohols, particularly cetyl and stearyl alcohols, are often used in place of the partial esters to make this type of cream base.[4] Beside

[2] For methods of identifying the amine soap formers in cosmetic creams, see Shupe, *J. Assoc. Offic. Agr. Chemists* **25**, 227 (1942).

[3] For discussions of emulsion types among cosmetic creams see: Redgrove, *Perfumery Essent. Oil Record* **31**, 165 (1940). Marriott, *Soap, Perfumery & Cosmetics* **13**, 307 (1940). King, *Mfg. Perfumer* **4**, 73 (1939). Jannaway, *Perfumery Essent. Oil Record* **31**, 81 (1940). Helfer, *Drug & Cosmetic Ind.* **45**, 42 (1939).

[4] Redgrove, *Am. Perfumer* **38**, No. 2, 32 (1939). U. S. Pats. 2,173,203; 2,302,121; to Benjamin R. Harris. French Pat. 690,330, to *Th. Goldschmidt A.-G.*

the two types of O/W emulsifying surface active agents mentioned in the preceding paragraph, several others have been used. The coconut fatty diethanolamide type (Ninol type) is quite popular among cosmetic formulators. The phosphated alcohols, sulfated oleic and ricinoleic esters, mineral oil sulfonates, and fatty monoglyceride sulfates have also been mentioned for making creams.[5]

Vanishing creams are O/W emulsions and contain little or no hydrocarbon oils or triglycerides. The basic ingredients are normally stearic acid, potassium stearate, water, and glycerol. Glycerol monostearate is often substituted for part or all of the stearic acid. Medicated creams for sunburn, insect bites, etc. usually are of the vanishing-cream type and contain menthol, camphor, or natural phenolic oils. Depilatory creams usually contain an alkaline earth sulfide or thioglycollate. Ability to withstand the prolonged action of these chemicals is an important prerequisite to be considered in choosing emulsifying agents for use in depilatory creams.

Antiperspirant creams contain from 3 to 15% of aluminum chloride or aluminum sulfate as the active ingredient. They have a pH in the neighborhood of 3 and this degree of acidity imposes a severe burden on the emulsifying agent. The base cream in these materials is usually made with glycerol monostearate and/or a higher fatty alcohol with a cationic or a fatty alkyl sulfate type O/W emulsifier.

The water-soluble surface active agents serve primarily as emulsifying agents in the types of creams we have been considering. They also serve the secondary purpose of making the cream more easily removable by water. This is of greater importance in make-up preparations than in general-purpose creams. The water-insoluble surface active agents (glycerol monostearate, fatty alcohols, etc.), which in many instances constitute the main non-aqueous ingredient of the cream, serve to aid the formation and stability of the emulsion regardless of whether it is O/W or W/O.

III. Lotions, Make-Up, and Facial Preparations

Lotions can generally be classed as emulsified or non-emulsified. The latter are clear solutions, which may be formulated either with thickening gums and mucilages or with considerable amounts of alcohol in case a non-viscous product is desired. The non-emulsified lotions usually do not contain surface active agents, although some formulas specify inclusion of small amounts of sodium lauryl sulfate, triethanolamine stearate, or various

[5] Redgrove, *Soap, Perfumery & Cosmetics* **16,** 38 (1943). *Am. Perfumer.* **38,** No. 3, 35 (1939). Allen, *Soap, Perfumery & Cosmetics* **14,** 705 (1941). U. S. Pat. 2,086,479, Schrauth to *Unichem.* Canadian Pat. 426,100, to *Colgate-Palmolive-Peet Co.* U. S. Pat. 2,216,485, Brandt to *Colgate-Palmolive-Peet Co.*

other types to facilitate spreading on the skin. The inclusion of such substances in clear lotions appears to be on the increase, as is evidenced by the fact that older cosmetic formularies do not mention them.

The emulsified lotions or "liquid creams" have recently enjoyed a great increase in popularity. This is due in no small measure to the shortage of alcohol during the war and the consequent intensive promotion of non-alcoholic preparations. The emulsified lotions are usually dilute O/W emulsions of glycerol monostearate or other partial fatty esters of polyhydric alcohols. The emulsifying agent may be a soap, such as triethanolamine oleate, or a fatty alkyl sulfate. Medicated skin lotions containing solid materials in suspension, such as calamine or powdered sulfur, often contain a small percentage of a surface active agent to aid in maintaining suspension.

The main items of make-up are powder, rouge, and lipstick. The ordinary solid powders and cake powders contain no surface active agents. Liquid powders and such items as leg make-up (quite popular during the war to replace hosiery) very rarely make use of surface active agents, and it is doubtful if any of the more successful formulations contain them. Bentonite, aluminum hydroxide gel, and magnesium silicate gel are usually used as the binding and suspending agents in this type of preparation.

Modern lipsticks consist essentially of an intimate mixture of powdered solid pigment in a base of mixed fats and oils, together with an eosine dye capable of staining mucous membrane. It is this dye which gives the "indelible" or "kissproof" property to the lipstick. Lecithin is often added to lipstick bases to improve the spreading properties and the pigment dispersion. Other oil-soluble, non-aqueous surface active agents may also be used, but lecithin is apparently the most successful. Small amounts of water-soluble surface active agents such as Aerosol OT have been recommended to aid in dispersing the eosine-type dyes.

Cake rouge is essentially a pigmented cake powder and requires no surface active agent. Paste rouge or cream rouge is usually a cold cream or a vanishing cream with the necessary pigment. Cream powders fall into the same category. Facial masks and mud packs, designed as skin stimulants and pore cleansers, often contain sulfonated oils, glycerol monostearate, etc., as bonding and blending agents. None of the cosmetic items mentioned in this paragraph, however, requires any considerable quantities of surface active agents.[6]

[6] For uses of various surface active agents in powder, rouge, and make-up see: U. S. Pats. 2,218,586, Quaedvlieg to *Winthrop Chemical Co.*; 2,101,843, Factor and Fisher to *Max Factor & Co.* French Pat. 786,828, to S. Sabetay. Brit. Pat. 464,400, to *Max Factor & Co.*

IV. Shampoos

Shampoo preparations form one of the major outlets for synthetic surface active agents in spite of the fact that the soap type shampoo is still the most widely used. The main requisites of a shampoo are as follows: (*1*) It must cleanse the hair and scalp thoroughly, preferably with one application, and leave the hair lustrous. (*2*) It must leave the hair soft rather than harsh and dry. (*3*) It must rinse off easily and not leave any residue which might interfere with subsequent waving or dressing treatments. (*4*) It must not be irritating. (*5*) It must be psychologically satisfying to the customer and the beauty shop operator with regard to color, odor, thickness, uniformity, degree of clarity or opacity, and lathering power. Most shampoos are sold in the form of clear viscous solutions, liquid creams (opaque viscous solutions), or paste creams. Some powdered shampoos are sold, but these are more popular on the European continent than in this country.

The usual soap shampoo is a clear viscous solution containing about 20% potash soap. In order to improve the properties of the shampoo, part of the potash may be replaced by triethanolamine or other amines. Thickening ingredients, such as small amounts of inorganic salt, and water-softening ingredients such as polyphosphates may also be included. A matter of decisive importance with regard to sales is the ability of a shampoo to provide a rich copious foam, which most customers regard as the mark of a good shampoo. This need to ensure good foam properties (in order to avoid excessive sales resistance) is an important consideration in selection of fats and oils used in preparing shampoos. Straight coconut oil soap gives the most desirable type of foam but it tends to be too irritating to the skin. Such irritating effect has been traced, with a high degree of certainty, to the presence of C_8 and C_{10} fatty acids in soaps derived from coconut oil. Very recently, shampoo manufacturers have been offered fractionated coconut oil fatty acids, from which these harmful lower fatty acids have been removed. The usual way of improving the mildness of coconut oil soap, however, consists of diluting it with oleic or ricinoleic soaps by using coconut oil together with olive oil, other high oleic acid oils, or castor oil as starting materials when making shampoos.

One of the greatest disadvantages of soap shampoos is their lack of resistance to hard water. The precipitated lime and magnesia salts of the fatty acids cling to the hair tenaciously and form a dull residual film. The difference in luster between hair shampooed with soap and with a lime-resistant synthetic detergent is quite marked even to the inexperienced eye. The usual way of overcoming this defect of soap shampoos is to rinse the hair with a mild acid such as citric or acetic (lemon rinse or vinegar rinse).

This decomposes the lime soap film and leaves a lustrous film of free fatty acid in its place. Another method is the inclusion in the shampoo composition of a lime-sequestering agent such as polyphosphate, or a lime soap disperser such as Igepon T, sulfonated castor oil, etc.[7] The efficacy of such additions is questionable since the deposition of lime soap on the hair seems to occur largely during rinsing rather than during actual washing.

In spite of the necessity for an acid rinse to obtain optimum results, soap continues to enjoy great favor as a shampoo. The acid rinse leaves a film of free fatty acid on the hair which prevents it from feeling too "dry" or "harsh." The dry, harsh feel is quite probably characteristic of completely cleaned hair.[8]

The shampoos formulated with synthetic detergents are of two main types, foamless oil and foaming, of which the latter is by far the more important. The foamless oil type is essentially a low-sulfated vegetable oil, (i.e., a mixture of sulfated and unsulfated oil), usually prepared from castor oil or oils containing oleic glycerides as starting materials. The oil shampoo is applied to the hair with little or no water and, after working, is rinsed off with excess warm water. The sulfated oil content is high enough to render the composition and its burden of dissolved and suspended dirt easily dispersible and removable by the rinse water. This type of shampoo leaves a considerable deposit of residual oil on the hair and enjoys only limited popularity. It is sometimes used as a preliminary treatment and is followed by a regular foaming shampoo. Such combination shampooing is quite effective where the hair and scalp have become excessively dirty.

A great number of synthetic detergents have been proposed and used as foaming shampoos but relatively few have attained a high degree of acceptance. In this country the highly sulfated fatty oils and fatty acid esters, the fatty alkyl sulfates, and the sulfated fatty acyl monoglycerides are the most widely used. The latter two types are synthesized with fatty acyl or alkyl radicals derived from coconut oil, while sodium or triethanolamine serve as the cationic salt-forming constituent. The products vary in active ingredient content from about 5 to about 30%, with an average of about 12–15%. A synthetic detergent, if it is to be used successfully, must have good solubility, must produce the desired type of foam, and must be free of harshening or irritating effects. It should also be relatively free

[7] French Pat. 797,083, to *Firma Hans Schwarzkopf.* Compare Gattefosse, *Parfumerie moderne,* **29,** 421 (1935).

[8] For a review of soap shampoo formulations see Jannaway, *Perfumery Essent. Oil Record* **36,** 179 (1945). For shampoo powders see Ekmann, *Riechstoff Ind. u. Kosmetik* **12,** 201 (1937).

of inorganic salts. In formulating clear liquid shampoos various thickening and solubilizing agents are often added, to render the product psychologically more satisfying to the user. In most shampoos a certain amount of neutral oil is added to combat the "drying" effect. This drying effect, which may be actual skin irritation caused by the lower alkyl radicals or may merely be the result of too thorough a cleansing, has been the chief customer objection to the synthetics. This has largely been overcome and today the largest selling single brand of shampoo is a synthetic-based product of the coconut fatty alkyl sulfate type.

Liquid cream shampoos may be formulated with a soap or synthetic detergent as principle cleansing agent. Such shampoos contain a suspended opacifying agent, which usually is sodium stearate, although the glycerol monostearate type of compound may also be used for this purpose. The paste cream shampoos are similarly constituted but are more concentrated both in active ingredient and in thickening agent. The most popular types of paste cream shampoo currently employ sodium lauryl sulfate as the detergent and sodium stearate as the opacifying, cream-forming agent.[9]

Aside from the three types of synthetic detergent mentioned above, the following are also stated to be very effective in shampoos: The Igepon A and Igepon T types, formerly widely used in Germany, the lauryl sulfoacetate and lauryl sulfoacetamide type, sulfated fatty acyl monoethanolamide, and cationic agents of the Sapamine class. The alkyl aryl sulfonates are generally considered too drying for use as the major ingredient, but they have been used in mixtures with other synthetic substances. Among thickening agents, low-molecular-weight cationic agents and detergents of the Ninol type have been used. More generally, however, either an emulsifiable constituent of the glycerol monostearate type, an inorganic salt, or a soluble polymer such as methyl cellulose is used.[10]

V. Other Preparations for the Hair

In the permanent waving of hair, wetting agents are used to insure thorough uniform contact between the hair and the active waving lotion.

[9] Compare DeNavarre, *Am. Perfumer* **49**, 269 (March, 1947); Beach, *Soap Sanit. Chemicals* **22**, No. 8, 46 (1946), for formulations of cream shampoos.

[10] For general reviews of shampoo formulation see Thomssen, *Soap Sanit. Chemicals* **22**, No. 3, 37; No. 4, 47 (1946). Harris, *Am. Perfumer* **48**, 54 (Nov., 1946). For specific detergents and non-detergent formulating agents see: U. S. Pat. 2,289,004; French Pat. 786,533; to Erhardt Franz. Swiss Pat. 172,509, to *A. Gubser-Knoch*. Lachampt, *Parfumerie* **1**, 76 (1943). U. S. Pats. 2,316,194; 2,401,726; Flett and Toone to *Allied Chemical and Dye Corp.* U. S. Pats. 2,166,127; 2,307,047; Katzman and Cahn to *Emulsol Corp.* U. S. Pats. 2,237,629, to John W. Orelup; 2,353,081, Robinson and Davis to *National Oil Products Co.*; 2,321,270, Bacon and Greminger to *Dow Chemical Co.* Brit. Pat. 529,040; French Pat. 847,001; to *Sandoz Ltd.* Brit. Pat. 470,494, to *Chemische Fabrik Stockhausen.*

Wetting agents are also sometimes used to suspend or emulsify emolient oils in such lotions.[11]

An interesting use of cationic agents in hair dyeing has recently been described. The cationic detergent is applied to the hair, which is subsequently treated with an acid dye. The dyestuff forms an adherent insoluble lake with the adsorbed detergent and is thus fixed to the hair.[12] The glycerol monostearate type of product has also been used to thicken the conventional p-phenylenediamine hair dyes.[13] Sulfonated oils and similar preparations are also used in removing colors from the hair.[14]

Hair-bleaching preparations based on hydrogen peroxide often include wetting agents and/or soaps.[15] The same is true of cream hair dressings, tonics, and brilliantines. In such products, surface active agents serve primarily as emulsifying agents for the active oily and fatty ingredients, or as suspending agents for medicated ingredients.[16]

VI. Bath Preparations

Aside from the special detergents which are used by soap-sensitive individuals, and which will be discussed under hand-cleaning compositions, considerable quantities of synthetic detergents are used in bubble bath preparations. The main requisite of these products is, of course, the formation of copious stable foam. A second desirable property is the ability to retain this foam even when soap is used in the bath. The more usual synthetic detergent types do not meet this specification, rapidly losing their foam when the bather starts using soap. A third requirement of bubble bath detergents is that they prevent the precipitation of lime soaps around the sides of the tub, *i.e.*, prevent the formation of a ring around the bathtub. The rising popularity of bubble bath compositions is probably due to this latter property as much as to any other single factor.[17]

VII. Shaving Preparations

Shaving soaps and creams, like shampoos, form a relatively large outlet for various types of surface active agents. They fall into two categories,

[11] U. S. Pat. 2,310,687, Friedman and Goldfarb to *Lawrence R. Bruce Inc.* Brit. Pats. 428,932 to *Zotos Corp.*; 468,845, to *Coriolan-G.m.b.H.*

[12] French Pat. 851,487, to John W. Orelup.

[13] Brit. Pats. 463,481, to *Franz Stroher A.-G.*; 524,293, to *Clairol N. V.* and Meyer.

[14] U. S. Pat. 2,236,970, Goldfarb to *Lawrence R. Bruce, Inc.*

[15] U. S. Pat. 2,283,350, Baum to McLaughlin and Wallenstein. Ger. Pat. 709,932 Stroher to *Franz Stroher A.-G.* French Pat. 796,709, to *Firma Hans Schwarzkopf.*

[16] Brit. Pats. 500,136, to *Osborne, Garrett and Co. Ltd.*; 492,412, to *Heinrich Macknachf.* Ger. Pat. 697,394, to Keil.

[17] For typical bubble bath formulations, see Augustin, *Seifensieder-Ztg.* **66,** 315 (1939). Molteni, *Am. Perfumer* **49,** 266 (March, 1947).

the brush creams and the brushless creams, of which the latter type is newer and currently more popular. The brush creams are for the most part straight soap. The fatty acids are blended to give the desired small-bubbled stable creamy lather; a mixture of soda and potash is usually used as the alkali. Phosphates or borates are sometimes added to the cream soaps, but seldom to the stick or mug soaps. The synthetic detergents have not been received with favor because they tend to remove too much of the natural oil from the skin during shaving. Certain synthetics, such as the sulfonated oils and the Lamepon type products have been proposed as emolient additives to soap shaving creams.[18]

The brushless shaving creams are similar to vanishing cream in their general composition, but usually contain more fixed oil and more emulsifying agent. Glycerol monostearate and similar substances are widely used in this type of preparation. As in the vanishing creams, the soaps used as emulsifying agents may be replaced to a greater or less degree by synthetic surface active agents. The latter substances also aid in establishing good contact between the cream and the beard hairs, and in making the cream more easily removable by water after shaving. The fatty alkyl sulfates and sulfated oleic esters are excellent for this purpose.[19]

VIII. Dentrifices

Most modern toothpastes contain a mild abrasive such as precipitated chalk, together with glycerol, sorbitol or other binder, and either soap or a synthetic detergent. The fatty alkyl sulfates, sulfated oils, sulfated oleic esters, and sulfated fatty monoglycerides are most widely used, although lauryl sulfoacetate and others have also proved successful. They function both as detergents and foaming agents, and have the advantage over soap of being useful in the neutral or acid pH range. Detergents to be used in toothpaste must be specially purified and certified and must be inherently non-toxic. The fatty alkyl sulfates have received recognition by the American Dental Association as acceptable toothpaste ingredients. Only 2 to 5% of the detergent is normally included in toothpastes based on fatty

---

[18] U. S. Pat. 2,100,090, Sommer and Nassau to *Chemische Fabrik Grünau*. Augustin, *Seifensieder-Ztg.* **63,** 525 (1936). Welwart, *ibid.* **61,** 274 (1934). Canadian Pats. 426,100; 426,101; to *Colgate-Palmolive-Peet Co.* For typical shaving-cream formulas see Kalish, *Drug & Cosmetic Ind.* **40,** 658 (1937).

[19] U. S. Pat. 2,195,713; Canadian Pat. 361,734; Kritchevsky to *Rit Products Corp.* Canadian Pat. 356,798 Ferguson to *Procter and Gamble Co.* Ger. Pat. 604,774, to Heckt, Hall and Marchant.

alkyl sulfates.[20]   In toothpastes which are based on soap about 5 to 10% of soap is used.[21]

Mouthwashes sometimes employ surface active agents also, although this is not common practice.[22]

IX. Toilet Soaps and Hand Cleaners

The ordinary toilet soaps are more appropriately discussed in treatises on soap than in a consideration of the synthetic surface active agents. There are, however, several preparations for special personal cleaning and toilet use which are composed partly or wholly of the synthetic products. These can be classed roughly into four groups, the first of which consists of soaps for general personal use which have been improved by the addition of various synthetic detergents.   The second group includes soapless synthetic detergents made into bar form for toilet use, and is closely related to the third group, which consists of synthetic detergent cleaners for persons allergic to soap.   The fourth and last group consists of industrial hand cleaners.

Ordinary toilet soaps, from a chemical point of view, practically always are mixtures of soaps derived from different fatty acids.   Correct proportioning of different raw material stocks is an important part of the art[23] of making toilet soap.   The nature and proportions of the individual components influence lathering properties,[24] detergent properties,[25] and action on the skin.[26]   In order to insure high lathering power, toilet soap stocks usually contain considerable proportions of coconut oil.   Unfortunately, the presence of lower fatty acid components from coconut oil not only improves the lathering power of toilet soap, but also tends to increase the

[20] DeNavarre, *Am. Perfumer* **39**, No. 3, 29 (1939); Compare *J. Am. Dental Assoc.* **27**, 953 (1940).

[21] For dentifrices soaps see Kranich, *Soap* **13**, No. 9, 19 (1937).   For synthetic detergents in dentrifices, see: Canadian Pat. 426,103, to *Colgate-Palmolive-Peet Co.* U. S. Pat. 2,236,828, Muncie to *Colgate-Palmolive-Peet Co.*   Ger. Pat. 630,344 to Behr. U. S. Pats. 2,359,326 Moss and Schilb to *Monsanto Chemical Co.*; 2,052,694 Breivogel to *Wm. R. Warner Co.*   French Pat. 807,271, to *Société anonyme substantia.*   Swiss Pat. 168,633-4, to *Ciba Co.*   Welwart, *Seifensieder-Ztg.* **62**, 22 (1935).   Redgrove, *Mfg. Perfumer* **3**, 267 (1938).

[22] French Pat. 826,426, to Rhein.   U. S. Pat. 2,222,830, Moss to *Monsanto Chemical Co.*

[23] Treffler, *Soap* **20**, No. 4, 29 (1944).

[24] Smith, *Am. Perfumer* **43**, No. 1, 55 (1941).

[25] Ruf and Renger, *Fette u. Seifen* **47**, 590 (1940).

[26] Matthews, *Perfumery Essent. Oil Rev.* **36**, 251 (1945).   Parkhurst, *Arch. Dermatol. and Syphilol.* **43**, 299 (1941).

possibility of skin irritation.   In order to make toilet soaps blander and less irritating, various "superfatting" agents such as lanolin, cholesterol, fatty alcohols, etc. are often used.[27]   The same purpose has also been served by additions of egg albumen[28] or of protein hydrolyzates.[29]   Another method of making blander toilet soaps consists of eliminating the lower fatty acids of the coconut oil and replacing them with high-foaming synthetic detergents and hydrogenated rosin stock.[30]   The addition of bentonite or colloidal clay is also said to decrease the irritating properties of soap, as well as to increase the foaming and cleaning action.[31]

There has been considerable controversy over the harmful effects of soap to the skin.   Although the vast majority of persons develop no sensitivity even on excessive use of soap, there is evidence that the relatively high pH of soap can cause sensitization in some individuals.[32]   The sulfonated oils are highly recommended as personal cleaners for soap-sensitive individuals. When used for this purpose the sulfonated oils, as well as the other products listed below, are preferably adjusted to a pH value between about 5 and 7. Other detergents which have been used include Igepon T, the sulfated higher fatty alcohols,[33] the sulfosuccinic esters,[34] non-ionic agents of the Igepal type,[35] the sulfonated polyoxyethylene ethers of alkyl phenols (Triton 720 type),[36] and the fatty alkyl sulfoacetates.   The latter have also been made into bar or tablet form by mixing with urea or thiourea,[37] and compacting into cakes under heavy pressure.

Most of the more common synthetic detergents are unsuitable from the practical point of view for forming into cakes or bars.   This is a distinct disadvantage in introducing them for personal use.   Numerous methods have been described for mixing them with various inert solid diluents and

[27] Augustin *Seifensieder-Ztg.* **64,** 536 (1937).   Housman, *Soap* **13,** No. 9, 23 (1937).

[28] U. S. Pat. 2,379,851, Morgan and Lowe to *Cities Service Corp.*

[29] U. S. Pat. 2,048,797, to Küller.

[30] U. S. Pats. 2,414,452, Cunder to *National Oil Products Co.*; 2,388,767, Safrin to *Wilson & Co.*

[31] Cass, *Am. Perfumer* **30,** 243 (1935).

[32] Parkhurst, *Arch. Dermatol. and Syphilol.* **43,** 299 (1941).   Peukert, *Arch. Dermatol. und Syphilis* **181,** 417 (1940).   Hebestreit, *Fette u. Seifen* **48,** 491 (1941).   For an authoritative discussion of this problem see L. Schwartz and Peck, *Cosmetics and Dermatitis,* Hoeber, New York, 1946.

[33] U. S. Pat. 2,091,704, Duncan and McAllister to *Procter and Gamble Co.*

[34] U. S. Pat. 2,373,863, Vitalis to *American Cyanamid Co.*

[35] French Pat. 679,227, to *H. Th. Böhme A.-G.*

[36] U. S. Pat. 2,303,932, Guild.

[37] U. S. Pats. 2,374,187, Flett, and 2,374,544, Hoyt, to *Allied Chemical and Dye Corp.*

pressing them into bars, but to date such compositions have had no appreciable commercial success. Among the diluents which have been used are clay, glycerol monostearate, and similar products, boric acid, urea, etc.[38]

Industrial hand cleaners may be of powdered, paste, or liquid types. The liquid types necessarily contain no abrasive material, but often contain large amounts of solvents, together with a solvent-soluble surface active agent such as mahogany sulfonate, sulfated oils, etc. The solvent loosens and dissolves the grease, which is then flushed away by the rinsing water.[39] A refined mixture of this type was used during the war for removing fuel oil from the burns of seamen who were rescued from burning ships.[40] Powder and paste hand cleaners usually include a mild abrasive such as corn meal or pumice together with soap or any one of a number of synthetic substances. The sulfonated oils are favored in these compositions, but many others have been used.[41] A variety of other specialized hand cleaners have also been described, and a number of detailed reviews are available.[42]

[38] U. S. Pats. 2,296,767, Caryl to *American Cyanamid Co.*; 2,175,285, Duncan to *Procter and Gamble Co.*; 2,169,829, to Beckers.

[39] Compare Brit. Pat. 460,839, to *J. Halden and Co. Ltd.*

[40] Stratton, *Pharm J.* **146,** 250 (1941).

[41] U. S. Pats. 2,035,940, to Berresford; 2,392,779, Showalter to *Standard Oil Development Co.*

[42] Lane and Blank, *J. Am. Med. Assoc.* **118,** 804 (1942). Rayner, *Soap, Perfumery & Cosmetics* **13,** 437 (1940). Davidsohn and Davidsohn, *ibid.* **15,** 382 (1942).

CHAPTER 20

Pharmaceutical, Germicidal, Fungicidal, and Disinfectant Uses

The germicidal effect of surface active agents has already been considered briefly as one of the specific gross effects. It varies markedly from product to product and seems to have only a very limited relationship to the usual surface and bulk properties. Surface active agents may be used by themselves as germicides and disinfectants, or they may be used as solubilizing, emulsifying, and carrying media for other types of germicidal compounds. In the latter case the surface active agent may be inert or it may act as a synergist for the active ingredient, increasing its germ-killing efficiency. In some combinations the surface active agent may reduce the potency of the active germicide and such combinations are, of course, to be avoided. Various micro-organisms differ greatly in their resistance to surface active agents as well as to other germicides. Furthermore, it has been shown recently that the established methods of testing non-surface active germicides do not always give accurate results when applied to surface active products. For these reasons the broad claims of high germicidal activity which are sometimes made in descriptive brochures and patents should be carefully checked. Test methods which simulate actual usage conditions are recommended for this class of materials.[1]

Of those surface active agents which are in themselves destructive of micro-organisms the most potent and most widely used are the cationic detergents. Their high potency and usefulness in this field was first pointed out by Domagk in a series of articles and patents beginning in 1935. Since then literally hundreds of new cationic surface active agents have been synthesized and their use as germicides and/or fungicides disclosed.[2] Among those which have been most widely accepted are the alkyl pyridinium halides and the alkyl trimethylammonium halides. In both these

---

[1] Bernstein, Epstein, and Wolk, *Soap Sanit. Chemicals* **22,** No. 9, 131 (1946).

[2] Among recent review articles and general discussions on cationic germicides, see Klarmann and Wright, *Soap Sanit. Chemicals* **22,** No. 1, 125 (1946). Klarmann, *ibid.* **22,** No. 2, 125 (1946); No. 3, 133 (1946). du Bois, *ibid.* **23,** No. 5, 139 (1947). Taub. *Merck Rept.* **53,** No. 3, 28 (1944). Hopkins, *Perfumery Essent. Oil Record* **36,** 71 (1945).

452

series the cetyl radical is said to give optimum results.[3] When the nature of the hydrophobic group is held constant and the smaller groups around the quaternary nitrogen atom are varied, the germicidal potency is also affected. Domagk found that excellent results are obtained when any of a wide variety of hydrophobic groups are combined with the dimethyl-benzyl ammonium group, the presence of which characterizes several of the most successful commercial cationic germicides. Thus Roccal and Zephiran contain the coconut fatty alkyl group, and the Hyamines contain the p-octylphenoxyethoxyethyl group, linked in both cases to dimethyl-benzyl ammonium. Numerous other attempts have been made to correlate chemical structure and germicidal activity within limited series of cationic surface active agents. None, however, have as yet brought forth any correlations of wide scope. Similarly there appears to be no simple relationship between germicidal activity and detergency or other surface effects.[4] Of interest in this connection is a recent observation[5] that the sodium salt of penicillin strongly reduces the surface tension of water, whereas in contrast, highly purified streptomycin calcium chloride complex slightly raises the surface tension. Both antibiotics exist, however, as hydrosols in aqueous solution. Many interesting relationships remain to be explored in this field.

The cationic surface active agents are used for such diverse disinfecting purposes as sanitization of fabrics in laundries, sterilization of dishes and glassware in restaurants, sterilization of instruments and skin in surgery, and general disinfection in military and naval establishments. A closely related use is for the treatment of fungus infections in dermatology. Cationic surface active agents appear to have the same general order of toxicity toward fungi as toward bacteria.

Although the cationic agents have received much wider acclaim there are also numerous effective germicides among the various classes of anionic surface active agents. Adams and co-workers studied the effect of a large number of fatty acid soaps on the bacilli causing leprosy and tuberculosis. The most effective were found to be those having branched chains or cyclo-

[3] Kolloff et al., J. Am. Pharm. Assoc. **31,** 51 (1942). Green and Birkeland, Proc. Soc. Exptl. Biol. Med. **51,** 55 (1942). Huyck, Am. J. Pharm. **116,** 50 (1944). Hagan, Maguire, and Miller, Arch. Surg. **52,** 149 (1946). Shelton et al., J. Am. Chem. Soc. **68,** 753 (1946). Barnes, Lancet **1942, I,** 531.

[4] Epstein, Harris, Katzman, and Epstein, Oil and Soap **20,** 171 (1943). Rawlins, Sweet, and Joslyn, J. Am. Pharm. Assoc. **32,** 11 (1943).

[5] Hauser, Phillips, and Phillips, Science **106,** 616 (1947).

aliphatic ring structures at the end of the chain.[6] Fatty acids of the latter type are found in the chaulmoogra oil. Although the ordinary C_{12} to C_{18} normal saturated fatty acids are not significantly toxic to the usual test organisms the C_8 and C_{10} members of this series have appreciable activity. Rosin soaps are stated to be highly effective against streptococci and staphylococci.[7] Bayliss has found that oleic soaps and the more highly unsaturated soaps are effective against pneumococci and certain strains of streptococci. Sodium undecylenate has a similar action. Bayliss also found that none of these soaps were as powerful as the fatty alkyl sulfates.[8] The sodium salts of hexylic and heptylic acids are reported to be toxic to human tubercle bacilli.[9]

The fungicidal effects of the naphthenic acid soaps have been fully demonstrated in connection with the mildewproofing of fabrics. These as well as the soaps of undecylenic acid have been used to treat fungus infections of the skin. Very recently it has been shown that the fatty acids with an odd number of carbon atoms from C_7 to C_{13} are highly fungistatic.[10] These acids have been shown to be present in the skin secretion of human adults and to be responsible for their immunity to ringworm of the scalp which infects children. The soaps of fatty acids obtained by the oxidation of paraffin wax are also claimed to be good germicides and fungicides.[11] Of the normal saturated fatty acids, capric, the C_{10} member of the series, is the most effective against fungi.[12]

It is interesting that the straight-chain ordinary soaps are much more destructive to influenza virus than either the fatty alkyl sulfates or the cationic Zephiran. It has been reported that this virus can be purified of contaminating bacteria by washing with 1:20,000 Zephiran solution. The bacteria are destroyed by this reagent leaving the virus unharmed.[13]

[6] R. Adams, Stanley, Ford, and Peterson, J. Am. Chem. Soc. 49, 2934–40 (1927): Arvin and R. Adams, ibid. 2940–2. R. Adams, Stanley, and Stearns, ibid. 50, 1475–8 (1928). Yore and R. Adams, ibid. 1503–8. U. S. Pat. 1,873,732; 1,917,681; R. Adams to Abbott Laboratories.

[7] Klarmann and Shternov, Soap Sanit. Chemicals 17, No. 1, 23 (1941). Stuart and Pohle, ibid. 17, No. 2, 34; No. 3, 34 (1941).

[8] Bayliss, J. Bact. 31, 489 (1936).

[9] Drea, J. Bact. 51, 507 (1946).

[10] Rothman, Weitkamp, Smiljanic, and Shapiro, J. Investigative Dermatol. 8, 81 (1947); Proc. Soc. Exptl. Biol. Med. 60, 394 (1945); Science 104, 201 (1946); J. Am. Chem. Soc. 69, 1936 (1947).

[11] U. S. Pat. 1,955,052 Burwell to Alox Chemical Corp. U. S. Pat. 2,100,469, to Burwell.

[12] Hoffman, Schweitzer, and Dalby, Food Research 4, 539 (1939).

[13] Krueger, U. S. Navy Med. Bull. 40, 622 (1942); Science 96, 543 (1942). Compare Stock and Francis, J. Exptl. Med. 71, 661 (1940); J. Immunol. 47, 309 (1943).

Among the synthetic anionic detergents the fatty alkyl sulfates have been most widely studied. The straight-chain primary C_{12} to C_{16} compounds have been found to be effective against Gram-positive bacteria but much less so against Gram-negative types.[14] Sodium lauryl sulfate has been suggested as the basis for commercial disinfectants.[15]

Secondary alkyl sulfates of the type of the Tergitols (q.v.) are stated to have excellent bactericidal powers, and similar claims have been made for the phosphates and sulfates of chaulmoogryl and hydnocarpyl alcohols.[16] Aerosol OT (q.v.) is effective against *Staphylococcus aureus*, particularly at lower pH values. It is twenty times as effective at pH 4.0 as at pH 6.6.[17]

The long-chain alkyl aromatic sulfonates of the type of Nacconol and Santomerse are effective germicides, but their action is slower than that of the usual standard non-surface active germicides. Like the other non-carboxylic anionic agents, the alkyl aromatic sulfonates are more effective in acid than in alkaline solutions.[18]

Both soaps and the various classes of synthetic detergents have been used in mixtures with non-surface active germicides such as phenols, chlorine derivatives, heavy-metal ions, etc. The effects of sulfonated oils, Aerosol OT, and a large number of other detergents on the activity of these germicides have been reported by Gershenfeld and co-workers.[19]

Since surface active agents greatly reduce the size of the drop taken for inoculation on the loop in the standard Food and Drug phenol coefficient method, Tobie and Orr recommend taking 0.02 cc. of the test mixture in sterile pipettes. Otherwise, the potentiation of phenol, cresol, phenylmercuric nitrate, etc. by the surface active agent may appear to be larger than it actually is.[20]

Soap–phenol mixtures have been widely used as disinfectants and have retained their popularity over a period of years, in spite of considerable controversy as to the effect of soaps on the germicidal power of phenols. It has recently been established that soap increases rather than hinders the

[14] Birkeland and Steinhaus, *Proc. Exptl. Biol. Med.* **40**, 86 (1939). Baker, Harrison, and Miller, *J. Exptl. Med.* **73**, 249 (1941); **74**, 621 (1941).

[15] U. S. Pat. 2,183,037, Bayliss, Wilson, and Ordal to *Economics Laboratory Inc.* Janke *et al.*, *Collegium* **1938**, 545.

[16] U. S. Pat. 2,380,011, Baker and Miller. Pollard, *Science* **103**, 758 (1946). Austrian Pat. 157,958, Stefanovic; *Chem. Abstracts* **36**, 2086.

[17] Gershenfeld and Perlstein, *Am. J. Pharm.* **113**, 88 (1941).

[18] Flett, *Oil and Soap* **22**, 245 (1945). Gonzalez and Madeiros, *Rev. med. y aliment.* (Santiago, Chile) **5**, 232 (1942–3); *Chem. Abstracts* **39**, 325.

[19] Gershenfeld and Witlin, *Am. J. Pharm.* **113**, 215 (1941); **11**⁴, 314–18 (1939). Gershenfeld and Perlstein, *ibid.* **113**, 237–55 (1941).

[20] Tobie and Orr, *J. Lab. Clin. Med.* **29**, 767 (1944).

germicidal action of phenols, and a large number of the new phenolic germicides are being formulated with soaps.[21] The fatty alkyl sulfates,[22] the sulfonated oils,[23] and the mahogany sulfonates,[24] are examples of anionic synthetic substances which have been used with phenolic germicides. The cationic agents, which are strongly germicidal in themselves and sufficiently surface active to be solubilizers and emulsifiers, have also been used with phenols.[25]

Among the active chlorine type of germicides, Chloramine T and azochlorinamide have been used with soaps.[26] Calcium hypochlorite has been used with a variety of lime-resistant anionic and cationic agents.[27]

Polysulfides and tetraethylthiuram monosulfide have been used with soaps and synthetic anionic agents as fungicides and germicides. The latter compound is said to be effective in treating scabies.[28]

Germicidal and fungicidal compositions have also been prepared by incorporating compounds of mercury, gold, silver, or copper into soaps and a variety of anionic detergents.[29] The sulfonated and sulfated synthetic detergents have also been used with hydrogen peroxide to give germicidal cleaning agents.[30]

One of the major uses for surface active agents in the pharmaceutical

[21] Cade, *Soap* **11**, No. 9, 27 (1935). Hueter and Engelbrecht, *Die Chemie* **55**, 329 (1942). Ordal, Wilson, and Borg, *J. Bact.* **42**, 117 (1941). Halvorson, Bayliss, Ordal, and Wilson, *Soap* **11**, No. 5, 25 (1935). U. S. Pats. 2,138,805, Halvorson *et al.* to *Economics Laboratory, Inc.*; 2,251,934, Hartung to *Sharp and Dohme*; 2,267,101, Hueter and Engelbrecht to *Unichem*. U. S. Pats. 2,353,724, Gump to *Burton T. Bush, Inc.*; 2,353,735, Kunz and Gump to *Burton T. Bush, Inc.*

[22] Canadian Pat. 402,938, Rawlins to *Parke, Davis and Co.* Brit. Pat. 515,386, Rawlins to *Parke, Davis and Co.* U. S. Pat. 2,191,405, Hueter, Neu, and Engelbrecht to *Unichem*.

[23] U. S. Pats. 2,289,476 Badertscher to *McCormick and Co.*; 2,297,388, Böhler, vested in Alien Property Custodian. Gershenfeld and Witlin, *Am. J. Pharm.* **112**, 45 (1940).

[24] U. S Pats. 2,079,772, Schuler to *Stanco, Inc.*; 2,228,407, Schuler and Archibald to *Stanco, Inc.*

[25] U. S. Pats. 2,309,592, Hueter to *Patchem A.-G.*; 2,264,150, Hueter and Neu to *Patchem A.-G.* Hornung, *Fortschr. Therap.* **16**, 384 (1940); *Chem. Abstracts* **37**, 1563.

[26] U. S. Pats. 2,296,121 to J. A. Smith; 2,263,948, Halvorson, Ordal, and Wilson to *Economics Laboratory, Inc.*

[27] Brit. Pat. 530,040; Canadian Pat. 403,649; French Pat. 852,048; to *Mathieson Alkali Works*.

[28] Ger. Pats. 689,373; 642,532, to *I. G. Farbenindustrie*. Brit. Med. *J.* **1**, 803 (1945).

[29] Brit. Pats. 427,324; 432,689; to *Lever Bros.* Brit. Pats. 489,222 to *Imperial Chemical Industries Ltd.*; 493,148, to Large and *Boots Pure Drug Co. Ltd.* Ger. Pat. 657,116; Brit. Pat. 407,039, to *I. G. Farbenindustrie*.

[30] U. S. Pat. 2,371,545, Riggs and Lehmann to *Pennsylvania Salt Mfg. Co.*

field is in the formation of various emulsions. Cod liver oil and other vitamin oils are ordinarily emulsified with the natural gums, pectins, etc. Synthetic agents of the type of sorbitol oleate have also been used.[31] Antiseptics such as proflavine and acriflavine have also been used advantageously in emulsion form.[32] Emulsions of castor oil, mineral oil, etc. for internal use are often made with proteins, lecithin, or with the glycerol monostearate type of emulsifier.[33]

The absorption bases already mentioned in the section on cosmetics are used as media for a variety of medicaments such as Chloramine T, ammoniated mercury, etc.[34] Recently ointment bases containing hydrophilic O/W emulsifiers such as sulfonated oils and fatty alkyl sulfates have been described. These have the advantage of being readily removed by washing with water.[35]

Sodium alkyl sulfates have been found to be effective in treatment of gastric and duodenal ulcers. The alkyl sulfates are said to exert beneficial action by inhibiting the action of pepsin rather than by decreasing the corrosive effects of hydrochloric acid.[36] A study of the relative effectiveness of various commercial detergents in inhibiting peptic activity revealed, among other things, that in the alkyl sulfate series the decyl and dodecyl derivatives are most effective.[37] The effectiveness of detergents in decreasing peptic activity is impaired to varying degrees by various substances, e.g., triacetin, sodium taurocholate, cream, butter, ethyl laurate.[38] A 2% ointment of sodium lauryl sulfate has been used to promote healing of wounds and underlying tissues following fistulas of the stomach, intestines, etc.[39] Related use of such products to inhibit adhesions and as a thrombotic and sclerotic agent in varicose veins has also been reported.[40]

The use of various synthetic surface active agents in preparing ointments for dermatologic therapy has been reviewed recently by Duemling and by

---

[31] U. S. Pat. 2,285,422, to Epstein and Harris. Goldner, *J. Am. Pharm. Assoc. Pract. Pharm. Ed.* **3**, 324 (1942).

[32] Heggie, Gerrard, and Heggie, *Lancet* **1942**, I, 347. Lum, *Pharm. J.* **135**, 457 (1935).

[33] Münzel, *Pharm. Acta Helv.* **18**, 32, 99 (1943); *Chem. Abstracts* **38**, 215. Chatterjee, *J. Indian Chem. Soc.* **13**, 563 (1936). Swallow, *Pharm J.* **147**, 226 (1941).

[34] Hopf, *Fette u. Seifen* **46**, 144 (1939). Soulsby, *Brit. J. Dermatol. Syphilis* **52**, 25 (1940). Bamber, *ibid.* **52**, 21 (1940). Gershenfeld, *Am. J. Pharm.* **112**, 281 (1940).

[35] U. S. Pat. 2,300,780 to Fiero; Fiero, *J. Am. Pharm. Assoc.* **30**, 145 (1941). Beeler, *Bull. Natl. Formulary Comm.* **11**, 27 (1943).

[36] Fogelson and Shoch, *Arch. Internal Med.* **73**, 212 (1944).

[37] Kirsner and Spitz, *Gastroenterology* **2**, 270 (1943).

[38] Kirsner and Wolff, *Gastroenterology* **2**, 348 (1943).

[39] Portis, Block, and Necheles, *Gastroenterology* **3**, 106 (1944).

[40] Whigham, *Lancet* **1944**, 646.

Mumford.[41] A specialized but important type of cream or ointment is used to protect industrial workers from agents which might cause contact dermatitis. The use of these protective ointments has expanded greatly within recent years.[42]

[41] Duemling, *Arch. Dermatol. and Syphilol.* **43,** 264 (1941). Mumford, *Brit. J. Dermatol. and Syphilis* **50,** 540 (1938).

[42] Klauder, *Ind. Med.* **9,** 221 (1940). L. Schwartz, Peck, and Tulipan, *Occupational Diseases of the Skin,* 2nd ed., Lea & Febiger, Philadelphia, 1947, p. 100 *et seq.*

CHAPTER 21

Household, Laundering, Dry Cleaning, and General Cleaning Uses

The four major types of soap products used in the household are: (*1*) bar soaps for bath and personal use; (*2*) bar soaps for laundering and scrubbing; (*3*) scouring powders; (*4*) powdered or flaked soaps for dishwashing, laundry, and general cleaning. Although, in the first three fields, the synthetic detergents have not as yet been used to any appreciable extent in this country, synthetic detergent preparations in powdered, flaked, and liquid form are rapidly gaining in popularity and are displacing soap powders to a considerable extent. At the present time a greater poundage of synthetic detergents is sold in packaged form for household use than for any other single purpose.

The two main uses for powdered soaps or detergents in the household are for dishwashing and laundering. Lesser quantities are also used for scrubbing tile and linoleum surfaces, painted woodwork, etc. There are two types of flaked or powdered household soaps in general use. One type, consisting of practically pure soap with little if any builder, is sold at a premium and is recommended for its mildness on hands and clothes. Most housewives prefer the pure soap powders for fine laundery and some prefer them for dishwashing. The other type of soap powder contains a substantial quantity of builders such as soda ash, sodium sesquicarbonate, phosphates, borax, etc., together with the soap. The built powders are recommended for heavy and machine laundering, dishwashing, and general cleaning. Both types of soap powder are today facing competition from washing powders based on synthetic detergents.

The washing of fabric materials in the household will be considered in two categories, machine washing and hand washing. Articles normally handled include table linen, sheets, flatware, shirts, house dresses, cotton or rayon hosiery and undergarments, cotton children's wear, etc. In short, machine washing is normally confined to sturdy fabrics of cotton or rayon or mixtures of the two. Hand laundering is used for the more delicate fabrics, including hose, lingerie, woolen sweaters and other garments of wool, silk, nylon, or the finer rayons. It is advantageous to wash these fabrics in relatively cool water with a minimum of rubbing or rough handling.

Among the large number of synthetic anionic agents ideally suited for fine laundering, three different types — (*1*) an alkyl aryl sulfonate, (*2*) a fatty alkyl sulfate, and (*3*) a fatty monoglyceride sulfate — are most widely used at present in various brands of soapless detergent powders. These usually contain approximately 25–40% organic detergent ingredients, the remainder being a non-alkaline, inorganic salt like sodium sulfate or sodium chloride. Sodium bicarbonate is also used in some instances. These detergent powders, which are intended to compete with the pure unbuilt soaps rather than with built soap powders, have the advantage of affording good detergency at low bath temperatures. Because of their lime resistance they are particularly valuable in hard water areas.

Polyethylene ether non-ionic agents, in liquid form diluted with water, are also being introduced into the fine laundering field.

The synthetic detergent powders, mentioned above as well suited for fine laundering, are also used very widely for household dishwashing where their lime resistance and high wetting power make them greatly superior to soap. If a dish or a glass is washed in soap and allowed to dry without wiping, the water will at first drain off unevenly, leaving droplets suspended on the vessel. When drying is complete, a dull film, or even streaks of lime soap will be left on the surface. This will occur even where the water is moderately soft. When the synthetic detergents are used, however, the dish will drain evenly and dry much more rapidly because of complete drainage. Furthermore, due to the absence of lime precipitates, the surface will be clean and sparkling. A further advantage of the more effective synthetic products is that they disperse grease and soil and prevent it from accumulating on the walls of the dish pan or sink. With soap (unless a special water-softening agent is added) an agglomerate of lime precipitate and grease is usually deposited on the ceramic surfaces, and must be removed by scouring.

A different type of synthetic detergent powder is intended to compete with the built soap powders for dishwashing, heavy laundering, and general cleaning. These products vary considerably in composition as well as in the nature of the base detergents. One inexpensive formula includes about 10% active alkyl aryl sulfonate, up to 30% soda ash, and the remainder sodium sulfate or sodium bicarbonate. Some similar formulas include smaller quantities of phosphate builders, borax, and sometimes bentonite or clay. The fatty alkyl sulfates, Igepons, and sulfated fatty acyl monoethanolamides have also been used as base detergents. In Germany powders of this class were based on Mersolat (a Reed reaction product) and often contained small amounts of Tylose (carboxymethylcellulose) and sodium metasilicate. Many of these heavily built powders have not been well received by American consumers. They are admittedly inferior to soap for laundering and many of them contain insufficient active detergent materials.

Two recently introduced products, intended for heavy laundry as well as general household use, appear to be more carefully designed. One contains about 25% active fatty alkyl sulfate together with a high percentage of pyrophosphate. The other contains about 20% of a polyethylene oxide non-ionic detergent as the active ingredient and also a large proportion of pyrophosphate. These powders appear able to match soap in cotton washing and they have all the advantages of high wetting power, lime resistance, etc., associated with the synthetic detergents.

Because of the extra cost involved, synthetic detergents have been seldom used in scouring powders. Most scouring powders contain about 5% soap, some phosphate or silicate builder, with the remainder an abrasive such as diatomaceous earth or ground pumice.[1]

Bars of soap – synthetic detergent mixtures were extensively used by the armed services during the war for washing clothes in hard water or even sea water. The alkyl aryl sulfonates, sulfated fatty alcohols, and fatty glyceride sulfates were the synthetic detergent types most widely used in such compositions. The mixed soap–synthetic bars have not as yet been well received for household use.[2] J. C. Harris reports that the effectiveness of Santomerse (an alkyl benzene sulfonate detergent) alone in very hard water and sea water is poor, but is improved by addition of trisodium phosphate, soda ash, sodium pyrophosphate, or soap made from tallow or tallow with coconut oil.[3] Methods for evaluating salt water detergents have been studied by Ruckman, Hughes, and Clark.[4] Sea water soaps consisting of mixtures of saturated higher alkyl sulfates such as palmityl or stearyl sulfate and unsaturated higher alkyl sulfates, e.g., oleyl sulfate, together with minor amounts of soap have been described.[5]

Commercial dishwashing and glassware washing is often done by machine using a cleaning agent which consists simply of a mixture of alkalies. Small amounts of a synthetic detergent, usually an alkyl aryl sulfonate or fatty alkyl sulfate, are sometimes added. In some restaurants, bars, and soda

[1] Compare *Soap Sanit. Chemicals* **21**, No. 6, 44 (1945). U. S. Pat. 2,041,744, Cummins to *Johns-Manville Corp.*

[2] Brit. Pat. 406,565, to *Deutsche Hydrierwerke A.-G.* U. S. Pats. 2,390,295, Flett to *Allied Chemical and Dye Corp.*; 2,303,212, Kise and Vitcha to *Solvay Process Co.*; 2,385,614, Dreger and Bell to *Colgate-Palmolive-Peet Co.*; 2,356,903, Wood to *Procter and Gamble Co.*

[3] J. C. Harris, *Rayon Textile Monthly* **26**, 77 (1945).

[4] Ruckman, Hughes, and Clark, *Oil and Soap* **19**, 3 (1942).

[5] Brit. Pat. 406,565, to *Deutsche Hydrierwerke A.-G.*

fountains, a special rinsing bath containing a cationic germicide is often used to sterilize dishes after washing.[6] In the dairy industry, metal equipment, bottles, and cans are often cleaned with the aid of synthetic anionic detergents, which, when used in acid solution with phosphoric acid or organic acids, are highly effective in removing and preventing the formation of milkstone.[7]

A number of compositions containing detergents of high wetting power are prepared especially for cleaning painted surfaces, mirrors and windows, ceramic surfaces, etc.   Such products may contain alcohol as well as water. The detergents most often used are the alkyl aryl sulfonates, petroleum sulfonates and sulfated oils.[8]   Other cleaning operations, such as the washing of streets, automobiles, railroad equipment, etc. may be facilitated by the use of surface active agents.   Thus, 5 pounds of an alkyl aryl sulfonate (Nacconol NR) in 2000 gallons of water produced superior results when used in a sprinkler truck to wash a mile of city street.[9]

Rug-shampooing compositions for use in the household usually contain a high-foaming synthetic anionic agent such as sodium lauryl sulfate. Commercial establishments for cleaning rugs usually use a solvent cleaner or a pulverulent cleaner.[10]

A considerable poundage of synthetic surface active agents is used in the preparation of floor waxes, furniture polishes, shoe creams, household solvent cleaners, and impregnated polishing cloths.   Floor waxes are usually emulsions of vegetable wax with an amine soap, but the anionic sulfates and sulfonates as well as the oil-soluble non-ionic emulsifying agents have been used.[11]   Furniture polishes and shoe polishes usually employ sulfated

[6] U. S. Pats. 2,035,652, Hall to *Hall Laboratories*; 2,142,870–1, Hall and Griffith to *Griffith Laboratories*.   Liddiard, *Chemistry & Industry* **1941**, 480.   Hileman, *9th Ann. Rept. N. Y. State Assoc. Dairy Milk Inspectors* **1935**, 71; *Chem. Abstracts* **30**, 3907. Mallmann, *Am. J. Pub. Health* **27**, 464 (1937).   Lesser, *Soap Sanit. Chemicals* **23**, No. 6, 43 (1947).

[7] U. S. Pat. 2,338,688, Parker and Bonewitz to *Rex Co.*   Parker, *Milk Plant Monthly* **1941**, No. 9, 49; *Chem. Abstracts* **36**, 7157.   Scales and Kemp, *Assoc. Bull. Intern. Assoc. Milk Dealers* **33**, 589 (1941); *Chem. Abstracts* **36**, 6680.

[8] Compare Lesser, *Soap Sanit. Chemicals* **21**, 37 (Dec., 1945).   Specification P-C-431, Federal Specifications Board, Bureau of Federal Supply, U. S. Treasury Dept. U. S. Pats. 2,241,790, Rembert to *Johns-Manville Corp.*; 2,079,793, Donlan to *Stanco Inc.*; 2,051,435, Colleran.   Ger. Pat. 542,441, to *Wollnerwerke G.m.b.H.* and Max Dittmer.

[9] Flett, *private communication*.

[10] Compare Brit. Pat. 526,647, to *Mathieson Alkali Works*.   U. S. Pat. 2,344,268, Rench to *Bigelow Sanford Carpet Co.*

[11] U. S. Pats. 2,349,326, Wilson to *Carbide and Carbon Chemicals Corp.*; 2,374,474, Dolian to *Commercial Solvents Corp.*; 2,009,345, Schrauth to *Unichem*.   Brit. Pat. 440,642, to *Henkel et Cie*.   U. S. Pat. 2,088,795, Kline to *Western Union Telegraph Co.* Becher, *Seifensieder-Ztg.* **67**, 321 (1940).

oils as the cleaning and emulsifying ingredient. Sulfated oils are also used in solvent cleaners and cleaning pads. The non-ionic detergents, petroleum sulfonates, and the oil-soluble amine salts of the fatty alkyl sulfates have also been used in these compositions.[12]

The synthetic detergents have hardly been used at all in commercial laundering in this country. They were very widely used for this purpose in Germany during the war. The non-ionic agents based on alkyl phenols, the Alipals (q.v.) and the Mersolats (Reed reaction products) appear to have been most successful, particularly when used with Tylose. Under normal circumstances, however, soap plus the usual alkaline builders is more economical for cotton washing.[13] The polyethylene oxide non-ionic agents and their sulfated derivatives are considered to have better detergency for cotton than the simpler sulfates or sulfonates, and they are as effective as soap under favorable conditions. It is possible that these products might be introduced into commercial laundry work if their cost relationships become favorable. Quite recently there has been an increased interest in synthetic detergents for laundry work, due in part to the increasing cost of soap. Successful pilot runs in commercial laundries have been reported for alkyl aryl sulfonates and alkyl sulfates, built with poly- or pyrophosphates and carboxymethylcellulose. Similarly built synergistic mixtures of alkyl aryl sulfonates with polyethylene oxide non-ionics have also given excellent results.

Modern dry-cleaning establishments generally use a detergent in the cleaning bath. The detergent is incorporated into the naphtha or chlorinated solvent together with a small amount of water and alcohol. The water is solubilized and the resulting bath is clear and non-foaming. The purpose of adding detergent is twofold. It serves primarily to suspend or peptize the insoluble soil removed from the garment and to prevent its redeposition. The water which is present also effects a cleaning action on any aqueous or water-soluble stains which may be present. Both soaps and synthetic detergents have been used in dry-cleaning baths. The usual soaps are potassium, ammonium, or alkylolamine salts of low-titer fatty acids, i.e., acids with a high oleic and/or linoleic content. Some excess free fatty acid should be present to obtain good solubilization and good cleaning results.

For studies and reviews on the action of dry cleaning soaps see: Bird, *J. Soc. Dyers Colourists* **48**, 30 (1932); **49**, 316 (1933); **50**, 389 (1934). French Pat. 798,796, to

---

[12] Schoenholz, *Chem. Inds.* **61**, No. 2, 228 (1947). Brit. Pat. 460,543 to *J. Halden and Co.* U. S. Pat. 2,209,785, Manierre. U. S. Pat. 1,986,936; Brit. Pat. 492,742, to *E. I. du Pont de Nemours & Co.* Smither, *Natl. Bur. Standards (U. S.) Circ.*, No. C424 (1939).

[13] Hoyt, *PB 3865*, Office of Technical Services, Dept. of Commerce, Washington, D. C.

F. G. C. Niedmann. Putnam, *Soap* **12**, No. 1, 25 (1936). Tyler, *ibid.* **12**, No. 5, 13 (1936). Treffler, *Soap Sanit. Chemicals* **20**, No. 12, 36 (1944). Stinson, *Laundry Dry Cleaning J. Can.* **19**, No. 10, 14 (1939). Mann, *Seifensieder-Ztg.* **64**, 102 (1937). Fabry *Soap Sanit. Chemicals* **22**, No. 7, 48 (1946).

Several different types of synthetic detergents have been proposed for use in dry-cleaning baths. They are all incorporated in the same way as the soaps, *i.e.*, with water and, if necessary, a coupling solvent such as alcohol. Among these surface active agents are:

(*1*) The long-chain alkyl phenol or alkyl benzene sulfonates.[14,15]

(*2*) Sulfated fatty alcohols, alkyl sulfoacetates, alkyl sulfosuccinates, and sulfated oleic or ricinoleic esters.[16]

(*3*) Petroleum sulfonates.[17]

(*4*) Cetylpyridinium bromide and other cationic agents.[18]

(*5*) Esters of long-chain fatty acids with a low-molecular-weight hydroxycarboxylic acid, such as stearic acid tartrate.[19]

(*6*) Oil-soluble non-ionic agents.[20]

At the present time most dry-cleaning plants use the soaps rather than the synthetic agents. One reason for this is that the soaps have been more thoroughly studied and much more is known about their behavior in both normal and unusual circumstances.

Special spot-cleaning compositions sometimes include synthetic detergents but more often if any surface active agent is used it is a soap.[21]

[14] U. S. Pats. 2,271,635; 2,317,986; Canadian Pat. 369,508; Flett to *Allied Chemical and Dye Corp.*

[15] Morgan, *Dyestuffs* **34**, 69 (1936); *Chem. Abstracts* **30**, 6573.

[16] U. S. Pats. 2,327,182–3; 2,326,772; 2,290,870; Flett to *Allied Chemical and Dye Corp.* U. S. Pat. 2,091,121, Lenher and Arnold to *E. I. du Pont de Nemours & Co.*

[17] U. S. Pats. 2,053,007, Parkhurst to *Standard Oil Co. of Indiana*; 2,084,483, Donlan to *Standard Oil Development Co.*

[18] U. S. Pat. 2,165,356, Dunbar to *Imperial Chemical Industries Ltd.*

[19] U. S. Pat. 2,251,694, Tucker to *Procter and Gamble Co.*

[20] U. S. Pats. 2,317,112, Pings to *E. I. du Pont de Nemours & Co.*; 2,251,691, Richardson to *Procter and Gamble Co.*

[21] Lesser, *Soap Sanit. Chemicals* **22**, 33 (March, 1946). Reich, *Chem. Inds.* **60**, 423 (1947). U. S. Pat. 2,052,891, Merrill to *Union Oil Co. of Calif.* Tyler, *Soap* **11**, 19 (1935). Meissner, *Deut. Wollen-Gewerbe* **70**, 381 (1938); *Chem. Abstracts* **32**, 4245.

CHAPTER 22

Metal Technology

Surface active agents, particularly the synthetic products rather than the soaps, are widely used in the various processes of metal fabrication. The types of utilization in this field can be classified as follows: (*1*) special cleaning operations and pickling; (*2*) rust and corrosion inhibition; (*3*) electroplating; (*4*) cutting lubricants; (*5*) drawing and buffing compositions, soldering fluxes, and other miscellaneous uses.

I. Special Cleaning Operations and Pickling

Fabricated metal parts which are to be electroplated, painted, or otherwise finished must be thoroughly cleaned of residual grease and dirt. This is usually accomplished with solvents, aqueous detergent solutions, or emulsions. Cleaning is normally carried out in two stages. In the first stage the main portion of grease and dirt is removed, leaving only a small amount of grease as a residual surface film. This operation is called degreasing and is carried out with solvents or with aqueous emulsions of solvents. Although the metal parts may be dipped into an emulsion already formed, it is more usual to treat the parts with a solvent containing a dissolved oil-soluble O/W emulsifier, such as oil-soluble soaps. Sulfonated oils and petroleum sulfonates are also excellent for this purpose, although they are less commonly used. Another variation of this type of metal-degreasing process involves first dipping the metal parts into a solvent bath which contains a dissolved soap-forming fatty acid, *e.g.*, oleic acid. The parts are then dipped in an aqueous alkaline bath which converts the fatty acid to its soap. This soap, formed *in situ*, removes the grease-laden solvent from the metal surfaces by virtue of its powerful emulsifying action. The solvents used in emulsion degreasing often contain cresylic acid. This component is particularly effective in increasing the solvent power and emulsifiability of the kerosene which is often used as the major solvent ingredient.[1]

---

[1] For examples and disclosures on emulsion degreasing see Mitchell, *Proc. Am. Electroplaters' Soc.* (June) **1938,** 238. Strachan, *Products Finishing* **2,** No. 11, 21 (1938). U. S. Pats. 2,032,174, Johnson to *Oakite Products, Inc.*; 2,208,524, Darsey and McVey to *Parker Rust-Proof Co.* Brit. Pat. 444,818; Ger. Pat. 608,928; to *Alexander Wacker Gesellschaft für Elektrochem. Ind.* Brit. Pat. 516,218, to *Bennet (Hyde) Ltd.*

Degreased metal surfaces are often cleaned with aqueous solutions of detergents to remove the last traces of grease and oil. This is usually done in strongly alkaline solutions containing silicates, phosphates, soda ash, or caustic soda. These are sometimes used alone but more often with soap or a synthetic detergent.[2] Sodium lauryl sulfate and the alkyl aryl sulfonates give excellent results when used in these alkaline baths. The efficiency of grease removal has been determined by measuring the ultraviolet fluorescence of the residual grease. Alkaline mixtures containing added detergent are stated to be about five times as effective as the straight alkalis.[3]

Before metals are painted or enameled they are often cleaned in baths containing phosphoric acid. This treatment converts oxide coatings on the metal to corresponding phosphates which aid in preventing corrosion. Alkyl aryl sulfonates, fatty alkyl sulfates, and phosphates, and the polyethylene oxide non-ionic agents have been recommended as surface active ingredients in these acid cleaning baths.[4] The soaps, of course, cannot be so used. The lime- and acid-resistant surface active agents are also used to remove flux residues from soldered metal parts, although this operation is usually carried out in strongly alkaline baths.[5]

The cleaning of aluminum prior to spot welding assumed great importance in the aircraft industry during the war. In this process it is necessary to remove the tenacious oxide film normally present on aluminum. Strongly acid solutions are used for this purpose and an acid-resistant detergent such as an alkyl aryl sulfonate is added. Degreasing and alkaline pre-cleaning are carried out in the usual way.[6]

The process of removing the oxide scale from ferrous metals by treatment with a dilute acid bath is termed pickling. This essential pretreatment for parts that are to be tinned, galvanized, or plated is usually carried out by immersion in 5% sulfuric acid at about 70°C. In order to minimize the attack of the acid on the underlying unoxidized metal, special agents

[2] Wernick, *Ind. Chemist* **9**, 275, 308 (1933). Promisel, *Monthly Rev. Am. Electroplaters' Soc.* (June) **1935**; *Iron Age* **155**, No. 15, 62 (1945).

[3] Morgan and Lankler, *Ind. Eng. Chem.* **34**, 1158 (1942); *Ind. Eng. Chem., Anal. Ed.* **14**, 725 (1942); *Proc. Am. Electroplaters' Soc.* (June) **1938**, 33. Peterson, *Metal Cleaning Finishing* **6**, 127 (1934). Harris, *ASTM Bull.* **133**, 23 (1945).

[4] Brit. Pat. 403,373, to J. H. Gravell. U. S. Pat. 2,304,299; Brit. Pat. 517,916; Canadian Pat. 356,078, to C. L. Boyle. U. S. Pat. 2,396,776, Douty and Heller to *Am. Chemical Paint Co.*

[5] Brit. Pat. 403,374, to J. H. Gravell.

[6] Harris, *Aluminum and Magnesium* **1**, No. 7, 28 (1945). Strow, *Monthly Rev. Am. Electroplaters' Soc.* **29**, 995 (1942). Morris, *Aluminum and Magnesium* **1**, No. 2, 26 (1944). Hess, Wyant, and Averbach, *Welding J. N.Y.* **23**, 417-S (1944).

called inhibitors, are added to the dilute-acid pickling bath. Although the acid-resistant surface active agents have little inhibiting power, they are sometimes added to pickling baths together with the inhibitor as assistants and dispersing agents.

In some instances electrolytic cleaning is used to remove traces of grease and oil. In this process the object to be cleaned is made one of the electrodes in an electrolytic bath.[7] Electrolytic pickling is more strictly a cleaning process in which the oxide scale may be only partly removed. The piece to be cleaned is made the cathode in a strong alkaline bath and a current in the range of 10 amperes per square foot is applied. The vigorous evolution of hydrogen accomplishes the cleaning.[8]

II. Rust and Corrosion Inhibition

Various treatments for protecting metals against rust, tarnish, and corrosion often make use of surface active agents as assistants. In some instances, the surface active agent itself is the active rust-preventing ingredient. In others it is merely an auxiliary, promoting the adhesion, solubility, or even distribution of the active agent. One of the best-known treatments for inhibiting the corrosion of iron and steel utilizes phosphoric acid or acid phosphates to form a thin protective or passivating coating on the metal. The alkyl aryl sulfonates and fatty alkyl sulfates have been recommended as auxiliaries to the treating bath since they promote wetting and afford a more even treatment. They have also been used in water-soluble type coatings containing nitrites or alkanolamine soaps as the active rust-inhibiting ingredients.[9]

The tarnishing of silverware can be inhibited by treating the surface with aqueous solutions of higher fatty amine salts. In this case the surface active agent itself is the inhibitor.[10]

One method of preventing the corrosion of machinery which runs in contact with water (radiators, engine blocks, etc.) consists of treating the water itself with an inhibitor. Surface active agents are sometimes added to radiator coolants together with an active inhibitor.[11]

Unpainted and unplated metal parts which are to be shipped or stored

[7] Mitchell, *Metal Cleaning Finishing* 7, 9 (1935). Griggs, *Can. Chem. Met.* 20, 260 (1936).

[8] Machu, *Korrosion u. Metallschutz.* 15, 105 (1939). Hartshorn, *Metal Finishing* 39, 561 (1941).

[9] U. S. Pats. 2,227,945, to H. R. Neilson; 2,398,212, Durgin to *Monsanto Chemical Co.*; 2,402,793, White, Schwartz, and Rouault to *Standard Oil Co. of Indiana.*

[10] U. S. Pats. 2,400,784–5, Rust to *Montclair Research Corp.*

[11] Canadian Pat. 362,601, Krekeler to *Shell Development Co.*

are usually protected from rusting by the application of a coating of grease or oil. These oils, which are specially treated to increase their rust-preventing powers, are known as slushing oils or rust-preventive oils. The addition of one or more special agents to such oils is practiced to accomplish two purposes: first, to increase the adhesion of the oil to the metal so that water and moisture cannot easily contact the metal; secondly, to directly inhibit corrosion even if water does come in contact with the metal. Of these two effects the first is probably the more important. The added reagents may be classed as oil-soluble surface active agents which enable the oil to displace water from the metal surface, *i.e.*, they *increase* the contact angle (measured in the water phase) at the oil–water–metal boundary line. In many cases these substances also impart positive corrosion-inhibiting properties to the oil. Large quantities of lanolin were used during the war in protective oil coatings for steel equipment to be shipped overseas. Among other substances used for this purpose are: (*1*) special long-chain fatty acids and/or their salts with higher fatty amines[12]; (*2*) higher alkyl phosphates[13]; and (*3*) higher alkyl ammonium hydroxides and chromates.[14] A large number of other compounds which have been used for this purpose will be described in the section on lubricants.[15]

III. Electroplating

Surface active agents have been added to electroplating baths to achieve a variety of effects in different individual plating operations. One important effect is the widening of the operating range with respect to pH, temperature, and current density. Another effect is the modification of the crystal size of the deposited metal, thus an aiding improving the brightness of the resultant plate. By reducing the surface tension of the electroplating solution, detachment of gas bubbles from the cathode is facilitated, and pitting and pinholing are avoided. The use of surface active agents in cleaning the metal surface preparatory to plating has already been mentioned.[16]

[12] U. S. Pats. 2,371,143, Barnum and Zublin to *Shell Development Co.*; 2,371,207, Zublin, Barnum, and White to *Shell Development Co.*; 2,330,524, Shields to *Alox Corp.*; 2,308,282, Howland and Ter Horst to *U. S. Rubber Co.*

[13] U. S. Pat. 2,080,299, Benning, Benton and Clarkson to *E. I. du Pont de Nemours & Co.*

[14] U. S. Pats. 2,340,996, Smyers to *Standard Oil Development Co.*; 2,270,386, Sloan to *E. I. du Pont de Nemours & Co.*

[15] Compare Pilz and Farley, *Ind. Eng. Chem.* **38,** 601 (1946).

[16] Lyons, *Trans. Electrochem. Soc.* **88,** 10 pp. (preprint) (1945). Silhan, *J. Electrodepositors' Tech. Soc.* **19,** 131 (1944). Kushner, *Metal Progress* **38,** 781 (1940). Hartshorn, *Metal Finishing* **38,** 476 (1940).

A considerable amount of study has been devoted to the effect of surface active agents in nickel-plating baths. The best deposits are said to be obtained in baths having the lowest surface tension, indicating that the beneficial effects are probably due primarily to a lowering of surface tension. No further useful effects are obtained by using more surface active agent than is required to lower surface tension to the minimum point. Those wetting agents which produce a precipitate in the plating bath are unsatisfactory.

Although in electroplating the wetting agents reduce the crystal size of the deposited nickel they do not give a true bright plate. A bright plate is one which requires no buffing and is produced by the addition of reagents which apparently cause an orientation of the deposited metal crystals. Although wetting agents may be used as brightening auxiliaries in bright-plating baths, their main function in electroplating is the suppression of pitting by releasing the bubbles of hydrogen which form at the cathode. The favored reagents for nickel-plating baths are those which reduce the surface tension to a very low value, are fully soluble in the bath, and are fully resistant to the chemical and electrolytic action of the cell. The fatty alkyl sulfates, alkyl aryl sulfonates, and Igepon have been most prominently mentioned.[17] Since plating baths are used for long periods of time and the same bath may be used for plating under different conditions, it is sometimes desirable selectively to remove the wetting agent from the plating bath. This may be done by adsorption on activated charcoal or clay.[18]

Surface active agents have also been recommended in the electrolytic baths used for anodizing aluminum.[19] In this important process, a highly resistant film of oxide is formed on the aluminum surface by making it the anode in an electrolytic cell.

Silver may be plated from acidic solutions of its nitrate or alkaline solutions of its cyanide. Taft and Hiebert have made an extensive study of the effect of various additives, including surface active agents, on plating from nitrate baths. They found that dibutyl sulfosuccinate, diglycol

---

[17] Young, *Proc. Am. Electroplaters' Soc.* (June) **1939**, 187. Davis, Wolfe, and France, *Ind. Eng. Chem.* **33**, 1546 (1941). Newark Branch of the A.E.S., *Proc. Am. Electroplaters' Soc.* (June) **1938**, 225. Stocker, *Monthly Rev. Am. Electroplaters' Soc.* **26**, 111 (1939). U. S. Pats. 2,195,409, Flett to *Allied Chemical and Dye Corp.*; 2,254,161, Waite and Martin to *McGean Chemical Co.* U. S. Pat. 2,207,989; Brit. Pats. 438,412; 528,498; Long to *Harshaw Chemical Co.* U. S. Pats. 2,389,179–81; Broun to *Udylite Corp.*

[18] Helbig, *Proc. Am. Electroplaters' Soc.* **1941**, 68. Canadian Pat. 406,030, Morgan to *Allied Chemical and Dye Corp.*

[19] French Pat. 788,874, to *Peintal S.A.*

stearate, and the ammonium soaps effectively decreased the crystal size of the deposit. There was no obvious relationship between the constitution of the wetting agent and the type of deposit obtained.[20]

In plating silver, as well as gold and copper, from alkaline solutions of the double cyanides addition of various surface active agents to the bath shortens or even eliminates the buffing operation otherwise required to produce a bright polished finish. It has been disclosed that this technically important purpose may be accomplished by adding soaps, sulfated oils, alkyl aryl sulfonates, or the betaine type of quaternary ammonium compounds to plating baths. High-silica sodium silicates such as ordinary water glass are also stated to be effective.[21]

In the plating of tin from acid solutions, only those surface active agents usually considered to be less powerful are said to be effective. These include some of the proteins, sulfonated cresols, and lower fatty alkyl sulfates.[22]

IV. Cutting Oils

By far the largest poundage of surface active agents used by the metal fabricating industry goes into the manufacture of so-called soluble cutting oils. These are used as lubricants and coolants in a great many machining operations and, as the name suggests, are applied in the form of an aqueous emulsion. For the heavier types of machine work, however, it is customary to use the non-aqueous type of cutting oil. All good cutting oils must form extremely tenacious adherent films on the metal to be tooled. Special ingredients which are added to impart this property, can rightfully be regarded as non-aqueous surface active agents. They are most often formed *in situ* by sulfurizing an appropriate petroleum oil. The sulfur compounds which are formed in this process act as powerful film formers and enable the oil to maintain contact with the metal in spite of extreme pressures at the cutting zone. Other additives designed to increase the "oiliness" or "load bearing capacity," as this property is variously called, are sometimes added. Among them are lard oil, fatty acids, alkyl phosphates, phenolates, etc. These will be considered in greater detail in the section on lubrication. The soluble cutting oils contain in addition an oil-soluble O/W emulsifying agent in sufficient quantity so that the clear

[20] Taft and Hiebert, *Trans. Kansas Acad. Sci.* **46**, 142 (1943).

[21] Brit. Pat. 450,979; U. S. Pat. 2,113,517, Powell and Davies to *Johnson, Matthey and Co.* Brit. Pat. 443,428, to Max Schlötter. U. S. Pat. 2,255,057, Holt to *E. I. du Pont de Nemours & Co.*

[22] Brit. Pat. 528,240; Canadian Pat. 401,204; to John S. Nachtman. French Pat. 790,884, to Max Schlötter. Taft and Fincke, *Trans. Kansas Acad Sci.* **45**, 173 (1942).

oil will instantly form a stable milky emulsion when added to water. In this respect they resemble the oils used for the emulsion degreasing of metals.[23]

The usual emulsifying agents for soluble cutting oils include the alkylolamine soaps of fatty acids, rosin acids, and naphthenic acids, the mahogany sulfonates, and the sulfated oils. These are most often used in mixtures rather than singly, and the mahogany sulfonates are probably used in larger quantities than the other products mentioned. In addition to the emulsifying agent and the extreme-pressure agent there is usually added a coupling agent or mutual solvent to aid in solubilizing the additives. For this purpose the glycol ethers, terpene solvents, butyl and amyl alcohols, etc. are usually used. Anti-foaming agents such as the ester waxes are often added to suppress foaming during actual use.[24]

Aside from the emulsifying agents mentioned above a number of other products have been used. These include the Igepon A class of sulfonates, the soaps of carboxylic acids from oxidized paraffin wax, and the polyethylene oxide type of non-ionic agent.[25] Two noteworthy products were developed by the Germans during the war for this purpose. The first was an alkyl xylene sulfonate potassium salt with a C_{12} side chain derived from propene tetramer. The second and more highly regarded product, called Emulphor STH was an alkyl sulfonylaminoacetic acid of formula:

$$RSO_2—NH—CH_2COONa$$

This was made from a Reed reaction sulfonyl chloride by first converting it to the amide and then condensing the latter with chloroacetic acid.

[23] Miller, *J. Inst. Petroleum Technol.* **24,** 645 (1938). Halls, *Ind. Chemist.* **10,** 45 (1934). Oldacre, *Mech. Eng.* **68,** 339 (1946). For typical disclosures on the newer non-aqueous cutting oils see: U. S. Pats. 2,285,853–5, Downing, Benning, and Johnson to *E. I. du Pont de Nemours & Co.*; 2,254,756, Segessemann to *National Oil Products Co.*; 2,296,037, Kaufman to *Texas Co.*

[24] For typical recent disclosures on soluble cutting oils see: U. S. Pats. 2,396,718, Moscowitz to *L. Sonneborn Sons*; 2,338,522, Liberthson to *L. Sonneborn Sons*; 2,391,-087, Donlan and Gathman to *Standard Oil Development Co.*; 2,265,799, Carlson and Cyphers to *Standard Oil Development Co.*; 2,231,214, Nelson and Langer to *Texas Co.*; 2,211,250, Anderson and Moir to *Pure Oil Co.*; 2,145,239, Flaxman to *Union Oil Co. of Calif.*; 2,289,536, Bradley to *Union Oil Co. of Calif.*; 2,214,634, Dombrow to *National Oil Products Co.*; 2,036,470, Gay to *Standard Oil Co. of Calif.*; 2,355,994, Morgan and Lowe to *Cities Service Oil Co.* Ger. Pat. 708,331, to *Standard Oil Development Co.* Brit. Pats. 557,755, to *C. C. Wakefield and Co. Ltd.*; 468,688 to Herrmann and Roehner. French Pat. 788,858, to *Standard Oil Development Co.* Canadian Pat. 362,261, to *General Chemical Co.*

[25] U. S. Pats. 2,231,228, to F. Singer; 2,043,922, Burwell and Kempe to *Alox Corp.*

Emulphor STH is said to have excellent rust-preventive and load-bearing characteristics as well as emulsifying properties.   Products of this type have not been used to date in this country, so far as the writers are aware.

V. Miscellaneous Uses

The lubricating properties of wire-drawing and cold-rolling lubricants have been improved by addition of alkyl aryl sulfonates and mahogany sulfonates.[26]   A more extensive use of such surface active agents (and also of the sulfated oils and fatty alkyl sulfates) is their application together with conventional components, *viz.*, solvents and finely divided abrasive solids, in formulating polishing and buffing compositions.[27]   Soaps have been used as protective temporary coatings during the hot working of magnesium, and as ingredients in wire-lacquering compositions.[28]   The acid-resistant water-soluble wetting agents have also been used as ingredients in soldering fluxes.   They are stated to improve the wetting and spreading of the molten solder, *i.e.*, the fluxing action itself.[29]

[26] U. S. Pats. 1,789,054, Sharp to *Standard Oil Co. of Indiana*; 2,089,883, Ellsworth to *American Steel and Wire Co. of N. J.*

[27] U. S. Pats. 2,196,992, to E. W. Keller; 1,986,388, Calcott and Clarkson to *E. I. du Pont de Nemours & Co.*; 2,032,482, to S. M. Hull.

[28] U. S. Pats. 2,229,236, Beck, Siebel, and Nachtigall to *Magnesium Development Corp.*; 2,216,376, Rimmel to *M. B. Suydam Co.*

[29] Canadian Pat. 356,079, to C. L. Boyle.

CHAPTER 23

Paints, Lacquers, Inks, and Pigments

Paints, lacquers, and similar protective or decorative coating compositions generally consist of a solid finely divided pigment suspended in a
liquid vehicle. Since the area of contact between pigment and vehicle is
extremely high as compared with the mass of the whole mixture, the interfacial relationships are of great importance. It is not surprising, therefore,
that surface active agents are widely used in the technology of coatings.
In this field, the oil-soluble, water-insoluble surface active agents are used
more widely than other types, since most paints, enamels, lacquers, and
printing inks and similar pigmented products are non-aqueous systems.
The water-soluble detergents and wetting agents, however, are also used for
several applications in the widely diversified operations of this industry.

One of the most important uses of the water-soluble types is in the
formulation of water paints and emulsion paints. Most of the so-called
water paints used today are really O/W emulsions containing at least a
moderate proportion of drying oil. There is relatively little true water
paint used any more. Emulsion paints may be of two types, O/W or W/O,
of which the first type is by far the most prevalent. This type may be
cheaply and conveniently thinned with water, thus avoiding both the
expense and the hazards of the ordinary solvent thinners. In O/W emulsion paints the pigment is usually not contained within the oil droplets
but is wet by the water and separately suspended in the water. In other
words, the droplets of oil and the particles of pigment exist side by side
in the aqueous phase. When the paint is applied the water evaporates
and the oil droplets coalesce to form a film, which wets and encloses the
pigment particles. The film so formed undergoes a final drying phase
similar to the drying of a film of conventional oil paint.

Emulsified paints are commonly formulated to contain soaps of fatty acids
with the volatile amines, such as morpholine. With such emulsifying
agents a protective colloid such as casein is almost always used. In addition to these ingredients, an oil-soluble non-ionic surface active agent such
as glycol or glycerol monostearate is often used to improve the brushing-
out qualities of the paint and to improve the emulsion stability. A strong
wetting agent of the sulfate or sulfonate class is often added to insure

complete wetting and dispersion of the pigment. The pigments for O/W emulsion paints must in themselves be carefully chosen. We have seen that finely divided solids are powerful emulsifying agents and may favor either type of emulsion according to the hydrophilic or hydrophobic character of the pigment surface. In the formulation of O/W emulsion paints, pigments which promote W/O emulsification should be avoided. Bentonite, a powerful O/W emulsifier, is often added in considerable quantities to these compositions.

Those alkyd type resins which are soluble in aqueous alkaline solutions, i.e., those which have a high acid number and a relatively low degree of polymerization, have been used as emulsifying agents. The quaternary ammonium compounds such as cetyltrimethylammonium chloride have also been prominently mentioned for this purpose.[1]

Although pigmented W/O emulsions are seldom used as paints, they are finding wide application in textile printing, not only because of their lower cost in comparison with the usual aqueous dyestuff compositions, but also because of their ability to produce sharp, light–fast prints. Pigmented W/O emulsions used in textile printing are usually formulated with an alkyd–urea–formaldehyde resin as the principle ingredient of the vehicle, which must dry quickly to form a flexible adherent film. The purpose of emulsifying water into such products is to impart workable flow and viscosity characteristics. The vehicle itself acts as a W/O emulsifying agent, but special agents of the hydrophobic type are sometimes added.

Most printing inks are non-aqueous in nature, and consist of a pigmented oily or resinous vehicle with relatively little thinner present. Aqueous printing inks, however, have been described as particularly suitable for high-speed intaglio printing. These products use a sulfated vegetable oil as the vehicle, and water-insoluble, oil-soluble coal tar dyestuffs as the pigments.[2] Wetting agents of the sulfate or sulfonate type have also been used in writing inks to improve the dispersion and flow properties.

The water-insoluble quaternary ammonium wetting agents such as laurylpyridinium chloride have been added in small quantities to nitrocellulose lacquers. They are soluble in the lacquer solvents and are said to aid in forming dust-resistant surfaces.[3]

[1] Jones, *Paint Manuf.* **4**, 277 (1934). Chubb, *J. Oil Colour Chemists' Assoc.* **25**, 211 (1942). Butler, *Paint Oil Chem. Rev.* **105**, No. 1, 10 (1943). U. S. Pats. 2,245,100, Bernstein to *Interchemical Corp.*; 2,391,041, Stamberger.

[2] Hadert, *Paint Manuf.* **9**, 83 (1939). U. S. Pat. 2,170,198, Hadert to *Miehle Printing Press and Mfg. Co.*

[3] U. S. Pat. 2,357,458, Clough to *E. I. du Pont de Nemours & Co.*

Several operations in the preparation of pigments offer opportunities for the advantageous application of surface active agents. Practically all the organic lakes and toners as well as the synthetic inorganic pigments are prepared or processed in aqueous media, and are initially recovered as a pulp or paste. They may then be filtered, dried, and ground, or they may be transferred directly from the aqueous medium to the oil medium in which they are used. This transfer process is called flushing. Since flushing consists essentially in a preferential wetting of the pigment by the oil phase, it is apparent that only those pigments whose surface is relatively more hydrophobic and more oleophilic will flush. The process itself consists of adding the desired quantity of oil to the aqueous pigment paste, and kneading or working the mass thoroughly. The pigment passes into the oil, while the water separates as a well-defined layer that may be poured or drawn off. Some pigments flush well without any added reagents. It is often necessary, however, to add a surface active substance which will be adsorbed on the pigment surface and make it more hydrophobic. Certain metal soaps or sulfonates precipitated *in situ* will accomplish this effect. The cationic long-chain quaternary ammonium salts are also said to be very effective flushing agents for a wide variety of pigments.[4]

In the preparation of the aqueous pigment pulps themselves, as well as in the dry grinding of pigments intended for water dispersion, surface active agents are often used. Their main purpose in these applications is the improvement of the degree of dispersion. The naphthalene sulfonates, proteins, lignin sulfonates, fatty alkyl sulfates, and quaternary ammonium compounds have all been used for this purpose.[5] Pigments are often coated with an oleophilic layer in order to make them more readily compatible with the vehicle in which they are to be used. Fatty acids and their soaps are the most commonly used agents for this purpose.[6]

The grinding of dry pigments into the oily or oleoresinous vehicle is one of the most important and troublesome steps in the manufacture of coating materials. The physical chemistry of this process has been intensively studied from both the practical and theoretical points of view, and surface active agents have found greater use here than in any other single process

---

[4] Brit. Pat. 469,559, to *United Color and Pigment Co.* Brit. Pat. 517,476; U. S. Pat. 2,219,395; Moilliet to *Imperial Chemical Industries Ltd.* U. S. Pats. 2,192,953–5, Sloan and Patterson to *E. I. du Pont de Nemours & Co.*; 2,238,275, Martone to *E. I. du Pont de Nemours & Co.*

[5] U. S. Pats. 2,160,119, Broderson to *General Aniline and Film Corp.*; 2,212,629, Alessandroni to *National Lead Co.*; 2,378,283, Bucher to *Ciba Co.*

[6] Compare Canadian Pat. 314,017, Endres to *Goodyear Tire and Rubber Co.* Park and Morris, *Ind. Eng. Chem.* **27**, 582 (1935). U. S. Pats. 2,245,104, Greubel to *Interchemical Corp.*; 2,067,060, Minor to *Industrial Process Corp.*

of the surface-coating industry. Dry pigment, as received at the grinding mill, consists of small, non-uniform agglomerates of the ultimate pigment particles. In the finished oil dispersion, the ultimate particles, which are at most a few microns thick and may range down to fractions of a micron, should be well separated, dispersed, and suspended. During the grinding operation the agglomerates are broken up, and the ultimate particles are thoroughly wetted by the vehicle. It is apparent that when the pigment is easily wet the agglomerates will be more easily disintegrated and the whole process will take place in much shorter time. The degree of dispersion, which is also markedly affected by surface active agents, has a strong influence on the rheological properties of the final paint as well as on its covering power. Specifically, the degree of dispersion can be correlated with the plastic viscosity, yield point, sedimentation rate, and sedimentation volume of the suspension.[7]

The so-called pigment-wetting and -dispersing agents used in paints are for the most part oil soluble and water insoluble. In particular, the heavy-metal soaps of stearic, naphthenic, and rosin acids, especially the zinc salt, have proved to be generally effective. Lecithin and the oil-soluble non-ionic agents of the glycol stearate type have also been used in pigment grinding. In spite of their water solubility the aqueous surface active agents such as sodium lauryl sulfate and the alkyl aryl sulfonates have been found useful in special cases, but the oil-soluble products appear to be preferred for most grinding applications. There appears to be no general relationship between the chemical and physical properties of the wetting agents and their effects on various pigments. The optimum wetting agent for each pigment–oil combination must be discovered by empirical methods.[8]

[7] Fischer and Jerome, *Ind. Eng. Chem.* **35,** 336 (1943). Carver, *Paint Manuf.* **14,** 45 (1944). Bartlett, *Paint Ind. Mag.* **60,** 48 (1945). Carr, *Paint Manuf.* **15,** 123 (1945). Burton, *Am. Ink Maker* **15,** No. 3, 21 (1937). Selden and Selker, *Official Digest Federation Paint & Varnish Production Clubs,* No. 195, 194 (1940).

[8] Young, *Paint Ind. Mag.* **57,** 336 (1942). Demant and Petzold, *Oil Colour Trades J.* **95,** 1279 (1939). Allen *et al., Am. Paint J.* **19,** Convention Daily, 14 (1935). Detroit Paint and Varnish Production Club, *Am. Paint J.* **25,** Convention Daily, 19 (1940). St. Louis Paint and Varnish Production Club, *Paint Oil and Chem. Rev.* **96,** No. 22, 87 (1934). U. S. Pats. 2,236,296, Minich and Levinson to *Nuodex Products Co.;* 2,296,382, Fischer, Gluck and Reynolds to *Interchemical Corp.;* 2,347,633, Kosolapoff to *Monsanto Chemical Co.*

CHAPTER 24

Leather Technology

The most important basic processes in the conversion of hides to leather include curing, soaking, fleshing, liming and depilation, degreasing, bating, pickling, tanning, fat-liquoring or oiling, and finishing. These operations are usually carried out in the order named, but not all of them are necessary for any single type of leather. In at least two of these processes, degreasing and fat-liquoring, surface active agents are of considerable importance. They find more or less limited use in most of the other processes with the exception of curing.

The freshly flayed skins are usually cured at the slaughterhouse by salting and are received by the tannery in the raw salted state. Goat skins and some other hides from foreign sources are merely dried instead of being salted. Whether the hides are dried or salt-cured the first stage in processing is a soaking operation. This serves to remove salt, if present, as well as water-soluble substances such as serum, etc. Wetting agents are sometimes added in the soaking of dried skins to speed up the penetration and aid in the removal or solution of encrusted material. They are not normally used, however, in the soaking of salt cured hides.

After the soaking operation heavier skins, such as cow hide, horsehide and calf skin, are fleshed. This is a mechanical operation in which the adherent adipose and connective tissue on the inner or flesh side of the skin is removed by rotating knives. Thick hides of this type are sufficiently cleaned of fatty matter in this operation and do not require any subsequent degreasing. After fleshing they are ready to be unhaired. The lime and alkalis used in unhairing remove any residual fatty matter from the skins.

Sheep skins, pig skins, reptile skins, and furs are usually degreased by washing rather than by fleshing. The degreasing process is ordinarily carried out just after unhairing, but may be carried out in other stages before bating and tanning. The skins may be degreased in aqueous baths containing a relatively high percentage of detergent. Another method of treatment is the application of fat solvents to the skin and the subsequent washing out of solvent and dissolved grease with the detergent solution. In either event, the detergents should be good emulsifying agents, should function well in the presence of lime, or salts, and should be effective over a wide pH range. The fatty alkyl sulfates, the alkyl aryl sulfonates, and, more recently, the water-soluble polyethyleneoxide non-ionic agents have proved satisfactory in this connection.[1]

Hair is removed from the hides by treating with alkaline reagents such as lime, methylamines and/or sodium or calcium sulfide solutions. Wetting agents may be added to the bath to speed the penetration of the depilating agent but they are usually not considered necessary.[2]

Bating is essentially a process of treating the leather with a mildly proteolytic enzyme. In modern practice clean, purified enzyme preparations are used almost exclusively. Wetting agents such as the sulfated fatty esters and sulfated fatty alcohols are sometimes used to promote penetration of the bating compositions, but are apparently not of primary importance in this process.[3]

The tanning process itself may be carried out in a number of different ways, and with reagents so apparently unrelated as chromium sulfate, formaldehyde, vegetable tannins, etc. Although tanning is a highly complex physicochemical process it involves essentially a physical change in the nature of the skin protein (coagulation and denaturation) plus a physical or chemical combination of the protein with the tanning agent. There is a whole group of tanning agents, generally called "the synthetic tanning agents" or "syntans," which have become increasingly important in the past two decades and which resemble the aqueous surface active agents in many superficial respects. It is fairly well established, however, that whatever surface activity the syntans may possess is purely incidental to their tanning power. The synthetic tanning agents are complex aromatic sulfonates in which several aromatic rings may be linked together with methylene bridges or sulfide links. Such products invariably contain phenolic hydroxyl groups and/or naphthalene nuclei, and are generally made by combining a phenol (complex phenolic compounds such as lignins may be used) with formaldehyde and a sulfonating agent. Sulfur or sulfur chloride may be used in place of formaldehyde to link the aromatic nuclei and increase the molecular weight.[4] Many of the aqueous detergents con-

---

[1] Blancher, *J. Intern. Soc. Leather Trades' Chemists* **22**, 425 (1938). Smith, *Hide Leather Shoes* **111**, No. 18, 15 (1946). U. S. Pats. 1,844,769, to Isermann and Kalscheuer; 2,283,421, to F. C. Castendyk. Ger. Pats. 696,735, to *Böhme Fettchemie G.m.b.H.*; 645,511, to *Tanning Process Co.*

[2] Ger. Pat. 673,649, to *Studiengesellschaft der deutschen Lederindustrie.*

[3] Brit. Pat. 491,801, to Otto Röhm. Smith, *Chem. Met. Eng.* **46**, 72 (1939).

[4] For discussions and typical disclosures on synthetic tanning agents see: Cheshire, *J. Intern. Soc. Leather Trades' Chemists* **27**, 123 (1943). Thuau, *Cuir tech.* **30**, 160

taining sulfate or sulfonate groups, particularly the alkyl phenol and alkyl naphthalene sulfonates, have a certain degree of tanning action but it does not approach that of the better synthetic tanning agents. The syntans are often added to vegetable tanning baths where they promote penetration and speed the tanning process, functioning both as surface active agents and tanning agents. The syntans and the alkyl aryl sulfonate detergents accordingly overlap in function as well as in molecular structure.

The sulfate and sulfonate wetting agents are often used as an adjunct in vegetable tanning liquors to promote their penetration. Wetting agents also exert a dispersing, solubilizing action on the tannins, thus counteracting the latter's tendency to precipitate and sediment out during the long course of the tanning process. Wetting agents are often used as assistants in chrome and alum tannages, particularly in the production of white leather with alum and formaldehyde.[5]

After the hide is tanned, oils must be worked into it in order to impart flexibility, fullness, and softness. If the oil is applied in the form of an aqueous emulsion the process is called fat-liquoring. Otherwise, it is called oiling or stuffing. In fat-liquoring the choice of both oil and emulsifying agent has an important bearing on the quality of leather produced. The oils most generally used are cod liver oil, neat's-foot oil, and tallow. Moellon and degras (spent cod oils from the oil tannage process, not to be confused with wool grease which is also called degras) as well as some synthetic higher esters are also used.[6] Mineral oils, in spite of their low cost, are not used, since they do not impart the desired properties to the leather.

The emulsifying agents used in fat-liquoring are typical water-soluble surface active products which yield O/W emulsions. The soaps which were formerly extensively used have largely been supplanted by the sulfated fatty oils. The petroleum mahogany sulfonates and the alkyl aromatic

---

(1942). Turley, *J. Am. Leather Chemists' Assoc.* **40**, 58 (1945). U. S. Pats. 2,136,997, Asch, Janz, Penner, and Voss to *I. G. Farbenindustrie*; 2,038,529, Schütte and Alles to *I. G. Farbenindustrie*; 2,099,717, Alles to *I. G. Farbenindustrie*. Ger. Pat. 631,017; French Pats. 794,078; 813,521; Brit. Pats. 490,296; 491,817; to *I. G. Farbenindustrie*. U. S. Pats. 2,289,278, Voss and Selle to *General Aniline and Film Corp.*; 2,282,264, Stiasny and Schuette to *General Aniline and Film Corp.*; 2,282,536, Swain and Adams to *American Cyanamid Co.*; 2,191,818, Stiasny to *Rohm and Haas Co.*; 2,220,867, Kirk to *E. I. du Pont de Nemours & Co.* Canadian Pat. 424,386, Robinson and Beach to *National Oil Products Co.*

[5] U. S. Pat. 2,226,579, to G. R. Pensel.

[6] Schindler, *J. Am. Leather Chemists' Assoc.* **34**, 414 (1939). Herfeld *et al.*, *Collegium* **1942**, 81; **1943**, 176. U. S. Pat. 1,949,990, Brodersen and Wagner to *I. G. Farbenindustrie*. Brit. Pat. 490,714, Jaeger to *American Cyanamid Co.*

sulfonates are also widely used in fat-liquoring.   Among other emulsifying agents which have been used for the same process are: (*1*) the fatty alkyl sulfates; (*2*) the Igepons; (*3*) the sulfated polyethylene ethers of alkyl phenols; (*4*) phosphated oils and alcohols; and (*5*) the cationic detergents.[7]

Fat-liquoring, although it is a simple technical process, involves a rather complicated physicochemical mechanism.   The emulsifying agent and part of the oil apparently combine with the leather and are held by fairly strong adsorption or chemical forces.   The main portion of the raw oil is mechanically held in the interfibrillar spaces and is responsible for the actual lubrication.   The combined oil is stated to contribute to the stretch and elasticity of the leather.[8]

Surface active agents are extensively used in the dyeing of both leather and furs.   The sulfates and sulfonates of high surface activity and also the syntans are effective in increasing the penetration and evenness of dyeing.   The action is apparently similar to that which occurs in the case of wool, the surface active agent competing with the dye for positions on the protein molecule.   The quaternary ammonium, phosphonium, and sulfonium salts have also been used as leather dyeing assistants.[9]

Leather goods may be washed effectively with the same type of detergents used for emulsifying fat liquors.   If the washing is carried out in acid medium and in the presence of a tanning substance such as trivalent chromium ions, no loss of tanning substance is said to occur.   Emulsions of oils in the usual sulfated oil or alcohol type detergent are also used as dressings for leather and patent leather.[10]

[7] U. S. Pat. 1,883,042, to I. Somerville.

[8] MacLaughlin and Theis, *The Chemistry of Leather Manufacture*, Reinhold, New York, 1945.   Koppenhoefer and Retzsch, *J. Am. Leather Chemists' Assoc.* **35,** 78 (1940).   Nelles, *J. Intern. Soc. Leather Trades' Chemists* **23,** 211 (1939).   Stather and Lauffmann, *Collegium* **1932,** 391.   Merrill, *J. Am. Leather Chemists' Assoc.* **27,** 201 (1932).   Theis and Serfass, *J. Am. Leather Chemists' Assoc.* **31,** 120 (1936).   Newbury, *Oil and Soap* **17,** 43 (1940).

[9] Nestler and Royer, *J. Am. Leather Chemists' Assoc.* **40,** 40 (1945).   U. S. Pat. 2,044,099, Piggott and White to *Imperial Chemical Industries Ltd*.

[10] Lesser, *Soap Sanit. Chemicals* **20,** No. 9, 29, 131 (1944).   Brit. Pat. 495,082 to *I. G. Farbenindustrie*.   Ger. Pat. 693,373, to *Zschimmer and Schwarz Chemische Fabrik*.

CHAPTER 25

Petroleum and Lubricants

Surface active agents of both the aqueous and non-aqueous types are used in numerous operations of the petroleum industry, from the production of crude oil to the formulation of finished products. In this industry, the production of greases and lubricants probably consumes the largest poundage of surface active agents. The surface active agents used in greases and lubricating oils, however, are for the most part of the oil-soluble, water-insoluble type. The emulsifiable cutting oils, which have already been considered in the section on metal technology, utilize the water-soluble W/O emulsifying agents and form the most notable exception to this rule in the lubricant field.

One of the important uses of surface active agents is in breaking crude oil–water emulsions. When petroleum is pumped from the wells it quite frequently is mixed with water or naturally occurring brines in the form of a more or less stable emulsion. Before the oil can be refined it is necessary to break the emulsion and remove the salt and water. This can sometimes be accomplished by heating or even by cold settling. More often, however, a special reagent is added to break the emulsion. Since most naturally occurring petroleum–brine emulsions are usually of the W/O type, the emulsion breakers usually are water-soluble O/W emulsifying agents. These must be used in carefully controlled amounts, lest an excess cause reversal of the emulsion with formation of a stable W/O emulsion.

There are several hundred patents covering surface active compounds which can be used for breaking crude-oil emulsions and the number of actual compounds and compositions disclosed for this purpose runs into the thousands. They range from simple soaps, sulfates, and sulfonates to complex products of uncertain structure defined by the methods of preparation. They include anionic, cationic, and non-ionic agents, and even mixtures of the different major classes of surface active agents. They range in properties from low molecular, highly soluble compounds to resinous or polymeric products of low dispersibility.[1] The reason for

---

[1] By far the most prolific workers in the field of breaking petroleum emulsions are M. de Groote and his collaborators at *Tretolite Co.* and *Petrolite Corp.* The majority of patents on this subject have been granted to this group. A few of their typical

481

482 25. PETROLEUM AND LUBRICANTS

this diversity in compositions is not difficult to understand. Emulsions are delicately balanced systems. In order to break them efficiently and completely it is necessary to neutralize with great exactness the forces and agents responsible for their formation. Crude-oil emulsions are often quite complex in nature. They may have several different components and phases and may be of the dual or multiple type. Since emulsions from different oil fields or even from different wells in the same field will vary widely in stability and composition, there should be an optimum demulsifying composition for each naturally occurring emulsion. In actual practice empirical methods are used to ascertain which demulsifying composition will break difficultly separable crude emulsions effectively and economically. Some demulsifiers have proved to be generally effective whereas others are only effective on the particular crude emulsions for which they were developed.

Surface active agents have been used to a certain extent in other phases of crude-oil production. Various types have been mentioned as constituents of drilling muds and drilling fluids. Drilling mud is the suspension of finely divided minerals in water circulated down the hollow drill stem and up through the annular space between the drill and the wall of the hole. It serves as a hydrostatic seal to prevent blowouts when a deposit of oil or brine under high pressure is tapped. It also serves to carry the drill cuttings to the surface where they are allowed to settle out before the mud is recirculated. A good drilling mud must have high density and very low viscosity while it is being worked. These desirable properties may be imparted by adding various typical dispersing agents to the mud. Alkaline tannins such as a mixture of soda and quebracho extract are widely used and are very effective. More recently the higher polyphosphates of the Calgon type have been used. Anti-foaming agents such as glycol stearate are sometimes added to drilling muds to prevent the formation of gas bubbles, which are detrimental since they reduce the

---

disclosures, illustrating the wide variations in compositions which are possible, are: 1,943,815; 1,954,585; 1,961,963; 1,977,146; 1,978,227; 1,979,347; 1,984,633; 1,985,692; 1,985,720; 1,994,758; 2,023,976; 2,023,979; 2,025,803–4; 2,026,195; 2,034,941; 2,039,063; 2,050,923–7; 2,052,281; 2,153,744; 2,171,328; 2,246,842; 2,246,856; 2,262,357–8; 2,266,960; 2,273,181; 2,278,838; 2,281,419; 2,282,644. All U. S. Patent.

Among other patents in this field are: U. S. Pats. 1,858,186, Canfield to *Empire Oil and Refining Co.*; 1,860,562, to T. B. Wayne; 1,944,021 and 1,863,143, Walker to *Empire Oil and Refining Co.*; 1,980,118, Tyler to *E. I. du Pont de Nemours & Co.*; 1,981,292, Todd and Hailwood to *Imperial Chemical Industries Ltd.*; 2,153,560, Hershman to *Petro Chemical Co.*; 2,252,110, Armendt to *Standard Oil Development Co.*; 2,265,081, to G. L. Monighan; 2,355,778, Berger and Goodloe to *Socony-Vacuum Oil Co.* Brit. Pat. 493,221; Ger. Pat. 702,012; to *I. G. Farbenindustrie.*

density of the mass and impede circulation.[2] Drilling muds tend to be contaminated by clay, brines, etc. encountered during the drilling, and therefore they frequently must be fortified or reconstituted. Sodium silicate and protective colloids such as starch, casein, and gums are often added.

Non-aqueous drilling fluids have been described in which the suspending agent is a fatty acid soap.

The rate of flow of oil into a well, *i.e.*, the productivity of the well, is often impeded by formations of limestone, dolomite, and other acid-soluble rock. A widely used method for increasing the yield from such wells consists of injecting acids which will dissolve the rock and open up channels through which the oil can flow into the well. The most commonly used reagent is hydrochloric acid of about 15% strength. Inhibitors are normally added to prevent the acid from attacking the metal shaft of the well. Wetting agents of the conventional type resistant to the action of the acid, such as the alkyl aryl sulfonates, are often added to increase the penetrating action of the acid. Acid – wetting agent combinations have also been used to remove mud sheaths as well as solid lime rock from bores during drilling.[3]

Soap – hydrocarbon oil – oxygenated solvent mixtures have also been proposed for cleaning out solid waxes and other accumulations from petroleum pipe lines and well shafts.[4]

Still another use for surface active agents in crude-oil production is in flushing out oil from sand formations, where water or brine occurs together with the oil in a sandy matrix. The interfacial relationships among the three phases determine whether oil or water will flow more rapidly into the well. It has been found that reduction of the surface tension of the aqueous phase causes more oil and less water to flow through the sand. The fatty acid soaps, naphthenates, fatty alkyl sulfates, and alkyl aryl sulfonates have been used for this purpose.[5]

The surface active agents apparently find little application in the usual

---

[2] Garrison, *Am. Inst. Mining Met. Engrs. Tech. Pub.*, No. 1027 (1939). U. S. Pats. 1,999,766 to Lawton, Loomis, and Ambrose; 2,216,865, to Wayne; 2,297,660, Mazee to *Shell Development Co.*; 2,349,585, Bond and Botsford to *Pure Oil Co.*

[3] U. S. Pats. 2,204,580, de Groote to *Dow Chemical Co.*; 2,038,720, de Groote to *Tretolite Co.* Flood, *Petroleum Engr.* **12**, No. 1, 46 (1940). U. S. Pat. 2,234,790, Zacher to *Tide Water Associated Oil Co.* Morris, *Oil Gas J.* **35**, No. 48, 49 (1937).

[4] U. S. Pat. 2,356,254, Lehmann and Blair to *Petrolite Corp.*

[5] Plummer, Hunter, and Timmerman, *Oil Gas J.* **35**, No. 47, 42 (1937). Livingston, *Am. Inst. Mining Met. Engrs. Tech. Pub.*, No. 1526 (1942); No. 1001 (1938). U. S. Pat. 2,369,831, Jones and Flaxman to *Union Oil Co. of Calif.* Canadian Pat. 407,729, Moore and Cannon to *Standard Oil Development Co.*

refining processes for petroleum.   Sulfonated castor oil has been suggested as a dispersing agent in the aluminum chloride process for desulfuring oils.[6]

Large quantities of metallic soaps are used in making lubricating greases as well as jellied fuels for military flame throwers and fire bombs.   They can hardly be considered as surface active agents when used for this purpose.   Their ability to form gels when mixed with hydrocarbon oils depends on their solubility and crystallization behavior and not on the surface properties of their solutions.   The same considerations apply to the sulfonates, such as sulfonated rosin and mahogany sulfonates, which have been used in grease making.[7]

Hydrocarbon oils constitute the most widely used and most important type of lubricant.   Oil-soluble surface active agents as well as a variety of other addition agents are used with lubricating oils.   These addition agents may be classified as follows: (1) anti-oxidants and additives which inhibit sludge formation; (2) anti-corrosives, which prevent corrosion and fouling of the machine parts being lubricated; (3) detergents or sludge-suspending agents, which prevent the sludge from depositing and forming hard carbonaceous cakes; (4) anti-foaming agents; (5) extreme-pressure additives; (6) viscosity index improvers; (7) pour point improvers.   These functions often overlap one another and single additives may fulfill more than one purpose.   Thus the distinction between a sludge inhibitor and a sludge-dispersing agent cannot always be clearly drawn, although the first product may be an anti-oxidant and the second a dispersing agent.   Surface active properties are 'of predominant importance in the action of the dispersing agent, but of no importance in the anti-oxidant action.   Nevertheless, the operational results of both products, as observed on a test engine, may be very similar.   The extreme pressure additives are powerfully surface active, and exert their effect by virtue of being strongly adsorbed at the oil–metal interface.   Certain chemical structures which make good extreme pressure additives also are good pour point depressors, i.e., they inhibit the crystallization and separation of wax and reduce the low-temperature viscosity of the oil.   This relationship, however, may be considered fortuitous since surface activity is not of major importance in the lowering of pour point.   Corrosion and rust-inhibiting agents might be regarded as borderline cases, due to the fact that their adsorption at the oil–metal interface, although necessary to protect the metal surface, does not of itself

[6] U. S. Pat. 2,239,859, to Records and Louttit.

[7] U. S. Pats. 2,235,926, Hasselstrom to *G. and A. Laboratories, Inc.*; 2,358,939, Nelson, Moore, and Faust to *Texas Co.*; 2,265,791, Zimmer and Morway to *Standard Oil Development Co.*

ensure their effectiveness. Many compounds which are strongly adsorbed (*i.e.*, surface active) are not good corrosion inhibitors. Those additives which may be classed unequivocally as non-aqueous surface active agents include the extreme-pressure agents, the detergents, and the anti-foaming agents.[8]

The large number of compounds which have been proposed as extreme-pressure additives precludes any complete survey of the field in this book. Most of the successful products have the same general structural features as the aqueous surface active agents with the one exception that they are water insoluble. They are essentially molecules of chain-like structure in which one end is polar and the other end non-polar. The term "polar," in its broadest sense, means that the molecular grouping in question is surrounded by a relatively strong force field. This is usually considered to be coulombic in nature, but may be due to the stray forces associated with non-ionic chemical bonding. When metal surfaces are brought into contact with lubricating oils containing dissolved extreme-pressure additives, the latter are strongly adsorbed on the metal surfaces. In this way intimate metal-to-metal contact of adjacent metal surfaces is prevented and a lubricating film is maintained even under severe loading of such machine parts as bearings, gears, etc.

Free carboxylic acids and the fatty glyceride oils from which they are derived were among the first extreme-pressure additives to be used. One of the early compounds specially synthesized for this purpose was dichloro-stearic acid methyl ester. More modern developments include the higher trialkyl phosphates and products containing sulfide sulfur or metals. The actual mechanism by which extreme-pressure additives exert their effect is not completely understood but strong adsorption is certainly one of the most important factors in the process. The patent literature in this field is voluminous, but only relatively few of the products disclosed have found actual commercial use. Aside from favorable economic potentialities, a successful extreme-pressure additive must have the requisite stability to heat, pressure, and the catalytic effect of the metal surfaces being lubricated. It must also be soluble and non-corrosive and must not interfere with the performance characteristics of the oil in which it is used. Extreme-pressure additives are not generally used in automotive crankcase oils, although they may be used in the crankcases of heavy-duty Diesel

[8] Lincoln, Byrkit, and Nelson, *Oil and Soap* **17**, 73 (1940). Pritzker; *Natl. Petroleum News* **37**, R793 (1945). Strohmeier, *Power* **89**, No. 5, 99 (1945). Evans, *J. Inst. Petroleum* **29**, 333 (1943). Schnurman, *Chem. Products* **7**, 56 (1944). Musgrave, *Iron Steel Engr.* **20**, No. 5, 40 (1943).

engines or aircraft engines. They are used in hypoid-gear boxes, metal rolling machinery, and other places where good lubrication must be maintained under severe load conditions.[9]

The anti-rust and anti-corrosion additives for lubricating oils are quite often designed with long alkyl chains and invariably have a polar – non-polar structure. Like the corrosion inhibitors used in the aqueous pickling of steel, they must be adsorbed on the metal surface in order to exert their effect. Many of the substances described as anti-corrosive agents also function as extreme-pressure agents and vice versa. Among the types which have been mentioned are: heavy-metal salts of long-chain phthalamic or succinamic acids, higher fatty amine salts of phosphoric acid or alkyl acid phosphates, metal salts of alkyl thiophosphoric acids, heavy-metal salts of mahogany sulfonates, dialkyl phenol sulfides, higher amine salts of fatty acids, naphthenic acids, alkylated aromatic carboxy acids, and metal salts of alkylated phenols.[10]

Under service conditions, lubricating oils deteriorate by oxidation, pyrolysis, and/or polymerization to form insoluble gummy precipitates known as sludge. The sludge accumulates on the motor parts, particularly around the piston rings, and forms hard carbonaceous deposits which interfere with the efficient operation of the engines. The undesirable effects of sludge formation can be combatted in two ways, either by an additive which inhibits sludge formation, or by an additive which disperses the sludge after its formation and prevents it from forming hard deposits. The sludge inhibitors may act purely as anti-oxidants or they may act by poisoning the sludge-forming catalytic action of the metal surfaces. In the latter case the inhibitors must be adsorbed on the metal surface.

Conceptually there is a sharp distinction between a sludge inhibitor and a sludge-dispersing agent. Actually this distinction is often disregarded in both the patent and journal literature, because it is difficult to recognize in practice. A motor oil with an effective sludge inhibitor

[9] Beeck et al., Proc. Roy. Soc. London **A177**, 90 (1940). Iranth and Neyman, Petroleum Z. **31**, No. 49 (1935). Bowden, Gregory, and Tabor, Nature **156**, 97 (1945). U. S. Pats. 2,158,096, Werntz to E. I. du Pont de Nemours & Co.; 2,391,631, Kingerley to E. I. du Pont de Nemours & Co.; 2,366,539, McCleary and Fields to Texas Co.

[10] U. S. Pats. 2,378,443; 2,400,611; Smith, Cantrell, and Peters to Gulf Oil Co. U. S. Pats. 2,387,537; 2,371,854; 2,371,851; 2,371,656; Smith and Cantrell to Gulf Oil Co. U. S. Pats. 2,191,996, Shoemaker and Loane to Standard Oil Co. of Indiana; 2,125,305, Murphy to L. Sonneborn Sons; 2,125,934, Liberthson to L. Sonneborn Sons; 2,349,785, Faust to L. Sonneborn Sons; 2,374,559, Morris and McCleary to Texas Co.; 2,361,804, Wilson to Union Oil Co. of Calif.; 2,349,044, Jahn to Shell Development Co.; 2,366,013, Duncan to Standard Oil Development Co.; 2,366,074, Wasson and Wilson to Standard Oil Development Co.

should not become black and fouled in service, nor should it deposit sediment on the metal parts, *i.e.*, no decomposition products should be apparent. With a sludge-dispersing agent the metal parts should remain clean but the oil itself should become contaminated with a well-suspended black precipitate of insoluble material. Actually no such clean-cut distinction is encountered except in rare cases and the same types of compounds are used for sludge inhibitors as are used for sludge dispersants.

Among the polar – non-polar types of compounds which are said to be effective sludge inhibitors are alkyl thiophosphates, lecithin, heavy-metal salts of long-chain salicylic esters, long-chain alkyl maleates, and alkenyl succinic acids.[11] Insofar as they can be distinguished from the inhibitors, the sludge-dispersing agents, which have recently received wide and favorable attention, are excellent examples of true non-aqueous detergents. Their action in removing sludge from the metal surfaces (or preventing its deposition) and keeping it suspended in the liquid medium is strictly analogous to the action of aqueous detergents in washing soiled fabrics. These products are in fact commonly referred to as lubricating-oil detergents.[12] They have been used for several years in Diesel engine oils to prevent ring sticking and are now being introduced into automotive crankcase oils.

The most successful of the many compounds described as lubricating detergents are heavy-metal salts of carboxylic or sulfonic acids and metal alcoholates or phenolates. In every case the anionic constituent has a large hydrocarbon-soluble radical. The alkaline-earth metals are most widely used, particularly calcium. In some cases aluminum, zinc, cobalt, and other salts are used. Mahogany sulfonic acids, alkyl phosphoric acids, alkyl phenols, phthalic monoesters, alkyl aryl sulfonamides, long-chain phenolic ether alcohols, naphthenic acids, amino-substituted fatty acids, chlorinated fatty acids, and acids containing thioether groups are among the many anionic constituents which have been used. Among non-metallic sludge dispersants are the lecithins and various long-chain esters of polyhydric alcohols.[13]

---

[11] U. S. Pats. 2,342,431–3, Smith and Cantrell to *Gulf Oil Co.*; 2,366,817, Towne to *Texas Co.* U. S. Pats. 2,356,043; 2,347,546–7, Finley to *Sinclair Refining Co.* U. S. Pats. 2,349,817, Clayton and Etzler to *Standard Oil Co. of Calif.*; 2,292,308, Watkins to *Sinclair Refining Co.*

[12] Georgi, *Petroleum Refiner* **23**, 504 (1944). Talley and Larsen, *Ind. Eng. Chem., Anal. Ed.* **15**, 91 (1943).

[13] U. S. Pat. 2,395,105, Cook and Thomas to *American Cyanamid Co.* U. S. Pats. 2,372,955, Johnston; 2,371,333, Johnston; 2,351,380, White; 2,372,411, Van Ess and Bergstrom; 2,352,669, Van Ess and White; to *Shell Development Co.* U. S. Pats. 2,391,099, McNab, Wilson, and Winning to *Standard Oil Development Co.*; 2,234,005, Loane and Shoemaker to *Standard Oil Co. of Indiana*; 2,337,868, Burwell and Camel-

Foaming is highly detrimental in lubricants since it reduces the quantity of oil fed to the moving parts and therefore impedes proper lubricating action. Among the anti-foamers which have been described are sulfonated fatty oils and their amine salts, soaps such as potassium oleate, fluorinated hydrocarbons, and silicones.[14]

Surface active agents have been used to stabilize suspensions of normally insoluble additives in both fuels and lubricants. Thus sulfated fatty oils have been suggested for stabilizing tetraethyllead, and mineral oil sulfonates for stabilizing graphite in fuels. Sodium lauryl sulfate has been used to stabilize aryl amine anti-oxidants in lubricating oils.[15]

A final use for surface active agents in the petroleum industry is in the formulation of rinsing or flushing oils. These are light mineral oils used to flush crankcases or clean other engine parts which have become fouled with sludge. Oxygenated solvents are often used as additives in flushing oils but the mahogany sulfonates, fatty alkyl sulfates, and sulfated fatty oils have also been used for this purpose.[16]

ford to *Alox Corp.* U. S. Pats. 2,369,908, McCleary, and 2,345,156, Roberts, to *Texas Co.* U. S. Pats. 2,225,365, and 2,361,803, Wilson, to *Union Oil Co. of California.* U. S. Pat. 2,322,307, Neely and Kavanagh to *Standard Oil Co. of Calif.*

[14] U. S. Pats. 2,377,654, Smith to *Gulf Oil Corp.*; 2,353,587, Rill to *Socony-Vacuum Oil Co.*; 2,375,007, Larsen and Diamond to *Shell Development Co.* U. S. Pats. 2,394,596, Davis and Zimmer; 2,355,255, Zimmer and Seitz to *Standard Oil Development Co.*

[15] U. S. Pats. 2,323,670, Musselman to *Standard Oil Co. of Ohio*; 2,176,879, Bartell to *Acheson Colloids Corp.*; 2,292,352, Cook and Thomas to *American Cyanamid Co.*

[16] Brit. Pat. 436,044, to *Standard Oil Development Co.* U. S. Pats. 2,169,344, to C. S. Kimball; 2,355,591, Flaxman to *Union Oil Co. of Calif.*

CHAPTER 26

Ore Flotation

Aside from the textile-processing field, a wider variety of surface active agents has been used in the beneficiation of minerals than in any other single industry. The whole process of ore flotation depends on surface and interfacial phenomena in the aqueous medium employed. As the result of an enormous amount of empirical experimentation it is possible today to separate by flotation almost any naturally occurring combination of minerals, in spite of the fact that there is still considerable disagreement as to the fundamental mechanisms which are involved.

It is quite rare for valuable metal-bearing minerals to be found in massive formations. They occur for the most part as relatively small particles embedded in a matrix of more or less worthless rock, clay, sand, or other mineral matter. The whole conglomerate is generally known as ore, and its value depends on the percentage of the desired mineral present. Some ores are rich enough so that they can be smelted directly. With most ores, however, it is necessary to separate at least a part of the worthless material or gangue from the desired mineral before smelting. This process is known as ore dressing or concentration and may be accomplished in numerous ways. If the desired mineral is present in sufficiently large chunks it may be picked out by hand. There are also various mechanical devices such as screens, hydraulic classifiers, thickeners, magnetic separators, etc. by which ores may be enriched. Within recent years the flotation method for concentrating ores has been intensively developed and extended with the result that processing costs on many important ores have been considerably lowered. Exploitation of some mineral deposits would not be economically feasible if flotation methods were not used.

In order to render an ore amenable to flotation, it is necessary first to crush the whole ore, including the desired mineral and the gangue, to a relatively fine powder. The particle size of this powder should be such that the individual crystals of mineral are substantially detached from gangue and are small enough to be floatable by attached air bubbles. In practice this size will vary from about 30 mesh down to 200 mesh or less. Since grinding is a costly operation the coarser grinds are used wherever possible. The crushed ore is fed to a flotation cell where it is mixed or "pulped" with

water and the appropriate flotation reagents.   The flotation cell consists
of a tank fitted with an agitator and/or an air-blowing device which can
keep the pulp vigorously stirred and can form a supernatant foam or froth
of air bubbles.   Within a few minutes after starting a well-operated cell,
a mineral-laden foam is formed and can be skimmed off.   Ordinarily the
desired mineral is withdrawn with the foam, while the gangue is left behind
within the aqueous liquor in the cell.   However, in some operations the
gangue is floated and the valuable mineral left as a residue.   In any case,
the desired separation is effected by causing one mineral to become firmly
attached to the air bubbles of the foam, and thus to float to the top despite
its high density, while the other minerals do not become attached to air
bubbles and therefore remain either suspended or sedimented in the liquor.

A mineral particle becomes attached to an air bubble when its surface
is hydrophobic, and conversely a particle with a hydrophilic surface will
not become attached to an air bubble.   This fact is apparent from a theoretical consideration of the contact angle relationships at the mineral
surface–water–air boundary, and it has been verified experimentally in a
large number of cases.   Two types of bubble attachment have been observed in actual flotation practice.   In one type the mineral particle is
located in the liquid air interface with part of it actually projecting into the
air space.   This is the type of attachment which would be predicted from
theory.   In a second type of attachment the mineral particle is entirely
in the liquid but is contiguous to the bubble.   In this case it is probable
that the mineral is held to the adsorbed interfacial layer between the air
and the liquid.   In either case the floated mineral has a hydrophobic surface whereas the unfloated one is hydrophilic.

Practically all minerals as they occur in the natural state possess a hydrophilic surface.   It is the function of one class of flotation agents
selectively to convert the surface of the mineral to be floated from a hydrophilic to a hydrophobic state.   At the same time it must not alter the
hydrophilic surface of the gangue or mineral to be left in the liquor.
Reagents which can accomplish this effect are called promoters or collectors.
There has been considerable controversy over the actual mechanism by
which the collector exerts its amazing selective action.   One school con-
tends that in most cases there is a stoichiometric chemical reaction between
the collector and the mineral, occurring only on the surface layers of the
insoluble mineral and resulting in a hydrophobic compound.   The other
contention is that the collector is adsorbed and oriented on the mineral
surface to form a hydrophobic film.   With the feeling that such contro-
versies often boil down to a matter of semantics rather than of physical
chemistry we shall use the term "adsorption" to denote the interaction between the collector and the mineral.

The choice of a collector depends primarily on the mixture of minerals to be separated, *i.e.*, on the component minerals of the particular geological formation which is being mined. Collectors which have been used include almost the full range of polar – non-polar surface active agents as well as many water-soluble compounds which are not generally thought of as being surface active. Conspicuous among the latter are the xanthates of the lower alcohols, which are probably used in greater tonnage than any other collecting agents. Neutral hydrocarbon oils have also been used as collectors. Due to the insolubility of such oils in the water, they are applied with the aid of an emulsifying agent which may in itself aid the collecting action.

In general, the action of a collector may be hindered or retarded by the presence of secondary additives generally known as modifiers or conditioners. A conditioner which promotes collection is called an activator, while one which inhibits collection is called a depressor. Conditioners may act either on the surface of the mineral or they may indirectly influence the action of the primary collector. As might be expected, the *p*H of the pulp often is an important factor in collection, and acids or alkalis are often added to pulps to bring the *p*H to an optimum value for the particular collector being employed. Straight acids and alkalis are generally considered to be in a different class from other types of conditioners. Summarizing, we may regard collectors and conditioners together as forming one class of flotation reagents whose broad function is selectively to modify a mineral surface so that it becomes attachable to an air bubble.[1]

The second class of flotation reagents are the frothing agents. It is important in flotation practice that the froth have the correct characteristics. It must be stable enough to resist disruption by the circulating mineral particles and yet it should be easy to break down after removal from the flotation cell. It should not be too copious. The individual bubbles should be of optimum size to match both the chemical nature and particle size of the mineral being floated. It is preferable that the frothing agent should not be a collector under the conditions of use, although sometimes the same chemical may be used as a frother in one system and a collector in

---

[1] For detailed discussions of the theories of flotation and particularly of collector action see: Hassialis, *Ann. N. Y. Acad. Sci.* **46**, 495 (1946). Dean and Ambrose, *U. S. Bur. Mines Bull.*, No. 449 (1944). Sproule, *Can. Mining J.* **57**, 582 (1936). DeWitt, *Ind. Eng. Chem.* **32**, 652 (1940). Kellogg, *Mineral Ind.* **14**, No. 3, 1 (1944). Taggard and Arbiter, *Am. Inst. Mining Met. Engrs., Tech. Pub.*, No. 1585 (1943); No. 1685 (1944). Wark *et al.*, *Z. physik. Chem.* **A173**, 265 (1935). Taggart *et al.*, *Am. Inst. Mining Met. Engrs. Tech. Pub.*, No. 2022 (1946). *Trans. Am. Inst. Mining Met. Engrs.* **112**, 348 (1934). Freundlich, *Wetting and Detergency Symposium*, Feb. 19–20, 1937, *Brit. Sec. Intern. Soc. Leather Trades Chem.*, 1–10. Ostwald, *Z. physik. Chem.* **A173**, 393–400 (1935). Volkova *et al.*, *Kolloid-Z.* **72**, 82 (1935).

another.   Control problems are easier of solution if the operator, by use of separate agents, is able to control the frothing and the collecting independently of each other.   The well-known foaming surface active agents such as the fatty alkyl sulfates and alkyl aryl sulfonates are sometimes used as frothers but the most commonly used substances are pine oil, cresols, and the alcohols of medium molecular weight, usually C_5 to C_8.   These good frothers are effective in extremely low concentrations well within their solubility range.   A typical pulp may contain 2 tons of water, 1 ton of ore, and only 0.1 pound of frothing agent.

In view of the large number of mineral combinations which make up the various workable ores, there is almost an infinite number of frother–collector–conditioner combinations which may be used advantageously in separating valuable minerals from one another.   The number of flotation agents which have actually been used on a large scale is still relatively small when one considers the need for high selectivity.   It is probable that this number will be greatly increased as further exploratory work is carried out.   As existing mines become exhausted and new deposits are opened up it is expected that new and different flotation agents will come into use.   For instance, we have the current interest in enrichment of low-grade iron ores. The presently worked iron deposits in the U. S., notably the famous ranges along Lake Superior, consist of almost pure hematite; the ores can be used in the blast furnace directly as they come from the pit.   The threat that these mines may soon become exhausted has led to intensive research on economical methods for enriching the inexhaustable poorer deposits. At the present time, however, there are probably not more than a hundred reagents and reagent combinations which are used to any significant extent in flotation.[2]

Several processes for the separation of siliceous and calcareous gangue from hematite utilize the fatty acid soaps and lime as collectors for the silica, while the iron oxide is depressed with polyphosphates, lignin sulfonate, or quebracho.   This is an example of the anionic flotation of silica gangue in an alkaline circuit.   The fatty amines have also been used as collectors in this type of alkaline flotation, being introduced in the form of their hydrochlorides and liberated by the excess of alkali present in the pulp.[3]   Hematite has also been separated by flotation in an acid pulp of

---

[2] A useful compilation of flotation reagents and technical data has been published by *Denver Equipment Co.*, Denver, Colorado, as their Bulletin No. R 1-B.   Many of the reagents are identified by their trade name only but in most cases the composition is indicated.

[3] Clemmer *et al.*, *U. S. Bur. Mines Repts. Invest.*, No. 3799 (1945).   U. S. Pats. 2,383,467, Clemmer and Clemmons to Secretary of the Interior; 2,403,481, Clemmer and Rampacek to Secretary of the Interior.   Scott, Richardson, and Arbiter, *Am. Inst. Mining Met. Engrs. Tech. Pub.*, No. 1902 (1945).

*p*H 5. In this case an emulsion of oleic acid in Turkey-red oil and Monopole soap is used as the collector.[4] Both neutral oils and fatty acids are often used as collectors together with an emulsifying agent which facilitates their application. Among examples are: tall oil soap and mineral oil,[5] lower fatty acids (C_3 to C_{10}) and mineral oils,[6] sulfonated terpenephenols or sulfated fatty esters and mineral oil,[7] and sulfonated fatty oil with free fatty acid.[8] Molybdenite has been successfully floated using an emulsion of a neutral mineral oil with Syntex, a sulfated fatty monoglyceride, as the emulsifying agent.[9]

The long-chain fatty amines and quaternary ammonium salts have been widely used in the flotation of acidic minerals, particularly in the removal of siliceous gangue from non-metallic minerals such as phosphate rock. Numerous different cationic surface active agents have been described as useful for this purpose.[10] The cationic agents have also been used for the flotation of iron and manganese oxides,[11] tin-bearing minerals,[12] zinc and lead sulfides,[13] chalcopyrite, a copper iron sulfide,[14] micas,[15] and many other minerals.[16]

Numerous combinations of soluble surface active agents with insoluble, polar, oily materials have been disclosed as flotation agents. Typical of these are: a cationic surface active agent with a higher fatty nitrile,[17] a cationic agent with a partial fatty ester of a polyhydric alcohol,[18] a Ninol type detergent (*q.v.*) with a higher fatty acid,[19] ammonium tall oil soap with a lower aliphatic alcohol,[20] and a higher fatty alkyl sulfate with a higher fatty alcohol.[21]

[4] Bring, *Jernkontorets Ann.* **124,** 562 (1940); *Chem. Abstracts* **35,** 1735.

[5] U. S. Pat. 2,202,601, Ried to *Separation Process Co.*

[6] U. S. Pat. 2,125,852, Ralston and Pool to *Armour and Co.*

[7] U. S. Pat. 2,286,374, Ray to *Hercules Powder Co.*

[8] Brit. Pat. 515,606, to *F. L. Smidth and Co. Aktieselskab.*

[9] Cuthbertson, *Am. Inst. Mining Met. Engrs. Tech. Pub.*, No. 1675 (1944).

[10] Brit. Pats. 463,261, to *E. I. du Pont de Nemours & Co.*; 546,059, to *American Cyanamid Co.* U. S. Pats. 2,278,107; 2,312,387; 2,336,015; 2,364,272; Jayne *et al.* to *American Cyanamid Co.*

[11] Brit. Pat. 536,868, to *General Electric Co.*

[12] U. S. Pat. 2,381,662, to Gaudin.

[13] U. S. Pat. 2,287,274, Ralston and Segebrecht to *Armour and Co.*

[14] U. S. Pat. 2,289,996, Ralston and Segebrecht to *Armour and Co.*

[15] Norman and O'Meara, *U. S. Bur. Mines Repts. Invest.*, No. 3558 (1941).

[16] U. S. Pat. 2,221,485, Kirby and Gillson to *E. I. du Pont de Nemours & Co.* U. S. Pats. 2,177,985; 2,255,252, to Benj. R. Harris.

[17] U. S. Pat. 2,298,281, Corley, Ralston, and Segebrecht to *Armour and Co.*

[18] U. S. Pat. 2,362,432, Cahn to *Emulsol Corp.*

[19] U. S. Pat. 2,173,909, Kritchevsky to *Ninol, Inc.*

[20] U. S. Pat. 2,238,021, Jayne to *American Cyanamid Co.*

[21] U. S. Pat. 2,012,609, Lenher to *E. I. du Pont de Nemours & Co.*

Since the action of any individual collector can generally be drastically modified by the use of appropriate conditioners, many patents relating to flotation reagents do not specify any one ore for which the reagent is best adapted. Quite often, however, a new reagent is developed as the answer to a particular field problem. Such disclosures usually give detailed information on practical performance. The mahogany sulfonates have been described as suitable for oxidized metallic sulfide ores as well as some non-metallic ores.[22] The long-chain alkyl phenol sulfonates are claimed to be excellent frothing agents in the flotation of galena.[23] The titanium ore, ilmenite, has been separated by a two-stage flotation process using a cationic surface active agent made from β-chloroethyl oleate and thiourea in the first stage and using oleic acid in the second stage.[24] Lignin sulfonate has been described specifically as a depressor for carbonaceous material in the flotation of various minerals.[25] Oleic acid and tall oil acids have been found to be the most effective collectors for the aluminum ores, gibbsite and bauxite.[26]

A very interesting application of flotation methods is in the concentration of potassium chloride from its naturally occurring mixtures with sodium chloride and/or other salts such as polyhalite (a mixed calcium magnesium potassium sulfate). The pulping liquor consists of a brine saturated with respect to the soluble constituents. The flotation agent varies depending on whether the sylvinite (potassium chloride) is to be floated or left as a residue. In one case a carboxylic acid soap or sodium lauryl sulfate is used together with a soluble lead salt such as lead nitrate as a conditioner. Alternatively a cationic collector such as lauryl amine hydrochloride is used. The octyl and dodecyl esters of glycine are claimed to be very efficient collectors for separating potassium and sodium chlorides.[27]

Wetting agents have also been used in a few of the metal-winning processes other than flotation. As an example, copper may be recovered from solutions of its sulfate by adding iron powder. If an effective wetting agent is present the process is said to be considerably more efficient.[28]

[22] U. S. Pats. 2,310,240; 2,373,688, to Keck.

[23] U. S. Pat. 2,230,565, Gaylor to *Standard Oil Development Co.*

[24] U. S. Pat. 2,387,856, Pickens to *American Cyanamid Co.*

[25] Brit. Pat. 515,592, to *F. L. Smidth and Co. Aktieselskab.*

[26] Clemmer, Clemmons, and Stacy, *U. S. Bur. Mines Repts. Invest.*, No. 3586 (1941).

[27] U. S. Pats. 2,222,330, Weinig to *Potash Co. of America*; 2,382,360, Weiner to *Bonneville Ltd.* French Pats. 814,282; 814,526; Brit. Pat. 482,579; to *Potash Co. of America.*

[28] U. S. Pat. 2,245,217, Mowlds to *Glidden Co.*

CHAPTER 27

Other Industries Utilizing a Variety of Surface Active Agents

I. Agricultural and Insecticidal Uses

The surface active agents are used in agriculture for two major purposes. The more important of these is as an ingredient in a wide variety of spraying compositions for combating insects, fungi, and other enemies of plant life. The second use is in washing fruits and vegetables before marketing in order to remove hazardous spray residues.

There are several ways in which surface active agents may function when used in parasiticidal compositions. When the effective toxicant is an oily water-insoluble liquid it is usually employed in emulsion form and a good emulsifier is needed. If the toxicant is an insoluble powder, such as sulfur, aqueous suspension is very often the most effective and convenient form of application. This type of insecticide composition often is formulated to include a surface active agent because of the desirable effects resulting from improved suspension and dispersion of the toxicant. When the toxicant is water soluble, a wetting or spreading agent may be added to insure more even coverage. In all cases the surface active agent should help (or at least not hinder) the deposition and retention of the killing ingredient. Certain surface active compounds have been found to be in themselves toxic to plant parasites or synergistic with non-surface active toxicants.

Mineral oil emulsions are among the most widely used insecticidal sprays. A wide variety of water-soluble emulsifying agents and oil-soluble emulsifying auxiliaries have been used in these oil-based sprays. Soaps, alkyl aryl sulfonates, alkyl phenol sulfonates, lignin sulfonates, the fatty alkyl sulfates and their amine salts, mahogany sulfonates, sulfated oils, and naphthenic soaps are included in a long list of primary emulsifiers.[1] Bentonite, fatty alcohols, and the partial fatty esters of polyhydric alcohols are typical of the products used as auxiliaries. The latter compounds, as well

[1] U. S. Pats. 2,243,254, to Linstaedt; 2,210,964, Tuennermann to *Dow Chemical Co.*; 2,114,125, Kaufmann to *Tide Water Associated Oil Co.* Martin, *Chem. Trade J.* **95,** 385 (1934). Canadian Pat. 405,964, Hansberry to *Shell Development Co.* Ginsburg, *N. J. Agr. Expt. Sta. Circ.*, No. 382 (1939). U. S. Pat. 2,010,443, Sibley to *Rubber Service Laboratories Co.*

as being emulsifying aids,[2] have strong insecticidal powers of their own. One of the special properties which a good insecticidal emulsifier should possess is the ability to aid in the spreading, sticking, and deposition of the oil or other toxicant on leaf and plant surfaces.   In general, a surface active agent which imparts high spreading and wetting powers to an insecticidal emulsion tends also to impart poor sticking power to the emulsion, i.e., less oil will be deposited on the plant from the same quantity of emulsion if the more powerful wetting agents are used as emulsifiers. Agents such as gum arabic and casein aid deposition.   Bentonite, the amine soaps, and the fatty esters of pentaerythritol are also said to be effective in increasing the deposit.[3]   The deposition of oil is influenced by the state of the emulsion as well as by the wetting power of the external phase. Emulsions having large oil droplets or clusters of droplets and emulsions which are near the breaking or inversion point by virtue of their composition give better deposits than the more stable emulsions.   Another method of aiding deposition consists of decreasing the negative charge on the oil particles or charging them positively by the addition of adsorbable cations. Since wet foliage is usually negatively charged, it will accordingly attract and coagulate the oil.[4]

In a great many insecticidal emulsions a more highly active toxicant such as a copper compound, nicotine, pyrethrum, etc. is incorporated.

The spreading power of insecticidal oils intended for use without being emulsified has been improved in some instances by addition of small amounts of surface active agents such as sulfonated castor oil, cresols, mahogany sulfonates, etc.[5]

A number of the more important insecticides are solid materials.   They may be applied by dusting or, in the form of an aqueous dispersion, by spraying.   The surface active materials used in forming the dispersions must perform the same general functions required of the analogous emulsi-

[2] U. S. Pats. 2,091,062, Yates to *Shell Development Co.*; 1,949,798, Knight *et al.* to *Emulsoids, Inc.*; 2,258,832, Weitkamp to *Standard Oil Co. of Indiana*; 1,949,799, Knight to *Standard Oil Co. of Indiana*; 2,264,761, Knight to *Shell Development Co.*; 2,327,105, Guy to *E. I. du Pont de Nemours & Co.*; 2,244,685, Fritz and Beach to *National Oil Products Co.*

[3] U. S. Pat. 2,369,429, Boïssonou to *Shell Development Co.*   Dawsey, *U. S. Dept. Agr., Circ.*, No. 568 (1940).   Fajans and Martin, *J. Pomol. Hort. Sci.* **15**, 1 (1937). Smith, *J. Econ. Entomol.* **24**, 985 (1931).   Pearce, Avens, and Chapman, *ibid.* **34**, 202 (1941).

[4] U. S. Pats. 2,362,760, Maxwell to *Shell Development Co.*; 2,190,173, Knight to *Shell Development Co.*   Knight, *J. Econ. Entomol.* **35**, 330 (1942).

[5] French Pat. 798,734, to *California Spray-Chemical Corp.*   U. S. Pats. 2,171,598, Parker to *California Spray-Chemical Corp.*; 2,135,987, Murphy to *Rohm and Haas Co.* Murray, *Bull. Entomol. Research* **30**, 211 (1939).

fying agents, *i.e.*, they must aid not only in dispersion but in deposition and sticking. Soaps, sodium lauryl sulfate, alkyl aryl sulfonates, sulfated oils and esters, quaternary ammonium compounds, and others have been described as useful dispersers. They are used with derris, cube, lead arsenate, copper compounds, sulfur, fluosilicates, and various synthetic organic toxicants.[6] Small amounts of certain solid surface active agents have been incorporated into insecticidal dusts to improve the sticking property. In this case, it seems likely that the surface active products in question, because of their gummy nature in the solid state, act merely as adhesives while their surface active properties play, at most, a minor role.[7]

Wetting agents are often used in aqueous sprays where the active toxicant is a water-soluble substance such as nicotine or its salts. In this case, improved wetting and spreading appear to be the only favorable effects of the wetting agent, whose presence ensures that the insect is thoroughly drenched and wetted by the spray. Such improved contact with the insect makes it possible to destroy the insect with lower concentrations of toxicant in the spray. Soap as well as a great number of the synthetic wetting agents have been described as suitable for this type of application.[8]

A surprisingly large number of detergents have been discovered to have valuable insecticidal powers even when used without added toxic substances. Although in some instances such toxicity appears closely related to wetting power, nevertheless it is not possible to correlate surface active properties closely with toxicity to insects. Soaps, particularly potassium soaps of the C_{10} to C_{14} fatty acids, are effective against aphids and other soft-bodied insects. Sodium lauryl sulfate is equally effective against these and similar pests. Non-ionic agents of the fatty acid hexitol ester type, the alkyl aryl sulfonates, and the long-chain cationic agents have also been described, along with many others, as surface active toxicants for a variety of insects.[9] A number of water-soluble insecticides containing a

[6] Martin, *J. Pomol. Hort. Sci.* **18**, 34 (1940). McKenna and Hartzell, *Contrib. Boyce Thompson Inst.* **11**, 465 (1941). Fulton and Howard, *J. Econ. Entomol.* **35**, 867 (1942). Brit. Pat. 529,244, Wilfred Sexton and *Imperial Chemical Industries Ltd.* Canadian Pat. 355,888, Sharp to *Canadian Industries Ltd.* U. S. Pat. 2,264,762, Knight to *Shell Development Co.*

[7] Rainwater, *J. Econ. Entomol.* **32**, 700 (1939). Dudley, Bronson, and Carroll, *U. S. Dept. Agr., Bur. Entomol. Plant Quarantine Bull.*, No. E-400 (1937).

[8] Yothers and Griffin, *J. Econ. Entomol.* **33**, 800 (1940). Greenslade, *22nd Ann. Rept. East Malling Research Sta. Kent* **1934**; *Chem. Abstracts* **30**, 225. Ger. Pat. 708,615 to *Clorox Chemical Co.*

[9] Ginsburg and Kent, *J. N.Y. Entomol. Soc.* **45**, 109 (1937). Dills and Menusan, *Contrib. Boyce Thompson Inst.* **7**, 63 (1935). Dozier, *J. Econ. Entomol.* **30**, 968 (1937). Cory and Langford, *ibid.* **28**, 257 (1935). Ger. Pat. 693,827, to *Böhme Fettchemie G.m.b.H.* Canadian Pat. 410,205, Brown to *Atlas Powder Co.* U. S. Pats. 2,374,918, Brown to *Atlas Powder Co.*; 2,375,095, Flett to *Allied Chemical and Dye Corp.*

surface active radical and a toxic radical in the same molecule have been specially synthesized. Examples of these are the nicotine salts of alkyl aryl sulfonates,[10] the long-chain alkyl nicotinium halides,[11] and the mercuriated derivatives of sulfosuccinic esters.[12]

Aside from insecticides the agriculturist uses fungicides, seed disinfectants, and herbicides in the form of dusts or sprays. The surface active agents are used with these compositions, as with the insecticides, to facilitate their application and to improve their retention. Many of the patents and specifications in this field indicate that the wetting and spreading powers of the auxiliary agent are of primary importance. Chemically, the surface active agent must be stable and non-reactive toward the active toxicant. In formulating arsenical herbicides and fungicides the fatty alkyl sulfates, soaps, and alkyl aryl sulfonates have been used extensively. The cationic agents have been used with chlorinated organic herbicides, and the secondary alcohol sulfates with herbicides containing sodium chlorate as the active ingredient.[13] Fish oil soap has been recommended as the best depositing and sticking agent for copper phosphate fungicides.[14] Certain surface active agents, notably the sulfated terpenes and the quaternary ammonium compounds, are in themselves active fungicides.[15]

Fruit which has been sprayed with insecticides which are poisonous to humans must often be cleaned before being marketed. Normal weathering between the time of applying the insecticide and the time of harvesting will generally reduce the concentration of toxicant on the fruit surface to a low level but in the cases of lead, arsenic, copper, and others only extremely small amounts of residue can be tolerated.

Lead residues are usually removed by washing in 1 to 2% hydrochloric acid solutions. There is some difference of opinion as to whether the addition of a detergent or wetting agent to the bath is significantly advantageous. It is doubtful, however, if enough products have been tried under favorable conditions to arrive at a final decision as to their actual usefulness. At least one type, the sulfated oleic esters, is claimed to be very effective.

Arsenic and lime–Bordeaux mixture residues are apparently best re-

[10] U. S. Pat. 2,232,662, Hockenyos to *Monsanto Chemical Co.*

[11] U. S. Pat. 2,048,885, Oakeshott to *Imperial Chemical Industries Ltd.*

[12] U. S. Pat. 2,325,411, Lynch to *American Cyanamid Co.*

[13] Stetson, *Soap Sanit. Chemicals* **21**, No. 7, 104 (1945). U. S. Pats. 2,228,262, Engels and Stevens to *Merck and Co.*; 2,140,519, Elston to *E. I. du Pont de Nemours & Co.*; 2,218,787, to Catchings, Padget, and Dawsey. Ger. Pat. 543,193, to *I. G. Farbenindustrie.* Offord and Winslow, *Northwest Sci.* **12**, 95 (1938). Martin and Salmon, *J. Agr. Sci.* **21**, 638 (1931). Tisdale, *Rev. Applied Mycol.* **14**, 598 (1935).

[14] Green and Goldsworthy, *Phytopathology* **27**, 957 (1937).

[15] Bertin, *Compt. rend. acad. agr. France* **24**, 735 (1938); *Chem. Abstracts* **32**, 7651.

moved by alkaline washing agents. For this purpose wide use is made of alkali mixtures containing sodium silicate or the sodium phosphates as the major ingredients. Soaps and a variety of anionic sulfates and sulfonates have been found to be very effective in such cleaning baths.[16]

II. Uses in the Paper Industry

In the paper industry, surface active agents are used much more extensively in coating and impregnating operations rather than in pulping or in making the initial unfinished sheet. Surface active agents are of particular importance in the manufacture of special-purpose papers. The application of commercially available surface active agents in paper fabrication has been surveyed recently by Miskel.[17]

The synthetic lime-resistant detergents have been used to a certain extent in the cooking of rag stock. This is essentially a grease- and wax-removing operation, carried out with lime or caustic soda, and is somewhat analogous to kier boiling in the cotton industry. Within recent years waste paper itself has become an important raw material for pulping. An essential step in this operation is de-inking, for which the surface active agents, mostly the less-expensive synthetic sulfonates and sulfates, have proven to be valuable aids.

Surface active agents may be used in the beater to aid in the dispersion and deposition of fillers and rosin sizes. Similarly, the dispersion and application either of pigment colors or of the basic dyes in the beater may be aided by surface active agents. They serve the same purpose in tub sizing and coloring.

The fatty alkyl sulfates and alkyl aryl sulfonates have proven to be excellent agents for washing the broad endless fabric belts or blankets which are used in paper-making machines to carry the newly formed web of paper pulp from the wire screen through squeeze rolls. During paper mill operation these belts or blankets, which are made of finely woven wool and which are very expensive, must be washed either at frequent intervals or continuously to remove sizing materials and other extraneous matter with which they become contaminated. Such washing operations may be

[16] U. S. Pat. 2,223,168, Dombrow and Rosenberg to *National Oil Products Co.* Overly *et al., Proc. 30th Ann. Meeting Washington State Hort. Assoc.* **1934**, 77; *Wash. State Agr. Expt. Sta., Tech. Bull.*, No. 286 (1933). Ellenwood, *Proc. 69th Ann. Meeting Ohio State Hort. Soc.* **1936**, 26. Haller *et al., Md. Agr. Expt. Sta. Bull.*, No. 368, 121 (1934). Ruth and Kadow, *Trans. Ill. State Hort. Soc.* **68**, 191 (1934). Robinson and Hatch, *Oregon Agr. Expt. Sta. Bull.* No. 341 (1935). Pentzer, *N. Y. State Hort. Soc., Proc. 79th Ann. Meeting* **1934**, 48. Karr, *J. Econ. Entomol.* **34**, 676 (1941). Frear and Worthley, *J. Agr. Research* **51**, 61 (1935).

[17] Miskel, *Paper Trade J.* **118**, No. 26, 27–32 (1944).

carried out either in acid, alkaline, or neutral solution, depending on the mill conditions and the type of paper being made.

The surface active agents which have good rewetting properties are often added to paper towelling, particularly when it has been resin treated to impart high wet strength. The rewetting agent counteracts the tendency of the resin treatment to impair the absorbency of the paper.

A wide range of papers and paper products are made by applying coating and impregnating materials in emulsified form. Surface active agents, whose use is indispensible in such operations, serve not only to effect the emulsification but also to promote the deposition of the coating and impregnating materials. The choice of an appropriate surface active agent is governed by the type of impregnating material and the conditions of operation. The impregnating materials include asphalt, metal soaps, waxes, synthetic resins of various types, casein, oils, etc. The sulfated oils and esters are probably the most widely used single class of auxiliaries, but most of the other types have been at least mentioned in the patent literature.[18]

III. Concrete and Building Materials

The major uses of surface active agents in the construction industries are: (1) In making wallboards, roofings, sidings, and other similar materials not based on Portland cement. (2) In various types of Portland cement compositions. (3) In the construction of roads, principally as an ingredient in bitumen emulsions.

The rosin soaps, fatty alkyl sulfates, and lignin sulfonates have been used as ingredients in plaster, gypsum boards, and mineral wool insulating blocks to control the amount of aeration during preparation and setting.[19] Various surface active agents have also been used to aid the impregnation and adhesion of bonding resinous materials in the manufacture of fiberboard as well as boards based on ground cork, crushed stone, and glass fiber.[20]

Wetting agents of the Nekal type, i.e., lower alkylated naphthalene sulfonates, have been known for many years as valuable addition agents to

[18] Kinney, *Paper Ind. and Paper World* **25**, 50 (1941). Gleklen, *Am. Dyestuff Reptr.* **28**, 437 (1939). Getty, *Paper Trade J.* **114**, No. 8, 130 (1942). Brit. Pat. 540,650, to A. Müller. Canadian Pat. 387,131, Kress and Johnson to *Institute of Paper Chemistry*. U. S. Pats. 1,995,623, Richter to *Brown Co.*; 2,223,158, Licata and Nothum to *National Oil Products Co.*; 2,030,226, Rankin and Uhler to *E. I. du Pont de Nemours & Co.*

[19] U. S. Pats. 2,370,058, Maguire to *Hercules Powder Co.*; 2,382,561, to T. G. Gregory; 2,162,386, to B. Neuhof.

[20] U. S. Pat. 2,283,192, Ditto to *Emulsions Process Corp.* Brit. Pat. 528,947, to K. Oesterreich. French Pat. 845,489, to *I. G. Farbenindustrie.*

concrete mixes. Denseness, strength, and frost resistance are said to be improved. Although it is not known with certainty how such improvements are accomplished, it has been suggested that the entrainment of small amounts of air due to the presence of the surface active agent may be responsible. The non-ionic agents and lignin sulfonates have also been suggested for use in concrete. Porous and foamed concrete structures are made by adding larger amounts of a foaming surface active agent and beating air into the mixture. The percentage of entrained air in this case is much higher and the resulting hardened mass has a fine honeycomb structure.[21]

Surface active agents are also of importance in connection with the very extensive use of asphalts and bitumens in road constructing and paving. In such work, it is desirable that the bituminous binder material should thoroughly wet the sand and gravel in order to ensure optimum firmness and strength of the resulting structure. Bitumens and asphalts are often used in emulsified form to improve their spreading and wetting properties. The extensive literature on bitumen emulsions for road building mentions use of a large number of surface active agents as auxiliaries. They function not only as emulsifying agents but also as differential wetting agents to displace water from the rock and sand, thus affording intimate contact with the bitumen. Both the O/W and W/O type emulsions have been used, and, as might be expected, their preparation has involved use of both the water-soluble and oil-soluble types of surface active agents.

Relatively concentrated bituminous emulsions are ordinarily used in paving and road construction. More dilute emulsions are used for soil stabilization, particularly in making dust-free dirt roads, as well as for impregnation of paper, fabrics, etc. Formulation is, in general, similar to that of the more concentrated road emulsions.

The emulsifying agents usually employed in preparing the above-mentioned bituminous emulsions are the soaps, including rosin and naphthenic soaps. Inorganic modifying agents such as trisodium phosphate or calcium chloride are often incorporated to adjust the stability and viscosity of the emulsion. The petroleum sulfonates and the soluble proteins such as albumin or sodium caseinate are also effective bitumen emulsifiers. Heavy-metal soaps such as lead oleate or lead naphthenate have been used in the emulsions to increase the adhesion of the bitumen to rock.

Unemulsified oils are also used in road construction. Lecithin and sim-

---

[21] Ger. Pats. 574,793; 588,196; 555,893; 703,338; to *I. G. Farbenindustrie.* U. S. Pats. 2,240,622, Lawson to *E. I. du Pont de Nemours & Co.*; 2,052,586, Tucker to *Dewey and Almy Chemical Co.* Kennedy, *J. Am. Concrete Inst.* **14**, 529 (1943). Ernsberger and France, *Ind. Eng. Chem.* **37**, 598 (1945).

ilar oil-soluble surface active agents may be added to such road oils to improve their ability to displace water from the rock and sand and thus effect a firmer bond. Similar results may be obtained by first treating rock and sand with a cationic wetting agent. Long-chain fatty amines, heavy-metal soaps and heavy-metal phenolates are among other substances which may be added to road oils to increase their adhesive power.[22]

An unusual use for cationic surface active agents in connection with excavation work has recently been described. In order to prevent landslides when bentonite strata are encountered the loose seam is flooded with a low-viscosity mineral oil containing cationic detergent. This treatment is said to stabilize the formation and prevent dangerous slippage.[23]

IV. Photography, Printing, and Graphic Arts

Surface active agents have proved to be of value in a number of operations connected with photography. Wetting agents such as sodium lauryl sulfate are now quite commonly added to developing baths, particularly in processing motion picture film. Complete wetting of the film is thereby insured and flaws due to adherent air bubbles are eliminated. Furthermore, when hard water is used in preparing developer solutions, precipitation of lime salts is prevented, a result that can also be achieved by using the polyphosphates. A large number of wetting agents, other than sodium lauryl sulfate, have been used successfully in developing baths, although some of them are said to promote fogging or to slow down the developing process. With hydrazine developers the cationic agents and the carboxy soaps are claimed to be particularly effective.[24] Dodecylpyridinium bromide is said to accelerate development, particularly when hydroquinone developers are used.[25]

Surface active agents have been added to the silver halide emulsions which make up the sensitive photographic layer. The cationic and particularly the non-ionic agents of the hexitol ester – polyoxyethylene ether

[22] Nicholson et al., Proc. Assoc. Asphalt Paving Technol. **12**, 3 (1940). Weiss' Allgem. Oel- u. Fett- Ztg. **30**, No. 12, 623 (1933). Ger. Pats. 531,157; 542,634; to N. V. Bataafsche Petroleum Maatschappij. Brit. Pat. 427,720 to International Bitumen Emulsions Corp. French Pats. 807,958, to T. I. C. Research Co. Ltd.; 848,911 to Standard française des petroles; 760,544 to Bitumen Investments. U. S. Pats. 2,003,860, McConnaughay to Pre Cote Corp.; 2,298,612, Carr to Union Oil Co. of Calif.; 2,372,658, Buckley and Bly to American Bitumuls Co.; 2,326,610, Borglin to Hercules Powder Co.

[23] U. S. Pat. 2,348,458, Endersby to Shell Development Co.

[24] Canadian Pat. 435,334, Smith and Stauffer to Canadian Kodak Co. Ltd. U. S. Pat. 2,220,929, Kirby to E. I. du Pont de Nemours & Co. Kieser, Phot. Ind. **34**, 1166 (1936).

[25] Lottermoser and Steudel, Kolloid- Z. **82**, 319 (1938).

type are said to improve the speed, contrast, and stability of the emulsion.[26]

In the application of anti-halation layers, filter layers, or other supercoats to photographic films various wetting agents have proved useful. They facilitate smooth deposition and adhesion of the added layer.[27] Surface active agents have also been used in the recovery of silver from waste photographic solutions.[28]

We have already discussed how the preparation of printing inks is facilitated by the use of surface active agents. They are also used as wetting agents in the cleaning and etching baths in the preparation of photoengraved printing plates. Wetting agents such as dioctyl sulfosuccinate have been used in ink-repellent aqueous compositions on offset printing plates.[29] Similar surface active types have been added to the water-sensitive layers of decalcomania papers.[30] The sulfated fatty oils and agents of the fatty monoglyceride type have been used as ingredients in the coatings on stencil sheets, hectograph blankets, and paper-backed copy pads. Several of the more powerful wetting agents have been employed as re-wetters in the copy papers designed for use with hectograph machines.[31]

V. Foods

Probably the largest use for surface active agents in the foodstuffs industry is in the manufacture of margarine and shortening. When these fats are used for frying an annoying amount of spattering is usually encountered. This is due to the superheating and sudden volatilization of water which may be present in the fat and/or in the food which is being fried. If the correct type of surface active agent is present in small amounts the water will boil out of the hot fat smoothly and without spattering. One of the widely used anti-spattering agents is the sodium salt of sulfated stearic monoglyceride. Other agents which are added to margarine or shortening in order to improve the emulsifying power for water, the creaming properties, and the ability to produce light pastry include the

[26] Brit. Pat. 557,178, to *Kodak Ltd.* U. S. Pat. 2,400,532, Blake and Baldsiefen to *E. I. du Pont de Nemours & Co.*

[27] U. S. Pats. 2,203,767, Baldsiefen to *Du Pont Film Mfg. Corp.*; 2,366,439, Chilton and Woosley to *Ilford Ltd.*

[28] U. S. Pat. 2,221,018, Bachman and Pool to *Eastman Kodak Co.*

[29] U. S. Pats. 2,126,181 Field to *Pyrene Co. Ltd.*; 2,240,486, Beckley to *Western Electric Co.*

[30] U. S. Pat. 2,273,694, Davis and Tuukkanen to *McLaurin-Jones Co.*

[31] U. S. Pat. 2,052,291, to B. Hagg. U. S. Pats. 2,208,980–1; 2,098,042–3; to S. Horii. U. S. Pats. 2,368,583, Tatge to *Ditto, Inc.*; 2,240,032, Bour to *Ditto, Inc.*; 2,098,662, Hoskins to *Ditto, Inc.*; 2,240,031, Bour to *Ditto, Inc.*

fatty esters of polyhydric alcohols and similar non-ionic water-dispersible agents[32]

Various types of surface active agents have been used in emulsified foods such as mayonnaise or other salad dressings. Surface active agents have also been used to disperse the oil-soluble vitamins in fortified foods or beverages.[33] Lecithin is very widely used in food preparations as an emulsifying, blending and smoothing agent. The sulfated and phosphated fatty alcohols and fatty esters containing free hydroxy groups have been used to improve the wettability of cocoa and preparations based on powdered milk. Phosphated fats and oils have also been used to reduce the viscosity of chocolate coating mixes.[34]

Experiments with Aerosol OT and sodium lauryl sulfates added to bread dough indicate that beneficial results are obtained by using critical small amounts.[35]

One of the more recently developed uses for the powerful aqueous wetting agents is in the chemical peeling of fruits and vegetables. Peaches for canning and potatoes for making chips are most efficiently peeled by treating with warm dilute hydrochloric acid or caustic soda. The addition of wetting agents to the bath speeds up the process greatly and gives more uniform results.[36]

VI. Rubber and Resins

In the rubber and plastics industries, the surface active agents play a particularly important role (1) in emulsion polymerization (i.e., polymerization of emulsified reactive compounds of low molecular weight), and (2) in preparation of emulsions of materials that are already in the polymeric form. The emulsion polymerization of butadiene–styrene mixtures and butadiene–acrylonitrile mixtures to make synthetic rubber consumes large quantities of surface active agents. In this country, the carboxylic soaps, both of fatty acids and of rosin, are most widely used. In Germany, dibutylnaphthalene sulfonate (Nekal BX), was the favored agent for this purpose.

The enormous amount of study devoted to the theoretical and practical aspects of emulsion polymerization cannot be reviewed adequately in this

[32] French Pats. 832,196; 848,515, to H. Schou. U. S. Pats. 2,098,010, Newton et al. to *Industrial Patents Corp.*; 2,095,955, to H. Bennett and F. Braude; 2,377,610, L. C. Brown to *Industrial Patents Corp.*

[33] Compare U. S. Pat. 2,236,516, Cahn and Harris to *Emulsol Corp.*

[34] Brit. Pats. 514,721, to *Emulsol. Corp.*; 514,848, to *Ross and Rowe, Inc.* U. S. Pat. 2,185,592, to S. Jordan.

[35] Swanson and Johnson, *Cereal Chem.* **21,** 222–32 (1944).

[36] Olsen, *Food Inds.* **13,** No. 4, 51 (1941). Lankler and Morgan, *ibid.* **16,** 888–91, 940 (1944).

book because of space limitations, but a few of the salient points may be mentioned. A particularly important point is the fact that mere emulsification is not the only important effect exerted by emulsifiers. It is true that a large number of water-soluble surface active agents can be used effectively, but rates of polymerization vary somewhat depending on the emulsifier used. Using any single emulsifier the polymerization rate increases with the concentration. The present consensus of workers in this field is that the monomer is solubilized in the micelles of the emulsifying agent and that at least a part of the polymerization actually takes place in the solubilized state.[37] Emulsion polymerization cannot be carried out using solid, insoluble emulsifiers such as bentonite, even though they form stable, well-dispersed emulsions of the monomer.

Hydrogen ion concentration, which is known to affect the solubilizing power of emulsifiers, plays an important role in emulsion polymerization. Anionic emulsifiers appear to give best results in slightly alkaline media. The cationic agents are more effective in acidic media. Non-ionic agents of the polyoxyethylene ether type are also claimed to be more effective if the pH is kept below 7.[38]

A particularly important effect of pH, however, is concerned not with the emulsifying agent, but rather with the "regulator" or "modifier." The function of a regulator is to control the degree of polymerization in the emulsion polymerization of GR-S rubber (butadiene-styrene co-polymer). The presence of the regulator narrows the molecular-weight range of the product and reduces the proportion of polymer having a very high molecular weight. Aliphatic mercaptans having alkyl groups in the C_8 to C_{16} range are the most widely used and most effective regulators. Since these mercaptans are insoluble in water and apparently contribute little or nothing to the solubilization effect, it is doubtful if their effect on polymerization has anything to do with their inherent surface active properties. According to recent theory, the regulators control polymerization by a mechanism involving the breaking of the polymerization reaction chain, and the SH radical rather than the hydrocarbon group of the mercaptan is believed to be the effective chain breaker. The chain breaking reaction occurs at the locus of polymerization, i.e., in the material held in the solubilized state within the soap micelles. The relative rates at which

[37] Yurzhenko et al., Compt. rend. acad. sci. U. S. S. R. 47, 103, 348 (1945). J. Gen. Chem. U. R. S. S. 16, No. 8, 1171 (1946). Fryling and Harrington, Ind. Eng. Chem. 35, 114 (1944). Hohenstein and Mark, J. Polymer Sci. 1, 127, 549 (1946). Starkweather et al., Ind. Eng. Chem. 39, 210 (1947). Salomon and Konigsberger, J. Polymer Sci. 1, 364 (1946).

[38] Ger. Pat. 711,840, to I. G. Farbenindustrie.

monomers and regulator diffuse from the emulsified "oil" droplets into the interior of the micelle will therefore control the regulating action. The diffusion rate of the regulator is in turn controlled by its molecular weight, *i.e.*, the size of the hydrocarbon chain. It is fortuitous that C_8 to C_{16} chains are optimum under practical operating conditions. A medium of high pH favors the formation of mercaptide anions which diffuse more rapidly through the aqueous medium into the micelles. Hence the pH has a pronounced effect on regulator action.[39]

Although the soaps and the Nekals are, by a wide margin, the most commonly used emulsifiers for commercial emulsion polymerization, a large number of others have been disclosed as particularly advantageous in one respect or another. The fatty alkyl sulfates, the sulfated or sulfonated derivatives of petroleum-derived olefins, and the alkyl benzene sulfonates have been prominently mentioned among the anionic products.[40] The Reed reaction products (alkane sulfonates) appear to have been used to a limited extent in Germany. The cationic agents which may be used in emulsion polymerization include a variety of long-chain quaternary ammonium and pyridinium salts, as well as the acid salts of long-chain primary, secondary, and tertiary amines. Products such as the N-diethylamino-ethyl ester of oleic acid or the analogous product having an amide instead of an ester linkage have apparently given very promising results. Rubbers characterized by low dielectric loss are said to be produced by emulsion polymerization when special soaps consisting of fatty acids neutralized with lower aliphatic or cycloaliphatic amines are used as emulsifiers. The polyethylene oxide non-ionic agents have been used in admixture with both soaps and cationic emulsifiers to improve the stability of the resulting latex.[41]

The emulsion polymerization technique can also be applied with advantage to the polymerization of 2-chloro-1,3-butadiene to form the neoprene type of synthetic elastomer. The fatty alkyl sulfates and cationic agents of the Sapamine type (*q.v.*) have been used as emulsifiers. Cetyltrimethylammonium bromide and stearylpyridinium bromide have also been recommended.[42]

[39] Smith, *J. Am. Chem. Soc.* **68**, 2059, 2064 (1946). Harkins, *J. Chem. Phys.* **13**, 381 (1945). U. S. Pat. 2,281,613, Wollthan and Becker to *Jasco, Inc.*
[40] Canadian Pat. 402,352, Starkweather to *E. I. du Pont de Nemours & Co.* U. S. Pat. 2,384,969, Serniuk to *Standard Oil Development Co.*
[41] U. S. Pats. 2,386,764, Zwicker to *B. F. Goodrich Co.*; 2,393,133, White to *U. S. Rubber Co.* Ger. Pat. 702,749, to *I. G. Farbenindustrie*. Rainard, *India Rubber World* **114**, 67 (1946). Weidlein, *Chem. Eng. News.* **24**, 771 (1946).
[42] Brit. Pats. 497,706; 461,279; French Pat. 853,478, to *E. I. du Pont de Nemours & Co.*

Non-elastomeric polymers are often prepared by emulsion polymerization processes which are similar in theory and practice to those employed for the synthetic rubbers. The end product is a dispersion or "latex" of the polymer, which may be used as such or may be coagulated and worked up into massive forms.

The polymerization of vinyl monomers such as vinyl halides, acetates, ketones, ethers, thioethers, and mixtures of these may be carried out with a variety of individual emulsifying agents. Among those which have been used are the lower alkylated naphthalene sulfonates, the carboxylic soaps, fatty alkyl sulfates, alkyl benzene sulfonates, Reed reaction sulfonates, and dodecylamine hydrochloride.[43] Mixed emulsifying agents sometimes afford advantages in the polymerization of vinyl monomers. In this case one component may be a strong surface tension depressant such as sodium dioctyl sulfosuccinate, while a companion emulsifying agent may be a compound of the "protective colloid" type such as methyl cellulose, water-soluble polyvinyl alcohol, the vegetable gums, or soluble starches. Oil-soluble emulsion stabilizers such as the higher fatty alcohols have also been included in these mixtures.[44]

The emulsion polymerization of acrylic monomers can be carried out in a similar manner with different types and combinations of emulsifying agents. Plasticizers and solvents may be added to the emulsion before polymerization in order to modify the working properties and film-forming ability of the final polymer dispersion. Among the anionic emulsifying agents which have been disclosed are the aliphatic hydroxysulfonates, amine soaps, and butylated naphthalene sulfonates.[45] The use of higher alkyl amine salts such as cetyldimethylamine acetate is stated to result in viscous thixotropic dispersions when esters of acrylic or methacrylic acids are polymerized.[46] The effect of emulsifying agents on the stability of polymerized acrylate dispersions has recently been studied by Mast, Smith, and Fisher.[47] They find that a typical long-chain quaternary ammonium

[43] Brit. Pat. 445,434 to *I. G. Farbenindustrie*. Brit. Pat. 545,703; U. S. Pat. 2,364,227; Lewis, Morgan, and Watts to *Imperial Chemical Industries Ltd*. U. S. Pat. 2,379,292, Gleason to *Standard Oil Development Co.*

[44] Brit. Pat. 514,116, to *Alexander Wacker Gesellschaft für elektrochemische Industrie G.m.b.H.* U. S. Pats. 2,383,782, Dreisbach to *Dow Chemical Co.*; 2,388,600, Collins to *Shawinigan Chemicals, Ltd.*

[45] Brit. Pat. 422,697, to *Triplex Safety Glass Co.* Ger. Pats. 679,825; 663,469; French Pat. 797,686; to *I. G. Farbenindustrie.* Australian Pat. 107,431; Brit. Pat. 501,810; to *Imperial Chemical Industries Ltd.* U. S. Pat. 2,296,403, Renfrew and Gates to *Imperial Chemical Industries Ltd.*

[46] U. S. Pat. 2,372,108, Neher and Conn to *Rohm and Haas Co.*

[47] Mast, Smith, and Fisher, *Ind. Eng. Chem.* **37,** 365 (1945).

compound, Triton K-60, gives emulsions of only moderate stability. A combination of fatty alkyl sulfates and alkyl aryl polyoxyethylene ether sulfonates (Tergitol 4 and Triton 720) gives emulsions of high stability to electrolytes. Casein and alginates can be used to modify both the stability and viscosity of polymer emulsions finally produced.

A large number of other polymers and copolymers have been made by the emulsion technique. Styrene and the vinyl ketones may be polymerized in emulsions of soap, casein, or alkyl naphthalene sulfonates.[48] Similar emulsifying agents may be used in producing polymers and copolymers from fumaric and maleic ester polymers.[49] The emulsion polymerization of ethylene has also been thoroughly studied and reported.[50]

Emulsion polymerization, as ordinarily practiced, produces a more or less stable suspension of solid polymer in which the individual particles are very finely divided. It is often desirable to obtain the polymer in the form of relatively large granules. This may be done in either of two ways. Polymerization may be carried out in the usual manner and the resulting dispersion may be partially coagulated but not completely broken. This requires very careful control. Alternatively the polymerization itself may be carried out while the monomer is in a coarsely dispersed form. The reaction mechanism in the latter case probably is different from true emulsion polymerization and more nearly resembles bulk polymerization.[51]

Although emulsion polymerization is the simplest method for making dispersions of addition type polymers, it is sometimes expedient or necessary to disperse an already formed polymeric material. Dispersions of almost all condensation type polymers are so prepared. Dispersion of such products may be accomplished in several diffcrent ways and a wide variety of surface active dispersing agents may be used. One method involves dissolving the polymer in an appropriate water-immiscible solvent and emulsifying the non-aqueous solution in water. The solvent may be subsequently removed by evaporation if so desired. The butanol-soluble urea-formaldehyde condensates have been dispersed in this manner, using methylcellulose and cetyldimethylbenzylammonium chloride as emulsifying agents.[52] The condensation products of phenol sulfonic acid with formalde-

[48] U. S. Pat. 2,005,295, Meisenburg to *I. G. Farbenindustrie.*

[49] Ger. Pat. 699,445; French Pat. 833,459; to *I. G. Farbenindustrie.*

[50] Hopff and Kern, *Modern Plastics* **23,** No. 10, 153 (June, 1946).

[51] Hohenstein, Vingiello, and Mark, *India Rubber World* **110,** 291 (1944). U. S. Pats. 2,357,861, Wilson to *B. F. Goodrich Co.;* 2,191,520, Crawford and McGrath to *Imperial Chemical Industries Ltd.*

[52] U. S. Pat. 2,196,367, Thackston to *Rohm and Haas Co.* Brit. Pat. 525,190, to W. W. Triggs.

hyde have been used to emulsify phenol-formaldehyde resins.[53] Solvent solutions of the melamine-formaldehyde resins may be emulsified with amine soaps of fatty acids, fatty alkyl sulfates, cationic agents of the Sapamine type, etc.[54] The polyvinyl acetals may be emulsified by first dissolving in methylene chloride and then dispersing with the aid of butylated naphthalene sulfonates.[55] Polyethylene, polyisobutene, cellulose esters and ethers, and reclaimed rubber are among other polymeric substances which have been converted into aqueous dispersions by similar techniques.[56]

Aside from emulsification there are several other uses for surface active agents in rubber technology. They have been used as foaming agents in the manufacture of foamed latex and sponge rubber articles. Both anionic and cationic agents have been used to control the stability and working properties of latex itself. The more powerful wetting agents are particularly effective in promoting impregnation of fabrics with latex. The cationic agents, in particular, are said to afford a better bond between rubber and fabric by conferring a positive charge on the latex particles. The water-insoluble dispersing agents such as fatty alcohols, partial fatty esters of polyhydric alcohols, etc. have also been used to facilitate the compounding of pigments and fillers with rubber.[57]

[53] Brit. Pat. 523,222, to *Catalin Ltd.*

[54] Brit. Pat. 539,288, to *American Cyanamid Co.*

[55] U. S. Pat. 2,143,228, Orthner and Selle to *I. G. Farbenindustrie.*

[56] Brit. Pats. 539,476, to *I. G. Farbenindustrie*; 515,582, to *Imperial Chemical Industries Ltd.* Dagaev, *Legkaya Prom.* **1944**, No. 7/8, 21; *Chem. Abstracts* **39**, 4515.

[57] U. S. Pat. 2,105,234, Rodman to *E. I. du Pont de Nemours & Co.*

CHAPTER 28

Miscellaneous Uses of Surface Active Agents

As was pointed out in Chapter 17, the various gross effects of surface active agents may be used as a basis for classifying their almost innumerable specific uses. This basis of classification appears particularly well-suited as an aid in discussing certain typical applications which do not belong in any well-defined industrial category. In this final chapter various uses which are grouped according to the following gross effects will be discussed: wetting and spreading, emulsifying, dispersing and solubilizing, chemical effects, foaming and anti-foaming, and, in conclusion, miscellaneous uses not involving surface activity.

I. Wetting and Spreading

One of the uses for surface active agents which depends primarily on wetting power is the removal of dusts from air or other gases by "scrubbing," *i.e.*, by passing a spray of water or other appropriate liquid through the dust-laden gas. In this process droplets of liquid spray must wet the solid dust particles in order that they may settle rapidly. This is facilitated, and the efficiency of the scrubbing operation is improved, by adding a wetting agent to the water or other spray liquid.[1] This technique has been used to recover chemicals which are produced in dust form, one example being the recovery of phthalic anhydride from converter exit gases.[2] Dust and fumes have also been removed from their supporting atmospheres by means of foams. The solid particles are wet by foam films in the same manner as they may be wet by sprayed liquid droplets. In this application the wetting agent may also serve the accessory purpose of stabilizing the foam.[3] Dust in coal mines constitutes a serious hazard and wetting agents have proved very effective in combatting it. The dusty areas are sprayed or otherwise wet down with dilute aqueous solutions of a powerful wetting agent. Treatment with water alone is much less effective, due to the hydrophobic nature of coal dust.[4]

[1] Brit. Pat. 511,599, to *Calmic Ltd.* and Wm. C. Peek. Ger. Pat. 718,565, Höfer to *Kali-Forschungsanstalt G.m.b.H.*

[2] U. S. Pat. 2,190,001, Talbert to *National Aniline and Chemical Co.*

[3] Luchinskii, *J. Phys. Chem. U. S. S. R.* **13**, 302 (1939).

[4] Ger. Pat. 681,968, Daimler to *I. G. Farbenindustrie.* Compare Bryan and Smellie, *J. Roy. Tech. Coll. Glasgow* **4**, 178 (1937); *Chem. Abstracts* **31**, 3590.

The formulation of adhesives is another large-scale application in which wetting power is of primary importance. In water-soluble adhesives, such as glue, dextrins, and mucilages, the function of the wetting agent is primarily to insure intimate rapid contact with the pieces to be joined. The dried layer of adhesive may also be plasticized. Sulfated alcohols and oils are often used in water-soluble adhesives.[5] In the case of emulsified adhesives the wetting agent may also promote emulsification. As an example, the polyvinyl ethers may be emulsified with dibutylnaphthalene sulfonate and the resulting composition used as an adhesive.[6] In adhesives based on natural rubber latex a wetting agent is particularly useful since latex by itself has poor wetting properties. More precisely, the rubber globules in the latex emulsion do not easily make intimate contact with the fabrics or other materials which are usually treated. The soaps, sulfated oils, and other anionic types of wetting agents have been used in latex adhesives, but the cationic agents appear to offer distinct advantages. This is especially true when latex is applied to cotton or rayon fabrics.[7]

The re-wetting and spreading powers of sulfated oils, sulfated esters, and other surface active agents are utilized in the formulation of anti-fogging compositions for glass windows, windshields, spectacles, etc. Such compositions are applied as aqueous solutions and are allowed to dry so as to form a thin film of wetting agent. This film is very effective in causing condensed moisture to spread out evenly into a clear transparent film. On untreated glass condensed moisture usually forms discrete droplets with the result that visibility may be reduced seriously.[8]

Wetting agents with strong surface-tension-lowering properties have been used as ingredients of embalming fluids to facilitate the passage of such fluids through the capillaries. Wetting agents suitable for this purpose must not undergo undesirable reaction with the formaldehyde and other preservative materials which are present in rather high concentrations.[9]

For the removal of wallpaper or paper labels, a variety of surface active materials have been used. They are all characterized by high wetting and penetrating power.[10]

Wetting agents are often employed to facilitate the mixing of such widely

[5] U. S. Pat. 2,215,848, Bauer to *Stein-Hall Mfg. Co.* French Pat. 762,881, to *Deutsche Hydrierwerke A.-G.* Ger. Pat. 639,821 to *Henkel et Cie.*

[6] Ger. Pat. 705,394; French Pat. 835,831, to *I. G. Farbenindustrie.*

[7] U. S. Pats. 2,152,012, Albion to *Providence Braid Co.*; 2,256,194, Crawford to *B. F. Goodrich Co.*; 2,255,834, Taylor and Schaefer to *North American Rayon Corp.*

[8] Brit. Pats. 494,113, to J. S. Banks; 524,987, to J. E. Ramsbottom and J. D. Main-Smith.

[9] U. S. Pats. 2,085,806; 2,219,927; 2,253,625, Jones to *National Selected Morticians.*

[10] U. S. Pats. 2,221,960 Abramowitz to *National Oil Products Co.*; 2,067,326, to M. Leatherman. Brit. Pat. 445,191, to P. W. Reid.

different compositions as pencil leads, brake linings, wallboards, linoleum, etc.[11] An example of an unusual and interesting application is in the making of compost. A wetting agent which will overcome the natural water-repellency of the leaves but will not interfere with fermentation is used.[12] Wetting agents of high penetrating power have been added to the water stream used in fighting fires, particularly those occurring in paper warehouses or granaries. Penetration of the water into the hot smoldering mass of burning material is facilitated. The wetting agents may be injected into the water stream either as it leaves the nozzle or at the pump.

II. Emulsifying, Dispersing, and Solubilizing

Numerous examples of O/W emulsions have already been discussed in previous chapters. Wax emulsions made with the usual wide array of surface active agents may be used for impregnating paper, fabrics, leather, concrete, and other materials to render them water proof.[13] O/W emulsions have been used for impregnating wood, for incorporating bonding agents in foundry sand cores, and for numerous other purposes where non-aqueous compositions are ordinarily used.[14] Resin–wax emulsions have been pumped through gas distribution systems for the purpose of sealing leaks. Such emulsions must break easily when forced into the small openings responsible for leakage in order to effect the desired deposition of resinous sealing material.[15] Replacement of normally homogeneous aqueous systems by O/W emulsions is sometimes practiced. An example is the use of dilute mineral oil – petroleum sulfonate emulsions in engine cooling systems. These emulsions are said to be effective inhibitors of corrosion and fouling.[16]

The solubilizing effect of surface active agents is used to incorporate insoluble perfume oils into aqueous media. The soaps are particularly effective for this important application, but the acid-resistant sulfates and sulfonates are used where pH conditions make soap ineffective. In some instances it is desirable to solubilize small amounts of water in an oil.

[11] Brit. Pat. 415,930, to J. S. Staedtler. U. S. Pat. 2,185,333, Denman to *Detroit Gasket and Mfg. Co.*
[12] U. S. Pat. 2,218,695, to M. Leatherman.
[13] Brit. Pat. 406,862, to *Deutsche Hydrierwerke A.-G.* U. S. Pat. 2,172,392, Kress and Johnson to *Institute of Paper Chemistry.* Brit. Pat. 528,308, to *Bakelite Building Products Co.* U. S. Pat. 2,280,830, Johnson to *United Shoe Machinery Corp.* Brit. Pat. 477,280, to Gilder and *F. W. Gilder and Co. Ltd.*
[14] Anon, *Fette u. Seifen* **48**, 412 (1941). U. S. Pat. 2,203,471, Ray to *Hercules Powder Co.*
[15] Brit. Pat. 514,262, to *Standard Oil Development Co.*
[16] U. S. Pat. 2,023,367, Krekeler to *Shell Development Co.*

Products such as the oil-soluble mahogany sulfonates are useful in this connection.[17]

III. Chemical Effects

The Twitchell method for splitting fats is the most important process which utilizes the catalytic effect of surface active agents. The optimum catalysts in this reaction appear to be the long-chain alkyl aromatic sulfonic acids. Twitchell's original reagent dates from 1898 and was made by condensing oleic acid with naphthalene or benzene and sulfonating the product. Since that time numerous other acid-stable surface active agents have been investigated but relatively few superior products have been found.[18] It is noteworthy that there is little direct correlation between emulsifying power and catalytic effect, although none of the good catalysts are poor emulsifiers. The saponification reactions take place in the interfacial layer but the actual nature of the catalytic action is not thoroughly understood.[19]

The emulsifying power of certain surface active agents has been utilized effectively in the laboratory to facilitate chemical reactions between two immiscible liquid phases. In some instances true catalytic action may be involved. Little attention appears to have been given this possibility except in the case of hydrolysis reactions.[20] Dodecyl sulfate and the fatty acyl methyltaurines are said to catalyze the alkylation of isobutene by isobutane in the presence of sulfuric acid. The reagents are kept in the liquid phase by operating at high pressures and/or low temperatures.[21] Products of the Friedel-Crafts reaction, containing hydrocarbons and aluminum chloride complexes, may be decomposed more efficiently and rapidly if a wetting agent such as dodecylamine hydrochloride or dioctyl sulfosuccinate is added to the water used for hydrolysis.[22] The long-chain fatty esters of alkylolamines, which are typical cationic detergents, are said to have a promoting effect in the "doctor" sweetening of petroleum products. This process is directed to the removal of undesirable sulfur compounds from gasoline or oil by treatment with a solution containing sodium plum-

[17] French Pat. 722,752, to *Twitchell Process Co.* Albert and Wyburn, *Soap Perfumery Cosmetics* **12**, 498 (1939).

[18] U. S. Pats. 628,503; 1,082,662; 1,170,468, to Twitchell. Twitchell, *J. Am. Chem. Soc.* **22**, 22 (1900); **28**, 196 (1906). Toi, *J. Chem. Soc. Japan* **61**, 1279 (1940); *Chem. Abstracts* **37**, 4000. Nishizawa *et al.*, *J. Soc. Chem. Ind. Japan* **39**, 488 (1936); *Chem. Abstracts* **31**, 4839. Schlutius, *J. prakt. Chem.* **142**, 49 (1935). Brit. Pat. 349,527, to *N.-V. Chemische Fabriek Servo.*

[19] Talmud *et al.*, *J. Phys. Chem. U. S. S. R.* **4**, 793 (1933).

[20] Colonge, *Bull. Soc. Chim.* **3**, 501 (1936).

[21] U. S. Pat. 2,393,152, Ellis to *Standard Oil Development Co.*

[22] U. S. Pat. 2,189,383, Ralston and Vander Wal to *Armour and Co.*

bite and elemental sulfur, *i.e.*, the normal doctor solution. Addition of the cationic surface active agent to the oil before treatment is said to accelerate elimination of the undesirable sulfur compounds.[23] A final illustration of the use of surface active agents in chemical reactions is afforded by one of the processes for making butadiene. In this procedure vinylacetylene is reduced with metallic zinc in an aqueous caustic soda solution to which sodium dibutylnaphthalene sulfonate has been added as a promoter. The main function of the surface active agent probably is to bring about a more extensive and intimate contact between the hydrocarbon gas and the aqueous liquid.[24]

IV. Foaming and Anti-foaming Effects

One of the most important uses of foams is in fire-fighting systems. Such use involves one of two types of foams. The first type of foam is made up to contain an inert fire-extinguishing gas such as carbon dioxide. When so constituted, the foam serves primarily as a carrier for the inert gas. Alternatively the foam may contain air, in which case the liquid film walls of the foam cells are in fact the active fire-fighting elements and the foam functions as a means for spreading a blanket of water over an extended area. The steam and other combustion products from the fire are blanketed by the stable foam and the fire is effectively smothered. With either type of fire-extinguishing foam, exceptional stability to mechanical and thermal disruption is essential. Formation of carbon dioxide foams may often be brought about by generating carbon dioxide, from sodium bicarbonate and an acid, even in the absence of a foaming agent. The surface active agents which may be added in this case serve primarily as foam stabilizers.[25] The surface active agents used with air foams must promote their formation as well as increase their stability.

A large number of surface active substances have been used in preparing both chemical (carbon dioxide) and mechanical (air) fire-fighting foams. Chemical foams may be stabilized by free fatty acids, glues and other hydrolyzed proteins, silicates and colloidal silica, or natural saponins. The latter are derived from vegetable extracts such as licorice or sassafras. Certain synthetic detergents which are strong foam stabilizers have also been used.[26]

---

[23] U. S. Pat. 2,366,545 Morris to *Petrolite Corp. Ltd.*

[24] French Pat. 834,111, to *I. G. Farbenindustrie.*

[25] Müller, *Gasschutz u. Luftschutz* **6,** 41 (1936); *Chem. Abstracts* **30,** 2665. Maerklin, *Oel u. Kohle* **38,** 291 (1942). Wilmoth, *J. Inst. Petroleum* **32,** 1 (1946).

[26] French Pat. 796,615, to L. Maugé. Brit. Pats. 560,354, to *M. and P. Colloid Stabilizers, Ltd.* and Jack May, 369,012, to R. A. Blakeborough and W. R. Garratt.

Partially hydrolyzed proteins, particularly glues, albumens, and soybean proteins, are very effective stabilizers for mechanical (air) foams. Such stabilizers may be used alone, but more generally a primary foam former is added. The alkyl aromatic sulfonates, fatty alkyl sulfates, sulfated oils and esters, and carboxylic soaps have all been described as suitable foam initiators in these compositions.[27]

Oleic methyltauride, when used together with a solvent of the glycol ether type, is said to give excellent fire-fighting air foams.[28] Sodium lauryl sulfate and sodium lauryl phosphate are said to be very effective when the foam is generated by aspirating air into a rapidly flowing stream of water.[29] The alkyl aromatic sulfonates and the petroleum sulfonates have been used in the same manner. Water-soluble organic solvents such as glycol or ethanol may be added to insure rapid dispersion and solution of the foaming agent in the water.[30] Several cationic detergents have been mentioned specifically as good foaming ingredients for fire extinguishers. Among them are included both quaternary compounds, such as dodecyltrimethylammonium chloride, and amine salts such as cetyldimethylamine acetate, etc.[31]

Aside from fire extinguishers relatively few other important uses for foaming agents remain to be considered. They are used in latex and solvent solutions of rubber to make spongy rubber goods, and they have been used also to make spongy articles from other plastics such as urea–formaldehyde.[32] Foaming agents have also been added to paperboard adhesives of the sodium silicate type in order to give them better drying characteristics.[33]

In chemical industry, the disruption of troublesome or unwanted foams

---

U. S. Pats. 1,954,154 and 2,269,958, Urquhart to *National Foam System Inc.*; 2,269,426, to F. L. Boyd.

[27] Brit. Pat. 460,596, to *I. G. Farbenindustrie*; 469,325, to *Imperial Chemical Industries Ltd.*; 539,579, to J. C. Ferree. French Pats. 764,138, to *Minimax S.-A.*; 853,709, to von Blanquet. U. S. Pat. 2,368,623, to *Pyrene Development Co.* French Pat. 805,831, to *I. G. Farbenindustrie.*

[28] French Pat. 781,818; U. S. Pat. 2,088,085; Gross *et al.* to *I. G. Farbenindustrie.*

[29] U. S. Pats. 2,086,711, to W. Friedrich; 2,138,133, Betzler to *Pyrene Minimax Corp.*; 2,193,541, Timpson to *Pyrene Minimax Corp.*

[30] U. S. Pats. 2,089,646, to W. Friedrich; 2,165,997, Daimler *et al.* to *I. G. Farbenindustrie*; 2,166,008, Holter and Stewart to *Texas Co.*

[31] U. S. Pat. 2,136,963, Bertsch to *Böhme Fettchemie G.m.b.H.* Fr. Pat. 789,327; Ger. Pat. 666,782, to *I. G. Farbenindustrie.* Ger. Pat. 650,919, to *Henkel et Cie.*

[32] Scheuermann, *Z. ges. Kälte-Ind.* **47**, 105 (1940); *Chem. Abstracts* **36**, 5580. French Pat. 729,311, to *Dunlop Rubber Co. Ltd.* and *Anode Rubber Co. Ltd.* U. S. Pats. 2,321,111, Stamberger to *International Latex Corp.*; 2,309,005, Ogilby to *U. S. Rubber Co.*

[33] U. S. Pat. 2,347,419, Lander to *Diamond Alkali Co.*

is at least as important as the generation of useful foams. Anti-foaming agents fall into two different classes, one of which consists of water-soluble polar – non-polar compounds similar to the general run of aqueous surface active agents. These products are usually effective only within a very narrow range of conditions. Under other conditions they may promote foaming. They are analogous in this respect to the water-soluble emulsion breakers which can also function as emulsifying agents. The more commonly used type of anti-foaming agent includes the water-insoluble oils of low volatility and strong spreading power. The glyceride oils and fatty acids fall into this category and are noted in history for their power to still "troubled waters." The soluble type of anti-foaming agent is often used merely as a carrier for the insoluble type.[34]

Among insoluble foam killers, the octyl alcohols have proved to be very effective under a wide range of conditions, e.g., in paper pulp slurries, electroplating baths, and glue solutions. Both octanol-2 and 2-ethylhexanol are effective. These two isomeric octanols are produced on a large scale and are available at reasonably low cost.[35] Other higher alcohols, including cyclohexanol, lauryl and cetyl alcohols, and the higher by-product alcohols from methanol synthesis, are also effective. They may be used either alone or in mixtures containing non-polar oils or such products as aluminum stearate.[36] The higher 1,2 and 1,3 glycols are said to be fully as effective as the monohydric alcohols.[37] The water-insoluble esters of phosphoric acid and the vegetable oils, particularly castor oil, have been used to prevent foaming in textile-finishing baths.[38] Ethyl oleate has been added as an anti-foaming agent in the alcohol-based automotive anti-freeze compositions. Anti-foaming agents are desirable in this connection because alcohol–water mixtures have a marked tendency to foam when agitated by the engine's circulating pumps.[39] Sulfated shark liver oil is claimed to prevent foaming in cutting-oil emulsions and mahogany sulfonate–stearic acid mixtures have been used for the same purpose. Mahogany sulfonate–

[34] For a review of patent literature on anti-foaming agents see Philipp, *Allgem. Oel- u. Fett-Ztg.* **39**, 167 (1942).

[35] Morehouse, *Paper Trade J.* **120**, No. 10, 53 (1945). Kennedy, *Paper Ind. Paper World* **20**, 1170 (1939). French Pat. 762,851, to *Deutsche Hydrierwerke A.-G.* U. S. Pat. 2,147,415, Tucker to *Eastman Kodak Co.*

[36] French Pat. 793,173, to *F. Schacht, Kommanditgesellschaft.* U. S. Pat. 1,947,725, Macarthur and Stewart to *Imperial Chemical Industries Ltd.* Brit. Pat. 429,423, to *Deutsche Hydrierwerke A.-G.*

[37] Ger. Pats. 700,677, to *Henkel et Cie.*; 704,862, to *I. G. Farbenindustrie.*

[38] French Pat. 851,842, to *I. G. Farbenindustrie.* Ohl, *Monatsh. Seide u. Kunstseide, Zellwolle* **43**, 411 (1938). Vorobev, *Chimie & industrie* **32**, 1405 (1934). Gleim and Lissaievitch, *ibid.* **39**, 1097 (1937).

[39] U. S. Pat. 2,298,465, Clapsadle to *Carbide and Carbon Chemicals Corp.*

white oil mixtures are effective in settling the foams encountered in the liming operations of sugar refining.[40] Anti-foaming agents can also be added to non-aqueous systems. Thus, for example, the amount of free fatty acid in vegetable oils is known to have a marked effect on their foaming tendencies. The use of agents such as the sulfated oils and esters to inhibit foam formation in engine lubricants has been discussed above.[41]

V. Miscellaneous Uses Not Involving Surface Activity

Many uses of surface active compounds involve their surface activity to a very minor degree and sometimes not at all. Thus, for example, the mechanical properties of soaps, sulfonated oils, and similar products make them useful as binder ingredients or matrixes in various compositions made from ground cork, sawdust, or similar comminuted substances. The sulfonated oils and other surface active sulfates have been used as plasticizers for glue, starch sizings, adhesives, and similar materials, a function which has little if any connection with their surface activity. The alkyl aromatic sulfonates in the surface active range also happen to have the power of precipitating glue and gelatin from solution, and have been used for this purpose. The long-chain cationic agents, by virtue of their high molecular weight and positive charge, can precipitate or "fix" the sulfonic acid dyestuffs on fibers. Various heavy-metal soaps, in particular the linoleates of lead, cobalt, and manganese, have been used as driers in paints, varnishes, etc. The soaps of these metals are used because they furnish a convenient means for introducing catalytically active metal into such coating compositions. As implied by the term "drier," the desired effect is an acceleration of the oxidation–polymerization process by which the protective film is formed. The catalysis of oxidation and polymerization by the metal is of primary importance in this connection, while the previously mentioned surface activity of heavy-metal soaps in non-aqueous systems appears to be of no importance. Aqueous solutions of soap, sulfonated oils, and other surface active products are often used as lubricants because of their slippery, viscous nature. It is doubtful if their utilization for this purpose has any connection with their surface active properties.

Uses of this nature are regarded as lying outside the field of interest of this book. They are mentioned only to point out that substances having surface activity may be primarily useful in other connections.

[40] U. S. Pats. 2,052,164, Buc. to *Standard Oil Development Co.*; 1,964,641, Mathias to *Standard Oil Co. of Indiana*; 2,285,940, Nörring to *Shell Development Co.*

[41] Robinson, Black, and Mitchell, *Oil and Soap* **17**, 208 (1940). U. S. Pats. 2,390,-491–2, Bennett and Marshall to *Mid-Continent Petroleum Co.*

AUTHOR INDEX

A

Abbott, 293, 294
Abbott Laboratories, 454
Abbruscato, 171
Abramowitz, 53, 75, 511
Abribat, 264
Aceta Gesellschaft, 425
Acheson Colloids Corp., 488
Achmatov, 264
Ackermann, 228
Ackley, 196, 316, 400
Adam, N. K., 3, 247, 257, 263, 270, 273, 281, 284, 287, 302, 336, 362
Adams, E. W., 233
Adams, J. C., 268
Adams, P., 479
Adams, R., 141, 454
Addison, 284, 286, 287, 348
Adkins, 56, 158
Adler, 144
Aelony, 101
Aickin, 286, 289, 348, 363, 377
Alba Pharmaceutical Co., 162, 187, 189, 191
Albert, A., 513
Albert, O., 249
Albion, 511
Albrecht, 42, 131, 138, 168, 179, 180, 436
Aldinger, 334
Alessandroni, 475
Alexander, A. E., 280, 281, 282, 284, 287, 348
Alframine Corp., 75, 172, 173, 217
Alien Property Custodian, 42, 65, 129, 143, 161, 183, 186, 210, 211, 215, 236, 377, 379, 456
Allen, A. O., 476
Allen, C. H., 442
Alles, 479
Allied Chemical & Dye Corp., 98, 100, 111, 117, 122, 123, 124, 128, 131, 133, 234, 235, 236, 379, 446, 450, 461, 464, 469, 496, 497. See also *National Aniline & Chemical Co.*, and *Solvay Process Co.*
Alox Chemical Co., 160, 454, 468, 471, 488
Alrose Chemical Co., 129, 135, 212, 213
Alvarado, 83
Ambrose, H. A., 483
Ambrose, P. M., 491
American Association of Textile Chemists & Colorists, 12

American Bitumuls Co., 502
American Chemical Paint Co., 466
American Cyanamid Co., 60, 64, 99, 100, 101, 106, 117, 129, 145, 165, 172, 173, 174, 179, 182, 184, 187, 194, 197, 206, 226, 235, 450, 451, 479, 487, 488, 493, 494, 498, 509
American Dental Association, 448
American Enka Co., 420, 424, 429
American Hyalsol Corp., 51, 55, 56, 58, 59, 60, 63, 64, 87, 145
American Steel & Wire Co. of New Jersey, 472
Amott, 93
Amsel, 334
Anderson, 263, 471
Andes, 264
Andreas, 264
Andrews, 57
Anode Rubber Co. Ltd., 515
Antonoff, 251, 264
Apgar, 161
Arbiter, 491, 492
Archer, 115
Archibald, 93, 456
Arkansas Co., 49, 52, 116
Armendt, 482
Armour & Co., 57, 130, 164, 224, 424, 493, 513
Arnold, 60, 436, 464
Arnold, Hoffman & Co., 173
Arrowsmith, 70
Arvin, 454
Asch, 479
Asinger, 89
Askew, 281
Atlantic Refining Co., 123
Atlas Electric Devices Co., 355
Atlas Powder Co., 209, 214, 437, 497
Aubry, 287
Augustin, 447, 448, 450
Auerbach, 343
Avens, 496
Averbach, 466
Averill, 93

B

Bachman, 503
Bachrach, 29
Bacon, J. C., 101
Bacon, K. D., 446
Bacon, L. R., 359, 362, 368, 375

519

SUBJECT INDEX

A

Abietic acid. See *Rosin*.
Abietyl alcohol sulfate, 61
Absorbed films. See *Monofilms*.
Absorption base creams, 441
Acetic acid, isooctylphenoxy-, condensed with amino ethyl sulfuric ester, 76
alkoxy, by nitrile synthesis from lauryl chloromethyl ether, 41
from higher alcohols of carbon monoxide hydrogenation, 41–42
fatty alkoxy-, reduced to alcohol and sulfated, 78
p-benzylphenoxy-, perhydrogenated and alkylated, 42
phenoxy-, alkylated with diisobutene, 42
Acetophenone, containing higher alkyl group, oxidation to carboxy acid, 33
Acids as components in synthetic detergent cleaners, 235
Acrylic acid polymers as detergent builders, 379–80
Acrylic esters, sulfonation with bisulfite, 99
Acrylic monomer polymerization in emulsions containing surface active agents, 507
Acrylonitrile, fatty alcohol added on double bond, 187
Activated aluminum to reduce fatty acids to fatty alcohols, 57
Activators in ore flotation, 491
Acyloins, sulfated, 61
Additives, for lubricant, 229–230, 485
for detergents and detergent baths, 233, 374–380. See also *Builders*.
for wetting agents 323
Adhesion, and interfacial tension, 4
between two liquids, 251
Adhesions treated with surface active agents, 457
Adhesion tensions. See *Contact angle* and *"Spreading Pressure."*
Adhesives, containing foaming agents, 515
containing polyvinyl ether and dibutyl naphthalene sulfonates, 511
containing surface active agents, 511
Adsorption, at phase boundary, molecular picture, 272ff
at surface of solution, experimental detection, 273–274

Adsorption compounds as factor in detergency, 361
Adsorption phenomena in detergency, 371
Aerosol(s), 100. See also *Sulfosuccinic esters*.
freezing point lowering of solutions, 299–300
solubilization of Orange OT, 310
Aerosol 18, 106
Aerosol 22, 106.
Aerosol IB, AY, MA, OT, 100
Aerosol OS, 117.
Aerosol OT as germicides, 455
in bread dough, 504
After-soaping of dyeings, 430
After-washing, textile prints, 433
Agglomerates of pigments, disintegration, 476
Agglomeration and its counteraction, 324 ff.
Air entrainment and concrete, 501
Alcohols, complex hydrophobic, mono-etherified with polyethylene glycol, 205
fatty. See *Fatty alcohol* and *Higher alcohol*.
Aldimines, as intermediates in conversion of nitriles to amines, 157
of aldehydes to amines, 161
Alframine, 74
Alginate fibers, detergents in spinning baths, 422
Alicyclic hydrocarbons and Reed reaction, 88
Alipal, 78
Alipon OT, 103
Alkaline builders, effect on soap detergency, 374–375
for synthetic detergents, 232–233
in synthetic detergent powders, 460
Alkaline inorganic detergents in drilling muds, 482–483
Alkane sulfonamide(s), reacted with ethylene oxide and sulfated, 81
with formaldehyde and aminocarboxy acid, 223
Alkane sulfonate(s), 82–110. See also *Mahogany sulfonates*, *Naphthene sulfonates*, *Petroleum sulfonates*, and *Reed reaction products*.
containing amide intermediate linkage, 102–107.